Convergence and Applications
of Newton-type Iterations

Ioannis K. Argyros

Convergence and Applications of Newton-type Iterations

Springer

Ioannis K. Argyros
Department of Mathematical Sciences
Cameron University
Lawton, OK 73505
USA
ioannisa@cameron.edu

ISBN 978-1-4419-2492-6
e-ISBN: 978-0-387-72743-1
DOI: 10.1007/978-0-387-72743-1

Mathematics Subject Classification (2000): 65G99, 65K10, 65B05, 90C30, 90C33, 47J20, 47H04, 49M15

Dedicated to my father Konstantinos

Contents

Introduction xi

1 Operators and Equations 1
   1.1 Operators on linear spaces 1
   1.2 Divided differences of operators 9
   1.3 Fixed points of operators 25
   1.4 Exercises 29

2 The Newton-Kantorovich (NK) Method 41
   2.1 Linearization of equations 41
   2.2 Semilocal convergence of the NK method 42
   2.3 New sufficient conditions for the secant method 54
   2.4 Concerning the "terra incognita" between convergence regions of
       two Newton methods 62
   2.5 Enlarging the convergence domain of the NK method under regular
       smoothness conditions 75
   2.6 Convergence of NK method and operators with values in a cone ... 80
   2.7 Convergence theorems involving center-Lipschitz conditions 84
   2.8 The radius of convergence for the NK method 90
   2.9 On a weak NK method 102
   2.10 Bounds on manifolds 103
   2.11 The radius of convergence and one-parameter operator embedding . 106
   2.12 NK method and Riemannian manifolds 110
   2.13 Computation of shadowing orbits 113
   2.14 Computation of continuation curves 116
   2.15 Gauss-Newton method 121
   2.16 Exercises 125

3 Applications of the Weaker Version of the NK Theorem 133
   3.1 Comparison of Kantorovich and Moore theorems 133
   3.2 Comparison of Kantorovich and Miranda theorems 137

3.3   The secant method and nonsmooth equations 142
3.4   Improvements on curve tracing of the homotopy method 153
3.5   Nonlinear finite element analysis 157
3.6   Convergence of the structured PSB update in Hilbert space 162
3.7   On the shadowing lemma for operators with chaotic behavior 166
3.8   The mesh independence principle and optimal shape design
      problems 170
3.9   The conditioning of semidefinite programs 180
3.10 Exercises 186

4   Special Methods 193
4.1   Broyden's method 193
4.2   Stirling's method 202
4.3   Steffensen's method 207
4.4   Computing zeros of operator satisfying autonomous differential
      equations 215
4.5   The method of tangent hyperbolas 219
4.6   A modified secant method and function optimization 230
4.7   Local convergence of a King-Werner-type method 233
4.8   Secant-type methods 235
4.9   Exercises 239

5   Newton-like Methods 261
5.1   Newton-like methods of "bounded deterioration" 261
5.2   Weak conditions for the convergence of a certain class of iterative
      methods 269
5.3   Unifying convergence analysis for two-point Newton methods 275
5.4   On a two-point method of convergent order two 290
5.5   Exercises 304

**6   Analytic Computational Complexity: We Are Concerned with the
      Choice of Initial Approximations** 325
6.1   The general problem 325
6.2   Obtaining good starting points for Newton's method 328
6.3   Exercises 336

7   Variational Inequalities 339
7.1   Variational inequalities and partially relaxed monotone mapping ... 339
7.2   Monotonicity and solvability of nonlinear variational inequalities .. 345
7.3   Generalized variational inequalities 352
7.4   Semilocal convergence 354
7.5   Results on generalized equations 358
7.6   Semilocal convergence for quasivariational inequalities 362
7.7   Generalized equations in Hilbert space 365
7.8   Exercises 371

8 Convergence Involving Operators with Outer or Generalized Inverses 379
   8.1 Convergence with no Lipschitz conditions 379
   8.2 Exercises 388

9 Convergence on Generalized Banach Spaces: Improving Error Bounds and Weakening of Convergence Conditions 395
   9.1 K-normed spaces 395
   9.2 Generalized Banach spaces 408
   9.3 Inexact Newton-like methods on Banach spaces with a convergence structure 417
   9.4 Exercises 436

10 Point-to-Set-Mappings 445
   10.1 Algorithmic models 445
   10.2 A general convergence theorem 449
   10.3 Convergence of k-step methods 451
   10.4 Convergence of single-step methods 454
   10.5 Convergence of single-step methods with differentiable iteration functions... 458
   10.6 Monotone convergence................................... 468
   10.7 Exercises 471

11 The Newton-Kantorovich Theorem and Mathematical Programming 475
   11.1 Case 1: Interior point methods 475
   11.2 Case 2: LP methods 482
   11.3 Exercises 489

References 493

Glossary of Symbols 503

Index 505

Introduction

Researchers in computational sciences are faced with the problem of solving a variety of equations. A large number of problems are solved by finding the solutions of certain equations. For example, dynamic systems are mathematically modeled by difference or differential equations, and their solutions represent usually the states of the systems. For the sake of simplicity, assume that a time-invariant system is driven by the equation $x' = f(x)$, where x is the state, then the equilibrium states are determined by solving the equations $f(x) = 0$. Similar equations are used in the case of discrete systems. The unknowns of engineering equations can be functions (difference, differential, integral equations), vectors (systems of linear or nonlinear algebraic equations), or real or complex numbers (single algebraic equations with single unknowns). Except special cases, the most commonly used solutions methods are iterative; when starting from one or several initial approximations, a sequence is constructed, which converges to a solution of the equation. Iteration methods are applied also for solving optimization problems. In such cases, the iteration sequences converge to an optimal solution of the problem in hand. Because all of these methods have the same recursive structure, they can be introduced and discussed in a general framework.

To complicate the matter further, many of these equations are nonlinear. However, all may be formulated in terms of operators mapping a linear space into another, the solutions being sought as points in the corresponding space. Consequently, computational methods that work in this general setting for the solution of equations apply to a large number of problems and lead directly to the development of suitable computer programs to obtain accurate approximate solutions to equations in the appropriate space.

This monograph is written with optimization considerations including the weakening of existing hypotheses for solving equations. It can also be used as a reference book for an advanced numerical-functional analysis course. The goal is to introduce these powerful concepts and techniques at the earliest possible stage. The reader is assumed to have had courses in numerical functional analysis and linear algebra.

We have divided the material into 11 chapters. Each chapter contains several new theoretical results and important applications in engineering, in dynamic economic systems, in input-output systems, in the solution of nonlinear and linear differential equations, and optimization problems. The applications appear in the form of Examples or Applications or Exercises or they are implied as our results improve (weaken) (extend the applicability of) earlier ones that have already been applied in concrete problems. Sections have been written as independent of each other as possible. Hence the interested reader can go directly to a certain section and understand the material without having to go back and forth in the whole textbook to find related material.

There are four basic problems connected with iterative methods.

Problem 1: Show that the iterates are well defined. For example, if the algorithm requires the evaluation of F at each x_n, it has to be guaranteed that the iterates remain in the domain of F. It is, in general, impossible to find the exact set of all initial data for which a given process is well defined, and we restrict ourselves to giving conditions that guarantee that an iteration sequence is well defined for certain specific initial guesses.

Problem 2: Concerns the convergence of the sequences generated by a process and the question of whether their limit points are, in fact, solutions of the equation. There are several types of such convergence results. The first, which we call a local convergence theorem, begins with the assumption that a particular solution x^* exists, and then asserts that there is a neighborhood U of x^* such that for all initial vectors in U the iterates generated by the process are well defined and converge to x^*. The second type of convergence theorem, which we call semilocal, does not require knowledge of the existence of a solution, but states that, starting from initial vectors for which certain—usually stringent—conditions are satisfied, convergence to some (generally nearby) solutions x^* is guaranteed. Moreover, theorems of this type usually include computable (at least in principle) estimates for the error $x_n - x^*$, a possibility not afforded by the local convergence theorems. Finally, the third and most elegant type of convergence result, the global theorem, asserts that starting anywhere in a linear space, or at least in a large part of it, convergence to a solution is ensured.

Problem 3: Concerns the economy of the entire operations and, in particular, the question of how fast a given sequence will converge. Here, there are two approaches, which correspond with the local and semilocal convergence theorems. As mentioned above, the analysis that leads to the semilocal type of theorem frequently produces error estimates, and these, in turn, may sometimes be reinterpreted as estimates of the rate of convergence of the sequence. Unfortunately, however, these are usually overly pessimistic. The second approach deals with the behavior of the sequence $\{x_n\}$ when n is large, and hence when x_n is near the solutions x^*. This behavior may then be determined, to a first approximation, by the properties of the iteration function near x^* and leads to so-called asymptotic rates of convergence.

Problem 4: Concerns with how to best choose a method, algorithm, or software program to solve a specific type of problem and its descriptions of when a given algorithm or method succeeds or fails.

We have included a variety of new results dealing with Problems 1–4.

This monograph is an outgrowth of research work undertaken by us and complements/updates earlier works of ours focusing on in-depth treatment of convergence theory for iterative methods [7]–[43]. Such a comprehensive study of optimal iterative procedures appears to be needed and should benefit not only those working in the field but also those interested in, or in need of, information about specific results or techniques. We have endeavored to make the main text as self-contained as possible, to prove all results in full detail, and to include a number of exercises throughout the monograph. In order to make the study useful as a reference source, we have complemented each section with a set of "Remarks" in which literature citations are given, other related results are discussed, and various possible extensions of the results of the text are indicated. For completion, the monograph ends with a comprehensive list of references. Because we believe our readers come from diverse backgrounds and have varied interests, we provide "recommended reading" throughout the textbook. Often a long textbook summarizes knowledge in a field. This monograph, however, may be viewed as a report on work in progress. We provide a foundation for a scientific field that is rapidly changing. Therefore we list numerous conjectures and open problems as well as alternative models that need to be explored.

The monograph is organized as follows:

Chapter 1: The essentials on the solution of equations are provided.

Newton-type methods and their implications/applications are covered in the rest of the chapters.

The Newton-Kantorovich Theorem 2.2.4 for solving nonlinear equations is one of the most important tools in nonlinear analysis and in classic numerical analysis. This theorem has been successfully used for obtaining optimal bounds for many iterative procedures. The original paper or Kantorovich [124] contains optimal a priori bounds for the Newton-Kantorovich (NK) method (2.1.3), albeit not in explicit form. Explicit forms of those a priori bounds were obtained independently by Ostrowski [155], Gragg and Tapia [102].

The paper of Gragg and Tapia [102] also contains sharp a posteriori bounds for the NK method. By using different techniques and/or different a posteriori information, these bounds were refined by others [6], [53], [58], [59], [64], [74], [76]–[78], [128], [135], [139]–[142], [154], [162], [167], [184], [191], [209]–[212], [214]–[216], [218]–[220], and us [11]–[43]. Various extensions of the NK theorem also have been used to obtain error bounds for Newton-like (or Newton-type) methods: Inexact Newton method, the secant method, Halley's method, etc. A survey of such methods can be found in [26], [43].

The NK theorem has also been used in concrete applications for proving existence and uniqueness of solutions for nonlinear equations arising in various fields. The spectrum of applications of this theorem is immense. An Internet search seeking "Newton-Kantorovich Theorem" leads to hundreds if not thousands of works related/based on this theorem.

The list given below is therefore incomplete. However, we have included diverse problems such as the NK method on a cone, Robinson [178] (Section 2.6); the weak NK method, Tapia [188] (Section 2.4); bounds on manifolds, Argyros [39],

Paardekooper [156] (Section 2.10); radius of convergence and one-parameter imbedding Meyer [139] (Section 2.11); NK method on Riemannian manifolds, Ferreira and Svaiter [94] (Section 2.12); shadowing orbits in dynamical systems, Hadeller [108] (Section 2.13); computation of continuation curves, Deuflhard, Pesh, Rentrop [77], Rheinboldt [176] (Section 2.14); Moore's theorem [143] from interval analysis, Rall [171], Neumaier and Shen [146], Zuhe and Wolfe [220] (Section 3.1); Miranda's theorem [142] for enclosing solutions of equations, Mayer [136] (Section 3.2); point-based approximation (PBA) used successfully by Robinson [179], [180] in Mathematical Programming (Section 3.3); curve tracing, Allgower [2], Chu [62], Rheinboldt [176] (Section 3.4); finite element analysis for boundary value problems, Tsuchiya [194], Pousin [168], Feinstauer-Zernicek [93] (Section 3.5); PSB updates in Hilbert spaces using quasi-NK method, Laumen [134] (Section 3.6); shadowing Lemma and chaotic behavior for nonlinear equations, Palmer [156], Stoffer [186] (Section 3.7); mesh independence principle for optimal design problems, Laumen [133], Allgower, Böhmer, Potra, Rheinboldt [2] (Section 3.8); conditioning of semidefinite programs, Nayakkankuppam [144], Alizadeh [1], Haeberly [109] (Section 3.9); analytic complexity/enlarging the set of initial guesses for the NK method, Kung [131], Traub [192] (Chapter 6, Sections 6.1, 6.2, 6.3); interior point methods, Potra [165] (Section 11.1); LP methods, Rheinboldt [177], Wang-Zhao [206], Renegar-Shub [174], Smale [184] (Section 11.2).

The foundation of the NK theorem is famous for its simplicity and clarity of NK hypothesis (2.2.17) (or (2.2.37) in affine invariant form).

This hypothesis is the crucial sufficient condition for the convergence of Newton's method. However, convergence of Newton's method can be obtained even if the NK hypothesis is violated (see, e.g., Example 2.2.14). Therefore weakening this condition is of extreme importance because the applicability of this powerful method will be extended. Recently we showed [39] by considering more precise majorizing sequences that the NK hypothesis can always be replaced by the weaker (2.2.52) (if $\ell_0 \neq \ell$) (see also Theorem 2.2.11) which doubles (at most if $\ell_0 = 0$) the applicability of this theorem. Note that the verification of condition (2.2.56) requires the same information and computational cost as (2.2.37) because in practice the computation of Lipschitz constant ℓ requires the evaluation of center-Lipschitz constant ℓ_0 too.

Moreover the following advantages hold (see Theorem 2.2.11 and the Remarks that follow): semilocal case: finer error estimates on the distances involved and an at least as precise information on the location of the solution; local case: finer error bounds and larger trust regions (radius of convergence).

The following advantages carry over if our approach is extended to related methods/hypotheses: Below we provide a list: secant method, Argyros [12], [43], Dennis [74], Potra [162], Hernandez [116], [117] (Section 2.3); "Terra Incognita" and Hölder continuity, Argyros [32], [35], Lysenko [135], Ciancarruso, De Pascale [64] (Section 2.4); NK method under regular smoothness conditions, Galperin [98], Galperin and Waksman [99] (Section 2.5); enlarging the radius of convergence for the NK method using hypotheses on the m ($m > 1$ an integer) Fréchet-differentiable operators, Argyros [27], [43], Ypma [216] (Section 2.8); Gauss-Newton method, Ben-Israel [46], Häussler [110] (Section 2.15); Broyden's method [52], Dennis [75]

(Section 4.1); Stirling's method [185], Rall [170] (Section 4.2); Steffenssen-Aitken method, Catinas [54], Pavaloiu [158], [159]; method of tangent hyperbolas, Kanno [123], Yamamoto [211] (Section 4.5); modified secant method with applications in function optimization, Amat, Busquier, Gutierrez [4], Bi, Ren, Wu [47], Ren [172] (Section 4.6); the King-Werner method, Ren [172]; Newton methods (including two-point), Argyros [34], [35], [43] Dennis [74], [75], Chen, Yamamoto, [58], [59], [60], (in Chapters 5 and 8); variational inequalities in Chapter 7, K-theory and convergence on generalized Banach spaces with a convergence structure, Caponetti, De Pascale, Zabrejko [53], Meyer [139]–[141] in Chapter 9, and extensions to set-to-set mappings in Chapter 10.

Earlier results by us or others are included in sections mentioned above directly or indirectly as special cases of our results. Note that revisiting all results to date that have used the NK hypothesis (2.2.37) and replacing (2.2.37) with our weaker hypothesis (2.2.56) is worth it for the reasons/benefits mentioned above. However, this will be an enormous or even impossible task. That is why in this monograph we decided to include only the above chapters and leave the rest for the motivated reader. Note that some results are also listed as exercises to reduce the size of the book.

Finally we state that although the refinement of majorizing sequences technique inaugurated by us in [39] is very recent, several authors have already succesfully used it: Amat, Busquier, Gutierrez [4] (see Section 4.8), Bi, Ren, Wu [47] (see Section 4.6), and Ren [172] (see Section 4.7).

1

Operators and Equations

The basic background for solving equations is introduced here.

1.1 Operators on linear spaces

Some mathematical operations have certain properties in common. These properties
are given in the following definition.

Definition 1.1.1. *An operator T that maps a linear space X into a linear space Y
over the same scalar field S is said to be additive if*

$$T(x + y) = T(x) + T(y), \qquad \text{for all } x, y \in X,$$

and homogeneous if

$$T(sx) = sT(x), \qquad \text{for all } x \in X, s \in S.$$

An operator that is additive and homogeneous is called a linear operator.

Many examples of linear operators exist.

Example 1.1.2. Define an operator T from a linear space X into itself by $T(x) = sx$,
$s \in S$. Then T is a linear operator.

Example 1.1.3. The operator $D = \frac{d}{dt}$ mapping $X = C^1[0, 1]$ into $Y = C[0, 1]$
given by

$$D(x) = \frac{dx}{dt} = y(t), \ 0 \leq t \leq 1,$$

is linear.

If X and Y are linear spaces over the same scalar field S, then the set $L(X, Y)$
containing all linear operators from X into Y is a linear space over S if addition is
defined by

I.K. Argyros, *Convergence and Applications of Newton-type Iterations*,
DOI: 10.1007/978-0-387-72743-1_1, © Springer Science+Business Media, LLC 2008

$$(T_1 + T_2)(x) = T_1(x) + T_2(x), \qquad \text{for all } x \in X,$$

and scalar multiplication by

$$(sT)(x) = s(T(x)), \qquad \text{for all } x \in X, \ s \in S.$$

We may also consider linear operators B mapping X into $L(X, Y)$. For an $x \in X$ we have

$$B(x) = T,$$

a linear operator from X into Y. Hence, we have

$$B(x_1, x_2) = (B(x_1))(x_2) = y \in Y.$$

B is called a bilinear operator from X into Y. The linear operators B from X into $L(X, Y)$ form a linear space $L(X, L(X, Y))$. This process can be repeated to generate j-linear operators ($j > 1$ an integer).

Definition 1.1.4. *A linear operator mapping a linear space X into its scalar S is called a linear functional in X.*

Definition 1.1.5. *An operator Q mapping a linear space X into a linear space Y is said to be nonlinear if it is not a linear operator from X into Y.*

Some metric concepts of importance are now introduced.

Definition 1.1.6. *An operator F from a Banach space X into a Banach space Y is continuous at $x = x^*$ if*

$$\lim_{n \to \infty} \|x_n - x^*\|_X = 0 \implies \lim_{n \to \infty} \|F(x_n) - F(x^*)\|_Y = 0$$

Theorem 1.1.7. *If a linear operator T from a Banach space X into a Banach space Y is continuous at $x^* = 0$, then it is continuous at every point x of space X.*

Proof. We have $T(0) = 0$, and from $\lim_{n \to \infty} \|x_n\| = 0$ we get $\lim_{n \to \infty} \|T(x_n)\| = 0$. If sequence $\{x_n\}$ ($n \geq 0$) converges to x^* in X, by setting $y_n = x_n - x^*$ we obtain $\lim_{n \to \infty} \|y_n\| = 0$. By hypothesis this implies that

$$\lim_{n \to \infty} \|T(x_n)\| = \lim_{n \to \infty} \|T(x_n - x^*)\| = \lim_{n \to \infty} \|T(x_n) - T(x^*)\| = 0.$$

Definition 1.1.8. *An operator F from a Banach space X into a Banach space Y is Lipschitz continuous on the set A in X if there exists a constant $c < \infty$ such that*

$$\|F(x) - F(y)\| \leq c \|x - y\|, \qquad \text{for all } x, y \in A.$$

The greatest lower bound (infimum) of numbers c satisfying the above inequality for $x \neq y$ is called the bound of F on A. An operator that is bounded on a ball (open) $U(z, r) = \{x \in X \mid \|x - z\| < r\}$ is continuous at z. It turns out that for linear operators, the converse is also true.

Theorem 1.1.9. *A continuous linear operator T from a Banach space X into a Banach space Y is bounded on X.*

Proof. By the continuity of T there exists $\varepsilon > 0$ such that $\|T(z)\| < 1$, if $\|z\| < \varepsilon$. For $0 \neq z \in X$

$$\|T(z)\| \leq \tfrac{1}{\varepsilon} \|z\|, \tag{1.1.1}$$

because $\|cz\| < \varepsilon$ for $|c| < \frac{\varepsilon}{\|z\|}$, and $\|T(cz)\| = |c| \cdot \|T(z)\| < 1$. Letting $z = x - y$ and $c = \varepsilon^{-1}$ in (1.1.1), we conclude that operator T is bounded on X.

The bound on X of a linear operator T denoted by $\|T\|_X$ or simply $\|T\|$ is called the norm of T. As in Theorem 1.1.9 we get

$$\|T\| = \sup_{\|x\|=1} \|T(x)\|. \tag{1.1.2}$$

Hence, for any bounded linear operator T

$$\|T(x)\| \leq \|T\| \cdot \|x\|, \qquad \text{for all } x \in X. \tag{1.1.3}$$

From now on, $L(X, Y)$ denotes the set of all bounded linear operators from a Banach space X into another Banach space Y. It also follows immediately that $L(X, Y)$ is a linear space if equipped with the rules of addition and scalar multiplication introduced in Definition 1.1.1.

The proof of the following result is left as an exercise (see also [119], [125]).

Theorem 1.1.10. *The set $L(X, Y)$ is a Banach space for the norm (1.1.2).*

In a Banach space X, solving a linear equation can be stated as follows: given a bounded linear operator T mapping X into itself and some $y \in X$, find an $x \in X$ such that

$$T(x) = y. \tag{1.1.4}$$

The point x (if it exists) is called a solution of Equation (1.1.4).

Definition 1.1.11. *If T is a bounded linear operator in X and a bounded linear operator T_1 exists such that*

$$T_1 T = T T_1 = I, \tag{1.1.5}$$

where I is the identity operator in X (i.e., $I(x) = x$ for all $x \in X$), then T_1 is called the inverse of T and we write $T_1 = T^{-1}$. That is,

$$T^{-1}T = T T^{-1} = I. \tag{1.1.6}$$

If T^{-1} exists, then Equation (1.1.4) has the unique solution

$$x = T^{-1}(y). \tag{1.1.7}$$

The proof of the following result is left as an exercise (see also [130]).

Theorem 1.1.12. *(Banach Lemma on Invertible Operators) [125]. If T is a bounded linear operator in X, T^{-1} exists if and only if there is a bounded linear operator P in X such that P^{-1} exists and*

$$\|I - PT\| < 1. \tag{1.1.8}$$

If T^{-1} exists, then

$$T^{-1} = \sum_{n=0}^{\infty} (I - PT)^n P \qquad \text{(Neumann Series)} \tag{1.1.9}$$

and

$$\left\| T^{-1} \right\| \le \frac{\|P\|}{1 - \|I - PT\|}. \tag{1.1.10}$$

Based on Theorem 1.1.12, we can immediately introduce a computational theory for Equation (1.1.4) composed by three factors:

(A) *Existence and Uniqueness.* Under the hypotheses of Theorem 1.1.12, Equation (1.1.4) has a unique solution x^*.

(B) *Approximation.* The iteration

$$x_{n+1} = P(y) + (I - PT)(x_n) \quad (n \ge 0) \tag{1.1.11}$$

gives a sequence $\{x_n\}$ $(n \ge 0)$ of successive approximations, which converges to x^* for any initial guess $x_0 \in X$.

(C) *Error Bounds.* Clearly the speed of convergence of iteration $\{x_n\}$ $(n \ge 0)$ to x^* is governed by the estimate:

$$\|x_n - x^*\| \le \frac{\|I - PT\|^n}{1 - \|I - PT\|} \|P(y)\| + \|I - PT\|^n \|x_0\|. \tag{1.1.12}$$

Let T be a bounded linear operator in X. One way to obtain an approximate inverse is to make use of an operator sufficiently close to T.

Theorem 1.1.13. *If T is a bounded linear operator in X, T^{-1} exists if and only if there is a bounded linear operator P_1 in X such that P_1^{-1} exists, and*

$$\|P_1 - T\| \le \left\| P_1^{-1} \right\|^{-1}. \tag{1.1.13}$$

If T^{-1} exists, then

$$T^{-1} = \sum_{n=0}^{\infty} \left(I - P_1^{-1} T \right)^n P_1^{-1} \tag{1.1.14}$$

and

$$\left\| T^{-1} \right\| \le \frac{\|P^{-1}\|}{1 - \left\| I - P_1^{-1} T \right\|} \le \frac{\left\| P_1^{-1} \right\|}{1 - \left\| P_1^{-1} \right\| \|P_1 - T\|}. \tag{1.1.15}$$

Proof. Let $P = P_1^{-1}$ in Theorem 1.1.12 and note that by (1.1.13)

$$\left\| I - P_1^{-1} T \right\| = \left\| P_1^{-1} (P_1 - T) \right\| \leq \left\| P_1^{-1} \right\| \cdot \| P_1 - T \| < 1. \qquad (1.1.16)$$

That is, (1.1.8) is satisfied. The bounds (1.1.15) follow from (1.1.10) and (1.1.16). That proves the sufficiency. The necessity is proved by setting $P_1 = T$, if T^{-1} exists.

The following result is equivalent to Theorem 1.1.12.

Theorem 1.1.14. *A bounded linear operator T in a Banach space X has an inverse T^{-1} if and only if linear operators P, P^{-1} exist such that the series*

$$\sum_{n=0}^{\infty} (I - PT)^n P \qquad (1.1.17)$$

converges. In this case we have

$$T^{-1} = \sum_{n=0}^{\infty} (I - PT)^n P.$$

Proof. If series (1.1.17) converges, then it converges to T^{-1} (see Theorem 1.1.12). The existence of P, P^{-1} and the convergence of series (1.1.17) is again established as in Theorem 1.1.12, by taking $P = T^{-1}$, when it exists.

Definition 1.1.15. *A linear operator N in a Banach space X is said to be nilpotent if*

$$N^m = 0, \qquad (1.1.18)$$

for some positive integer m.

Theorem 1.1.16. *A bounded linear operator T in a Banach space X has an inverse T^{-1} and only if there exist linear operators P, P^{-1} such that $I - PT$ is nilpotent.*

Proof. If P, P^{-1} exists and $I - PT$ is nilpotent, then series

$$\sum_{n=0}^{\infty} (I - PT)^n P = \sum_{n=0}^{m-1} (I - PT)^n P$$

converges to T^{-1} by Theorem 1.1.14. Moreover, if T^{-1} exists, then $P = T^{-1}$, $P^{-1} = T$ exists, and $I - PT = I - T^{-1}T = 0$ is nilpotent.

The computational techniques to be considered later make use of the derivative in the sense of Fréchet [125], [204].

Definition 1.1.17. *Let F be an operator mapping a Banach space X into a Banach space Y. If there exists a bounded linear operator L from X into Y such that*

$$\lim_{\|\Delta x\| \to 0} \frac{\|F(x_0 + \Delta x) - F(x_0) - L(\Delta x)\|}{\|\Delta x\|} = 0, \qquad (1.1.19)$$

then F is said to be Fréchet-differentiable at x_0, and the bounded linear operator

$$F'(x_0) = L \qquad (1.1.20)$$

is called the first Fréchet derivative of F at x_0. The limit in (1.1.19) is supposed to hold independently of the way that Δx approaches 0. Moreover, the Fréchet differential

$$\delta F(x_0, \Delta x) = F'(x_0) \Delta x \qquad (1.1.21)$$

is an arbitrary close approximation to the difference $F(x_0 + \Delta x) - F(x_0)$ relative to $\|\Delta x\|$, for $\|\Delta x\|$ small.

If F_1 and F_2 are differentiable at x_0, then

$$(F_1 + F_2)'(x_0) = F_1'(x_0) + F_2'(x_0). \qquad (1.1.22)$$

Moreover, if F_2 is an operator from a Banach space X into a Banach space Z, and F_1 is an operator from Z into a Banach space Y, their composition $F_1 \circ F_2$ is defined by

$$(F_1 \circ F_2)(x) = F_1(F_2(x)), \qquad \text{for all } x \in X. \qquad (1.1.23)$$

It follows from Definition 1.1.17 that $F_1 \circ F_2$ is differentiable at x_0 if F_2 is differentiable at x_0 and F_1 is differentiable at $F_2(x_0)$ of Z, with (chain rule):

$$(F_1 \circ F_2)'(x_0) = F_1'(F_2(x_0))F_2'(x_0). \qquad (1.1.24)$$

In order to differentiate an operator F we write:

$$F(x_0 + \Delta x) - F(x_0) = L(x_0, \Delta x)\Delta x + \eta(x_0, \Delta x), \qquad (1.1.25)$$

where $L(x_0, \Delta x)$ is a bounded linear operator for given x_0, Δx with

$$\lim_{\|\Delta x\| \to 0} L(x_0, \Delta x) = L, \qquad (1.1.26)$$

and

$$\lim_{\|\Delta x\| \to 0} \frac{\|\eta(x_0, \Delta x)\|}{\|\Delta x\|} = 0. \qquad (1.1.27)$$

Estimates (1.1.26) and (1.1.27) give

$$\lim_{\|\Delta x\| \to 0} L(x_0, \Delta x) = F'(x_0). \qquad (1.1.28)$$

If $L(x_0, \Delta x)$ is a continuous function of Δx in some ball $U(0, R)$ $(R > 0)$, then

$$L(x_0, 0) = F'(x_0). \qquad (1.1.29)$$

Higher-order derivatives can be defined by induction:

Definition 1.1.18. *If F is $(m - 1)$-times Fréchet-differentiable ($m \geq 2$ an integer), and an m-linear operator A from X into Y exists such that*

$$\lim_{\|\Delta x\| \to 0} \frac{\left\| F^{(m-1)}(x_0 + \Delta x) - F^{(m-1)}(x_0) - A(\Delta x) \right\|}{\|\Delta x\|} = 0, \qquad (1.1.30)$$

then A is called the m-Fréchet derivative of F at x_0, and

$$A = F^{(m)}(x_0) \qquad (1.1.31)$$

Higher partial derivatives in product spaces can be defined as follows: Define

$$X_{ij} = L(X_j, X_i), \qquad (1.1.32)$$

where $X_1, X_2, \ldots$ are Banach spaces and $L(X_j, X_i)$ is the space of bounded linear operators from X_j into X_i. The elements of X_{ij} are denoted by L_{ij}, etc. Similarly,

$$X_{ijm} = L(X_m, X_{ij}) = L(X_m, L(X_j, X_i)) \qquad (1.1.33)$$

denotes the space of bounded bilinear operators from X_k into X_{ij}. Finally, we write

$$X_{ij_1 j_2 \cdots j_m} = L\left(X_{jk}, X_{ij_1 j_2 \cdots j_{m-1}} \right), \qquad (1.1.34)$$

which denotes the space of bounded linear operators from X_{jm} into $X_{ij_1 j_2 \cdots j_{m-1}}$. The elements $A = A_{ij_1 j_2 \cdots j_m}$ of $X_{ij_1 j_2 \cdots j_m}$ are a generalization of m-linear operators [10], [125].

Consider an operator F_i from space

$$X = \prod_{p=1}^{n} X_{j_p} \qquad (1.1.35)$$

into X_i, and that F_i has partial derivatives of orders $1, 2, \ldots, m - 1$ in some ball $U(x_0, R)$, where $R > 0$ and

$$x_0 = \left(x_{j_1}^{(0)}, x_{j_2}^{(0)}, \ldots, x_{j_n}^{(0)} \right) \in X. \qquad (1.1.36)$$

For simplicity and without loss of generality we renumber the original spaces so that

$$j_1 = 1, j_2 = 2, \ldots, j_n = n. \qquad (1.1.37)$$

Hence, we write

$$x_0 = (x_1^{(0)}, x_2^{(0)}, \ldots, x_n^{(0)}). \qquad (1.1.38)$$

A partial derivative of order $(m - 1)$ of F_i at x_0 is an operator

$$A_{iq_1 q_2 \cdots q_{m-1}} = \frac{\partial^{(m-1)} F_i(x_0)}{\partial x_{q_1} \partial x_{q_2} \cdots \partial x_{q_{m-1}}} \qquad (1.1.39)$$

(in $X_{iq_1 q_2 \cdots q_{m-1}}$) where

$$1 \leq q_1, q_2, \ldots, q_{m-1} \leq n. \tag{1.1.40}$$

Let $P(X_{q_m})$ denote the operator from X_{q_m} into $X_{iq_1 q_2 \cdots q_{m-1}}$ obtained from (1.1.39) by letting

$$x_j = x_j^{(0)}, \quad j \neq q_m, \tag{1.1.41}$$

for some q_m, $1 \leq q_m \leq n$. Moreover, if

$$P'(x_{q_m}^{(0)}) = \frac{\partial}{\partial x_{q_m}} \cdot \frac{\partial^{m-1} F_i(x_0)}{\partial x_{q_1} \partial x_{q_2} \cdots \partial x_{q_{m-1}}} = \frac{\partial^m F_i(x_0)}{\partial x_{q_1} \cdots \partial x_{q_m}}, \tag{1.1.42}$$

exists, it will be called the partial Fréchet derivative of order m of F_i with respect to $x_{q_1}, \ldots, x_{q_m}$ at x_0.

Furthermore, if F_i is Fréchet-differentiable m times at x_0, then

$$\frac{\partial^m F_i(x_0)}{\partial x_{q_1} \cdots \partial x_{q_m}} x_{q_1} \cdots x_{q_m} = \frac{\partial^m F_i(x_0)}{\partial x_{s_1} \partial x_{s_2} \cdots \partial x_{s_m}} x_{s_1} \cdots x_{s_m} \tag{1.1.43}$$

for any permutation $s_1, s_2, \ldots, s_m$ of integers $q_1, q_2, \ldots, q_m$ and any choice of points $x_{q_1}, \ldots, x_{q_m}$, from $X_{q_1}, \ldots, X_{q_m}$ respectively. Hence, if $F = (F_1, \ldots, F_t)$ is an operator from $X = X_1 \times X_2 \times \cdots \times X_n$ into $Y = Y_1 \times Y_2 \times \cdots \times Y_t$, then

$$F^{(m)}(x_0) = \left(\frac{\partial^m F_i}{\partial x_{j_1} \cdots \partial x_{j_m}} \right)_{x=x_0} \tag{1.1.44}$$

$i = 1, 2, \ldots, t$, $j_1, j_2, \ldots, j_m = 1, 2, \ldots, n$, is called the m-Fréchet derivative of F at $x_0 = (x_1^{(0)}, x_2^{(0)}, \ldots, x_n^{(0)})$.

We now state results concerning the mean value theorem, Taylor's theorem, and Riemannian integration. The proofs are left out as exercises [125], [186].

The mean value theorem for differentiable real functions f:

$$f(b) - f(a) = f'(c)(b - a), \tag{1.1.45}$$

where $c \in (a, b)$, does not hold in a Banach space setting. However, if F is a differentiable operator between two Banach spaces X and Y, then

$$\|F(x) - F(y)\| \leq \sup_{\bar{x} \in L(x,y)} \|F'(\bar{x})\| \cdot \|x - y\|, \tag{1.1.46}$$

where

$$L(x, y) = \{z : z = \lambda y + (1 - \lambda)x, \ 0 \leq \lambda \leq 1\}. \tag{1.1.47}$$

Set

$$z(\lambda) = \lambda y + (1 - \lambda)x, \quad 0 \leq \lambda \leq 1, \tag{1.1.48}$$

and

$$F(\lambda) = F(z(\lambda)) = F(\lambda y + (1 - \lambda)x). \tag{1.1.49}$$

Divide the interval $0 \leq \lambda \leq 1$ into n subintervals of lengths $\Delta \lambda_i$, $i = 1, 2, \ldots, n$, choose points λ_i inside corresponding subintervals and as in the real Riemann integral consider sums

$$\sum_{\sigma} F(\lambda_i)\Delta\lambda_i = \sum_{i=1}^{n} F(\lambda_i)\Delta\lambda_i, \qquad (1.1.50)$$

where σ is the partition of the interval, and set

$$|\sigma| = \max_{(i)} \Delta\lambda_i. \qquad (1.1.51)$$

Definition 1.1.19. *If*

$$S = \lim_{|\sigma|\to 0} \sum_{\sigma} F(\lambda_i)\,\Delta\lambda_i \qquad (1.1.52)$$

exists, then it is called the Riemann integral from $F(\lambda)$ *from* 0 *and* 1, *denoted by*

$$S = \int_0^1 F(\lambda)\,d\lambda = \int_x^y F(\lambda)\,d\lambda. \qquad (1.1.53)$$

Note that a bounded operator $P(\lambda)$ on $[0, 1]$ such that the set of points of discontinuity is of measure zero is said to be integrable on $[0, 1]$.

We now state the famous Taylor theorem [103].

Theorem 1.1.20. *If* F *is* m-*times Fréchet-differentiable in* $U(x_0, R)$, $R > 0$, *and* $F^{(m)}(x)$ *is integrable from* x *to any* $y \in U(x_0, R)$, *then*

$$F(y) = F(x) + \sum_{n=1}^{m-1} \frac{1}{n!} F^{(n)}(x)(y-x)^n + R_m(x, y), \qquad (1.1.54)$$

$$\left\| F(y) - \sum_{n=0}^{m-1} \frac{1}{n!} F^{(n)}(x)(y-x)^n \right\| \le \sup_{\bar{x}\in L(x,y)} \left\| F^{(m)}(\bar{x}) \right\| \frac{\|y-x\|^m}{m!}, \qquad (1.1.55)$$

where

$$R_m(x, y) = \int_0^1 F^{(m)}\big(\lambda y + (1-\lambda)x\big)(y-x)^m \frac{(1-\lambda)^{m-1}}{(m-1)!}\,d\lambda. \qquad (1.1.56)$$

1.2 Divided differences of operators

This section introduces the fundamentals of the theory of divided differences of a nonlinear operator. Several results are also provided using differences as well as Fréchet derivatives satisfying Lipschitz or monotone-type conditions.

Let X be a linear space. We introduce the following definition:

Definition 1.2.1. *A partially ordered topological linear space (POTL-space) is a locally convex topological linear space* X *which has a closed proper convex cone.*

A proper convex cone is a subset K such that $K + K \subset K$, $\alpha K \subset K$ for $\alpha > 0$, and $K \cap (-K) = \{0\}$. Thus the order relation $\leq$, defined by $x \leq y$ if and only if $y - x \in K$, gives a partial ordering that is compatible with the linear structure of the space. The cone K that defines the ordering is called the positive cone as $K = \{x \in X \mid x \geq 0\}$. The fact that K is closed implies also that intervals, $[a, b] = \{z \in X \mid a \leq z \leq b\}$, are closed sets.

Example 1.2.2. Some simple examples of POTL-spaces are:

(1) $X = E^n$, n-dimensional Euclidean space, with

$$K = \left\{(x_1, x_2, ..., x_n) \in E^n \mid x_i \geq 0, \ i = 1, 2, ..., n\right\};$$

(2) $X = E^n$ with $K = \{(x_1, x_2, ..., x_n) \in E^n \mid x_i \geq 0, \ i = 1, 2, ..., n - 1, x_n = 0\}$;
(3) $X = C^n [0, 1]$, continuous functions, maximum norm topology, pointwise ordering;
(4) $X = C^n [0, 1]$, n-times continuously differentiable functions with

$$\|f\| = \sum_{k=0}^{n} \max \left| f^{(K)}(t) \right|, \text{ and pointwise ordering;}$$

(5) $C = L^p [0, 1]$, $0 \leq p \leq \infty$ usual topology,

$$K = \left\{f \in L^p [0, 1] \mid f(t) \leq 0 \text{ a.e.}\right\}.$$

Remark 1.2.3. Using the above examples, it is easy to see that the closedness of the positive cone is not, in general, a strong enough connection between the ordering and the topology. Consider, for example, the following properties of sequences of real numbers:

(1) $x_1 \leq x_2 \leq \cdots \leq x^*$, and $\sup \{x_n\} x^*$ implies $\lim_{n\to\infty} x_n = x^*$;
(2) $\lim_{n\to\infty} x_n = 0$ implies that there exists a sequence $\{y_n\}$ with $y_1 \geq y_2 \geq \cdots \geq 0$, $\inf \{y_n\} = 0$ and $-y_n \leq x_n \leq y_n$;
(3) $0 \leq x_n \leq y_n$, and $\lim_{n\to\infty} y_n = 0$ imply $\lim_{n\to\infty} x_n = 0$.

Unfortunately, these statements are not true for all POTL-spaces:

(a) In $X = C[0, 1]$ let $x_n(t) = -t^n$. Then $x_1 \leq x_2 \leq \cdots \leq 0$, and $\sup \{x_n\} = 0$, but $\|x_n\| = 1$ for all n, so $\lim_{n\to\infty} x_n$ does not exist. Hence (1) does not hold.
(b) In $X = L^1 [0, 1]$ let $x_n(t) = n$ for $\frac{1}{n+1} \leq t \leq \frac{1}{n}$ and zero elsewhere. Then $\lim_{n\to\infty} \|x_n\| = 0$ but clearly property (2) does not hold.
(c) In $X = C^1 [0, 1]$ let $x_n(t) = \frac{t^n}{n}$, $y_n(t) = \frac{1}{n}$. Then $0 \check{S} x_n \leq y_n$, and $\lim_{n\to\infty} y_n = 0$, but $\|x_n\| = \max \left|\frac{t^n}{n}\right| + \max \left|t^{n-1}\right| = \frac{1}{n} + 1 > 1$; hence x_n does not converge to zero.

We will now devote a brief discussion of certain types of POTL-spaces in which some of the above statements are true.

Definition 1.2.4. *A POTL-space is called regular if every order-bounded increasing sequence has a limit.*

Remark 1.2.5. Examples of regular POTL-spaces are E^n and L^p, $0 \leq p \leq \infty$, whereas $C[0, 1]$, $C^n[0, 1]$ and $L^\infty[0, 1]$ are not regular, as was shown in (a) of the above remark. If $\{x_n\}$ $n \geq 0$ is a monotone increasing sequence and $\lim_{n \to \infty} x_n = x^*$ exists, then for any k_0, $n \geq k_0$ implies $x_n \geq x_{k_0}$. Hence $x^* = \lim_{n \to \infty} x_n \geq x_{k_0}$, i.e., x^* is an upper bound on $\{x_n\}$ $n = 0$. Moreover, if y is any other upper bound, then $x_n \leq y$, and hence $x^* = \lim_{n \to \infty} x_n \leq y$, i.e., $x^* = \sup\{x_n\}$. This shows that in any POTL-space, the closedness of the positive cone guarantees that, if a monotone increasing sequence has a limit, then it is also a supremum. In a regular space, the converse of this is true; i.e., if a monotone increasing sequence has a supremum, then it also has a limit. It is important to note that the definition of regularity involves both an order concept (monotone boundedness) and a topological concept (limit).

Definition 1.2.6. *A POTL-space is called normal if, given a local base U for the topology, there exists a positive number η so that if $0 \leq x \in V \in U$ then $[0, x] \subseteq \eta^U$.*

Remark 1.2.7. If the topology of a POTL-space is given by a norm then this space is called a partially ordered normed space (PON)-space. If a PON-space is complete with respect to its topology then it is called a partially ordered Banach space (POB)-space. According to Definition 1.2.6. A PON-space is normal if and only if there exists a positive number α such that

$$\|x\| \leq \alpha \|y\| \quad \text{for all} \quad x, y \in X \quad \text{with} \quad 0 \leq x \leq y.$$

Let us note that any regular POB-space is normal. The converse is not true. For example, the space $C[0, 1]$, ordered by the cone of nonnegative functions, is normal but is not regular. All finite-dimensional POTL-spaces are both normal and regular.

Remark 1.2.8. Let us now define some special types of operators acting between two POTL-spaces. First we introduce some notation if X and Y are two linear spaces then we denote by (X, Y) the set of all operators from X into Y and by $L(X, Y)$ the set of all linear operators from X into Y. If X and Y are topological linear spaces, then we denote by $LB(X, Y)$ the set of all continuous linear operators from X into Y. For simplicity, the spaces $L(X, X)$ and $LB(X, X)$ will be denoted by $L(X)$ and $LB(X)$. Now let X and Y be two POTL-spaces and consider an operator $G \in (X, Y)$. G is called isotone (resp. antitone) if $x \geq y$ implies $G(x) \leq G(y)$ (resp. $G(x) \leq G(y)$). G is called nonnegative if $x \geq 0$ implies $G(x) \geq 0$. For linear operators, the nonnegativity is clearly equivalent with the isotony. Also, a linear operator is inverse nonnegative if and only if it is invertible and its inverse is nonnegative. If G is a nonnegative operator, then we write $G \geq 0$. If G and H are two operators from X into Y such that $H - G$ is nonnegative, then we write $G \leq H$. If Z is a linear space, then we denote by $I = I_z$ the identity operator in Z (i.e., $I(x) = x$

for all $x \in Z$). If Z is a POTL-space, then we have obviously $I \geq 0$. Suppose that X and Y are two POTL-spaces and consider the operators $T \in L(X, Y)$ and $S \in L(Y, X)$. If $ST \leq I_x$ (resp. $ST \geq I_x$), then S is called a left subinverse (resp. superinverse) of T and T is called a right subinverse (resp. superinverse) of S. We say that S is a subinverse of T if S is a left as well as a right subinverse of T.

We finally end this section by noting that for the theory of partially ordered linear spaces, the reader may consult M.A. Krasnosel'skii [128], [129], Vandergraft [199], or Argyros and Szidarovszky [43].

The concept of a divided difference of a nonlinear operator generalizes the usual notion of a divided difference of a scalar function in the same way in which the Fréchet derivative generalizes the notion of a derivative of a function.

Definition 1.2.9. *Let F be a nonlinear operator defined on a subset D of a linear space X with values in a linear space Y, i.e., $F \in (D, Y)$ and let x, y be two points of D. A linear operator from X into Y, denoted $[x, y]$, which satisfies the condition*

$$[x, y] (x - y) = F(x) - F(y) \tag{1.2.1}$$

is called a divided difference of F at the points x and y.

Remark 1.2.10. If X and Y are topological linear spaces, then we shall always assume the continuity of the linear operator $[x, y]$. (Generally, $[x, y] \in L(X, Y)$ if X, Y are POTL-spaces then $[x, y] \in LB(X, Y)$).

Obviously, condition (1.2.1) does not uniquely determine the divided difference, with the exception of the case when X is one-dimensional. An operator $[\cdot, \cdot]: D \times D \to L(X, Y)$ satisfying (1.2.1) is called a divided difference of F on D. If we fix the first variable, we get an operator

$$\left[x^0, \cdot\right]: D \to L(X, Y). \tag{1.2.2}$$

Let x^1, x^2 be two points of D. A divided difference of the operator (1.2.2) at the points x^1, x^2 will be called a divided difference of the second order of F at the points x^0, x^1, x^2 and will be denoted by $\left[x^0, x^1, x^2\right]$. We have by definition

$$\left[x^0, x^1, x^2\right] \left(x^1 - x^2\right) = \left[x^0, x^1\right] - \left[x^0, x^2\right]. \tag{1.2.3}$$

Obviously, $\left[x^0, x^1, x^2\right] \in L(X, L(X, Y))$.

Let us now state a well-known result due to Kantorovich concerning the location of fixed points, which will be used extensively later [125].

Theorem 1.2.11. *Let X be a regular POTL-space and let x, y be two points of X such that $x \leq y$. If $H: [x, y] \to X$ is a continuous isotone operator having the property that $x \leq H(x)$ and $y \geq H(y)$, then there exists a point $z \in [x, y]$ such that $H(z) = z$.*

We now assume that X and Y are Banach spaces. Accordingly we shall have $[x, y] \in LB(X, Y)$, $[x, y, z] \in LB(X, LB(X, Y))$. As we will see in later chapters, most convergence theorems in a Banach space require that the divided differences of F satisfy Lipschitz conditions of the form:

$$\|[x, y] - [x, z]\| \le c_0 \|y - z\| \tag{1.2.4}$$

$$\|[y, x] - [z, x]\| \le c_1 \|y - z\| \tag{1.2.5}$$

$$\|[x, y, z] - [u, y, z]\| \le c_2 \|x - y\| \quad \text{for all} \quad x, y, z, u \in D. \tag{1.2.6}$$

It is a simple exercise to show that if $[\cdot, \cdot]$ is a divided difference of F satisfying (1.2.4) or (1.2.5), then F is Fréchet-differentiable on D and we have

$$F'(x) = [x, x] \quad \text{for all} \quad x \in D. \tag{1.2.7}$$

Moreover, if (1.2.4) and (1.2.5) are both satisfied, then the Fréchet derivative F' is Lipschitz continuous on D with Lipschitz constant $I = c_0 + c_1$.

We shall also give an example of divided differences of the first and of the second order in the finite-dimensional case. We shall consider the space $|R^q$ equipped with the Chebysheff norm, which is given by

$$\|x\| = \max\{|x_i| \in \mathbf{R} : 1 \le I \le q\} \quad \text{for} \quad x = (x_1, x_2, ..., x_q) \in \mathbf{R}^q. \tag{1.2.8}$$

It follows that the norm of a linear operator $L \in LB(\mathbf{R}^q)$ represented by the matrix with entries I_{ij} is given by

$$\|L\| = \max\left\{\sum_{j=1}^{q} |I_{ij}| \mid |1 \le i \le q\right\}. \tag{1.2.9}$$

We cannot give a formula for the norm of a bilinear operator. However, if B is a bilinear operator with entries b_{ijk}, then we have the estimate

$$\|B\| \le \max\left\{\sum_{j=1}^{q} \sum_{k=1}^{q} |b_{ijk}| \mid |1 \le i \le q\right\}. \tag{1.2.10}$$

Let U be an open ball of $|R^q$ and let F be an operator defined on U with values in $\mathbf{R}^q$. We denote by $f_1, ..., f_q$ the components of F. For each $x \in U$ we have

$$F(x) = (f_1(x), ..., f_q(x))^T. \tag{1.2.11}$$

Moreover, we introduce the notation

$$D_j f_i(x) = \frac{\partial f(x)}{\partial x_j}, \quad D_{kj} f_i(x) = \frac{\partial^2 f_i(x)}{\partial x_j \partial x_k}. \tag{1.2.12}$$

Let x, y be two points of U and let us denote by $[x, y]$ the matrix with entries

$$[x, y]_{ij} = \frac{1}{x_j - y_j} \left(f_i(x_1, ..., x_j, y_{j+1}, ..., y_q) - f_i(x_1, ..., x_{j-1}, y_j, ..., y_q) \right). \tag{1.2.13}$$

The linear operator $[x, y] \in LB\,(\mathbf{R}^q)$ defined in this way obviously satisfies condition (1.2.1). If the partial derivatives $D_j f_i$ satisfy some Lipschitz conditions of the form

$$\left| D_j f_i\,(x_1, ..., x_k + t, ..., x_q) - D_j f_i\,(x_1, ..., x_k, ..., x_q) \right| \le p^i_{jk}\,|t| \qquad (1.2.14)$$

then condition (1.2.4) and (1.2.5) will be satisfied with

$$c_0 = \max \left\{ \frac{1}{2} \sum_{j=1}^{q} \left(p^i_{jj} + \sum_{k=j+1}^{q} p^i_{jk} \right) |1 \le i \le q \right\} \qquad (1.2.15)$$

and

$$c_1 = \max \left\{ \frac{1}{2} \sum_{j=1}^{q} \left(p^i_{jj} + \sum_{k=1}^{j-1} p^i_{jk} \right) |1 \le i \le q \right\}. \qquad (1.2.16)$$

We shall prove (1.2.4) only as (1.2.5) can be proved similarly.

Let x, y, z be three points of U. We shall have in turn

$$[x, y]_{ij} - [x, z] = \sum_{k=1}^{q} \left\{ [x, (y_1, ..., y_k, z_{k+1}, ..., z_q)]_{ij} \right.$$
$$\left. - [x\,(y_1, ..., y_{k-1}, z_k, ..., z_q)]_{ij} \right\} \quad \text{by (1.2.13). } (1.2.17)$$

If $k \le j$ then we have

$$[x, (y_1, ..., y_k, z_{k+1}, ..., z_q)]_{ij} - [x, (y_1, ..., y_{k-1}, z_k, ..., z_q)]_{ij}$$
$$= \frac{1}{x_j - z_j} \{ f_i\,(x_1, ..., x_j, z_{j+1}, ..., z_q) - f_i\,(x_1, ..., x_{j-1}, z_j, ..., z_q) \}$$
$$- \frac{1}{x_j - z_j} \{ f_i\,(x_1, ..., x_j, z_{j+1}, ..., z_q) - f_i\,(x_1, ..., x_{j-1}, z_j, ..., z_q) \} = 0.$$

For $k = j$ we have

$$\left| [x, (y_1, ..., y_j, z_{j+1}, ..., z_q)]_{ij} - [x, (y_1, ..., y_{j-1}, z_j, ..., z_q)_{ij}] \right|$$
$$= \left| \frac{1}{x_j - y_j} \{ f_i\,(x_1, ..., x_j, z_{j+1}, ..., z_q) - f_i\,(x_1, ..., x_{j-1}, y_i, z_{j+1}, ..., z_q) \} \right.$$
$$\left. - \frac{1}{x_j - y_j} \{ f_i\,(x_1, ..., x_j, z_{j+1}, ..., z_q) - f_i\,(x_1, ..., x_{j-1}, z_j, ..., z_q) \} \right|$$
$$= \left| \int_0^1 \{ D_j f_i\,(x_1, ..., x_j, y_j + t\,(x_j - y_j), z_{j+1}, ..., z_q) \right.$$
$$\left. - D_j f_i\,(x_1, ..., x_j, z_j + t\,(x_j - z_j), z_{j+1}, ..., z_q) \} dt \right|$$
$$\le |y_j - z_j|\,p^i_{jj} \int_0^1 t\,dt = \frac{1}{2} |x_j - z_j|\,p^i_{jj}$$

(by (1.2.14)).

Finally for $k > j$ we have using (1.2.13) and (1.2.17) again

$$\left|\left[x, (y_1, ..., y_k, z_{k+1}, ..., z_q)\right]_{ij} - \left[x, (y_1, ..., y_{k-1}, z_k, ..., z_q)\right]_{ij}\right|$$

$$= \left|\frac{1}{x_j - y_j}\left\{f_i\left(x_1, ..., x_j, y_{j+1}, ..., y_k, z_{k+1}, ..., z_q\right)\right.\right.$$

$$- f_i\left(x_1, ..., x_{j-1}, y_j, ..., y_k, z_{k+1}, ..., z_q\right)$$

$$- f_i\left(x_1, ..., x_j, y_{j+1}, ..., y_{k-1}, z_k, ..., z_q\right)$$

$$\left.+ f_i\left(x_1, ..., x_{j-1}, y_j, ..., y_{k-1}, z_k, ..., z_q\right)\right\}\right|$$

$$= \left|\int_0^1 \left\{f_i\left(x_1, ..., x_{j-1}, y_j + t\left(x_j - y_j\right), y_{j+1}, ..., y_k, z_{k+1}, ..., z_q\right)\right.\right.$$

$$\left.\left.- f_i\left(x_1, ..., x_{j-1}, y_j + t\left(x_j - y_j\right), y_{j+1}, ..., y_{k-1}, z_k, ..., z_q\right)\right\} dt\right|$$

$$\leq |y_k - z_k|\, p_{jk}^i.$$

By adding all the above, we get

$$\left|[x, y]_{ij} - [x, z]_{ij}\right| \leq \frac{1}{2}|y_j - z_j|\, p_{jj}^i + \sum_{k=j+1}^q |y_k - z_k|\, p_{jk}^i$$

$$\leq \|y - z\|\left\{\frac{1}{2}\sum_{j=1}^q \left(p_{jj}^i + \sum_{k=j+1}^q p_{jk}^i\right)\right\}.$$

Consequently, condition (1.2.4) is satisfied with c_0 given by (1.2.15). If each f_j has continuous second-order partial derivatives that are bounded on U, we have

$$p_{jk}^i = \sup\left\{\left|D_{jk} f_i\left(x\right)\right| \,|x \in U\right\}.$$

In this case $p_{jk}^i = p_{kj}^i$ so that $c_0 = c_1$.

Moreover, consider again three points x, y, z of U. Similarly with (1.2.17), the second divided difference of F at x, y, z is the bilinear operators defined by

$$[x, y, z]_{ijk} = \frac{1}{y_k - z_k}\left\{\left[x, (y_1, ..., y_k, z_{k+1}, ..., z_q)\right]_{ij}\right.$$

$$\left.- \left[x, (y_1, ..., y_{k-1}, z_k, ..., z_q)\right]_{ij}\right\}. \quad (1.2.18)$$

It is easy to see as before that $[x, y, z]_{ijk} = 0$ for $k < j$. For $k = j$ we have

$$[x, y, z]_{ijj} = \left[x_j, y_j, z_j\right]_t f_i\left(x_1, ..., x_{i-1}, t, z_{j+1}, ..., z_q\right) \quad (1.2.19)$$

where the right-hand side of (1.2.19) represents the divided difference of $f_i(x_1, ..., x_{j-1}, t, z_{j+1}, ..., z_q)$ as a function of t, at the points x_j, y_j, z_j. Using Genocchi's integral representation of divided differences of scalar functions [154], we get

$$[x, y, z]_{ijj} = \int_0^1 \int_0^1 t D_{jj} f_i \left(x_1, ..., x_{j-1}, x_j\right.$$
$$\left. + t \left(y_j - x_j\right) + ts \left(z_j - y_j\right), z_{j+1}, ..., z_q\right) ds dt. \quad (1.2.20)$$

Hence, for $k > j$ we obtain

$$[x, y, z]_{ijk} = \frac{1}{(y_k - z_k)(x_j - y_j)} \left\{ f_i \left(x_1, ..., x_j, y_{j+1}, ..., y_k, z_{k+1}, ..., z_q\right) \right.$$

$$- f_i \left(x_1, ..., x_j, x_{j+1}, ..., y_{k-1}, z_k, ..., z_q\right)$$

$$- f_i \left(x_1, ..., x_{j-1}, y_j, ..., y_k, z_{k+1}, ..., z_q\right)$$

$$\left. + f_i \left(x_1, ..., x_{j-1}, y_j, ..., y_{k-1}, z_k, ..., z_q\right) \right\}$$

$$\times \frac{1}{x_j - y_j} \int_0^1 \left\{ D_k f_i \left(x_1, ..., x_j, y_{j+1}, ..., y_{k-1}, z_k + t \left(y_k - z_k\right), z_{k+1}, ..., z_q\right) \right.$$

$$\left. - D_k f_i \left(x_1, ..., x_{j-1}, y_j, ..., y_{k-1}, z_k + t \left(y_k - z_k\right), z_{k+1}, ..., z_q\right) \right\} dt$$

$$= \int_0^1 \int_0^1 D_{kj} f_i \left(x_1, ..., x_{j-1}, y_j\right.$$

$$\left. + s \left(x_j - y_i\right), y_{j+1}, ..., y_{k-1}, z_k + t \left(y_k - z_k\right), z_{k+1}, ..., z_q\right) ds dt. \quad (1.2.21)$$

We now want to show that if

$$\left| D_{kj} f_i \left(v_1, ..., v_m + t, ..., v_q\right) - D_{kj} f_i \left(v_1, ..., v_m, ... v_q\right) \right| \le q_{km}^{ij} |t|$$
$$\text{for all} \quad v = \left(v_1, ..., v_q\right) \in U, \quad 1 \le i, j, k, m \le q, \quad (1.2.22)$$

then the divided difference of F of the second order defined by (1.2.18) satisfies condition (1.2.6) with the constant

$$c_2 = \max_{1 \le i \le q} \sum_{j=1}^q \left\{ \frac{1}{6} q_{jj}^{ij} + \frac{1}{2} \sum_{m=1}^{j-1} q_{jm}^{ij} + \frac{1}{2} \sum_{k=j+1}^q q_{kj}^{ij} + \sum_{k=j+1}^q \sum_{m=1}^{j-1} q_{km}^{ij} \right\}. \quad (1.2.23)$$

Let u, x, y, z be four points of U. Then using (1.2.18), we can easily have

$$[x, y, z]_{ijk} - [u, y, z]_{ijk} = \sum_{m=1}^q \left\{ \left[\left(x_1, ..., x_m, u_{m+1}, ..., u_q\right), y, z\right]_{ijk} \right.$$

$$\left. \left[\left(x_1, ..., x_{m-1}, u_m, ..., u_q\right), y, z\right]_{ijk} \right\}. \quad (1.2.24)$$

If $m = j$, the terms in (1.2.24) vanish so that using (1.2.21) and (1.2.22), we deduce that for $k > j$

$$\left| [x, y, z]_{ijk} - [u, y, z]_{ijk} \right|$$

$$= \left| \sum_{m=1}^{j-1} \int_0^1 \int_0^1 \left\{ D_{kj} f_i \left(x_1, ..., x_m, u_{m+1}, ..., u_{j-1}, y_j + s\,(x_j - y_j), \right.\right.$$

$$y_{j+1}, ..., y_{k-1}, z_k + t\,(y_k - z_k), z_{k+1}, ..., z_q \bigr)$$

$$- D_{kj} f_i \left(x_1, ..., x_{m-1}, u_m, ..., u_{j-1}, y_j + s\,(x_j - y_j), \right.$$

$$\left. y_{j+1}, ..., y_{k-1} z_k + t\,(y_k - z_k), z_{k+1}, ..., z_q \right) \bigr\} ds\,dt$$

$$+ \int_0^1 \int_0^1 \left\{ D_{kj} f_i \left(x_1, ..., x_{j-1}, y_j + s\,(x_j - y_j), \right.\right.$$

$$y_{j+1}, ..., y_{k-1}, z_k + t\,(y_k - z_k), z_{k+1}, ..., z_q \bigr)$$

$$- D_{kj} f_i \left(x_1, ..., x_{j-1}, y_j + s\,(u_j - y_j), \right.$$

$$\left. y_{j+1}, ..., y_{k-1}, z_k + t\,(y_k - z_k), z_{k+1}, ..., z_q \right) \bigr\} ds\,dt \right|$$

$$\leq \frac{1}{2} |x_j - u_j|\, q_{kj}^{ij} + \sum_{m=1}^{j-1} |x_m - u_m|\, q_{km}^{ij}.$$

Similarly for $k = j$, we obtain in turn

$$\left| [x, y, z]_{ijj} - [u, y, z]_{ijj} \right|$$

$$= \left| \int_0^1 \int_0^1 t \left\{ D_{jj} f_i \left(x_1, ..., x_{j-1}, x_j + t\,(y_j - x_j) + ts\,(z_j - y_j), z_{j+1}, ..., z_q \right) \right.\right.$$

$$\left. - D_{jj} f_i \left(x_1, ..., x_{j-1}, u_j + t\,(y_j - u_j) + ts\,(z_j - y_j), z_{j+1}, ..., z_q \right) \right\} ds\,dt$$

$$+ \sum_{m=1}^{j-1} \int_0^1 \int_0^1 t \left\{ D_{jj} f_i \left(x_1, ..., x_m, u_{m+1}, ..., u_{j-1}, \right.\right.$$

$$x_j + t\,(x_j - y_j) + ts\,(z_j - y)\,j, z_{j+1}, ..., z_q \bigr)$$

$$- D_{jj} f_i \left(x_1, ..., x_{m-1}, u_m, ..., u_{j-1}, \right.$$

$$\left. x_j + t\,(y_j - x_j) + ts\,(z_j - y_j), z_{j+1}, ..., z_q \right) \bigr\} ds\,dt \right|$$

$$\leq \frac{1}{6} |x_j - u_j|\, q_{jj}^{ij} + \frac{1}{2} \sum_{m=1}^{j-1} |x_m - u_m|\, q_{jm}^{ij}.$$

Finally using the estimate (1.2.10) of the norm of a bilinear operator, we deduce that condition (1.2.6) holds with c_2 given by (1.2.23).

We make an introduction to the problem of approximating a locally unique solution x^* of the nonlinear operator equation $F(x) = 0$, in a POTL-space X. In particular, consider an operator $F: D \subseteq X \to Y$ where X is a POTL-space with values in a POTL-space Y. Let x_0, y_0, y_{-1} be three points of D such that

$$x_0 \leq y_0 \leq y_{-1}, \quad [x_0, y_{-1}],$$

and denote

$$D_1 = \left\{ (x, y) \in X^2 \mid x_0 \leq x \leq y \leq y_0 \right\},$$

$$D_2 = \left\{ (y, y_{-1}) \in X^2 \mid x_0 \leq u \leq y_0 \right\},$$

$$D_3 = D_1 \cup D_2. \tag{1.2.25}$$

Assume there exist operators $A_0: D_3 \rightarrow LB(X, Y)$, $A: D_1 \rightarrow L(X, Y)$ such that:

(a)

$$F(y) - F(x) \leq A_0(w, z)(y - x)$$
$$\text{for all} \quad (x, y), (y, w) \in D_1, \ (w, z) \in D_3; \tag{1.2.26}$$

(b) the linear operator $A_0(u, v)$ has a continuous nonsingular nonnegative left subinverse;

(c)

$$F(y) - F(x) \geq A(x, y)(y - x) \quad \text{for all} \quad (x, y) \in D_1; \tag{1.2.27}$$

(d) the linear operator $A(x, y)$ has a nonnegative left superinverse for each $(x, y) \in D_1$

$$F(y) - F(x) \leq A_0(y, z)(y - x) \quad \text{for all} \quad x, y \in D_1, \ (y, z) \in D_3. \tag{1.2.28}$$

Moreover, let us define approximations

$$F(y_n) + A_0(y_n, y_{n-1})(y_{n+1} - y_n) = 0 \tag{1.2.29}$$

$$F(x_n) + A_0(y_n, y_{n-1})(x_{n+1} - x_n) = 0 \tag{1.2.30}$$

$$y_{n+1} = y_n - B_n F(y_n) n \geq 0 \tag{1.2.31}$$

$$x_{n+1} = x_n - B_n^1 F(x_n) \quad n \geq 0, \tag{1.2.32}$$

where B_n and B_n^1 are nonnegative subinverses of $A_0(y_n, y_{n-1}) n \geq 0$.

Under very natural conditions, hypotheses of the form (1.2.26) or (1.2.27) or (1.2.28) have been used extensively to show that the approximations (1.2.26) and (1.2.30) or (1.2.31) and (1.2.32) generate two sequences $\{x_n\} n \geq 1, \{y_n\} n \geq 1$ such that

$$x_0 \leq x_1 \leq \cdots \leq x_n \leq x_{n+1} \leq y_{n+1} \leq y_n \leq \cdots \leq y_1 \leq y_0 \tag{1.2.33}$$

$$\lim_{n \to \infty} x_n = x^* = y^* = \lim_{n \to \infty} y_n \quad \text{and} \quad F(x^*) = 0. \tag{1.2.34}$$

For a complete survey on these results, we refer to the works of Potra [164] and Argyros and Szidarovszky [42]–[44].

Here we will use similar conditions (i.e., like (1.2.26), (1.2.27), (1.2.28)) for two-point approximations of the form (1.2.29) and (1.2.30) or (1.2.31) and (1.2.32).

Consequently, a discussion must follow on the possible choices of the linear operators A_0 and A.

Remark 1.2.12. Let us now consider an operator $F: D \subseteq X \to Y$, where X, Y are both POTL-spaces. The operator F is called order-convex on an interval $[x_0, y_0] \subseteq D$ if

$$F(\lambda x + (1 - \lambda) y) \le \lambda F(x) + (1 - \lambda) F(y) \qquad (1.2.35)$$

for all comparable $x, y \in [x_0, y_0]$ and $\lambda \in [0, 1]$. If F has a linear G-derivative $F'(x)$ at each point $[x_0, y_0]$, then (1.2.35) holds if and only if

$$F'(x)(y - x) \le F(y) - F(x) \le F'(y)(y - x) \quad \text{for} \quad x_0 \le x \le y \le y_0. \quad (1.2.36)$$

(See, e.g., Ortega and Rheinboldt [154] or Exercise 1.4.65 for the properties of the Gâteaux derivative.)

Hence, for order-convex G-differentiable operators, conditions (1.2.27) and (1.2.28) are satisfied with $A_0(y, v) = A(y, v) = F'(u)$. In the unidimensional case, (1.2.36) is equivalent with the isotony of the operator $x \to F'(x)$ but in general the latter property is stronger. Assuming the isotony of the operator $x \to F'(x)$, it follows that

$$F(y) - F(x) \le F'(w)(y - x) \quad \text{for} \quad x_0 \le x \le y \le w \le y_0.$$

Hence, in this case condition (1.2.26) is satisfied for $A_0(w, z) = F'(w)$.

The above observations show one to choose A and A_0 for single- or two-step Newton methods. We note that the iterative algorithm (1.2.29)–(1.2.30) with

$$A_0(u, v) = F'(u)$$

is the algorithm proposed by Fourier in 1818 in the unidimensional case and extended by Baluev in 1952 in the general case. The idea of using an algorithm of the form (1.2.31)–(1.2.32) goes back to Slugin [183]. In Ortega and Rheinboldt [154], it is shown that with B_n properly chosen, (1.2.31) reduces to a general Newton-SOR algorithm. In particular, suppose (in the finite-dimensional case) that $F'(y_n)$ is an M-matrix and let $F'(y) = D_n - L_n - U_n$ be the partition of $F'(y_n)$ into diagonal, strictly lower- and strictly upper-triangular parts, respectively, for all $n \ge 0$. Consider an integer $m_n \ge 1$, a real parameter $w_n \in (0, 1]$ and denote

$$P_n = w_n^{-1}(D_n - w_n L_n), \qquad Q_n = w_n^{-1}[(1 - w_n) D_n + w_n U_n], \qquad (1.2.37)$$

$$H_n = P_n^{-1} Q_n, \qquad \text{and} \qquad B_n = \left(I + H_n + \cdots + H_n^{m_n - 1}\right) P_n^{-1}. \quad (1.2.38)$$

It can easily be seen that B_n $n \ge 0$ is a nonnegative subinverse of $F'(y_n)$ (see also [164]). If $f: [a, b] \to | R$ is a real function of a real variable, then f is (order) convex if and only if

$$\frac{f(x) - f(y)}{x - y} \le \frac{f(u) - f(v)}{u - v}$$

for all x, y, u, v from $[a, b]$ such that $x \le u$ and $y \le v$.

This fact motivates the notion of convexity with respect to a divided difference considered by J.W. Schmidt and H. Leonhardt [181]. Let $F: D \subseteq X \to Y$ be a nonlinear operator between the POTL-spaces X and Y. Assume that the nonlinear operator F has a divided difference $[\cdot, \cdot]$ on D. F is called convex with respect to the divided difference $[\cdot, \cdot]$ on D if

$$[x, y] \leq [u, v] \quad \text{for all } x, y, u, v \in D \text{ with } x \leq u \text{ and } y \leq v. \tag{1.2.39}$$

In the above quoted study, Schmidt and Leonhardt studied (1.2.29)–(1.2.30) with $A_0(u, v) = [u, v]$ in case the nonlinear operator F is convex with respect to $[\cdot, \cdot]$. Their result was extended by N. Schneider [43] who assumed the milder condition

$$[u, v](u - v) \geq F(u) - F(v) \quad \text{for all comparable } u, v \in D. \tag{1.2.40}$$

An operator $[\cdot, \cdot]: D \times D \to L(X, Y)$ satisfying (1.2.40) is called a generalized divided difference of F on D. If both (1.2.39) and (1.2.40) are satisfied, then we say that F is convex with respect to the generalized divided difference of $[\cdot, \cdot]$. It is easily seen that if (1.2.39) and (1.2.40) are satisfied on $D = [x_0, y_{-1}]$, then conditions (1.2.26) and (1.2.27) are satisfied with $A = A_0 = [\cdot, \cdot]$. Indeed for $x_0 \leq x \leq y \leq w \leq z \leq y_{-1}$ we have

$$[x, y](y - x) \leq F(y) - F(x) \leq [y, x](y - x) \leq [w, z](y - x).$$

The monotonicity results can also be used to study general secant-SOR methods, in case the generalized difference $[y_n, y_{n-1}]$ is an M-matrix and if B_n $n \geq 0$ is computed according to (1.2.37) and (1.2.38) where $[y_n, y_{n-1}] = D_n - L_n - U_n$ $n \geq 0$ is the partition of $[y_n, y_{n-1}]$ into its diagonal, strictly lower- and strictly upper-triangular parts.

We remark that an operator that is convex with respect to a generalized divided difference is also order-convex. To see that, consider $x, y \in D$, $x \leq y$, $\lambda \in [0, 1]$ and set $z = \lambda x + (I - \lambda) y$. Observing that $y - x = (1 - \lambda)^{-1}(z - x) = \lambda^{-1}(y - z)$ and applying (1.2.40), we have in turn:

$$\begin{aligned}
(1 - \lambda)^{-1}(F(z) - F(x)) &\leq (1 - \lambda)^{-1}[z, x](z - x) \\
&= [z, x](y - x) \leq [z, y](y - x) \\
&= \lambda^{-1}[z, y](y - z) \leq^1 (F(y) - F(z)).
\end{aligned}$$

By the first and last term we deduce that $F(z) \leq \lambda F(x) + (1 - \lambda) F(y)$. Thus, Schneider's result can be applied only to order-convex operators, and its importance resides in the fact that the use of a generalized divided difference instead of the G-derivative may be more advantageous from a numerical point of view. We note, however, that conditions (1.2.26) and (1.2.27) do not necessarily imply convexity. For example, if f is a real function of a real variable such that

$$\inf_{x, y \in [x_0, y_0]} \frac{f(x) - f(y)}{x - y} = m > 0, \qquad \sup_{x, y \in [x_0, y_0]} \frac{f(x) - f(y)}{x - y} = M < \infty$$

then (1.2.26) and (1.2.27) are satisfied for $A_0(u, v) = M$ and $A(u, v) = m$. It is not difficult to find examples of nonconvex operators in the finite- (or even in the infinite-) dimensional case satisfying a condition of the form

$$A(y - x) \leq F(y) - F(x) \leq A_0(y, x), \quad x_0 \leq x \leq y \leq y_0,$$

where A_0 and A are fixed linear operators. If A_0 has a continuous nonsingular nonnegative left subinverse and A has a nonnegative right superinverse, then convergence of the algorithm (1.2.29)–(1.2.30) can be discussed. This algorithm becomes extremely simple in this case. The monotone convergence of such an iterative procedure seems to have been first investigated by S. Slugin [183].

In the end of this section, we shall consider a class of nonconvex operators that satisfy condition (1.2.26) but do not necessarily satisfy condition (1.2.27). Consequently from convergence theorems involving (1.2.29) and (1.2.30), it will follow that Jacobi-Newton and the Jacobi-secant methods have monotonous convergence for operators belonging to this class (see also the elegant papers by F. Potra [163], [164]).

Let $F = (f_1, ..., f_q)^{T0}$ be an operator acting in the finite-dimensional space $\mathbf{R}^q$, endowed with the natural (componentwise) partial ordering. Let us denote by e_i the ith coordinate vector of $\mathbf{R}^q$. We say that F is off-diagonally antitone if the functions

$$g_{ij} : \mathbf{R} \to \mathbf{R}, \quad g_{ij}(t) = f_i(x + te_j), \quad i \neq j, \; i, j = 1, ..., q$$

are antitone. Suppose that at each point x belonging to an interval $U = [x_0, y_{-1}]$ the partial derivatives $\partial_i F_i(x)$, $i = 1, 2, ..., q$, exist and are positive. For any two points $x, y \in U$, we consider the quotients

$$[x, y]_i = \begin{cases} \frac{f_i(y) - f_i(y - e_i^T(y-x)e_i)}{e_i^T(y-x)}, & \text{if } e_i^T(y - x) \neq 0 \\ \partial_i F_i(x) & \text{if } e_i^T(y - x) = 0. \end{cases} \tag{1.2.41}$$

Let us denote $\Delta[x, y]$ the diagonal matrix having as elements the number $[x, y]_i$, $i = 1, 2, ..., q$. For the diagonal matrix $\Delta[x, y]$ formed by the partial derivatives $\partial_i f_i(x)$, $i = 1, 2, ..., q$, we shall also use the notation $DF(x)$.

Suppose now that F is off-diagonally antitone and that the operator $DF : U \to LB(\mathbf{R}^q)$ is isotone (i.e., all functions $\partial_i F_i : \mathbf{R} \to \mathbf{R}$ are isotone). In this case for all $x_0 \leq x \leq y \leq w \leq z \leq y_{-1}$ and all $i \in \{1, 2, ..., q\}$, there exist $\lambda, \mu \in [0, 1]$ such that

$$f_i(y) - f_i(x) \leq f_i(y) - f_i\left(y - e_i^T(y - x)e_i\right) = \partial_i f_i\left(y - \lambda e_i^T(y - x)e_i\right),$$

$$e_i^T(y - x) \leq \partial_i f_i(y) e_i^T(y - x) \leq \partial_i f_i(w) e_i^T(y - x)$$

$$\leq \partial_i f_i\left(z - \mu e_i^T(z - w)\right) e_i^T(y - x) = [w, z]_i e_i^T(y - x).$$

It follows that condition (1.2.26) is satisfied for $A_0(w, z) = DF(w)$ as well as for $A_0(w, z) = [\Delta w, z]$. With the choice $A_0(w, z) = \Delta[w, z]$, the iterative procedure (1.2.29) is a Jacobi-secant method whereas with the choice $A_0(w, z) = DF(w)$ it reduces to the Jacobi-Newton method. For some applications of the latter method, see W. Torning [190].

We now connect divided differences and Fréchet derivatives.

Let F be a nonlinear operator defined on an open subset D of a Banach space X with values in a Banach space Y.

Let x, y be two points in D and suppose that the segment

$$S = \{x + t(y - x) \mid t \in [0, 1]\} \subseteq D.$$

Let y' be a continuous linear functional, set $h = y - x$ and define

$$\varphi(t) = \big(F(x + th), y'\big).$$

If F is Fréchet-differentiable at each point of the segment S, then φ is differentiable on $[0, 1]$ and

$$\varphi(t) = \big(F'(x + th), y'\big).$$

Let us now suppose that

$$\alpha = \sup_{t \in [0,1]} \big\| F'(x + t(y - x)) \big\| < \infty;$$

then we have

$$\big\| (F(y) - F(x), y') \big\| = \| \varphi(1) - \varphi(0) \|$$
$$\leq \sup_{t \in [0,1]} \| \varphi'(t) \| \leq \alpha \, \| y' \| \, \| y - x \|.$$

But, we also have

$$\| F(y) - F(x) \| = \sup_{\| y' \| \leq 1} \big\| (F(y) - F(x), y') \big\|,$$

we deduce that

$$\| F(y) - F(x) \| \leq \sup_{t \in [0,1]} \big\| F'(x + t(y - x)) \big\| \cdot \| y - x \| \qquad (1.2.42)$$

so, we proved:

Theorem 1.2.13. *Let D be a convex subset of a Banach space X and $F: D \subseteq X \to Y$. If F is Fréchet-differentiable on D and if there exists a constant c such that*

$$\big\| F'(x) \big\| \leq M \quad \text{for all} \ \ x \in D \Rightarrow \| F(x) - F(y) \| \leq c \, \| x - y \| \quad \text{for all} \ \ x \in D. \qquad (1.2.43)$$

The estimate (1.2.42) is the analogue of the famous mean value formula from real analysis. If the operator F' is Riemann integrable on the segment S, we can give the following integral representation of the mean value formula

$$F(x) - F(y) = \int_0^1 F'(x + t(y - x))\, dt\,(x - y). \qquad (1.2.44)$$

Let now D be a convex open subset of X and let us suppose that we have associated to each pair (x, y) of distinct points from D a divided difference $[x, y]$ of F at these points. In applications, one often has to require that the operator $(x, y) \to [x, y]$ satisfies a Lipschitz condition. We suppose that there exists a nonnegative $c > 0$ such that

$$\|[x, y] - [x_1, y_1]\| \le c\,(\|x - x_1\| + \|y - y_1\|) \qquad (1.2.45)$$

for all $x, y, x_1, y_1 \in D$ with $x \ne y$ and $x_1 = y_1$.

We say in this case that F has a Lipschitz continuous difference on D. This condition allows us to extend by continuity the operator $(x, y) \to [x, y]$ to the whole Cartesian product $D \times D$. It follows that F is Fréchet-differentiable on D and that $[x, x] = F'(x)$. It also follows that

$$\left\| F'(x) - F'(y) \right\| \le c_1 \|x - y\| \quad \text{with} \quad c_1 = 2c \qquad (1.2.46)$$

and

$$\left\| [x, y] - F'(z) \right\| \le c\,(\|x - z\| + \|y - z\|) \qquad (1.2.47)$$

for all $x, y \in D$. Conversely, if we assume that F is Fréchet-differentiable on D and that its Fréchet derivative satisfies (1.2.46), then it follows that F has a Lipschitz continuous divided difference on D. We can certainly take

$$[x, y] = \int_0^1 F'(x + t(y - x))\, dt. \qquad (1.2.48)$$

We now want to give the definition of the second Fréchet derivative of F. We must first introduce the definition of bounded multilinear operators (which will also be used later).

Definition 1.2.14. *Let X and Y be two Banach spaces. An operator $A: X^n \to y$ ($n \in N$) will be called n-linear operator from X to Y if the following conditions are satisfied:*

(a) The operator $(x_1, ..., x_n) \to A(x_1, ..., x_n)$ is linear in each variable x_k $k = 1, 2, ..., n$.

(b) There exists a constant c such that

$$\|A(x_1, x_2, ..., x_n)\| \le c\,\|x_1\| ... \|x_n\|. \qquad (1.2.49)$$

The norm of a bounded n-linear operator can be defined by the formula

$$\|A\| = \sup\{\|A(x_1, ..., x_n)\| \mid \|x_n\| = 1\}. \qquad (1.2.50)$$

Set $LB^{(1)}(X, Y) = LB(X, Y)$ and define recursively

$$LB^{(k+1)}(X, Y) = LB\left(X, LB^{(k)}(X, Y)\right), \quad k \ge 0. \tag{1.2.51}$$

In this way we obtain a sequence of Banach spaces $LB^{(n)}(X, Y)$ $(n \ge 0)$. Every $A \in LB^{(n)}(X, Y)$ can be viewed as a bounded n-linear operator if one takes

$$A(x_1, ..., x_n) = (... (A(x_1)(x_2)(x_3)) ...)(x_n). \tag{1.2.52}$$

In the right-hand side of (1.2.52) we have

$$A(x_1) \in LB^{(n-1)}(X, Y), \quad (A(x_1))(x_2) \in LB^{(n-2)}(X, Y), \quad etc.$$

Conversely, any bounded n-linear operator A from X to Y can be interpreted as an element of $B^{(n)}(X, Y)$. Moreover, the norm of A as a bounded n-linear operator coincides with the norm as an element of the space $LB^{(n)}(X, Y)$. Thus we may identify this space with the space of all bounded n-linear operators from X to Y. In the sequel, we will identify $A(x, x, ..., x) = Ax^n$, and

$$A(x_1)(x_2) ... (x_n) = A(x_1, x_2, ..., x_n) = Ax_1x_2...x_n.$$

Let us now consider a nonlinear operator $F: D \subseteq X \to Y$ where D is open. Suppose that F is Fréchet-differentiable on D. Then we may consider the operator $F': D \to LB(X, Y)$ that associates to each point x the Fréchet derivative of F at x. If the operator F' is Fréchet-differentiable at a point $x_0 \in D$, then we say that F is twice Fréchet-differentiable at x_0. The Fréchet derivative of F' at x_0 will be denoted by $F''(x_0)$ and will be called the second Fréchet derivative of F at x_0. Note that $F''(x_0) \in LB^{(2)}(X, Y)$. Similarly, we can define Fréchet derivatives of higher order. Finally by analogy with (1.2.48)

$$[x_0, ..., x_k] = \int_0^1 \cdots \int_0^1 t_1^{k-1} t_2^{k-2} \cdots t_{k-1} F(x_0 + t_1(x_1 - x_0) + t_1 t_2(x_2 - x_1)$$
$$+ \cdots + t_1 t_2, ..., t_k(x_k - x_{k-1})) dt_1 dt_2 ... dt_k. \tag{1.2.53}$$

It is easy to see that the multilinear operators defined above verify

$$[x_0, ..., x_{k-1}, x_k, x_{k+1}](x_k - x_{k+1}) = [x_0, ..., x_{k-1}, x_{k+1}]. \tag{1.2.54}$$

We note that throughout this sequel, a 2-linear operator will also be called bilinear.

Finally, we will also need the definition of a *n*-linear symmetric operator. Given a *n*-linear operator $A: X^n \to Y$ and a permutation $i = (i_1, i_2, ..., i_n)$ of the integers $1, 2, ..., n$, the notation $A(i)$ (or $A_n(i)$ if we want to emphasize the *n*-linearity of A) can be used for the *n*-linear operator $A(i) = A_n(i)$ such that

$$A(i)(x_1, x_2, ..., x_n) = A_n(i)(x_1, x_2, ..., x_n) = A_n\left(x_{i_1}, x_{i_2}, ..., x_{i_n}\right) A_n x_{i_1} x_{i_2}...x_{i_n} \tag{1.2.55}$$

for all $x_1, x_2, ..., x_n \in X$. Thus, there are $n!$ *n*-linear operators $A(i) = A_n(i)$ associated with a given *n*-linear operator $A = A_n$.

Definition 1.2.15. *A n-linear operator $A = A_n: X^n \to Y$ is said to be symmetric if*

$$A = A_n = A_n\,(i) \tag{1.2.56}$$

for all i belonging in R_n, which denotes the set of all permutations of the integers 1, 2, ..., n. The symmetric n-linear operator

$$\overline{A} = \overline{A}_n = \frac{1}{n!} \sum_{i \in R_n} A_n\,(i) \tag{1.2.57}$$

is called the mean of $A = A_n$.

Notation 1.2.16. *The notation*

$$A_n X^P = A_n xx...x\ (p\text{-}times) \tag{1.2.58}$$

$p \leq n$, $A = A_n: X^n \to Y$, *for the result of applying A_n to $x \in X$ p times will be used. If $p < n$, then (1.2.58) will represent a $(n - p)$-linear operator. For $p = n$, note that*

$$A_k x^k = \overline{A}_k x^k = A_k\,(i)\,x^k \tag{1.2.59}$$

for all $i \in R_k$, $x \in X$. It follows from (1.2.59) that whenever we are dealing with an equation involving n-linear operators A_n, we may assume that they are symmetric without loss of generality, as each A_n may be replaced by $\overline{A}_n$ without changing the value of the expression at hand.

1.3 Fixed points of operators

The ideas of a contraction operator and its fixed points are fundamental to many questions in applied mathematics. We outline the essential ideas.

Definition 1.3.1. *Let F be an operator mapping a set X into itself. A point $x \in X$ is called a fixed point of F if*

$$x = F(x). \tag{1.3.1}$$

Equation (1.3.1) leads naturally to the construction of the method of successive approximations or substitutions or the Picard iteration

$$x_{n+1} = F\,(x_n)\ (n \geq 0)\ \ x_0 \in X. \tag{1.3.2}$$

If sequence $\{x_n\}$ $(n \geq 0)$ converges to some point $x^* \in X$ for some initial guess $x_0 \in X$, and F is a continuous operator in a Banach space X, we can have

$$x^* = \lim_{n \to \infty} x_{n+1} = \lim_{n \to \infty} F\,(x_n) = F\left(\lim_{n \to \infty} x_n\right) = F\,(x^*).$$

That is, x^* is a fixed point of operator F. Hence, we showed:

Theorem 1.3.2. *If F is a continuous operator in a Banach space X, and the sequence $\{x_n\}$ ($n \geq 0$) generated by (1.3.2) converges to some point $x^* \in X$ for some initial guess $x_0 \in X$, then x^* is a fixed point of operator F.*

We need information on the uniqueness of x^* and the distances $\|x_n - x^*\|$, $\|x_{n+1} - x_n\|$ ($n \geq 0$). That is why we introduce the concept:

Definition 1.3.3. *Let $(X, \| \, \|)$ be a metric space and F a mapping of X into itself. The operator F is said to be a contraction or a contraction mapping if there exists a real number c, $0 \leq c < 1$, such that*

$$\|F(x) - F(y)\| \leq c \|x - y\|, \quad \text{for all } x, y \in X. \tag{1.3.3}$$

It follows immediately from (1.3.3) that every contraction mapping F is uniformly continuous. Indeed, F is Lipschitz continuous with a Lipschitz constant c. The point c is called the contraction constant for F.

We now arrive at the Banach contraction mapping principle.

Theorem 1.3.4. *Let $(X, \|\cdot\|)$ be a Banach space and $F: X \to X$ be a contraction mapping. Then F has a unique fixed point.*

Proof. Uniqueness: Suppose there are two fixed points x, y of F. Because F is a contraction mapping,

$$\|x - y\| = \|F(x) - F(y)\| \leq c \|x - y\| < \|x - y\|,$$

which is impossible. That shows uniqueness of the fixed point of F.

Existence: Using (1.3.2) we obtain

$$\|x_2 - x_1\| \leq c \|x_1 - x_0\|$$
$$\|x_3 - x_2\| \leq x \|x_2 - x_1\| \leq c^2 \|x_1 - x_0\| \tag{1.3.4}$$
$$\vdots$$
$$\|x_{n+1} - x_n\| \leq c^n \|x_1 - x_0\|$$

and, so

$$\|x_{n+m} - x_n\| \leq \|x_{n+m} - x_{n+m-1}\| + \cdots + \|x_{n+1} - x_n\|$$

$$\leq \left(c^{m+1} + \cdots + c + 1 \right) c^n \|x_1 - x_0\|$$

$$\leq \frac{c^n}{1 - c} \|x_1 - x_0\|.$$

Hence, sequence $\{x_n\}$ ($n \geq 0$) is Cauchy in a Banach space X and such it converges to some x^*. The rest of the theorem follows from Theorem 1.3.2.

Remark 1.3.5. It follows from (1.3.1), (1.3.3) and $x^* = F(x^*)$ that

$$\left\|x_n - x^*\right\| = \left\|F(x_{n-1}) - F(x^*)\right\| \le c\left\|x_{n-1} - x^*\right\| \le c^n\left\|x_0 - x^*\right\| \quad (n \ge 1).$$
(1.3.5)

Inequality (1.3.5) describes the convergence rate. This is convenient only when an a priori estimate for $x_0 - x^*$ is available. Such an estimate can be derived from the inequality

$$\left\|x_0 - x^*\right\| \le \left\|x_0 - F(x_0)\right\| + \left\|F(x_0) - F(x^*)\right\| \le \left\|x_0 - F(x_0)\right\| + c\left\|x_0 - x^*\right\|,$$

which leads to

$$\left\|x_0 - x^*\right\| \le \frac{1}{1-c}\left\|x_0 - F(x_0)\right\|.$$
(1.3.6)

By (1.3.5) and (1.3.6), we obtain

$$\left\|x_n - x^*\right\| \le \frac{c^n}{1-c}\left\|x_0 - F(x_0)\right\| \quad (n \ge 1).$$
(1.3.7)

Estimates (1.3.5) and (1.3.7) can be used to determine the number of steps needed to solve Equation (1.3.1). For example if the error tolerance is $\varepsilon > 0$, that is, we use $\left\|x_n - x^*\right\| < \varepsilon$, then this will certainly hold if

$$n > \frac{1}{\ln c} \ln \frac{\varepsilon(1-c)}{\left\|x_0 - F(x_0)\right\|}.$$
(1.3.8)

Example 1.3.6. Let $F: \mathbf{R} \to \mathbf{R}$, $q > 1$, $F(x) = qx + 1$. Operator F is not a contraction, but it has a unique fixed point $x = (1-q)^{-1}$.

Example 1.3.7. Let $F: X \to X$, $x = \left(0, 1/\sqrt{6}\right]$, $F(x) = x^3$. We have

$$|F(x) - F(y)| = \left|x^3 - y^3\right| \le \left(|x|^2 + |x| \cdot |y| + |y|^2\right)|x - y| \le \tfrac{1}{2}|x - y|.$$

That is, F is a contraction with $c = \tfrac{1}{2}$, with no fixed points in X. This is not violating Theorem 1.3.4 because X is not a Banach space.

Example 1.3.8. Let $F: [a, b] \to [a, b]$, F differentiable at every $x \in (a, b)$ and $\left|F'(x)\right| \le c < 1$. By the mean value theorem, if $x, y \in [a, b]$ there exists a point z between x and y such that

$$F(x) - F(y) = F'(z)(z - y),$$

from which it follows that F is a contraction with constant c.

Example 1.3.9. Let $F: [a, b] \to \mathbf{R}$. Assume there exist constants p_1, p_2 such that $p_1 p_2 < 0$ and $0 < p_1 < F'(x) \le p_2^{-1}$ and assume that $F(a) < 0 < F(b)$. How can we find the zero of $F(x)$ guaranteed to exist by the intermediate value theorem? Define a function P by

$$P(x) = x - p_2 F(x).$$

Using the hypotheses, we obtain $P(a) = a - p_2 F(a) > a$, $P(b) = b - F(b) < b$, $P'(x) = 1 - p_2 F'(x) \geq 0$, and $P'(x) \leq 1 - p_1 p_2 < 1$. Hence P maps $[a, b]$ into itself and $|p'(x)| \leq 1$ for all $x \in [a, b]$. By Example 1.3.8, $P(x)$ is a contraction mapping. Hence, P has a fixed point that is clearly a zero of F.

Example 1.3.10. (Fredholm integral equation). Let $K(x, y)$ be a continuous function on $[a, b] \times [a, b]$, $f_0(x)$ be continuous on $[a, b]$ and consider the equation

$$f(x) = f_0(x) + \lambda \int_a^b K(x, y) f(y) \, dy.$$

Define the operator $P: C[a, b] \to C[a, b]$ by $p(f) = g$ given by

$$g(x) = f_0(x) + \lambda \int_a^b K(x, y) f(y) \, dy.$$

Note that a fixed point of P is a solution of the integral equation. We get from $p(f) = g$:

$$\|P(q_1) - P(q_2)\| = \sup_{x \in [a,b]} |p(q_1(x)) - P(q_2(x))|$$

$$= |\lambda| \sup_{x \in [a,b]} \left| \int_a^b k(x, y)(q_1(y) - q_2(y)) \, dy \right|$$

$$\leq |\lambda| \delta \left| \int_a^b (q_1(y) - q_2(y)) \, dy \right| \quad \text{(by)}$$

$$\leq |\lambda| \delta (b - a) \sup_{x \in [a,b]} |q_1(y) - q_2(y)|$$

$$\leq |\lambda| \delta (b - a) \|q_1 - q_2\|.$$

Hence, P is a contraction mapping if there exists a $c < 1$ such that

$$|\lambda| \delta (b - a) \leq c.$$

We need the definition:

Definition 1.3.11. *Let $\{x_n\}$ be a sequence in a normed space X. Then a nonnegative sequence $\{v_n\}$ for which*

$$\|x_{n+1} - x_n\| \leq v_{n+1} - v_n, \quad \forall n \geq 0, \tag{1.3.9}$$

holds is a majorizing sequence for $\{x_n\}$.

Note that any majorizing sequence is necessarily nondecreasing.
The following will be a frequently used result on majorization.

Lemma 1.3.12. *Let $\{v_n\}$ be a majorizing sequence for $\{x_n\}$ in X where X is a Banach space. Assume $\lim_{n \to \infty} v_n = v^* < \infty$ exists. Then $x^* = \lim_{n \to \infty} x_n$ exists and*

$$\|x^* - x_n\| \leq v^* - v_n, \quad \forall n \geq 0. \tag{1.3.10}$$

Proof. The estimate

$$\|x_{n+m} - x_n\| \le \sum_{j=n}^{n+m-1} \|x_{j+1} - x_j\| \le \sum_{j=n}^{n+m-1} (v_{j+1} - v_j) = v_{n+m} - v_n \quad (1.3.11)$$

shows that $\{x_n\}$ is a Cauchy sequence in the Banach space X and as such it converges to some x^* and the error estimates (1.3.10) follow from (1.3.11) as $m \to \infty$.

1.4 Exercises

1.4.1. Show that the operators introduced in Examples 1.1.2 and 1.1.3 are indeed linear.

1.4.2. Show that the Laplace transform

$$\Delta = \frac{\partial^2}{\partial x_1^2} + \frac{\partial^2}{\partial x_2^2} + \frac{\partial^2}{\partial x_3^2}$$

is a linear operator mapping the space of real functions $x = x(x_1, x_2, x_3)$ with continuous second derivatives on some subset D of $\mathbf{R}^3$ into the space of continuous real functions on D.

1.4.3. Define $T: C''[0, 1] \times C'[0, 1] \to C[0, 1]$ by

$$T(x, y) = \alpha \frac{d^2 x}{dt^2} + \beta \frac{dy}{dt}, \quad \text{for any real constants } \alpha, \beta \text{ and } 0 \le t \le 1.$$

Show that T is a linear operator.

1.4.4. In an inner product $\langle \cdot, \cdot \rangle$ space X show that for any fixed z in X

$$T(x) = \langle x, z \rangle$$

is a linear functional.

1.4.5. Show that an additive operator T from a real Banach space X into a real Banach space Y is homogeneous if it is continuous.

1.4.6. Show that matrix $A = \{a_{ij}\}$, $i, j = 1, 2, \dots, n$ has an inverse if

$$|a_{ii}| > \frac{1}{2} \sum_{j=1}^{n} |a_{ij}| > 0, \quad i = 1, 2, \dots, n.$$

1.4.7. Show that the linear integral equation of second Fredholm kind in $C[0, 1]$

$$x(s) - \lambda \int_0^1 K(s, t)x(t)dt = y(s), \quad 0 \le \lambda \le 1,$$

where $K(s, t)$ is continuous on $0 \le s, t \le 1$, has a unique solution $x(s)$ for $y(s) \in C[0, 1]$ if

$$|\lambda| < \left[\max_{[0,1]} \int_0^1 |K(s, t)|dt \right]^{-1}.$$

1.4.8. Prove Theorem 1.1.10.

1.4.9. Prove Theorem 1.1.12.

1.4.10. Show that the operators defined below are all linear.

(a) Identity operator. The identity operator $I_X: X \to X$ given by $I_X(x) = x$, for all $x \in X$.

(b) Zero operator. The zero operator $O: X \to Y$ given by $O(x) = 0$, for all $x \in X$.

(c) Integration. $T: C[a, b] \to C[a, b]$ given by $T(x(t)) = \int_0^t x(s)ds$, $t \in [a, b]$.

(d) Differentiation. Let X be the vector space of all polynomials on $[a, b]$. Define T on X by $T(x(t)) = x'(t)$.

(e) Vector algebra. The cross product with one factor kept fixed. Define $T_1: \mathbf{R}^3 \to \mathbf{R}^5$. Similarly, the dot product with one fixed factor. Define $T_2: \mathbf{R}^3 \to \mathbf{R}$.

(f) Matrices. A real matrix $A = \{a_{ij}\}$ with m rows and n columns. Define $T: \mathbf{R}^n \to \mathbf{R}^m$ given by $y = Ax$.

1.4.11. Let T be a linear operator. Show:

(a) the $R(T)$ (range of T) is a vector space;
(b) if $\dim(T) = n < \infty$, then $\dim R(T) \leq n$;
(c) the null/space $N(T)$ is a vector space.

1.4.12. Let X, Y be vector spaces, both real or both complex. Let $T: D(T) \to Y$ (domain of T) be a linear operator with $D(T) \subseteq X$ and $R(T) \subseteq Y$. Then, show:

(a) the inverse $T^{-1}: R(T) \to D(T)$ exists if and only if

$$T(x) = 0 \Rightarrow x = 0;$$

(b) if T^{-1} exists, it is a linear operator;
(c) if $\dim D(T) = n < \infty$ and T^{-1} exists, then $\dim R(T) = \dim D(T)$.

1.4.13. Let $T: X \to Y$, $P: Y \to Z$ be bijective linear operators, where X, Y, Z are vector spaces. Then, show: the inverse $(ST)^{-1}: Z \to X$ of the product ST exists, and
$$(ST)^{-1} = T^{-1}S^{-1}.$$

1.4.14. If the product (composite) of two linear operators exists, show that it is linear.

1.4.15. Let X be the vector space of all complex 2×2 matrices and define $T: X \to X$ by $T(x) = cx$, where $c \in X$ is fixed and cx denotes the usual product of matrices. Show that T is linear. Under what conditions does T^{-1} exist?

1.4.16. Let $T: X \to Y$ be a linear operator and $\dim X = \dim Y = n < \infty$. Show that $R(T) = Y$ if and only if T^{-1} exists.

1.4.17. Define the integral operator $T: C[0, 1] \to C[0, 1]$ by $y = T(x)$, where $y(t) = \int_0^1 k(x, s)x(s)ds$ and k is continuous on $[0, 1] \times [0, 1]$. Show that T is linear and bounded.

1.4.18. Show that the operator T defined in 1.4.10(f) is bounded.

1.4.19. If a normed space X is finite-dimensional, then show that every linear functional on X is bounded.

1.4.20. Let $T: D(T) \to Y$ be a linear operator, where $D(T) \subseteq X$ and X, Y are normed spaces. Show:

(a) T is continuous if and only if it is bounded;

(b) if T is continuous at a single point, it is continuous.

1.4.21. Let T be a bounded linear operator. Show:

(a) $x_n \to x$ (where $x_n, x \in D(T)$) $\Rightarrow T(x_n) \to T(x)$;

(b) the null space $N(T)$ is closed.

1.4.22. If $T \neq 0$ is a bounded linear operator, show that for any $x \in D(T)$ such that $\|x\| < 1$, we have $\|T(x)\| < \|T\|$.

1.4.23. Show that the operator $T: \ell^\infty \to \ell^\infty$ defined by $y = (y_i) = T(x)$, $y_i = \frac{x_i}{i}$, $x = (x_i)$, is linear and bounded.

1.4.24. Let $T: C[0, 1] \to C[0, 1]$ be defined by

$$y(t) = \int_0^t x(s)ds.$$

Find $R(T)$ and $T^{-1}: R(T) \to C[0, 1]$. Is T^{-1} linear and bounded?

1.4.25. Show that the functionals defined on $C[a, b]$ by

$$f_1(x) = \int_a^b x(t)y_0(t)dt \quad (y_0 \in C[a, b])$$

$$f_2(x) = c_1 x(a) + c_2 x(b) \quad (c_1, c_2 \text{ fixed})$$

are linear and bounded.

1.4.26. Find the norm of the linear functional f defined on $C[-1, 1]$ by

$$f(x) = \int_{-1}^0 x(t)dt - \int_0^1 x(t)dt.$$

1.4.27. Show that

$$f_1(x) = \max_{t \in J} x(t), \quad f_2(x) = \min_{t \in J} x(t), \quad J = [a, b]$$

define functionals on $C[a, b]$. Are they linear? Bounded?

1.4.28. Show that a function can be additive but not homogeneous. For example, let $z = x + iy$ denote a complex number, and let $T: \mathbb{C} \to \mathbb{C}$ be given by

$$T(z) = \bar{z} = x - iy.$$

1.4.29. Show that a function can be homogeneous but not additive. For example, consider the operator $T: \mathbf{R}^2 \to \mathbf{R}$ given by

$$T((x_1, x_2)) = \frac{x_1^2}{x_2}.$$

1.4.30. Let F be an operator in $C[0, 1]$ defined by

$$F(x, y) = x(s) \int_0^1 \frac{s}{s+t} x(t)dt, \quad 0 \leq \lambda \leq 1.$$

Show that for $x_0, z \in C[0, 1]$

$$F'(x_0)z = x_0(s) \int_0^1 \frac{s}{s+t} z(t)dt + z(s) \int_0^1 \frac{s}{s+t} x_0(t)dt.$$

1.4.31. Find the Fréchet derivative of the operator F in $\mathbf{R}_\infty^2$ given by

$$F\binom{x}{y} = \binom{x^2 + 7x + 2xy - 3}{x + y^3}.$$

1.4.32. Find the first and second Fréchet derivatives of the Uryson operator

$$U(x) = \int_0^1 k(s, t, x(t))dt$$

in $C[0, 1]$ at $x_0 = x_0(s)$.

1.4.33. Find the Fréchet derivative of the Riccati differential operator

$$R(z) = \frac{dz}{dt} + p(t)z^2 + q(t)z + r(t),$$

from $C'[0, s]$ into $C[0, s]$ at $z_0 = z_0(t)$ in $C'[0, s]$, for p, q, r being given differentiable functions on $[0, s]$.

1.4.34. Find the first two Fréchet derivatives of the operator

$$F\binom{x}{y} = \binom{x^2 + y^2 - 3}{x \sin y} \quad \text{in } \mathbf{R}^2.$$

1.4.35. Consider the partial differential operator

$$F(x) = \Delta x - x^2$$

from $C^2(I)$ into $C(I)$, the space of all continuous functions on the square $0 \leq \alpha, \beta \leq 1$. Show that

$$F'(x_0)z = \Delta z(\alpha, \beta) - 2x_0(\alpha, \beta)z(\alpha, \beta),$$

where Δ is the usual Laplace operator.

1.4.36. Let $F(L) = L^3$, in $L(x)$. Show:

$$F'(L_0) = L_0[\ ]L_0 + L_0^2[\ ] + [\ ]L_0.$$

1.4.37. Let $F(L) = L^{-1}$, in $L(x)$. Show:

$$F'(L_0) = -L_0^{-1}[\ ]L_0^{-1},$$

provided that L_0^{-1} exists.

1.4.38. Show estimates (1.1.45) and (1.1.46).

1.4.39. Show Taylor's Theorem 1.1.20.

1.4.40. Integrate the operator

$$F(L) = L^{-1} \text{ in } L(X)$$

from $L_0 = I$ to $L_1 = A$, where $\|I - A\| < 1$.

1.4.41. Show that the spaces defined in Examples 1.2.2 are POTL.

1.4.42. Show that any regular POB-space is normal but the converse is not necessarily true.

1.4.43. Prove Theorem 1.2.11.

1.4.44. Show that if (1.2.4) or (1.2.5) are satisfied then $F'(x) = [x, x]$ for all $x \in D$. Moreover show that if both (1.2.4) and (1.2.5) are satisfied, then F' is Lipschitz continuous with $I = c_0 + c_1$.

1.4.45. Find sufficient conditions so that estimates (1.2.33) and (1.2.34) are both satisfied.

1.4.46. Show that B_n $(n \geq 0)$ in (1.2.38) is a nonnegative subinverse of $F'(y_n)$ $(n \geq 0)$.

1.4.47. Let $x_0, x_1, ..., x_n$ be distinct real numbers, and let f be a given real-valued function. Show that:

$$[x_0, x_1, ..., x_n] = \sum_{j=0}^{n} \frac{f(x_j)}{g_n'(x_j)}$$

and

$$[x_0, x_1, ..., x_n](x_n - x_0) = [x_1, ..., x_n] - [x_0, ..., x_{n-1}]$$

where

$$g_n(x) = (x - x_0) ... (x - x_n).$$

1.4.48. Let $x_0, x_1, ..., x_n$ be distinct real numbers, and let f be n times continuously differentiable function on the interval $I\{x_0, x_1, ..., x_n\}$. Then show that

$$[x_0, x_1, ..., x_n] = \int_{\tau_n} \cdots \int f^{(n)}(t_0 x_0 + \cdots + t_n x_n)\, dt_1 \cdots dt_n$$

in which

$$\tau_n = \left\{ (t_1, ..., t_n) \,|\, t_1 \geq 0, ..., t_n \geq 0, \sum_{i=1}^{n} t_i \leq 1 \right\}$$

$$t_0 = 1 - \sum_{i=1}^{n} t_i.$$

1.4.49. If f is a real polynomial of degree m, then show:

$$[x_0, x_1, ..., x_n, x] = \begin{cases} \text{polynomial of degree } m - n - 1, & n \leq m - 1 \\ a_m & n = m - 1 \\ 0 & n > m - 1 \end{cases}$$

where $f(x) = a_m x^n +$ lower-degree terms.

1.4.50. The tensor product of two matrices $M, N \in L(\mathbf{R}^n)$ is defined as the $n^2 \times n^2$ matrix $M \times N = (m_{ij} N \mid i, j = 1, ..., n)$, where $M = (m_{ij})$. Consider two F-differentiable operators $H, K: L(\mathbf{R}^n) \to L(\mathbf{R}^n)$ and set $F(X) = H(X) K(X)$ for all $X \in L(\mathbf{R}^n)$. Show that $F'(X) = [H(X) \times I] K'(X) + [I \times K(X)^T] H'(X)$ for all $X \in L(\mathbf{R}^n)$.

1.4.51. Let $F: \mathbf{R}^2 \to \mathbf{R}^2$ be defined by $f_1(x) = x_1^3$, $f_2(x) = x_2^2$. Set $x = 0$ and $y = (1, 1)^T$. Show that there is no $z \in [x, y]$ such that $F(y) - F(x) = F'(z)(y - x)$.

1.4.52. Let $F: D \subset \mathbf{R}^n \to \mathbf{R}^m$ and assume that F is continuously differentiable on a convex set $D_0 \subset D$. For and $x, y \in D_0$, show that

$$\| F(y) - F(x) - F'(x)(y - x) \| \le \|y - x\| \, w(\|y - x\|),$$

where w is the modulus of continuity of F' on $[x, y]$. That is

$$w(t) = \sup \{ \|F'(x) - F'(y)\| \mid x, y \in D_0, \|x - y\| \le t \}.$$

1.4.53. Let $F: D \subset \mathbf{R}^n \to \mathbf{R}^m$. Show that F'' is continuous at $z \in D$ if and only if all second partial derivatives of the components $f_1, ..., f_m$ of F are continuous at z.

1.4.54. Let $F: D \subset \mathbf{R}^n \to \mathbf{R}^m$. Show that $F''(z)$ is symmetric if and only if each Hessian matrix $H_1(z), ..., H_m(z)$ is symmetric.

1.4.55. Let $M \in L(\mathbf{R}^n)$ be symmetric, and define $f: \mathbf{R}^n \to R$ by $f(x) = x^T M x$. Show, directly from the definition that f is convex if and only if M is positive semidefinite.

1.4.56. Show that $f: D \subset \mathbf{R}^n \to R$ is convex on the set D if and only if, for any $x, y \in D$, the function $g: [0, 1] \to R$, $g(t) = g(tx + (1 - t)y)$, is convex on $[0, 1]$.

1.4.57. Show that if $g_i: \mathbf{R}^n \to R$ is convex and $c_i \ge 0$, $i = 1, 2, ..., m$, then $g = \sum_{i=1}^{m} c_i g_i$ is convex.

1.4.58. Suppose that $g: D \subset \mathbf{R}^n \to R$ is continuous on a convex set $D_0 \subset D$ and satisfies

$$\frac{1}{2} g(x) + \frac{1}{2} g(y) - g\left(\frac{1}{2}(x + y)\right) \ge \gamma \|x - y\|^2$$

for all $x, y \in D_0$. Show that g is convex on D_0 if $\gamma = 0$.

1.4.59. Let $M \in L(\mathbf{R}^n)$. Show that M is a nonnegative matrix if and only if it is an isotone operator.

1.4.60. Let $M \in L(\mathbf{R}^n)$ be diagonal, nonsingular, and nonnegative. Show that $\|x\| = \|D(x)\|$ is a monotonic norm on $\mathbf{R}^n$.

1.4.61. Let $M \in L(\mathbf{R}^n)$. Show that M is invertible and $M^{-1} \ge 0$ if and only if there exist nonsingular, nonnegative matrices $M_1, M_2 \in L(\mathbf{R}^n)$ such that $M_1 M M_2 = 1$.

1.4.62. Let $[\cdot, \cdot]: D \times D$ be an operator satisfying conditions (1.2.1) and (1.2.44). The following two assertions are equivalent:

(a) Equality (1.2.48) holds for all $x, y \in D$.

(b) For all points $u, v \in D$ such that $2v - u \in D$ we have

$$[u, v] = 2[u, 2v - u] - [v, 2v - u].$$

1.4.63. If δF is a consistent approximation F' on D, show that each of the following four expressions

$$c_1 = h\left(\|x - u\| + \|y - u\| + \|u - v\|\right) \|x - y\|,$$
$$c_2 = h\left(\|x - v\| + \|y - v\| + \|u - v\|\right) \|x - y\|,$$
$$c_3 = h\left(\|x - y\| + \|y - u\| + \|y - v\|\right) \|x - y\|,$$

and

$$c_4 = h\left(\|x - y\| + \|x - u\| + \|x - v\|\right) \|x - y\|$$

is an estimate for

$$\|F(x) - F(y) - \delta F(u, v)(x - y)\|.$$

1.4.64. Show that the integral representation of $[x_0, ..., x_k]$ is indeed a divided difference of kth order of F. Let us assume that all divided differences have such an integral representation. In this case for $x_0 = x_1 = \cdots = x_k = x$, we shall have

$$\underbrace{[x, x, ..., x]}_{k+1 \text{ times}} = \frac{1}{k} f^{(k)}(x).$$

Suppose now that the nth Fréchet derivative of F is Lipschitz continuous on D, i.e., there exists a constant c_{n+1} such that

$$\left\|F^{(n)}(u) - F^{(n)}(v)\right\| \leq c_{n+1} \|u - v\|$$

for all $u, v \in D$. In this case, set

$$R_n(y) = \left([x_0, ..., x_{n-1}, y] - [x_0, ..., x_{n-1}, x_n]\right)(y - x_{n-1}), ..., (y - x_0)$$

and show that

$$\|R_n(y)\| \leq \frac{c_{n+1}}{(n+1)!} \|y - x_n\| \cdot \|y - x_{n-1}\| \cdots \|y - x_0\|$$

and

$$\left\|F(x + h) - \left(F(x) + F'(x)h + \frac{1}{2}F''(x)h^2 + \cdots + \frac{1}{n!}F^{(n)}(x)h^n\right)\right\|$$
$$\leq \frac{c_{n+1}}{(n+1)!} \|h\|^{n+1}.$$

1.4.65. We recall the definitions:

(a) An operator $F: D \subset \mathbf{R}^n \to \mathbf{R}^m$ is Gâteaux- (or G-) differentiable at an interior point x of D if there exists a linear operator $L \in L(\mathbf{R}^n, \mathbf{R}^m)$ such that, for any $h \in \mathbf{R}^n$

$$\lim_{t \to 0} \frac{1}{t} \|F(x + th) - F(x) - tL(h)\| = 0.$$

L is denoted by $F'(x)$ and called the G-derivative of F at x.

(b) An operator $F: D \subset \mathbf{R}^n \to \mathbf{R}^m$ is hemicontinuous at $x \in D$ if, for any $h \in \mathbf{R}^n$ and $\varepsilon > 0$, there is a $\delta = \delta(\varepsilon, h)$ so that whenever $|t| < \delta$ and $x + th \in D$, then $\|F(x + th) - F(x)\| < \varepsilon$.

(c) If $F: D \subset \mathbf{R}^n \to \mathbf{R}^m$ and if for some interior point x of D, and $h \in \mathbf{R}^n$, the limit

$$\lim_{t \to 0} \frac{1}{t} [F(x + th) - F(x)] = A(x, h)$$

exists, then F is said to have a Gâteaux differential at x in the direction h.

(d) If the G-differential exists at x for all h and if, in addition

$$\lim_{h \to 0} \frac{1}{\|h\|} \|f(x + H) - F(x) - A(x, h)\| = 0,$$

then F has a Fréchet differential at x.

Show:

(i)   The linear operator L is unique;

(ii)   If $F: D \subset \mathbf{R}^n \to \mathbf{R}^m$ is G-differentiable at $x \in D$, then F is hemicontinuous at x.

(iii) G-differential and "uniform in h" implies F-differential;

(iv) F-differential and "linear in h" implies F-derivative;

(v) G-differential and "linear in h" implies G-derivative;

(vi) G-derivative and "uniform in h" implies F-derivative;

Here "uniform in h" indicated the validity of (d).

Linear in h means that $A(x, h)$ exists for all $h \in \mathbf{R}^n$ and

$$A(x, h) = M(x)h, \quad \text{where} \quad M(x) \in L(\mathbf{R}^n, \mathbf{R}^m).$$

Define $F: \mathbf{R}^2 \to \mathbf{R}$ by $F(x) = \text{sgn}(x_2) \min(|x_1|, |x_2|)$.
Show that, for any $h \in \mathbf{R}^2$, $A(0, h) = F(h)$, but F does not have a G-derivative at 0.

(vii)   Define $F: \mathbf{R}^2 \to R$ by $F(0) = 0$ if $x = 0$ and

$$F(x) = x_2 \left(x_1^2 + x_2^2\right)^{\frac{3}{2}} \bigg/ \left[\left(x_1^2 + x_2^2\right)^2 + x_2^2\right], \quad \text{if} \quad x \neq 0.$$

Show that F has a G-derivative at 0, but not an F-derivative. Show, moreover, that the G-derivative is hemicontinuous at 0.

(viii) If the g-differential $A(x, h)$ exists for all x in an open neighborhood of an interior point x_0 of D and for all $h \in \mathbf{R}^n$, then F has an F-derivative at x_0 provided that for each fixed h, $A(x, h)$ is continuous in x at x_0.

(e) Assume that $F: D \subset \mathbf{R}^n \to \mathbf{R}^m$ has a G-derivative at each point of an open set $D_0 \subset D$. If the operator $F': D_0 \subset \mathbf{R}^n \to L(\mathbf{R}^n, \mathbf{R}^m)$ has a G-derivative at $x \in D_0$, then $\left(F'\right)'(x)$ is denoted by $F''(x)$ and called the second G-derivative of F at x.

Show:

(i) If $F: \mathbf{R}^n \to \mathbf{R}^m$ has a G-derivative at each point of an open neighborhood of x, then F' is continuous at x if and only if all partial derivatives $\partial_i F_i$ are continuous at x.

(ii) F'' is continuous at $x_0 \in D$ if and only if all second partial derivatives of the components $f_1, ..., f_m$ of F are continuous at x_0. $F''(x_0)$ is symmetric if and only if each Hessian matrix $H_1(x_0), ..., H_m(x_0)$ is symmetric.

1.4.66. Consider the problem of approximating a solution $y \in C'[0, t_0]$ of the non-linear ordinary differential equation

$$\frac{dy}{dt} = K(t, y(t)), \quad 0 \le t \le t_0, \ y(0) = y_0.$$

The above equation may be turned into a fixed point problem of the form

$$y(t) = y_0 + \int_0^t K(s, y(s))ds, \quad 0 \le t \le t_0.$$

Assume $K(x, y)$ is continuous on $[0, t_0] \times [0, t_0]$ and satisfies the Lipschitz condition

$$\max_{[0, t_0]} \left| K(s, q_1(s)) - K(s, q_2(s)) \right| \le M \|q_1 - q_2\|,$$

$$\text{for all } q_1, q_2 \in C[0, t_0].$$

Note that the integral equation above defines an operator P from $C[0, t_0]$ into itself. As in Example 1.3.10 find a sufficient condition for P to be a contraction mapping.

1.4.67. Let F be a contraction mapping on the ball $\bar{U}(x_0, r)$ in a Banach space X, and let

$$\|F(x_0) - x_0\| \le (1 - c)r.$$

Show F has a unique fixed point in $\bar{U}(x_0, r)$.

1.4.68. Under the assumptions of Theorem 1.3.4, show that the sequence generalized by (1.3.2) minimizes the functional

$$f(x) = \|x - F(x)\|$$

for any x_0 belonging to a closed set A such that $F(A) \subseteq A$.

1.4.69. Let the equation $F(x) = x$ have a unique solution in a closed subset A of a Banach space X. Assume that there exists an operator F_1 that $F_1(A) \subseteq A$ and F_1 commutes with F on A. Show the equation $x = F_1(x)$ has at least one solution in A.

1.4.70. Assume that operator F maps a closed set A into a compact subset of itself and satisfies

$$\|F(x) - F(y)\| < \|x - y\| \quad (x \neq y), \quad \text{for all } x, y \in A.$$

Show F has a unique fixed point in A. Apply these results to the mapping $F(x) = x - \frac{x^2}{2}$ of the interval $[0, 1]$ into itself.

1.4.71. Show that operator F defined by

$$F(x) = x + \frac{1}{x}$$

maps the half line $[1, \infty)$ into itself and satisfies

$$\|F(x) - F(y)\| < \|x - y\|$$

but has no fixed point in this set.

1.4.72. Consider condition

$$\|F(x) - F(y)\| \leq \|x - y\| \quad (x, y \in A).$$

Let A be either an interval $[a, b]$ or a disk $x^2 + y^2 \leq r^2$. Find conditions in both cases under which F has a fixed point.

1.4.73. Consider the set c_0 of null sequences $x = \{x_1, x_2, \ldots\}$ ($x_n \to 0$) equipped with the norm $\|x\| = \max_n |x_n|$. Define the operator F by

$$F(x) = \left\{ \tfrac{1}{2}(1 + \|x\|), \tfrac{3}{4}x_1, \tfrac{7}{8}x_2, \ldots, \left(1 - \tfrac{1}{2^{n+1}}\right) x_n, \ldots \right\}.$$

Show that $F: \bar{U}(0, 1) \to \bar{U}(0, 1)$ satisfies

$$\|F(x) - F(y)\| < \|x - y\|,$$

but has no fixed points.

1.4.74. Repeat Exercise 1.4.73 for the operator F defined in c_0 by

$$F(x) = \{y_1, \ldots, y_n, \ldots\},$$

where $y_n = \frac{n-1}{n}x_n + \frac{1}{n}\sin(n)$ ($n \geq 1$).

1.4.75. Repeat Exercise 1.4.73 for the operator F defined in $C[0, 1]$ by

$$Fx(t) = (1 - t)x(t) + t\sin\left(\frac{1}{t}\right).$$

1.4.76. Let F be a nonlinear operator on a Banach space X which satisfies (1.3.3) on $\bar{U}(0, r)$. Let $F(0) = 0$. Define the resolvent $R(x)$ of F by

$$F(x)f = xFR(x)f + f.$$

Show:

(a) $R(x)$ is defined on the ball $\|f\| \leq (1 - |x|c)r$ if $|x| < c^{-1}$;

(b) $\frac{1}{1+|x|c}\|f-g\| \le \|R(x)f - R(x)g\| \le \frac{1}{1-|x|c}\|f-g\|$;

(c) $\|R(x)f - R(y)f\| \le \frac{c\|f\|\,|x-y|}{(1-|x|c)(1-|y|c)}$.

1.4.77. Let A be an operator mapping a closed set A of a Banach space X into itself. Assume that there exists a positive integer m such that A^m is a contraction operator. Prove that sequence (1.3.2) converges to a unique fixed point of F in A.

1.4.78. Let F be an operator mapping a compact set $A \subseteq X$ into itself with $\|F(x) - F(y)\| < \|x - y\|$ $(x \ne y$, all $x, y \in A)$. Show that sequence (1.3.2) converges to a fixed point of (1.3.1).

1.4.79. Let F be a continuous function on $[0, 1]$ with $0 \le f(x) \le 1$ for all $x \in [0, 1]$. Define the sequence

$$x_{n+1} = x_n + \frac{1}{n+1}(F(x_n) - x_n).$$

Show that for any $x_0 \in [0, 1]$ sequence $\{x_n\}$ $(n \ge 0)$ converges to a fixed point of F.

1.4.80. Show:

(a) A system $x = Ax + b$ of n linear equations in n unknowns $x_1, x_2, \dots, x_n$ (the components of x) with $A = \{a_{jk}\}$, $j, k = 1, 2, \dots, n$, b given, has a unique solution x^* if

$$\sum_{k=1}^{n} |a_{jk}| < 1, \qquad j = 1, 2, \dots, n.$$

(b) The solution x^* can be obtained as the limit of the iteration $(x^{(0)}, x^{(1)}, x^{(2)}, \dots\}$, where $x^{(0)}$ is arbitrary and

$$x^{(m+1)} = Ax^{(m)} + b \quad (m \ge 0).$$

(c) The following error bounds hold:

$$\|x^{(m)} - x^*\| \le \frac{c}{1-c}\|x^{(m-1)} - x^{(m)}\| \le \frac{c^m}{1-c}\|x^{(0)} - x^{(1)}\|,$$

where

$$c = \max_j \sum_{k=1}^{n} |a_{jk}| \text{ and } \|x - z\| = \max_j |x_i - z_i|, \ j = 1, 2, \dots, n.$$

1.4.81. (Gershgorin's theorem: If λ is an eigenvalue of a square matrix $A = \{a_{jk}\}$, then for some j, where $1 \le j \le n$,

$$|a_{jj} - \lambda| \le \sum_{\substack{k=1 \\ k \ne j}}^{n} |a_{jk}|.)$$

Show that $x = Ax + b$ can be written $Bx = b$, where $B = I - A$, and $\sum_{k=1}^{n} |a_{jk}| < 1$ together with the theorem imply that 0 is not an eigenvalue of B and A has spectral radius less than 1.

1.4.82. Let (X, d), (X, d_1), (X, d_2) be metric spaces with $d(x, z) = \max_j |x_j - z_j|$, $j = 1, 2, \ldots, n$,

$$d_1(x, z) = \sum_{j=1}^{n} |x_j - z_j| \quad \text{and} \quad d_2(x, z) = \left[\sum_{j=1}^{n} (x_j - z_j)^2 \right]^{1/2},$$

respectively. Show that instead of $\sum_{k=1}^{n} |a_{jk}| < 1$, $j = 1, 2, \ldots, n$, we obtain the conditions

$$\sum_{j=1}^{n} |a_{jk}| < 1, \quad k = 1, 2, \ldots, n \quad \text{and} \quad \sum_{j=1}^{n} \sum_{k=1}^{n} a_{jk}^2 < 1.$$

1.4.83. Let us consider the ordinary differential equation of the first order (ODE)

$$x' = f(t, x), \quad x(t_0) = x_0,$$

where t_0 and x_0 are given real numbers. Assume:

$$|f(t, x)| \leq c_0$$

on $R = \{(t, x) \mid |t - t_0| \leq a, |x - x_0| \leq b\}$,

$$|f(t, x) - f(t, v)| \leq c_1 |x - v|, \quad \text{for all } (t, x), (t, v) \in R.$$

Then show: the (ODE) has a unique solution on $[t_0 - c_2, t_0 + c_2]$, where

$$c_2 < \min \left\{ a, \frac{b}{c_0}, \frac{1}{c_1} \right\}.$$

1.4.84. Show that f defined by $f(x, y) = |\sin y| + x$ satisfies a Lipschitz condition with respect to the second variable (on the whole xy-plane).

1.4.85. Does f defined by $f(t, x) = |x|^{1/2}$ satisfy a Lipschitz condition?

1.4.86. Apply Picard's iteration $x_{n+1}(t) = \int_{t_0}^{t} f(s, x_n(s)) ds$ used for the (ODE) $x' = f(t, x), x(t_0) = x_0 + 0, x' = 1 + x^2, x(0) = 0$. Verify that for x_3, the terms involving $t, t^2, \ldots, t^5$ are the same as those of the exact solution.

1.4.87. Show that $x' = 3x^{2/3}, x(0) = 0$ has infinitely many solutions x.

1.4.88. Assume that the hypotheses of the contraction mapping principle hold, then show that x^* is accessible from any point $\bar{U}(x_0, r_0)$.

1.4.89. Define the sequence $\{\bar{x}_n\}$ $(n \geq 0)$ by $\bar{x}_0 = x_0, \bar{x}_{n+1} = F(\bar{x}_n) + \varepsilon_n$ $(n \geq 0)$. Assume:

$$\|\varepsilon_n\| \leq \lambda^n \varepsilon \ (n \geq 0) \ (0 \leq \lambda < 1);$$

F is a c-contraction operator on $U(x_0, r)$. Then show sequence $\{\bar{x}_n\}$ $(n \geq 0)$ converges to the unique fixed point x^* of F in $\bar{U}(x_0, r)$ provided that

$$r \geq r_0 + \frac{\varepsilon}{1 - c}.$$

2

The Newton-Kantorovich (NK) Method

We study the problem of approximating locally unique solution of an equation in a Banach space. The Newton-Kantorovich method is undoubtedly the most popular method for solving such equations.

2.1 Linearization of equations

Let F be a Fréchet-differentiable operator mapping a convex subset D of a Banach space X into a Banach space Y. Consider the equation

$$F(x) = 0. \tag{2.1.1}$$

We will assume D is an open set, unless otherwise stated.

The principal method for constructing successive approximations x_n to a solution x^* (if it exists) of Equations (2.1.1) is based on successive linearization of the equation.

The interpretation of (2.1.1) is that we model F at the current iterate x_n with a linear function:

$$L_n(x) = F(x_n) + F'(x_n)(x - x_n). \tag{2.1.2}$$

L_n is called the local linear model. If $F'(x_n)^{-1} \in L(Y, X)$ the space of bounded linear operators from Y into X, then approximation x_{n+1}, which is the root of $L_n(x_{n+1}) = 0$, is given by

$$x_{n+1} = x_n - F'(x_n)^{-1} F(x_n) \quad (n \geq 0). \tag{2.1.3}$$

The iterative procedure generated by (2.1.3) is the famous Newton-Kantorovich (NK) method [125]. The geometric interpretation of this method is well-known, if F is a real function. In such a case, x_{n+1} is the point where the tangential line $y - F(x_n) = F'(x_n)(x - x_n)$ of function F at the point $(x, F(x_n))$ intersects the x-axis.

The basic defect of method (2.1.3) is that each step involves the solution of a linear equation with a different linear operator $F'(x_n)$. For this reason, one often

I.K. Argyros, *Convergence and Applications of Newton-type Iterations*,
DOI: 10.1007/978-0-387-72743-1_2, © Springer Science+Business Media, LLC 2008

constructs successive approximations that employ linear equations other than (2.1.2), though similar to them.

The most frequently used substitute for (2.1.2) is the equation

$$F(x_n) + F'(x_0)(x - x_n),\tag{2.1.4}$$

where x_0 is the initial approximation. The successive approximations are then defined by the recurrence relation

$$x_{n+1} = x_n - F'(x_0)^{-1} F(x_n) \quad (n \geq 0).\tag{2.1.5}$$

We will call this method the modified Newton-Kantorovich method (MNK).

We are concerned about the following aspects:

(a) finding effectively verifiable conditions for its applicability;
(b) computing convergence rates and a priori error estimates;
(c) choosing an initial approximation x_0 for which the method converges; and
(d) the degree of "stability" of the method.

2.2 Semilocal convergence of the NK method

Define the operator P by

$$P(x) = x - F'(x)^{-1} F(x)\tag{2.2.1}$$

Then the NK method (2.1.3) may be regarded as the usual iterative method

$$x_{n+1} = P(x_n) \quad (n \geq 0),\tag{2.2.2}$$

for approximating solution x^* of the equation

$$x = P(x)\tag{2.2.3}$$

Suppose that

$$\lim_{n \to \infty} x_n = x^*.\tag{2.2.4}$$

We would like to know under what conditions on F and F' the point x^* is a solution of Equation (2.1.1).

Proposition 2.2.1. *If F' is continuous at $x = x^*$, then we have*

$$F(x^*) = 0.\tag{2.2.5}$$

Proof. The approximations x_n satisfy the equation

$$F'(x_n)(x_{n+1} - x_n) = -F(x_n).\tag{2.2.6}$$

Because the continuity of F at x^* follows from the continuity of F', taking the limit as $n \to \infty$ in (2.2.6) we obtain (2.2.5).

Proposition 2.2.2. *If*
$$\|F'(x)\| \leq b \tag{2.2.7}$$
in some closed ball that contains $\{x_n\}$, *then* x^* *is a solution of* $F(x) = 0$.

Proof. By (2.2.7) we get
$$\lim_{n \to \infty} F(x_n) = F(x^*), \tag{2.2.8}$$
and as
$$\|F(x_n)\| \leq b \|x_{n+1} - x_n\|, \tag{2.2.9}$$
(2.2.5) is obtained by taking the limit as $n \to \infty$ in (2.2.4).

Proposition 2.2.3. *If*
$$\|F''(x)\| \leq K \tag{2.2.10}$$
in some closed ball $\overline{U}(x_0, r) = \{x \in X \mid \|x - x_0\| \leq r\}$, $0 < r < \infty$, *which contains* $\{x_n\}$, *then* x^* *is a solution of equation* $F(x) = 0$.

Proof. By (2.2.10)
$$\|F'(x) - F'(x_0)\| \leq K \|x - x_0\| \leq Kr \tag{2.2.11}$$
for all $x \in \overline{U}(x_0, r)$. Moreover we can write
$$\|F'(x)\| \leq \|F'(x_0)\| + \|F'(x) - F'(x_0)\|, \tag{2.2.12}$$
so the conditions of Proposition 2.2.2 hold with
$$b = \|F'(x_0)\| + Kr \tag{2.2.13}$$

Let us assume that the operator F is Fréchet-differentiable on D, where x_0 is an initial approximation for the NK method (2.1.3) and that the operator $F'(x)$ satisfies a Lipschitz condition
$$\|F'(x) - F'(y)\| \leq \ell \|x - y\|, \quad \text{for all } x, y \in D. \tag{2.2.14}$$

Throughout the sequel, we shall assume that the operator $\Gamma_0 = F'(x_0)^{-1}$ exists.

We shall now state and prove the famous Newton-Kantorovich theorem for approximating solutions of equation (2.1.1) [125]:

Theorem 2.2.4. *Assume that*
$$\Gamma_0 \leq b_0, \tag{2.2.15}$$
$$\|\Gamma_0 F(x_0)\| \leq \eta_0 = \eta \tag{2.2.16}$$
$$h_0 = b_0 \ell \eta_0 \leq \tfrac{1}{2} \tag{2.2.17}$$
$$r_0 = \frac{1 - \sqrt{1 - 2h_0}}{h_0} \eta_0, \tag{2.2.18}$$
and
$$\overline{U}(x_0, r) \subseteq D$$
then the NK method (2.1.3) converges to a solution x^* *of equation (2.1.1) in the ball* $U(x_0, r)$.

There are several proofs of this theorem; we present one due to Kantorovich [125].

Proof. Define a number sequence

$$b_{n+1} = \frac{b_n}{1-h_n}, \quad \eta_{n+1} = \frac{h_n}{2(1-h_n)}\eta_n, \quad h_{n+1} = b_{n+1}\ell\eta_{n+1}, \tag{2.2.19}$$

$$r_{n+1} = \frac{1-\sqrt{1-2h_{n+1}}}{h_{n+1}}\eta_{n+1}. \tag{2.2.20}$$

We claim that under the assumptions (2.2.15), (2.2.18) the successive approximations (2.1.3) exists; moreover

$$\|\Gamma(x_n)\| \le b_n, \quad \|\Gamma(x_n)F(x_n)\| \le \eta_n, \quad h_n \le \tfrac{1}{2} \tag{2.2.21}$$

and

$$U(x_n, r_n) \subset U(x_{n-1}, r_{n-1}). \tag{2.2.22}$$

The proof is by induction. Assume that (2.2.21) and (2.2.22) hold for $n = m$. Because $\|x_{m+1} - x_m\| = \|\Gamma(x_m)F(x_m)\| \le \eta_m$, it follows from the definition of r_m that $x_{m+1} \in U(x_m, r_m)$; a fortiori, $x_{m+1} \in D$. The derivative $F'(x_{m+1})$ therefore exists. By (2.2.14)

$$\|\Gamma(x_m)(F'(x_{m+1}) - F'(x_m))\| \le b_m\ell\|x_{m+1} - x_m\| \le h_m \le \tfrac{1}{2};$$

the operator $\Gamma(x_{m+1}) = F'(x_{m+1})^{-1}$ therefore exists, and has the representation

$$\Gamma(x_{m+1}) = \{I + \Gamma(x_m)[F'(x_{m+1}) - F'(x_m)]\}^{-1}\Gamma(x_m)$$

$$= \sum_{i=0}^{\infty}(-1)^i\{\Gamma(x_m)[F'(x_{m+1}) - F'(x_m)]\}^i\Gamma(x_m) \tag{2.2.23}$$

Hence

$$\|\Gamma(x_{m+1})\| \le \sum_{i=0}^{\infty}\|\Gamma(x_m)[F'(x_{m+1}) - F'(x_m)]\|^i b_m \le \frac{b_m}{1-h_m} = b_{m+1}. \tag{2.2.24}$$

Now consider the second inequality of (2.2.21) (for $n = m + 1$). It follows from the identity

$$F(x_{m+1}) = F(x_{m+1}) - F(x_m) - F'(x_m)(x_{m+1} - x_m) \tag{2.2.25}$$

that

$$\|F(x_{m+1})\| \le \tfrac{\ell}{2}\|x_{m+1} - x_m\|^2 \le \tfrac{\ell}{2}\eta_m^2, \tag{2.2.26}$$

and, by (2.2.24),

$$\|\Gamma(x_{m+1})F(x_{m+1})\| \le \frac{b_m\ell\eta_m^2}{2(1-h_m)} = \frac{h_m}{2(1-h_m)}\eta_m = \eta_{m+1}.$$

The third inequality of (2.2.21) is easily proved; by definition,

$$h_{m+1} = b_{m+1}\ell\eta_{m+1} = \frac{b_m}{1-h_m}\ell\frac{h_m}{2(1-h_m)}\eta_m = \frac{h_m^2}{2(1-h_m)^2} \leq \frac{1}{2}. \tag{2.2.27}$$

To prove the inclusion (2.2.22) it suffices to note that if $\|x - x_{k+1}\| \leq r_{k+1}$ then

$$\|x - x_k\| \leq \|x - x_{k+1}\| + \|x_{k+1} - x_k\| \leq r_{k+1} + \eta_k, \tag{2.2.28}$$

as the right-hand side is identically equal to r_k (the simple computation is left to the reader).

Thus the successive approximations (2.1.3) are well defined.

The third inequality of (2.2.21) implies that $\eta_{m+1} \leq \frac{1}{2}\eta_m$; therefore $r_n \to 0$ as $n \to \infty$. Thus the successive approximations converge to some point $x^* \in U(x_0, r_0)$. To complete the proof, it suffices to leave $m \to \infty$ in (2.2.26).

Remark 2.2.5. (a) It is clear from the proof that, under the assumptions of Theorem 2.2.4,

$$\|x_n - x^*\| \leq r_n \quad (n = 1, 2, ...).$$

(b) Under the assumptions of Theorem 2.2.4, one can easily prove that

$$\|x_n - x^*\| \leq \frac{1}{2^n}(2h_0)^{2^n - 1}\eta_0.$$

(c) In Exercise 2.16.9 we have provided a list of error bounds and the relationship between them.

It is natural to call a solution x^* of equation (2.1.1) a simple zero of the operator F if the operator $\Gamma(x^*)$ exists and is continuous.

Theorem 2.2.6. *If, under the assumptions of Theorem 2.2.4, $h_0 < \frac{1}{2}$, then the zero x^* of F to which the successive approximations (2.1.3) converge is simple.*

Proof. It suffices to note that $r_0 < (\ell b_0)^{-1}$ for $h_0 < \frac{1}{2}$, and that $\|x - x_0\| < (\ell b_0)^{-1}$ implies $\|\Gamma_0 F'(x) - I\| < 1$. Thus both operators $\Gamma_0 F'(x)$ and $F'(x)$ are invertible.

Note that when $h_0 = \frac{1}{2}$ the successive approximations may converge to a "multiple" zero. An example is the scalar equation $x^2 = 0$ for any $x_0 \neq 0$.

We now examine the convergence of the MNK method (2.1.5).

The method (2.1.5) coincides with the usual iterative method

$$x_{n+1} = Ax_n \quad (n = 0, 1, 2, ...) \tag{2.2.29}$$

for approximate solution of the equation

$$x = Ax \tag{2.2.30}$$

where

$$Ax = x - \Gamma_0 Fx. \tag{2.2.31}$$

Theorem 2.2.7. *Under the hypotheses of Theorem 2.2.4 with (2.2.15) holding as strict inequality the successive approximations (2.1.5) converge to a solution $x^* \in U(x_0, r_0)$ of equation (2.1.1).*

Proof. First note that equation (2.1.1) indeed has a solution in the ball $U(x_0, r_0)$—this follows from Theorem 2.2.4. Below we shall prove that the operator (2.2.31) satisfies the assumptions of the contractive mapping principle Theorem 1.3.4 in the ball $U(x_0, r_0)$. This will imply that the solution x^* in the ball $U(x_0, r_0)$ is unique, and that the approximations (2.1.5) converge.

Obviously, for any $x, y \in U(x_0, r)$ $(r \leq R)$,

$$Ax - Ay = x - y - \Gamma_0(Fx - Fy)$$

$$= \Gamma_0 \int_0^1 \left[F'(x_0) - F'(y + t(x - y)) \right] (x - y) \, dt. \qquad (2.2.32)$$

This identity, together with (2.2.14) implies the estimate

$$\| Ax - Ay \| \leq b_0 Lr \| x - y \| . \qquad (2.2.33)$$

Consequently, A is a contraction operator in the ball $U(x_0, r_0)$. To complete the proof, it remains to show that

$$AU(x_0, r_0) \subseteq U(x_0, r_0).$$

Let $x_0 \in D$. Then, by (2.2.32)

$$\| Ax - x_0 \| \leq \| Ax - Ax_0 \| + \| Ax_0 - x_0 \|$$

$$\leq \left\| \Gamma_0 \int_0^1 \left[F'(x_0) - F'(x_0 + t(x - x_0)) \right] (x - x_0) \, dt \right\| + \eta_0.$$

Therefore, when $\| x - x_0 \| \leq r_0$,

$$\| Ax - x_0 \| \leq \frac{b_0 Lr_0^2}{2} + \eta_0 = r_0.$$

Note that, by (2.2.33), the operator A satisfies a Lipschitz condition with constant $q = 1 - \sqrt{1 - 2h_0}$ (see also (1.3.3)).

The above analysis of the modified Newton-Kantorovich method relates the simplest case. More subtle arguments (see, e.g., Kantorovich and Akilov [67]) show that Theorem 2.2.7 remains valid if the sign $<$ in (2.2.17) is replaced by $\leq$.

If $D = U(x_0, R)$, consider the operator A defined by (2.2.31). Assume that the conditions of Theorem 2.2.6 hold, and set

$$\alpha(r) = \sup_{\|x - x_0\| \leq r} \| Ax - x_0 \| . \qquad (2.2.34)$$

The function $\alpha(r)$ is obviously continuous and nondecreasing. It was shown in the proof of Theorem 2.2.6 that

$$\| Ax - x_0 \| \leq \frac{b_0 L \| x - x_0 \|^2}{2} + \eta_0 \quad (\| x - x_0 \| \leq R). \qquad (2.2.35)$$

Hence it follows:

Lemma 2.2.8. *The function* $\alpha(r)$ *satisfies the inequality*

$$\alpha(r) \leq \frac{b_0 L r^2}{2} + \eta_0 \quad (r_0 \leq r \leq R).$$

Theorem 2.2.9. *If, under the assumptions of Theorem 2.2.4,*

$$r_0 = \frac{1 - \sqrt{1 - 2h_0}}{h_0} \eta_0 \leq R < \frac{1 + \sqrt{1 - 2h_0}}{h_0} \eta_0,$$

as the quadratic trinomial $\frac{1}{2} b_0 L r^2 - r + \eta_0$ *is negative in the interval*

$$\left(r_0, \frac{1 + \sqrt{1 - 2h_0}}{h_0} \eta_0 \right).$$

Remark 2.2.10. If one repeats the proofs of Theorem 2.2.4 and 2.2.7 using $F'(x_0)^{-1} F(x)$ instead of the operator $F(x)$, condition

$$\left\| F'(x_0)^{-1} \left(F'(x) - F'(y) \right) \right\| \leq \ell \|x - y\| \tag{2.2.36}$$

for all $x, y \in U(x_0, R)$ instead of (2.2.14), condition

$$h = \frac{1}{2} \ell \eta \tag{2.2.37}$$

instead of (2.2.17), and finally

$$s^* = \frac{1 - \sqrt{1 - 2h}}{h} \eta, \quad U(x_0, s^*) \subseteq D \tag{2.2.38}$$

instead of (2.2.38) then the results hold in an affine invariant setting. The advantages of such an approach have elegantly been explained in [78] and also in [43]. From now on we shall be referring to (2.2.37) as the famous for its simplicity and clarity Newton-Kantorovich hypothesis.

Note that we are using for simplicity the same symbol ℓ to denote the Lipschitz constant in both conditions (2.2.14) and (2.2.36).

It also turns out from the proof of Theorem 2.2.4 that the scalar sequence $\{s_n\}$ $(n \geq 0)$ given by

$$s_0 = 0, \; s_1 = \eta, \; s_{n+2} = s_{n+1} + \frac{\ell (s_{n+1} - s_n)^2}{2 (1 - \ell s_n)} \tag{2.2.39}$$

is a majorizing sequence for $\{x_n\}$ such that

$$0 \leq s_0 \leq s_1 \leq \ldots \leq s_n \leq \ldots \leq s^* \tag{2.2.40}$$

and

$$\lim_{n \to \infty} s_n = s^*$$

[43], [67], [96].

Moreover the following error estimates hold for all $n \geq 0$:

$$\|x_{n+1} - x_n\| \leq s_{n+1} - s_n \tag{2.2.41}$$

and

$$\|x_n - x^*\| \leq s^* - s_n. \tag{2.2.42}$$

In the rest of the book motivated by optimization considerations using the same information (F, x_0, ℓ, η) we attempt to weaken crucial condition (2.2.37) and also provide a finer majorizing sequence than $\{s_n\}$. We also investigate in applications how is this effecting results by others based on (2.2.37).

To achieve all the above we introduce the center-Lipschitz condition

$$\left\| F'(x_0)^{-1} \left(F'(x) - F'(x_0) \right) \right\| \leq \ell_0 \|x - x_0\| \tag{2.2.43}$$

for all $x \in D$, where D is an open convex subset of X.

We also define scalar sequence $\{t_n\}$ by

$$t_0 = 0, \quad t_1 = \eta, \quad t_{n+2} = t_{n+1} + \frac{\ell (t_{n+1} - t_n)^2}{2 (1 - \ell_0 t_{n+1})} \quad (n \geq 0). \tag{2.2.44}$$

In [24] we showed:

Theorem 2.2.11. *Let $F: D \subseteq X \to Y$ be a Fréchet-differentiable operator and for $x_0 \in D$, assume*

$$F'(x_0)^{-1} \in L(Y, X); \tag{2.2.45}$$

conditions (2.2.36), (2.2.43), and

$$\bar{U}(x_0, t^*) \subseteq D \tag{2.2.46}$$

hold, where

$$t^* = \lim_{n \to \infty} t_n. \tag{2.2.47}$$

Moreover, assume that the following conditions hold:

$$h_\delta = (\delta \ell_0 + \ell) \eta \leq \delta, \text{ for } \delta \in [0, 1] \tag{2.2.48}$$

or

$$h_\delta \leq \delta, \quad \frac{2 \ell_0 \eta}{2 - \delta} \leq 1, \quad \frac{\ell_0 \delta^2}{2 - \delta} \leq \ell, \text{ for } \delta \in [0, 2) \tag{2.2.49}$$

or

$$h_\delta \leq \delta, \quad \ell_0 \eta \leq 1 - \frac{1}{2} \delta, \text{ for } \delta \in [\delta_0, 2), \tag{2.2.50}$$

where

$$\delta_0 = \frac{-\frac{\ell}{\ell_0} + \sqrt{\left(\frac{\ell}{\ell_0}\right)^2 + 8 \frac{\ell}{\ell_0}}}{2} \quad (\ell_0 \neq 0) \tag{2.2.51}$$

Then sequence $\{x_n\}$ generated by the NK method (2.1.3) is well defined, remains in $\bar{U}(x_0, t^*)$ for all $n \geq 0$ and converges to a solution $x^* \in \bar{U}(x_0, t^*)$ of equation $F(x) = 0$.

Moreover the following error estimates hold for all $n \geq 0$:

$$\|x_{n+2} - x_{n+1}\| \leq \frac{\ell \|x_{n+1} - x_n\|^2}{2[1 - \ell_0 \|x_{n+1} - x_0\|]} \leq t_{n+2} - t_{n+1} \tag{2.2.52}$$

and

$$\|x_n - x^*\| \leq t^* - t_n, \tag{2.2.53}$$

where t_n, t^* are given by (2.2.44) and (2.2.47) respectively.

Furthermore, if there exists $t^{**} \geq t^*$ such that

$$U\left(x_0, t^{**}\right) \in D \tag{2.2.54}$$

and

$$\ell_0\left(t^* + t^{**}\right) \leq 2, \tag{2.2.55}$$

the solution x^* is unique in $U(x_0, t^{**})$.

Note that optimum condition is given by (2.2.50) for $\delta = \delta_0$. However, we will be mostly using condition (2.2.48) for $\delta = 1$, which is the simplest, in the rest of this book.

We now compare our results with the ones obtained in Theorem 2.2.4 for the NK method (2.1.3).

Remark 2.2.12. Let us set $\delta = 1$ in condition (2.2.48). That is, consider

$$h_1 = (\ell + \ell_0) \leq 1. \tag{2.2.56}$$

Although (2.2.56) is not the weakest condition among (2.2.48)–(2.2.50) we will only compare this one with condition (2.2.37), since it seems to be the simplest.

(a) Note that

$$\ell_0 \leq \ell \tag{2.2.57}$$

holds in general and $\frac{\ell}{\ell_0}$ can be arbitrarily large as the following example indicates:

Example 2.2.13. Let $X = Y = D = \mathbf{R}$, $x_0 = 0$ and define function F on D by

$$F(x) = c_0 + c_1 x + c_2 \sin e^{c_3 x} \tag{2.2.58}$$

where c_i, $i = 0, 1, 2, 3$ are given parameters. Then it can easily be seen that for c_3 large and c_2 sufficiently small, $\frac{\ell}{\ell_0}$ can be arbitrarily large.

(b) We have

$$h \leq \frac{1}{2} \Rightarrow h_1 \leq 1 \tag{2.2.59}$$

but not vice versa unless if $\ell = \ell_0$. Indeed, for $\ell = \ell_0$, the NK theorem 2.2.4 is a special case of our Theorem 2.2.11. Otherwise our Theorem 2.2.11 can double the applicability of the NK theorem 2.2.4 as $\ell_0 \in [0, \ell]$:

Example 2.2.14. Let $X = Y = \mathbf{R}$, $D = [a, 2 - a]$, $a \in \left[0, \frac{1}{2}\right)$, $x_0 = 1$, and define function F on D by

$$F(x) = x^3 - a. \tag{2.2.60}$$

Using (2.2.16), (2.2.36), and (2.2.43), we obtain

$$\eta = \frac{1}{3}(1 - a), \quad \ell = 2(2 - a) > \ell_0 = 3 - a. \tag{2.2.61}$$

The Newton-Kantorovich hypothesis (2.2.37) cannot hold since

$$h = \frac{2}{3}(1 - a)(2 - a) > \frac{1}{2}, \quad \text{for all } a \in \left[0, \frac{1}{2}\right). \tag{2.2.62}$$

That is, there is no guarantee that NK method (2.1.3) converges to a solution of equation $F(x) = 0$.

However our condition (2.2.56), which becomes

$$h_1 = \frac{1}{3}(1 - a)[3 - a + 2(2 - a)] \leq 1, \tag{2.2.63}$$

holds for all $a \in \left[\frac{5 - \sqrt{13}}{3}, \frac{1}{2}\right)$.

In fact we can do better if we use (2.2.50) for

$$.4505 < \frac{5 - \sqrt{13}}{3} = .464816242...,$$

as we get $\eta = .183166...$, $\ell_0 = 2.5495$, $\ell = 3.099$, and $\delta_0 = 1.0656867$.

Choose $\delta = \delta_0$. Then we get that the interval $\left[\frac{5 - \sqrt{13}}{3}, \frac{1}{2}\right)$ can be extended to $\left[.450339002, \frac{1}{2}\right)$.

(c) Using simple induction (see [24]) we showed:

$$t_n \leq s_n \tag{2.2.64}$$

$$t_{n+1} - t_n \leq s_{n+1} - s_n \tag{2.2.65}$$

$$t^* \leq s^* = \lim_{n \to \infty} s_n \tag{2.2.66}$$

and

$$t^* - t_n \leq s^* - s_n. \tag{2.2.67}$$

Note also that strict inequality holds in (2.2.64) and (2.2.65) if $\ell_0 < \ell$.

That is, in this case our error estimates are finer and the information on the location of the solution at least as precise.

Note that all the above advantages are obtained using the same information and with the same computational cost since in practice the evaluation of ℓ requires the evaluation of ℓ_0.

We now compare our results with the ones obtained in Theorem 2.2.7 for the MNK method (2.1.5).

Remark 2.2.15. (a) Conditions (2.2.36), (2.2.37), and (2.2.38) can be replaced by

$$h^0 = \ell_0 \eta \le \frac{1}{2},$$
(2.2.68)

$$s_0^* = \frac{1 - \sqrt{1 - 2h^0}}{h^0} \eta,$$
(2.2.69)

and (2.2.43), respectively.

Indeed the proof of Theorem 2.2.7 can simply be rewritten with the above changes as condition (2.2.36) is never used full strength. This observation is important in computational mathematics for the following reasons: condition (2.2.68) is weaker than (2.2.37) if $\ell_0 < \ell$. That increases the applicability of Theorem 2.2.7 (see also Example 2.2.13); the error estimates are finer since the ratio q becomes smaller for $\ell_0 < \ell$; it is easier to compute ℓ_0 than computing ℓ (see also the three examples that follow). Finally by comparing (2.2.38) with (2.2.69) for $\ell_0 < \ell$ we obtain

$$s_0^* < s^*.$$
(2.2.70)

That is, we also obtain a more precise information on the location of the solution s^*.

Example 2.2.16. Returning back to Example 2.2.14 we see that Theorem 2.2.7 cannot be applied as condition (2.2.37) is violated. However, our condition (2.2.68) which becomes

$$h^0 = \frac{1}{3}(1-a)(3-a) \le \frac{1}{2}$$
(2.2.71)

holds for $a \in \left[\frac{4-\sqrt{10}}{2}, \frac{1}{2}\right)$.

Our motivation for introducing condition (2.2.39) instead of (2.2.36) can also be seen in the following example:

Example 2.2.17. Let $X = Y = \mathbf{R}$, $D = [0, \infty)$, $x_0 = 1$ and define function F on D by

$$F(x) = \frac{x^{1+\frac{1}{i}}}{1 + \frac{1}{i}} + c_1 x + c_2,$$
(2.2.72)

where c_1, c_2 are real parameters and $i > 2$ an integer. Then $F'(x) = x^{\frac{1}{i}} + c_1$ is not Lipschitz on D. However, center-Lipschitz condition (2.2.43) holds for $\ell_0 = (1 + c_1)^{-1}$ $(c_1 \ne -1)$.

Indeed, we have

$$\left\| F'(x_0)^{-1} \left[F'(x) - F'(x_0) \right] \right\| = (1 + c_1)^{-1} \left| x^{\frac{1}{i}} - x_0^{\frac{1}{i}} \right|$$

$$= \frac{(1 + c_1)^{-1} |x - x_0|}{x_0^{\frac{i-1}{i}} + \ldots + x^{\frac{i-1}{i}}}$$

$$\le \ell_0 |x - x_0|.$$
(2.2.73)

Example 2.2.18. We consider the integral equation

$$u(s) = f(s) + \lambda \int_a^b G(s,t) u(t)^{1+\frac{1}{n}} dt, \quad n \in \mathbf{N}. \tag{2.2.74}$$

Here, f is a given continuous function satisfying $f(s) > 0$, $s \in [a,b]$, λ is a real number, and the kernel G is continuous and positive in $[a,b] \times [a,b]$.

For example, when $G(s,t)$ is the Green kernel, the corresponding integral equation is equivalent to the boundary value problem

$$u'' = \lambda u^{1+\frac{1}{n}},$$

$$u(a) = f(a), \quad u(b) = f(b).$$

These type of problems have been considered in [71].

Equations of the form (2.2.74) generalize equations of the form

$$u(s) = \int_a^b G(s,t) u(t)^n dt \tag{2.2.75}$$

studied in [45].

Instead of (2.2.74), we can try to solve the equation $F(u) = 0$ where

$$F: \Omega \subseteq C[a,b] \to C[a,b], \quad \Omega = \{u \in C[a,b] : u(s) \geq 0, s \in [a,b]\},$$

and

$$F(u)(s) = u(s) - f(s) - \lambda \int_a^b G(s,t) u(t)^{1+\frac{1}{n}} dt.$$

The norm we consider is the max-norm.

The derivative F' is given by

$$F'(u) v(s) = v(s) - \lambda \left(1 + \frac{1}{n}\right) \int_a^b G(s,t) u(t)^{\frac{1}{n}} v(t) dt, \quad v \in \Omega.$$

First of all, we notice that F' does not satisfy a Lipschitz-type condition in Ω. Let us consider, for instance, $[a,b] = [0,1]$, $G(s,t) = 1$ and $y(t) = 0$. Then $F'(y) v(s) = v(s)$ and

$$\|F'(x) - F'(y)\| = |\lambda| \left(1 + \frac{1}{n}\right) \int_0^1 x(t)^{\frac{1}{n}} dt.$$

If F' were a Lipschitz function, then

$$\|F'(x) - F'(y)\| \leq L_1 \|x - y\|,$$

or, equivalently, the inequality

$$\int_0^1 x(t)^{\frac{1}{n}} dt \leq L_2 \max_{x \in [0,1]} x(s), \tag{2.2.76}$$

would hold for all $x \in \Omega$ and for a constant L_2. But this is not true. Consider, for example, the functions

$$x_j(t) = \frac{t}{j}, \quad j \geq 1, \quad t \in [0, 1].$$

If these are substituted into (2.2.76)

$$\frac{1}{j^{1/n}\left(1 + \frac{1}{n}\right)} \leq \frac{L_2}{j} \iff j^{1-1/n} \leq L_2\left(1 + \frac{1}{n}\right), \quad \forall j \geq 1.$$

This inequality is not true when $j \to \infty$.

Therefore, condition (2.2.36) fails in this case. However, condition (2.2.43) holds. To show this, let $x_0(t) = f(t)$ and $\alpha = \min_{s\in[a,b]} f(s)$, $\alpha > 0$. Then, for $v \in \Omega$,

$$\left\|\left[F'(x) - F'(x_0)\right]v\right\|$$

$$= |\lambda|\left(1 + \frac{1}{n}\right)\max_{s\in[a,b]}\left|\int_a^b G(s,t)\left(x(t)^{\frac{1}{n}} - f(t)^{\frac{1}{n}}\right)v(t)\,dt\right|$$

$$\leq |\lambda|\left(1 + \frac{1}{n}\right)$$

$$\cdot \int_a^b \max_{s\in[a,b]} \frac{G(s,t)\,|x(t) - f(t)|}{x(t)^{(n-1)/n} + x(t)^{(n-2)/n}f(t)^{1/n} + \ldots + f(t)^{(n-1)/n}}\,dt\,\|v\|.$$

Hence,

$$\left\|F'(x) - F'(x_0)\right\| \leq \frac{|\lambda|\left(1 + \frac{1}{n}\right)}{\alpha^{(n-1)/n}}\max_{s\in[a,b]}\int_a^b G(s,t)\,dt\,\|x - x_0\|$$

$$\leq K\,\|x - x_0\|,$$

where $K = \frac{|\lambda|\left(1 + \frac{1}{n}\right)}{\alpha^{(n-1)/n}}N$ and $N = \max_{s\in[a,b]}\int_a^b G(s,t)\,dt$.

Set $\ell_0 = \left\|F'(x_0)^{-1}K\right\|$. Then condition (2.2.68) holds for sufficiently small λ.

Remark 2.2.19. (a) We showed above that although the convergence of NK method (2.1.5) is quadratic (for $h < \frac{1}{2}$) there are cases when MNK method is preferred over the NK method.

(b) Although $t^* \in [\eta, 2\eta]$ say if condition (2.2.56) holds, we do not have an explicit form for it like, e.g., (2.2.38).

In practice though we can handle this problem in several ways. It follows from (2.2.56) that condition (2.2.68) also holds. Therefore we know that the solution x^* is unique in $U(x_0, s_0^*)$. That is, there exists a finite $n_0 \geq 1$ such that if $n \geq n_0$ the sequence $\{x_n\}$ will enter the ball $U(x_0, s_0^*)$ and enjoy quadratic convergence according to Theorem 2.2.11. Note that if $t^* \leq s_0^*$ then we can take $n_0 = 1$. Moreover, if (2.2.37) also holds, then $t^* \in [\eta, s^*]$, with $s^* \leq 2\eta$.

2.3 New sufficient conditions for the secant method

It turns out that the ideas introduced in Section 2.2 for Newton's method can be extended to the method of chord or the secant method.

In this section, we are concerned with the problem of approximating a locally unique solution x^* of equation

$$F(x) = 0, \tag{2.3.1}$$

where F is a nonlinear operator defined on a convex subset D of a Banach space X with values in a Banach space Y.

We consider the secant method in the form

$$x_{n+1} = x_n - \delta F(x_{n-1}, x_n)^{-1} F(x_n) \quad (n \geq 0), \tag{2.3.2}$$

where $\delta F(x, y) \in L(X, Y)$ $(x, y \in D)$ is a consistent approximation of the Fréchet derivative of F, Dennis [74], Potra [162], Argyros [12], [43], Hernandez [116], [117], and others have provided sufficient convergence conditions for the secant method based on "Lipschitz-type" conditions on δF (see also Section 1.2). Here using "Lipschitz-type" and center-"Lipschitz-type" conditions, we provide a semilocal convergence analysis for (2.3.2). It turns out that our error bounds are more precise and our convergence conditions hold in cases where the corresponding hypotheses mentioned in earlier references mentioned above are violated.

We need the following result on majorizing sequences.

Lemma 2.3.1. *Assume there exist nonnegative parameters* $\ell, \ell_0, \eta, c,$ *and* $a \in [0, 1],$

$$\delta \in \begin{cases} \left[0, \dfrac{-1 + \sqrt{1 + 4a}}{2a} \right], & a \neq 0 \\[2mm] [0, 1), & a = 0 \end{cases} \tag{2.3.3}$$

such that:

$$(\ell + \delta \ell_0)(c + \eta) \leq \delta, \tag{2.3.4}$$

$$\eta \leq \delta c, \tag{2.3.5}$$

$$\ell_0 \leq a\ell. \tag{2.3.6}$$

Then,

(a) iteration $\{t_n\}$ $(n \geq -1)$ *given by*

$$t_{-1} = 0, \quad t_0 = c, \quad t_1 = c + \eta,$$

$$t_{n+2} = t_{n+1} + \frac{\ell(t_{n+1} - t_{n-1})}{1 - \ell_0 [t_{n+1} - t_0 + t_n]} (t_{n+1} - t_n) \quad (n \geq 0) \tag{2.3.7}$$

is nondecreasing, bounded above by

$$t^{**} = \frac{\eta}{1 - \delta} + c \tag{2.3.8}$$

and converges to some t^* such that

$$0 \le t^* \le t^{**}. \tag{2.3.9}$$

Moreover, the following estimates hold for all $n \ge 0$

$$0 \le t_{n+2} - t_{n+1} \le \delta(t_{n+1} - t_n) \le \delta^{n+1}\eta. \tag{2.3.10}$$

(b) Iteration $\{s_n\}$ ($n \ge 0$) given by

$$s_{-1} - s_0 = c, \quad s_0 - s_1 = \eta,$$

$$s_{n+1} - s_{n+2} = \frac{\ell(s_{n-1} - s_{n+1})}{1 - \ell_0 \left[(s_0 + s_{-1}) - (s_n + s_{n+1})\right]}(s_n - s_{n+1}) \quad (n \ge 0) \tag{2.3.11}$$

for $s_{-1}, s_0, s_1 \ge 0$ is nonincreasing, bounded below by

$$s^{**} = s_0 - \frac{\eta}{1 - \delta} \tag{2.3.12}$$

and converges to some s^* such that

$$0 \le s^{**} \le s^*. \tag{2.3.13}$$

Moreover, the following estimates hold for all $n \ge 0$

$$0 \le s_{n+1} - s_{n+2} \le \delta(s_n - s_{n+1}) \le \delta^{n+1}\eta. \tag{2.3.14}$$

Proof. (a) The result clearly holds if $\delta = 0$ or $\ell = 0$ or $c = 0$. Let us assume $\delta \ne 0$, $\ell \ne 0$ and $c \ne 0$. We must show for all $k \ge 0$:

$$\ell(t_{k+1} - t_{k-1}) + \delta\ell_0\left[(t_{k+1} - t_0) + t_k\right] \le \delta, \quad 1 - \ell_0\left[(t_{k+1} - t_0) + t_k\right] > 0. \tag{2.3.15}$$

Inequalities (2.3.15) hold for $k = 0$ by the initial conditions. But then (2.3.7) gives

$$0 \le t_2 - t_1 \le \delta(t_1 - t_0).$$

Let us assume (2.3.10) and (2.3.15) hold for all $k \le n+1$. By the induction hypotheses we can have in turn:

$$\ell(t_{k+2} - t_k) + \delta\ell_0\left[(t_{k+2} - t_0) + t_{k+1}\right]$$

$$\le \ell\left[(t_{k+2} - t_{k+1}) + (t_{k+1} - t_k)\right] + \delta\ell_0\left[\frac{1 - \delta^{k+2}}{1 - \delta} + \frac{1 - \delta^{k+1}}{1 - \delta}\right]\eta + \delta\ell_0 c$$

$$\le \ell(\delta^{k+1} + \delta^k)\eta + \frac{\delta\ell_0}{1 - \delta}(2 - \delta^{k+1} - \delta^{k+2})\eta + \delta\ell_0 c. \tag{2.3.16}$$

We must show that δ is the upper bound in (2.3.16). Instead by (2.3.5) we can show

$$\ell\delta^k(1 + \delta)\eta + \frac{\delta\ell_0}{1 - \delta}(2 - \delta^{k+2} - \delta^{k+1})\eta + \delta\ell_0 c \le (\ell + \delta\ell_0)(c + \eta)$$

or

$$\delta\ell_0\left[\frac{2-\delta^{k+2}-\delta^{k+1}}{1-\delta}-1\right]\eta \le \ell\left[c+\eta-\delta^k(1+\delta)\eta\right]$$

or

$$a\delta\ell\frac{1+\delta-\delta^{k+1}(1+\delta)}{1-\delta}\eta \le \ell\left[\frac{\eta}{\delta}+\eta-\delta^k(1+\delta)\eta\right]$$

or

$$a\delta^2(1+\delta)(1-\delta^{k+1}) \le (1-\delta)(1+\delta)(1-\delta^{k+1})$$

or

$$a\delta^2+\delta-1\le 0,$$

which is true by the choice of δ. Moreover, by (2.3.5) and (2.3.10)

$$\delta\ell_0\left[(t_{k+2}-t_0)+t_{k+1}\right] \le \frac{\delta\ell_0}{1-\delta}(2-\delta^{k+2}-\delta^{k+1})\eta + \delta\ell_0 c$$

$$< (\ell+\delta\ell_0)(c+\eta) \le \delta, \qquad (2.3.17)$$

which shows the second inequality in (2.3.15). We must also show:

$$t_k \le t^{**} \quad (k \ge -1). \qquad (2.3.18)$$

For $k = -1, 0, 1, 2$ we have $t_{-1} = 0 \le t^{**}$, $t_0 = \eta \le t^{**}$, $t_1 = \eta + c \le t^{**}$ by (2.3.8), and $t_2 = c + \eta + \delta\eta = c + (1+\delta)\eta \le t^{**}$ by the choice of δ. Assume (2.3.18) holds for all $k \le n + 1$. It follows from (2.3.10)

$$t_{k+2} \le t_{k+1} + \delta(t_{k+1}-t_k) \le t_k + \delta(t_k-t_{k-1}) + \delta(t_{k+1}-t_k)$$

$$\le \cdots \le t_1 + \delta(t_1-t_0) + \cdots + \delta(t_{k+1}-t_k)$$

$$\le c+\eta+\delta\eta+\cdots+\delta^{k+1}\eta = c + \frac{1-\delta^{k+2}}{1-\delta}\eta$$

$$< \frac{\eta}{1-\delta}+c = t^{**}.$$

That is $\{t_n\}$ $(n \ge -1)$ is bounded above by t^{**}. It also follows from (2.3.7) and (2.3.15) that it is also nondecreasing and as such it converges to some t^* satisfying (2.3.9).

(b) As in part (a) but we show $\{s_n\}$ $(n \ge -1)$ is nonincreasing and bounded below by s^{**}. Note that the inequality corresponding with (2.3.16) is

$$\ell(s_k - s_{k+2}) \le \delta\left[1 - \beta(s_0+s_{-1}) + \beta(s_{k+1}+s_{k+2})\right]$$

or

$$\ell\left[\delta^k(s_0-s_1) + \delta^{k+1}(s_0-s_1)\right]$$

$$\le \delta\left[1 - \ell_0(s_0+s_{-1}) + \ell_0\left(s_0 - \frac{1-\delta^{k+1}}{1-\delta}(s_0-s_1)\right) + \ell\left(s_0 - \frac{1-\delta^{k+2}}{1-\delta}(s_0-s_1)\right)\right]$$

or

$$\ell\delta^k(1+\delta)\eta + \delta\ell_0\left[\frac{2-\delta^{k+1}-\delta^{k+2}}{1-\delta}\eta + c\right]$$

must be bounded above by δ which was shown in part (a).

Remark 2.3.2. It follows from (2.3.16) and (2.3.17) that the conclusions of Lemma 2.3.1 hold if (2.3.3), (2.3.5), (2.3.6) are replaced by the weaker conditions:

for all $n \geq 0$ there exists $\delta \in [0, 1)$ such that:

$$\ell\delta^n(1+\delta)\eta + \frac{\delta\ell_0}{1-\delta}(2 - \delta^{n+2} - \delta^{n+1})\eta + \delta\ell_0 c \leq \delta,$$

and

$$\frac{\delta\ell_0}{1-\delta}(2 - \delta^{n+2} - \delta^{n+1})\eta + \delta\ell_0 c < 1.$$

The above conditions hold in many cases for all $n \geq 0$. One such stronger case is

$$\ell(1+\delta)\eta + \frac{2\delta\ell_0\eta}{1-\delta} + \delta\ell_0 c \leq \delta,$$

and

$$\frac{2\delta\ell_0\eta}{1-\delta} + \delta\ell_0 c < 1.$$

We shall study the iterative procedure (2.3.2) for triplets (F, x_{-1}, x_0) belonging to the class $C(\ell, \ell_0, \eta, c)$ defined as follows:

Definition 2.3.3. *Let ℓ, ℓ_0, η, c be nonnegative parameters satisfying the hypotheses of Lemma 2.3.1 or Remark 2.3.2 (including (2.3.4)).*
We say that a triplet (F, x_{-1}, x_0) belongs to the class $C(\ell, \ell_0, \eta, c)$ if:
(c_1) F is a nonlinear operator defined on a convex subset D of a Banach space X with values in a Banach space Y;
(c_2) x_{-1} and x_0 are two points belonging to the interior D^0 of D and satisfying the inequality

$$\|x_0 - x_{-1}\| \leq c; \tag{2.3.19}$$

(c_3) F is Fréchet-differentiable on D^0 and there exists an operator $\delta F: D^0 \times D^0 \to L(X, Y)$ such that:
the linear operator $A = \delta F(x_{-1}, x_0)$ is invertible, its inverse A^{-1} is bounded and:

$$\|A^{-1}F(x_0)\| \leq \eta; \tag{2.3.20}$$

$$\|A[\delta F(x, y) - F'(z)]\| \leq \ell(\|x - z\| + \|y - z\|), \tag{2.3.21}$$

$$\|A[\delta F(x, y) - F'(x_0)]\| \leq \ell_0(\|x - x_0\| + \|y - x_0\|) \tag{2.3.22}$$

for all $x, y, z \in D$.
(c_4) the set $D_c = \{x \in D; F \text{ is continuous at } x\}$ contains the closed ball $\bar{U}(x_0, s^*)$ where s^* is given in Lemma 2.3.1.

We present the following semilocal convergence theorem for secant method (2.3.2).

Theorem 2.3.4. *If $(F, x_{-1}, x_0) \in C(\ell, \ell_0, \eta, c)$ then sequence $\{x_n\}$ ($n \geq -1$) generated by secant method (2.3.2) is well defined, remains in $\bar{U}(x_0, s^*)$ for all $n \geq 0$ and converges to a solution $x^* \in \bar{U}(x_0, s^*)$ of equation $F(x) = 0$.*

Moreover the following estimates hold for all $n \geq 0$

$$\|x_{n+2} - x_{n+1}\| \leq s_{n+1} - s_{n+2}, \qquad (2.3.23)$$

$$\|x_n - x^*\| \leq \alpha_n \qquad (2.3.24)$$

and

$$\|x_n - x^*\| \geq \beta_n \qquad (2.3.25)$$

where,

$$s_{-1} = \frac{1 + \ell_0 c}{2\ell_0}, \quad s_0 = \frac{1 - \ell_0 c}{2\ell_0} \ \text{ for } \ell_0 \neq 0, \qquad (2.3.26)$$

sequence $\{s_n\}$ ($n \geq 0$) given by (2.3.11), α_n, β_n are respectively the nonnegative solutions of equations

$$\ell_0 t^2 - 2\ell_0(s_0 - \|x_n - x_0\|)t - \ell(\|x_n - x_{n-1}\| + \|x_{n-1} - x_{n-2}\|)\|x_n - x_{n-1}\| = 0, \qquad (2.3.27)$$

and

$$\ell t^2 + \left[\ell\|x_n - x_{n-1}\| + 1 - \ell_0(\|x_n - x_0\| + \|x_{n-1} - x_0\| + c)\right] t$$
$$+ \left[\ell_0(\|x_n - x_0\| + \|x_{n-1} - x_0\| + c) - 1\right]\|x_{n+1} - x_n\| = 0. \qquad (2.3.28)$$

Proof. We first show operator $L = \delta F(u, v)$ is invertible for all $u, v \in D^0$ with

$$\|u - x_0\| + \|v - x_0\| < 2s_0. \qquad (2.3.29)$$

It follows from (2.3.22) and (2.3.29)

$$\|I - A^{-1}L\| = \|A^{-1}(L - A)\| \leq \|A^{-1}(L - F'(x_0))\| + \|A^{-1}(F'(x_0) - A)\|$$
$$\leq \ell_0(\|u - x_0\| + \|v - x_0\| + \|x_0 - x_{-1}\|) < 1. \qquad (2.3.30)$$

According to the Banach Lemma on invertible operators and (2.3.30), L is invertible and

$$\|L^{-1}A\| \leq [1 - \ell_0(\|u - x_0\| + \|v - x_0\| + c)]^{-1}. \qquad (2.3.31)$$

Condition (2.3.21) implies the Lipschitz condition for F'

$$\|A^{-1}(F'(u) - F'(v))\| \leq 2\ell\|u - v\|, \quad u, v \in D^0. \qquad (2.3.32)$$

By the identity,

$$F(x) - F(y) = \int_0^1 F'(y + t(x - y))dt(x - y) \qquad (2.3.33)$$

we get

$$\|A_0^{-1}\left[F(x) - F(y) - F'(u)(x - y)\right]\| \leq \ell(\|x - u\| + \|y - u\|)\|x - y\| \qquad (2.3.34)$$

and

$$\|A_0^{-1}[F(x) - F(y) - \delta F(u, v)(x - y)]\| \leq \ell(\|x - v\| + \|y - v\| + \|u - v\|)\|x - y\|$$
$$(2.3.35)$$

for all $x, y, u, v \in D^0$. By a continuity argument (2.3.33)–(2.3.35) remain valid if x and/or y belong to D_c.

We first show (2.3.23). If (2.3.23) hold for all $\eta \leq k$ and if $\{x_n\}$ ($n \geq 0$) is well defined for $n = 0, 1, 2, \ldots, k$ then

$$\|x_0 - x_n\| \leq s_0 - s_n < s_0 - s^*, \quad n \leq k. \tag{2.3.36}$$

Hence (2.3.29) holds for $u = x_i$ and $v = x_j$ ($i, j \leq k$). That is (2.3.2) is well defined for $n = k + 1$. For $n = -1$ and $n = 0$, (2.3.23) reduces to $\|x_{-1} - x_0\| \leq c$ and $\|x_0 - x_1\| \leq \eta$. Suppose (2.3.23) holds for $n = -1, 0, 1, \ldots, k$ ($k \geq 0$). Using (2.3.31), (2.3.35) and

$$F(x_{k+1}) = F(x_{k+1}) - F(x_k) - \delta F(x_{k-1}, x_k)(x_{k+1} - x_k) \tag{2.3.37}$$

we obtain in turn

$$\|x_{k+2} - x_{k+1}\| = \|\delta F(x_k, x_{k+1})^{-1} F(x_{k+1})\|$$
$$\leq \|\delta F(x_k, x_{k+1})^{-1} A\| \, \|A^{-1} F(x_{k+1})\|$$
$$\leq \frac{\ell(\|x_{k+1} - x_k\| + \|x_k - x_{k-1}\|)}{1 - \ell_0[\|x_{k+1} - x_0\| + \|x_k - x_0\| + c]} \|x_{k+1} - x_k\|$$
$$\leq \frac{\ell(s_k - s_{k+1} + s_{k-1} - s_k)}{1 - \ell_0[s_0 - s_{k+1} + s_0 - s_k + s_{-1} - s_0]} (s_k - s_{k+1}) = s_{k+1} - s_{k+2}. \tag{2.3.38}$$

The induction for (2.3.23) is now complete. It follows from (2.3.23) and Lemma 2.3.1 that sequence $\{x_n\}$ ($n \geq -1$) is Cauchy in a Banach space X and as such it converges to some $x^* \in \bar{U}(x_0, s^*)$ (as $\bar{U}(x_0, s^*)$ is a closed set) so that

$$\|x_n - x^*\| \leq s_n - s^*. \tag{2.3.39}$$

By letting $k \to \infty$ in (2.3.38), we obtain $F(x^*) = 0$.

Set $x = x_n$ and $y = x^*$ in (2.3.33), $M = \int_0^1 F'(x^* + t(x_n - x^*))dt$. Using (2.3.23) and (2.3.39) we get in turn

$$\|x_n - x_0\| + \|x^* - x_0\| + \|x_0 - x_{-1}\| \leq 2\|x_n - x_0\| + \|x_n - x^*\| + c$$
$$< 2(\|x_n - x_0\| + \|x_n - x^*\|)$$
$$\leq 2(s_0 - s_n + s_n - s^*) + c \leq 2s_0 + c = \frac{1}{\ell_0}. \tag{2.3.40}$$

By (2.3.40) and the Banach Lemma on invertible operators we get

$$\|M^{-1} A\| \leq [1 - \ell_0(2\|x_n - x_0\| + \|x_n - x^*\| + c)]^{-1}. \tag{2.3.41}$$

It follows from (2.3.2) and (2.3.41)

$$\|x_n - x^*\| \leq \|M^{-1} A\| \cdot \|A^{-1} F(x_n)\|$$
$$\leq \frac{\ell[\|x_n - x_{n-1}\| + \|x_{n-1} - x_{n-2}\|]}{1 - \ell_0[2\|x_n - x_0\| + \|x_n - x^*\| + c]} \|x_n - x_{n-1}\|, \tag{2.3.42}$$

which shows (2.3.24).

Using the approximation

$$x_{n+1} - x^* = x^* - x_n + [A\delta F(x_{n-1}, x_n)]^{-1}$$
$$\cdot A[F(x^*) - F(x_n) - \delta F(x_{n-1}, x_n)(x^* - x_n)] \quad (2.3.43)$$

and estimates (2.3.30), (2.3.35) we get

$$\|x_{n+1} - x_n\| \le \frac{\ell[\|x^* - x_n\| + \|x_n - x_{n-1}\|]}{1 - \ell_0[\|x_n - x_0\| + \|x_{n-1} + x_0\| + c]} \|x_n - x^*\| + \|x_n - x^*\|, \quad (2.3.44)$$

which shows (2.3.25).

In the next result we examine the uniqueness of the solution x^*.

Theorem 2.3.5. *If $(F, x_{-1}, x_0) \in C(\ell, \ell_0, \eta, c)$ equation (2.3.1) has a solution $x^* \in \bar{U}(x_0, s^*)$. This solution is unique in the set $U_1 = \{x \in D_c \mid \|x - x_0\| < s_0 + \gamma\}$ if $\gamma > 0$ or in the set $U_2 = \{x \in D_c \mid \|x - x_0\| \le s_0\}$ if $\gamma = 0$.*

Proof. Case 1: $\gamma > 0$. Let $x^* \in \bar{U}(x_0, s^*)$ and $y^* \in U_1$ be solutions of equation $F(x) = 0$. Set $P = \int_0^1 F'(y + t(x - y))dt$. Using (2.3.22) we get

$$\|I - A^{-1}P\| = \|A^{-1}(A - P)\| \le \ell_0(\|y^* - x_0\| + \|x^* - x_0\| + \|x_0 - x_{-1}\|)$$
$$< \ell_0(s_0 + \gamma + s_0 - \gamma + c) = 1.$$

Hence, P is invertible and from (2.3.33) we get $x^* = y^*$.

Case 2: $\gamma = 0$. Consider the modified secant method

$$s_{n+1} = s_n - A^{-1}F(y_n) \quad (n \ge 0). \quad (2.3.45)$$

By Theorem 2.3.4 sequence $\{y_n\}$ $(n \ge 0)$ converges to x^* and

$$\|x_n - x_{n+1}\| \le \bar{s}_n - \bar{s}_{n+1} \quad (2.3.46)$$

where,

$$\bar{s}_0 = \sqrt{\tfrac{\eta}{\ell}}, \quad \bar{s}_{n+1} = \bar{s}_n - \ell s_n^2 \quad (n \ge 0), \quad \text{for } \ell > 0. \quad (2.3.47)$$

Using induction on $n \ge 0$ we get

$$\bar{s}_n \ge \frac{\sqrt{\tfrac{\eta}{\ell}}}{n+1} \quad (n \ge 0). \quad (2.3.48)$$

Let y^* be a solution of $F(x) = 0$. Set $P_n = \int_0^1 F'(y^* + t(x_n - y^*))dt$. It follows from (2.3.22), (2.3.33), (2.3.45), and (2.3.48)

$$\|x_{n+1} - y^*\| = \|A^{-1}(A - P_n)(x_n - y^*)\|$$
$$\le \ell(\|y^* - x_0\| + \|x_n - x_0\| + \|x_0 - x_{-1}\|)\|x_n - y^*\|$$
$$\le (1 - \ell\bar{s}_n)\|x_n - y^*\| \le \cdots \le \prod_{i=1}^{n}(1 - \ell\bar{s}_i)\|x_1 - y^*\|. \quad (2.3.49)$$

By (2.3.49), we get $\lim_{n \to \infty} \prod_{i=1}^{n}(1 - \ell\bar{s}_i) = 0$. Hence, we deduce $x^* = y^*$.

That completes the proof of the theorem.

Remark 2.3.6. The parameter s^* can be computed as the limit of sequence $\{s_n\}$ ($n \geq -1$) using (2.3.11). Simply set

$$s^* = \lim_{n \to \infty} s_n. \tag{2.3.50}$$

Remark 2.3.7. A similar convergence analysis can be provided if sequence $\{s_n\}$ is replaced by $\{t_n\}$. Indeed under the hypotheses of Theorem 2.3.4 we have for all $n \geq 0$

$$\|x_{n+2} - x_{n+1}\| \leq t_{n+2} - t_{n+1} \tag{2.3.51}$$

and

$$\|x^* - x_n\| \leq t^* - t_n. \tag{2.3.52}$$

In order for us to compare with earlier results we first need the definition:

Definition 2.3.8. *Let ℓ, η, c be three nonnegative numbers satisfying the inequality*

$$\ell c + 2\sqrt{\ell \eta} \leq 1. \tag{2.3.53}$$

We say that a triplet $(F, x_{-1}, x_0) \in C_1(\ell, \eta, c)$ ($\ell > 0$ if conditions (c_1)–(c_4) hold (excluding (2.3.22)). Define iteration $\{p_n\}$ ($n \geq -1$) by

$$p_{-1} = \frac{1+\ell c}{2\ell}, \quad p_0 = \frac{1-\ell c}{2\ell}, \quad p_{n+1} = p_n - \frac{p_n^2 - p^2}{p_n + p_{n-1}}, \tag{2.3.54}$$

where,

$$p = \frac{1}{2\ell}\sqrt{(1 - \ell c)^2 - 4\ell\eta}. \tag{2.3.55}$$

The proof of the following semilocal convergence theorem can be found in [164].

Theorem 2.3.9. *If $(F, x_{-1}, x_0) \in C_1(\ell, \eta, c)$ sequence $\{x_n\}$ ($n \geq -1$) generated by secant method (2.3.2) is well defined, remains in $\bar{U}(x_0, p)$ for all $n \geq 0$ and converges to a unique solution $x^* \in \bar{U}(x_0, p)$ of equation $F(x) = 0$.*

Moreover the following error bounds hold for all $n \geq 0$:

$$\|x_{n+1} - x_n\| \leq p_n - p_{n+1} \tag{2.3.56}$$

and

$$\|x_n - x^*\| \leq p_n - p. \tag{2.3.57}$$

Using induction on n we can easily show the following favorable comparison of error bounds between Theorems 2.3.4 and 2.3.9.

Proposition 2.3.10. *Under the hypotheses of Theorems 2.3.4 and 2.3.9 the following estimates hold for all $n \geq 0$*

$$p_n \leq s_n \tag{2.3.58}$$

$$s_n - s_{n+1} \leq p_n - p_{n+1} \tag{2.3.59}$$

and

$$s_n - s^* \leq p_n - p. \tag{2.3.60}$$

Remark 2.3.11. We cannot compare conditions (2.3.4) and (2.3.53) in general because of ℓ_0. However in the special case $\ell = \ell_0 \neq 0$, we can set $a = 1$ to obtain
$\delta = \frac{\sqrt{5}-1}{2}$. Condition (2.3.4) can be written

$$\ell c + \ell \eta \leq \beta = \frac{\delta}{1+\delta} = .381966011.$$

It can then easily be seen that if

$$0 < \ell c < 2\sqrt{\beta} - 1 = .236067977,$$

condition (2.3.4) holds but (2.3.53) is violated. That is, even in the special case of
$\ell = \ell_0$, our Theorem 2.3.4 can be applied in cases not covered by Theorem 2.3.9.

2.4 Concerning the "terra incognita" between convergence regions of two Newton methods

There is an unknown area, between the convergence regions ("terra incognita") of
the NK method, and the corresponding MNK method, when F' is an λ-Hölder continuous operator, $\lambda \in [0, 1)$. Note that according to Kantorovich theorems 2.2.4 and
2.2.7, these regions coincide when $\lambda = 1$. However, we already showed (see (2.2.70))
that this is not the case unless if $\ell_0 = \ell$. Here, we show how to investigate this region
and improve on earlier attempts in this direction for $\lambda \in [0, 1)$ [32], [35], [64].

To make the study as self-contained as possible, we briefly reintroduce some
results (until Remark 2.4.3) that can originally be found in [32], [64].

Let $x_0 \in D$ be such that $F'(x_0)^{-1} \in L(Y, X)$. Assume F' satisfies a center-
Hölder condition

$$\|F'(x_0)^{-1}(F'(x) - F'(x_0))\| \leq \ell_0 \|x - x_0\|^\lambda, \qquad (2.4.1)$$

and a Hölder condition

$$\|F'(x_0)^{-1}(F'(x) - F'(y))\| \leq \ell \|x - y\|^\lambda \qquad (2.4.2)$$

for all $x, y \in U(x_0, R) \subseteq D$.

The results in [64] were given in non-affine invariant form. Here we reproduce
them in affine invariant form. The advantages of such an approach have been well
explained in [43], [78].

Define:

$$h_0 = \ell_0 \eta^\lambda, \qquad (2.4.3)$$
$$h = \ell \eta^\lambda \qquad (2.4.4)$$

and function

$$\psi(r) = \frac{\ell}{1+\lambda} r^{1+\lambda} - r + \eta, \qquad (2.4.5)$$

where η is given by (2.2.16).

The first semilocal convergence result for methods NK and MNK under Hölder
conditions were given in [135]:

Theorem 2.4.1. *Assume:*

$$h \leq \left(\frac{\lambda}{1+\lambda}\right)^{\lambda} \tag{2.4.6}$$

and

$$r^* \leq R, \tag{2.4.7}$$

where r^* is the smallest positive zero of function ψ. Then sequence $\{x_n\}$ ($n \geq 0$) generated by MNK method is well defined, remains in $U(x_0, r^*)$ for all $n \geq 0$ and converges to a unique solution x^* of equation (2.1.1) in $U(x_0, r^*)$. If r^* is the unique zero of ψ on $[0, R]$ and $\psi(R) \leq 0$, then x^* is unique in $U(x_0, R)$.
    Moreover, if

$$h \leq h_\nu, \tag{2.4.8}$$

where h_ν is the unique solution in $(0, 1)$ of equation

$$\left(\frac{t}{1+\lambda}\right)^{\lambda} = (1 - t)^{1+\lambda} \tag{2.4.9}$$

method NK converges as well.

Theorem 2.4.1 holds [135] if condition (2.4.6) is replaced by the weaker

$$h \leq 2^{\lambda-1}\left(\frac{\lambda}{1+\lambda}\right)^{\lambda}. \tag{2.4.10}$$

Later in [64], (2.4.10) was replaced by an even weaker condition

$$h \leq \frac{1}{g(\lambda)}\left(\frac{\lambda}{1+\lambda}\right)^{\lambda}, \tag{2.4.11}$$

where,

$$g(\lambda) = \max_{t\geq 0} f(t), \tag{2.4.12}$$

$$f(t) = \frac{t^{1+\lambda}+(1+\lambda)t}{(1+t)^{1+\lambda}-1} \tag{2.4.13}$$

with

$$g(\lambda) < 2^{1-\lambda} \text{ for all } \lambda \in (0, 1). \tag{2.4.14}$$

Recently in [64], (2.4.11) was replaced by

$$h \leq \frac{1}{a(\lambda)}\left(\frac{\lambda}{1+\lambda}\right)^{\lambda}, \tag{2.4.15}$$

where,

$$a(\lambda) = \min\left\{b \geq 1: \max_{0\leq t\leq t(b)} f(t) \leq b\right\}, \tag{2.4.16}$$

$$t(b) = \frac{b\lambda^{\lambda}}{(1+\lambda)[b(1+\lambda)^{\lambda}-\lambda^{\lambda}]}. \tag{2.4.17}$$

The idea is to optimize b in the equation

$$\psi_b(r) = 0, \tag{2.4.18}$$

where,

$$\psi_b(r) = \frac{b\ell}{1+\lambda} r^{1+\lambda} - r + \eta \tag{2.4.19}$$

assuming

$$h \leq \frac{1}{b}\left(\frac{\lambda}{1+\lambda}\right)^{\lambda}. \tag{2.4.20}$$

Note that condition (2.4.20) guarantees that equation (2.4.18) is solvable (see Proposition 1.1 in [64]).

With the above notation it was shown in [64] (Theorem 2.2, p. 719):

Theorem 2.4.2. *Assume (2.4.15) holds and that $r^* \leq R$, where r^* is the smallest solution of the scalar equation*

$$\psi_a(r) = \frac{a(\lambda)\ell}{1+\lambda} r^{1+\lambda} - r + \eta = 0. \tag{2.4.21}$$

Then sequence $\{x_n\}$ $(n \geq 0)$ generated by NK method is well defined, remains in $U(x_0, r^*)$ for all $n \geq 0$, and converges to a unique solution x^* of equation $F(x) = 0$ in $U(x_0, r^*)$.

Moreover if sequence r_n is defined by

$$r_0 = 0, \quad r_n = r_{n-1} - \frac{\psi_a(r_{n-1})}{\psi'(r_{n-1})} \quad (n \geq 1) \tag{2.4.22}$$

then the following estimates hold for all $n \geq 1$:

$$\|x_n - x_{n-1}\| \leq r_n - r_{n-1} \tag{2.4.23}$$

and

$$\|x_n - x^*\| \leq r^* - r_n. \tag{2.4.24}$$

Remark 2.4.3. It was also shown in [64] (see Theorem 2.3) that

$$a(\lambda) < f(2) < g(\lambda) \quad \text{for all } \lambda \in (0, 1), \tag{2.4.25}$$

which shows that (2.4.15) is a real improvement over (2.4.10) and (2.4.11).

We can summarize as follows:

$$h_v < 2^{\lambda-1}\left(\frac{\lambda}{1+\lambda}\right)^{\lambda} < \frac{1}{g(\lambda)}\left(\frac{\lambda}{1+\lambda}\right)^{\lambda}$$

$$< \frac{1}{a(\lambda)}\left(\frac{\lambda}{1+\lambda}\right)^{\lambda} \leq \left(\frac{\lambda}{1+\lambda}\right)^{\lambda} = h_{exi}. \tag{2.4.26}$$

Below we present our contributions/improvements in the exploration of "terra incognita."

First of all, we have observed that the Vertgeim result given in Theorem 2.4.1 holds under weaker conditions. Indeed:

Theorem 2.4.4. *Assume:*

$$h_0 \le \left(\tfrac{\lambda}{1+\lambda}\right)^{\lambda} \tag{2.4.27}$$

replaces condition (2.4.6) in Theorem 2.4.1. Then under the rest of the hypotheses of Theorem 2.4.1, the conclusions for method (2.1.5) and equation (2.1.1) hold.

Proof. We note that (2.4.1) can be used instead of (2.4.2) in the proof of Theorem 1 given in [135].

Remark 2.4.5. Condition (2.4.27) is weaker than (2.4.6) because

$$h \le \left(\tfrac{\lambda}{1+\lambda}\right)^{\lambda} \Rightarrow h_0 \le \left(\tfrac{\lambda}{1+\lambda}\right)^{\lambda} \tag{2.4.28}$$

but not vice versa unless if $\ell = \ell_0$. Therefore our Theorem 2.4.4 improves the convergence region for MNK method under weaker conditions and cheaper computational cost.

It turns out that we can improve on the error bounds given in Theorem 2.4.2 under the same hypotheses and computational cost. Indeed:

Theorem 2.4.6. *Assume hypotheses of Theorem 2.4.1 and condition (2.4.1) hold.*
Then sequence $\{x_n\}$ $(n \ge 0)$ generated by NK method is well defined, remains in $U(x_0, r^*)$ for all $n \ge 0$, and converges to a unique solution x^* of equation $F(x) = 0$ in $U(x_0, r^*)$. Moreover, if scalar sequence s_n is defined by

$$s_0 = 0, \quad s_n = s_{n-1} - \frac{\psi_a(s_{n-1})}{a(\lambda)\ell_0 s_{n-1}^{\lambda} - 1} \quad (n \ge 1), \tag{2.4.29}$$

then the following estimates hold for all $n \ge 1$

$$\|x_n - x_{n-1}\| \le s_n - s_{n-1} \tag{2.4.30}$$

and

$$\|x_n - x^*\| \le r^* - s_n. \tag{2.4.31}$$

Furthermore, if $\ell_0 < \ell$, then we have:

$$s_n < r_n \quad (n \ge 2), \tag{2.4.32}$$

$$s_n - s_{n-1} < r_n - r_{n-1} \quad (n \ge 2), \tag{2.4.33}$$

and

$$s^* - s_n \le r^* - r_n \quad (n \ge 0). \tag{2.4.34}$$

Proof. We simply arrive at the more precise estimate

$$\|F'(x)^{-1}F'(x_0)\| \le \left[1 - \ell_0\|x - x_0\|^{\lambda}\right]^{-1} \tag{2.4.35}$$

instead of

$$\|F'(x)^{-1}F'(x_0)\| \le (1 - \ell\|x - x_0\|^{\lambda}) \tag{2.4.36}$$

used in the proof of Theorem 1.4.2 in [64, pp. 720], for all $x \in U(x_0, R)$. Moreover note that if $\ell_0 < \ell$, $\{s_n\}$ is a more precise majorizing sequence of $\{x_n\}$ than sequence $\{r_n\}$ otherwise $r_n = s_n$ $(n \geq 0)$). With the above changes, the proof of Theorem 2.4.2 can be utilized so we can reach until (2.4.31).

Using (2.4.22), (2.4.29), and simple induction on n, we immediately obtain (2.4.32) and (2.4.33), whereas (2.4.34) is obtained from (2.4.33) by using standard majorization techniques.

At this point we wonder if:

(a) condition (2.4.15) can be weakened, by using more precise majorizing sequences along the lines of the proof of Theorem 2.4.4;

(b) even more precise majorizing sequences than $\{s_n\}$ can be found.

We need the following result on majorizing sequences for the NK method.

Lemma 2.4.7. *Assume there exist parameters $\ell \geq 0$, $\ell_0 \geq 0$, $\eta \geq 0$, $\lambda \in [0, 1]$, and $q \in [0, 1)$ with η and λ not zero at the same time such that:*
(a)

$$\left[\ell + \frac{\delta \ell_0}{(1-q)^\lambda}\right] \eta^\lambda \leq \delta, \quad for \quad \delta = (1+\lambda)q \quad \lambda \in [0, 1), \tag{2.4.37}$$

or
(b)

$$(\ell + \bar{\delta}\ell_0)\eta \leq \bar{\delta}, \quad for \quad \lambda = 1, \quad \ell_0 \leq \ell, \quad and \quad \bar{\delta} \in [0, 1]. \tag{2.4.38}$$

Then, iteration $\{t_n\}$ $(n \geq 0)$ given by

$$t_0 = 0, \quad t_1 = \eta,$$

$$t_{n+2} = t_{n+1} + \frac{\ell}{(1+\lambda)\left[1 - \ell_0 t_{n+1}^\lambda\right]}(t_{n+1} - t_n)^{1+\lambda} \tag{2.4.39}$$

is nondecreasing, bounded above by

$$(a) \quad t^{**} = \frac{\eta}{1-q}, \quad or \quad (b) \quad t^{**} = \frac{2\eta}{2-\bar{\delta}}, \quad \bar{\delta} \in [0, 1] \tag{2.4.40}$$

and converges to some t^* such that

$$0 \leq t^* \leq t^{**}. \tag{2.4.41}$$

Moreover, the following estimates hold for all $n \geq 0$:

$$(a) \quad 0 \leq t_{n+2} - t_{n+1} \leq q(t_{n+1} - t_n) \leq q^{n+1}\eta, \tag{2.4.42}$$

or

$$(b) \quad 0 \leq t_{n+2} - t_{n+1} \leq \frac{\bar{\delta}}{2}(t_{n+1} - t_n) \leq \left(\frac{\bar{\delta}}{2}\right)^{n+1}\eta,$$

respectively.

Proof. (a) The result clearly holds if q or ℓ or n or $\ell_0 = 0$. Let us assume q, ℓ, η, $\ell_0 \neq 0$. We must show:

$$\ell(t_{k+1}-t_k)^\lambda + \delta\ell_0 t_{k+1}^\lambda \leq \delta, \quad t_{k+1}-t_k \geq 0, \quad 1-\ell_0 t_{k+1}^\lambda > 0 \text{ for all } k \geq 0. \quad (2.4.43)$$

Estimate (2.4.42) can then follow immediately from (2.4.39) and (2.4.43). Using induction on the integer k, we have for $k = 0$, $\ell\eta^\lambda + \delta\ell_0\eta^\lambda = (\ell + \delta\ell_0)\eta^\lambda \leq \delta$ (by (2.4.37)) and $1 - \ell_0\eta^p > 0$. But then (2.4.43) gives

$$0 \leq t_2 - t_1 \leq q(t_1 - t_0).$$

Assume (2.4.43) holds for all $k \leq n + 1$. We can have in turn

$$\ell(t_{k+2} - t_{k+1})^\lambda + \delta\ell_0^\lambda t_{k+2}$$

$$\leq \ell\eta^\lambda q^{k+1} + \delta\ell_0 \left[t_1 + q(t_1 - t_0) + q^2(t_1 - t_0) + \cdots + q^{k+1}(t_1 - t_0) \right]^\lambda$$

$$\leq \ell\eta^\lambda q^{(k+1)\lambda} + \delta\ell_0\eta^\lambda \left[\tfrac{1-q^{k+2}}{1-q} \right]^\lambda$$

$$= \left[\ell q^{(k+1)\lambda} + \tfrac{\delta\ell_0}{(1-q)^\lambda}(1 - q^{k+2})^\lambda \right]\eta^\lambda \leq \left[\ell + \tfrac{\delta\ell_0}{(1-q)^\lambda} \right]\eta^\lambda \quad (2.4.44)$$

which is smaller or equal to δ by (2.4.37). Hence, the first estimate in (2.4.43) holds for all $k \geq 0$. We must also show:

$$t_k \leq t^{**} \quad (k \geq 0). \quad (2.4.45)$$

For $k = 0, 1, 2$ we have

$$t_0 = \eta \leq t^{**}, \quad t_1 = \eta \leq t^{**} \quad \text{and} \quad t_2 \leq \eta + q\eta = (1 + q)\eta \leq t^{**}.$$

Assume (2.4.45) holds for all $k \leq n + 1$. We also can get

$$t_{k+2} \leq t_{k+1} + q(t_{k+1} - t_k) \leq t_k + q(t_k - t_{k-1}) + q(t_{k+1} - t_k)$$

$$\leq \cdots \leq t_1 + q(t_1 - t_0) + \cdots + q(t_k - t_{k-1}) + q(t_{k+1} - t_k)$$

$$\leq \eta + q\eta + q^2\eta + \cdots + q^{k+1}\eta = \tfrac{1-q^{k+2}}{1-q}\eta < \tfrac{\eta}{1-q} = t^{**}. \quad (2.4.46)$$

Moreover the second inequality in (2.4.43) holds since

$$\ell_0 t_{k+2}^\lambda \leq \ell_0 \left(\tfrac{\eta}{1-q} \right)^\lambda < 1 \quad \text{by (2.4.37)}.$$

Furthermore the third inequality in (2.4.43) holds by (2.4.39), (2.4.44), and (2.4.46). Hence (2.4.43) holds for all $k \geq 0$. Iteration $\{t_n\}$ is nondecreasing and bounded above by t^{**} and as such it converges to some t^* satisfying (2.4.41).

(b) See [39] and the proof of part (a).

We can show the main semilocal convergence theorem for the NK method:

Theorem 2.4.8. *Let $F: D \subseteq X \to Y$ be a Fréchet-differentiable operator. Assume: there exist a point $x_0 \in D$ and parameters $\eta \geq 0$, $\ell_0 \geq 0$, $\ell \geq 0$, $\lambda \in [0, 1]$, $q \in [0, 1)$, $\bar{\delta} \in [0, 1]$, $R \geq 0$ such that: conditions (2.4.1), (2.4.2), and hypotheses of Lemma 2.4.7 hold, and*

$$\overline{U}(x_0, t^*) \subseteq U(x_0, R). \tag{2.4.47}$$

Then, $\{x_n\}$ $(n \geq 0)$ generated by NK method is well defined, remains in $\overline{U}(x_0, t^*)$ for all $n \geq 0$ and converges to a unique solution $x^* \in \overline{U}(x_0, t^*)$ of equation $F(x) = 0$.
  Moreover the following estimates hold for all $n \geq 0$:

$$\|x_{n+2} - x_{n+1}\| \leq \frac{\ell \|x_{n+1} - x_n\|^{1+\lambda}}{(1+\lambda)[1 - \ell_0 \|x_{n+1} - x_0\|^\lambda]} \leq t_{n+2} - t_{n+1} \tag{2.4.48}$$

and

$$\|x_n - x^*\| \leq t^* - t_n, \tag{2.4.49}$$

where iteration $\{t_n\}$ $(n \geq 0)$ and point t^* are given in Lemma 2.4.7.
  Furthermore, if there exists $R > t^*$ such that

$$R_0 \leq R \tag{2.4.50}$$

and

$$\ell_0 \int_0^1 \left[\theta t^* + (1-\theta)R\right]^\lambda d\theta \leq 1, \tag{2.4.51}$$

the solution x^* is unique in $U(x_0, R_0)$.

Proof. We shall prove:

$$\|x_{k+1} - x_k\| \leq t_{k+1} - t_k, \tag{2.4.52}$$

and

$$\overline{U}(x_{k+1}, t^* - t_{k+1}) \subseteq \overline{U}(x_k, t^* - t_k) \tag{2.4.53}$$

hold for all $n \geq 0$.
  For every $z \in \overline{U}(x_1, t^* - t_1)$

$$\|z - x_0\| \leq \|z - x_1\| + \|x_1 - x_0\| \leq t^* - t_1 + t_1 = t^* - t_0$$

implies $z \in \overline{U}(x_0, t^* - t_0)$. Because also

$$\|x_1 - x_0\| = \|F'(x_0)^{-1} F(x_0)\| \leq \eta = t_1$$

(2.4.52) and (2.4.53) hold for $n = 0$. Given they hold for $n = 0, 1, \ldots, k$, then

$$\|x_{k+1} - x_0\| \leq \sum_{i=1}^{k+1} \|x_i - x_{i-1}\| \leq \sum_{i=1}^{k+1} (t_i - t_{i-1}) = t_{k+1} - t_0 = t_{k+1} \tag{2.4.54}$$

and

$$\|x_k + \theta(x_{k+1} - x_k) - x_0\| \leq t_k + \theta(t_{k+1} - t_k) < t^*, \quad \theta \in [0, 1]. \tag{2.4.55}$$

Using NK we obtain the approximation

$$F(x_{k+1}) = F(x_{k+1}) - F(x_k) - F'(x_k)(x_{k+1} - x_k)$$
$$= \int_0^1 \left[F'(x_k + \theta(x_{k+1} - x_k)) - F'(x_k) \right] (x_{k+1} - x_k)d\theta, \qquad (2.4.56)$$

and by (2.4.2)

$$\| F'(x_0)^{-1} F(x_{k+1}) \| \le$$
$$\le \int_0^1 \| F'(x_0)^{-1} \left[F'(x_k + \theta(x_{k+1} - x_k)) - F'(x_k) \right] \| d\theta \| x_{k+1} - x_k \|$$
$$\le \frac{\ell}{1+\lambda} \| x_{k+1} - x_k \|^{1+\lambda}. \qquad (2.4.57)$$

By (2.4.1), the estimate

$$\| F'(x_0)^{-1} \left[F'(x_{k+1}) - F'(x_0) \right] \| \le \ell_0 \| x_{k+1} - x_0 \|^\lambda \le \ell_0 t_{k+1}^\lambda < 1$$

and the Banach Lemma on invertible operators $F'(x_{k+1})^{-1}$ exists and

$$\| F'(x_0) F'(x_{k+1})^{-1} \| \le \frac{1}{1 - \ell_0 \| x_{k+1} - x_0 \|^\lambda} \le \frac{1}{1 - \ell_0 t_{k+1}^\lambda}. \qquad (2.4.58)$$

Therefore, by NK, (2.4.39), (2.4.57), and (2.4.58) we obtain in turn

$$\| x_{k+2} - x_{k+1} \| = \| F'(x_{k+1})^{-1} F(x_{k+1}) \|$$
$$\le \| F'(x_{k+1})^{-1} F'(x_0) \| \cdot \| F'(x_0)^{-1} F(x_{k+1}) \|$$
$$\le \frac{\ell \| x_{k+1} - x_k \|^{1+\lambda}}{(1+\lambda)[1 - \ell_0 \| x_{k+1} - x_0 \|^\lambda]}$$
$$\le \frac{\ell(t_{k+1} - t_k)^{1+\lambda}}{(1+\lambda)\left[1 - \ell_0 t_{k+1}^\lambda\right]} = t_{k+2} - t_{k+1}. \qquad (2.4.59)$$

Thus for every $z \in \overline{U}(x_{k+2}, t^* - t_{k+2})$, we have

$$\| z - x_{k+1} \| \le \| z - x_{k+2} \| + \| x_{k+2} - x_{k+2} \| \le t^* - t_{k+2} + t_{k+2} - t_{k+2} = t^* - t_{k+1}.$$

That is

$$z \in \overline{U}(x_{k+1}, t^* - t_{k+1}). \qquad (2.4.60)$$

Estimates (2.4.59) and (2.4.60) imply that (2.4.52) and (2.4.53) hold for $n = k + 1$. By induction the proof of (2.4.52) and (2.4.53) is completed.

Lemma 2.4.7 implies that $\{t_n\}$ $(n \ge 0)$ is a Cauchy sequence. From (2.4.52) and (2.4.53) $\{x_n\}$ $(n \ge 0)$ becomes a Cauchy sequence, too, and as such it converges to some $x^* \in \overline{U}(x_0, t^*)$ so that (2.4.49) holds.

The combination of (2.4.59) and (2.4.60) yields $F(x^*) = 0$. Finally to show uniqueness let y^* be a solution of equation $F(x) = 0$ in $U(x_0, R)$. It follows from (2.4.1), the estimate

$$\left\| F'(x_0)^{-1} \int_0^1 \left[F'(y^* + \theta(x^* - y^*)) - F'(x_0) \right] d\theta \right\| \le$$

$$\le \ell_0 \int_0^1 \| y^* + \theta(x^* - y^*) - x_0 \|^\lambda d\theta$$

$$\le \ell_0 \int_0^1 \left[\theta \| x^* - x_0 \| + (1 - \theta) \| y^* - x_0 \| \right]^\lambda d\theta$$

$$< \ell_0 \int_0^1 \left[\theta t^* + (1 - \theta) R_0 \right]^\lambda d\theta \le 1 \quad \text{(by (2.4.51))} \qquad (2.4.61)$$

and the Banach Lemma on invertible operators that linear operator

$$L = \int_0^1 F'(y^* + \theta(x^* - y^*)) d\theta \qquad (2.4.62)$$

is invertible.

Using the identity

$$0 = F(y^*) - F(x^*) = L(x^* - y^*) \qquad (2.4.63)$$

we deduce $x^* = y^*$. To show uniqueness in $\overline{U}(x_0, t^*)$ as in (2.4.61), we get:

$$\| F'(x_0)^{-1}(L - F'(x_0)) \| \le \tfrac{\ell_0}{1+\lambda}(t^*)^{1+\lambda} < 1 \quad \text{(by Lemma 2.4.7)}$$

which implies again $x^* = y^*$.

Remark 2.4.9. In the result that follows we show that our error bounds on the distances involved are finer and the location of the solution x^* at least as precise.

Proposition 2.4.10. *Under hypotheses of Theorems 2.4.6 and 2.4.8 with $\ell_0 < \ell$ the following estimates hold:*

$$r_0 = t_0 = s_0 = 0, \quad r_1 = t_1 = s_1 = \eta,$$

$$t_{n+1} < s_{n+1} < r_{n+1} \quad (n \ge 1), \qquad (2.4.64)$$

$$t_{n+1} - t_n < s_{n+1} - s_n < r_{n+1} - r_n \quad (n \ge 1), \qquad (2.4.65)$$

$$t^* - t_n \le s^* - s_n \le r^* - r_n \quad (n \ge 0), \qquad (2.4.66)$$

and

$$t^* \le s^* \le r^*. \qquad (2.4.67)$$

Proof. We use induction on the integer k to show the left-hand sides of (2.4.64) and (2.4.65) first. By (2.4.29) and (2.4.39), we obtain

$$t_2 - t_1 = \frac{\ell \eta^{1+\lambda}}{(1+\lambda)[1 - \ell_0 \eta^\lambda]} < \frac{\psi_a(s_1)}{(1+\lambda)[1 - \ell_0 \eta^\lambda]} = s_2 - s_1,$$

and

$$t_2 < s_2.$$

Assume:

$$t_{k+1} < s_{k+1}, \quad t_{k+1} - t_k < s_{k+1} - s_k \quad (k \leq n). \tag{2.4.68}$$

Using (2.4.29), and (2.4.39), we get

$$t_{k+2} - t_{k+1} = \frac{\ell(t_{k+1} - t_k)^{1+\lambda}}{(1+\lambda)\left[1 - \ell_0 t_{k+1}^\lambda\right]} < \frac{\ell(s_{k+1} - s_k)^{1+\lambda}}{(1+\lambda)\left[1 - \ell t_{k+1}^\lambda\right]} \leq s_{k+2} - s_{k+1},$$

(by the proof of Theorem 2.2 in [64], end of page 720 and first half of page 721) and

$$t_{k+2} < s_{k+2}.$$

Let $m \geq 0$, we can obtain

$$
\begin{aligned}
t_{k+m} - t_k &< (t_{k+m} - t_{k+m-1}) + (t_{k+m-1} - t_{k+m-2}) + \cdots + (t_{k+1} - t_k) \\
&< (s_{k+m} - s_{k+m-1}) + (s_{k+m-1} - s_{k+m-2}) + \cdots + (s_{k+1} - s_k) \\
&= s_{k+m} - s_k.
\end{aligned}
\tag{2.4.69}
$$

By letting $m \to \infty$ in (2.4.69) we obtain (2.4.66). For $n = 1$ in (2.4.66) we get (2.4.67).

That completes the proof of Proposition 2.4.10, as the right-hand side estimates in (2.4.65)–(2.4.67) were shown in Theorem 2.4.6.

In the next remark, we also show that our sufficient convergence conditions are weaker in general than the earlier ones (i.e., the Lipschitz case):

Remark 2.4.11. Case $\lambda = 1$. (see Section 2.2 of Chapter 2)

Case $\lambda = 0$. It was examined here but not in [64], [78], [135].

Case $\lambda \in (0, 1)$. We can compare condition (2.4.37) with (2.4.15) (or (2.4.11) or (2.4.10) or (2.4.6)). For example set

$$q = \frac{1}{\lambda + 1}. \tag{2.4.70}$$

Then for

$$\ell_0 = \ell d, \quad d \in [0, 1], \tag{2.4.71}$$

and

$$c(\lambda, d) = d + \left(\frac{\lambda}{1 + \lambda}\right)^\lambda, \tag{2.4.72}$$

condition (2.4.37) becomes:

$$h \leq \frac{1}{c(\lambda, d)} \left(\frac{\lambda}{1 + \lambda}\right)^\lambda. \tag{2.4.73}$$

(a) Choose $d = \frac{1}{2}$, then using Mathematica we compare the magnitude of $a(\lambda)$ with $c\left(\lambda, \frac{1}{2}\right)$ to obtain the following favorable for our approach table:

Comparison table

λ	.1	.2	.3	.4	.5	.6	.7	.8	.9
$a(\lambda)$	1.842	1.695	1.562	1.445	1.341	1.252	1.174	1.108	1.050
$c\left(\lambda, \frac{1}{2}\right)$	1.287	1.200	1.444	1.106	1.080	1.055	1.037	1.023	1.010

See also the corresponding table in [64, pp. 722].

(b) If $d = 1$ (i.e., $\ell = \ell_0$), say for $\lambda = .1$ we found

$$c(.1, 1) = 1.787 < a(.1) = 1.842.$$

(c) Because $\frac{\ell}{\ell_0}$ can be arbitrarily large (see Example 2.2.13) for

$$d = 1 - \left(\frac{\lambda}{1+\lambda}\right)^{\lambda} = p, \tag{2.4.74}$$

condition (2.4.73) reduces to (2.4.6), whereas for

$$0 \leq d < p \tag{2.4.75}$$

(2.4.73) improves (2.4.6), which is the weakest of all conditions given before (see [64]).

Other favorable comparisons can also be made when q is not necessarily given by (2.4.70). However we leave the details to the motivated reader.

We state the following local convergence result for the NK method.

Theorem 2.4.12. *Let $F: D \subseteq X \to Y$ be a Fréchet-differentiable operator. Assume:*
(a) there exist a simple zero $x^* \in D$ of equation $F(x) = 0$, parameters $\bar{\ell}_0 \geq 0$, $\ell \geq 0$, $\mu \in [0, 1]$ not all zero at the same time such that:

$$\|F'(x^*)^{-1}[F'(x) - F'(y)]\| \leq \bar{\ell}\|x - y\|^{\mu}, \tag{2.4.76}$$

$$\|F'(x^*)^{-1}[F'(x) - F'(x^*)]\| \leq \bar{\ell}_0\|x - x^*\|^{\mu} \tag{2.4.77}$$

for all $x, y \in \overline{U}(x_0, R) \subseteq D$ $(R \geq 0)$;
(b) Define:

$$q = \begin{cases} \left[\dfrac{1+\mu}{\bar{\ell} + (1+\mu)\bar{\ell}_0}\right]^{1/\mu} & \mu \neq 0 \\[2ex] R \quad \text{and} \quad \bar{\ell} + \bar{\ell}_0 \leq 1 & \text{for } \mu = 0 \end{cases} \tag{2.4.78}$$

and

$$q \leq R. \tag{2.4.79}$$

Then, sequence $\{x_n\}$ $(n \geq 0)$ generated by NK is well defined, remains in $U(x^*, q)$ for all $n \geq 0$ and converges to x^*, provided that $x_0 \in U(x^*, q)$. Moreover the following estimates hold for all $n \geq 0$:

$$\|x_{n+1} - x^*\| \leq \frac{\bar{\ell}\|x_n - x^*\|^{1+\mu}}{(1+\mu)\left[1 - \bar{\ell}_0\|x_n - x^*\|^{\mu}\right]}. \tag{2.4.80}$$

Proof. Inequality (2.4.80) follows from the approximation

$$x_{n+1} - x^* =$$
$$= x_n - x^* - F'(x_n)^{-1}F(x_n)$$
$$= -\left[F'(x_n)^{-1}F'(x^*)\right] \times$$
$$\times \left\{F'(x^*)^{-1}\int_0^1 \left[F'(x^* + t(x_n - x^*)) - F'(x_n)\right](x_n - x^*)dt\right\}, \quad (2.4.81)$$

and estimates

$$\|F'(x_n)^{-1}F'(x^*)\| \le \left[1 - \bar{\ell}_0\|x_n - x^*\|^\mu\right]^{-1} \quad \text{(see (2.4.58))} \quad (2.4.82)$$

$$\left\|F'(x^*)^{-1}\int_0^1 \left[F'(x^* + t(x_n - x^*)) - F'(x_n)\right](x_n - x^*)dt\right\| \le$$

$$\le \frac{\bar{\ell}}{1+\mu}\|x_n - x^*\|^{1+\mu}, \quad \text{(see (2.4.57))} \quad (2.4.83)$$

The rest follows using induction on the integer n, (2.4.81)–(2.4.83), and along the lines of the proof of Theorem 2.4.8.

The corresponding local result for the MNK method is:

Remark 2.4.13. Using only condition (2.4.76) and the approximation

$$y_{n+1} - x^* = F'(y_0)^{-1}\int_0^1 \left[F'(x^* + t(y_n - x^*)) - F'(y_0)\right](y_n - x^*)dt, \quad (2.4.84)$$

as in the proof of Theorem 2.4.12 we obtain the convergence radius

$$\bar{q}_0 = \begin{cases} \left[\dfrac{1+\mu}{(2^{1+\mu} - 1)\bar{\ell}}\right]^{1/\mu}, & \bar{\ell} \ne 0, \ \mu \ne 0 \\ R, & \mu = 0, \end{cases} \quad (2.4.85)$$

and the corresponding estimates

$$\|y_{n+1} - x^*\| \le \bar{\ell}\int_0^1 \left[\|x^* - y_0\| + t\|y_n - x^*\|\right]^\mu dt\|y_n - x^*\|$$

$$\le \frac{\bar{\ell}(2^{1+\mu} - 1)}{1+\mu}\bar{q}_0^\mu\|y_n - x^*\| \quad (n \ge 0). \quad (2.4.86)$$

Remark 2.4.14. As noted in [43] and [216], the local results obtained here can be used for projection methods such as Arnoldi's, the Generalized Minimum Residual method (GMRES), the generalized conjugate residual method (GCR), for combined Newton/finite-difference projection methods, and in connection with the mesh independence principle in order to develop the cheapest mesh refinement strategies.

Remark 2.4.15. The local results obtained here can also be used to solve equations of the form $F(x) = 0$, where F' satisfies the autonomous differential equation [71]:

$$F'(x) = T(F(x)), \tag{2.4.87}$$

where $T: Y \to X$ is a known continuous operator. Because $F'(x^*) = T(F(x^*)) = T(0)$, we can apply the results obtained here without actually knowing the solution x^* of the equation $F(x) = 0$.

We complete this section with a numerical example to show that through Theorem 2.4.6 we can obtain a wider choice of initial guesses x_0 than before.

Example 2.4.16. Let $X = Y = \mathbf{R}$, $D = U(0, 1)$ and define function F on D by

$$F(x) = e^x - 1. \tag{2.4.88}$$

Then it can easily be seen that we can set $T(x) = x + 1$ in [35]. Because $F'(x^*) = 1$, we get $\|F'(x) - F'(y)\| \le e\|x - y\|$. Hence we set $\bar{\ell} = e$, $\mu = 1$. Moreover, because $x^* = 0$, we obtain in turn

$$F'(x) - F'(x^*) = e^x - 1 = x + \frac{x^2}{2!} + \cdots + \frac{x^n}{n!} + \cdots$$

$$= \left(1 + \frac{x}{2!} + \cdots + \frac{x^{n-1}}{n!} + \cdots\right)(x - x^*)$$

and for $x \in U(0, 1)$,

$$\|F'(x) - F'(x^*)\| \le (e - 1)\|x - x^*\|.$$

That is, $\bar{\ell}_0 = e - 1$. Using (2.4.85) we obtain

$$r^* = .254028662.$$

Rheinboldt's radius [175] is given by

$$p = \frac{2}{3\bar{\ell}}.$$

Note that

$$p < r^* \quad (\text{as } \bar{\ell}_0 < \bar{\ell}).$$

In particular, in this case we obtain

$$p = .245252961.$$

Note also that our error estimates are finer as $\bar{\ell}_0 < \bar{\ell}$. That is our convergence radius r^* is larger than the corresponding one p due to Rheinboldt [175]. This observation is very important in computational mathematics (see Remark 2.4.15). Note also that local results were not given in [64].

The case $\mu \in [0, 1)$ was not covered in [64]. The "terra incognita" can be examined along the lines of the semilocal case studied above. However, we leave the details to the motivated reader.

2.5 Enlarging the convergence domain of the NK method under regular smoothness conditions

Sufficient convergence conditions such as the famous Newton-Kantorovich hypothesis (see Chapter 2, Sections 2.2 and 2.4) have been given under the hypotheses for all $x, y \in D$

$$\left\| F'(x) - F'(y) \right\| \leq L \left\| x - y \right\|^{\lambda} \quad \lambda \in [0, 1], \tag{2.5.1}$$

or the w-smoothness

$$\left\| F'(x) - F'(y) \right\| \leq w \left(\left\| x - y \right\| \right) \tag{2.5.2}$$

for some increasing conditions function $w: [0, \infty) \to [0, \infty)$ with $w(0) = 0$ [6], [35], [43], [58], [98], [99], [146].

Under (2.5.1) the error bound

$$\left\| F(y) - F(x) - F'(x)(y - x) \right\| \leq \tfrac{L}{1+\lambda} \left\| x - y \right\|^{1+\lambda} \tag{2.5.3}$$

crucial in any convergence analysis of method NK has been improved only if $\lambda \in (0, 1)$ under an even more flexible condition than (2.5.2) called w-regular smoothness (to be precised later).

Here motivated by the elegant works in [98], [99] but using more precise majorizing sequences and under the same computational cost, we provide a semilocal convergence analysis for NK method under w-regular smoothness conditions on F' with the following advantages:

(a) finer estimates on the distances

$$\left\| x_{n+1} - x_n \right\|, \quad \left\| x_n - x^* \right\| \quad (n \geq 0);$$

(b) an at least as precise information on the location of the solution; and

(c) a larger convergence domain.

Expressions $r \to cr^{\lambda}$, $\lambda \in (0, 1]$ are typical representations of the class C of nondecreasing functions $w: (0, \infty] \to (0, \infty]$ that are concave and vanishing at zero. By w^{-1} we denote the function whose closed epigraph $cl\{(s, t), \ s \geq 0, \text{ and } t \geq w^{-1}(s)\}$ is symmetrical to closure of the subgraph of w with respect to the axis $t = s$ [98], [99]. Consider $T \in L(X, Y)$ and denote $\underline{h}(T)$ the inf $\|T(x)\|$. Given an $w_0 \in C$ and $x_0 \in D$, we say that T is w_0-regularly continuous on D with respect to $x = x_0 \in D$ or, equivalently, that w_0 is a regular continuity modulus of T on D relative to x_0, if there exists $\underline{h} = \underline{h}(x_0) \in [0, \underline{h}(T)]$ such that for all $x \in D$:

$$w_0^{-1}\left(h_T(x_0, x) + \|T(x) - T(x_0)\| \right) - w_0^{-1}\left(h_T(x_0, x) \right) \leq \|x - x_0\|, \tag{2.5.4}$$

where

$$h_{0T}(x_0, x) = \|T(x)\| - \underline{h}_0.$$

Given $w \in C$, we say T is w-regularly continuous on D if there exists $\underline{h} \in [0, \underline{h}(T)]$ such that for all $x, y \in D$

$$w^{-1}\left(h_T\left(x, y\right) + \|T\left(y\right) - T\left(x\right)\|\right) - w^{-1}\left(h_T\left(x, y\right)\right) \le \|y - x\|, \qquad (2.5.5)$$

where

$$h_T\left(y, x\right) = \min\left\{\|T\left(x\right)\|, \|T\left(y\right)\|\right\} - \underline{h}.$$

The operator F is w_0-regularly smooth on D with respect to a given point $x_0 \in D$, if its Fréchet derivative F' is w_0-regularly continuous with respect to x_0. Operator F is w-regularly smooth on D, if its Fréchet derivative F' is w-regularly continuous there [98], [99].

Note that in general

$$w_0\left(r\right) \le w\left(r\right) \quad \text{for all} \quad r \in [0, \infty) \qquad (2.5.6)$$

holds.

Given $w \in C$, set $Q_0\left(t\right) = \int_0^t w\left(\theta\right) d\theta$ and define function Q by

$$Q\left(u, t\right) = \begin{cases} tw\left(u\right) - Q_0\left(u\right) + Q_0\left(u - t\right) & \text{for } t \in [0, u], u \ge 0 \\ uw\left(u\right) - 2Q_0\left(u\right) + Q_0\left(t\right) & \text{for } t \ge u, u \ge 0. \end{cases} \qquad (2.5.7)$$

Denote by the superscript$^+$ the nonnegative part of a real number

$$a^+ = \max\left\{a, 0\right\}. \qquad (2.5.8)$$

Given $x_0 \in D$, if the operator $F_0 = F'\left(x_0\right)^{-1} F$ is w_0-regularly smooth with respect to x_0 and w-regularly smooth on D, define the sequence $\bar{u}_n = \left(\bar{t}_n, \bar{\alpha}_n, \bar{\bar{\alpha}}_n, \bar{\varepsilon}_n\right)$ by

$$\bar{t}_n = \|x_n - x_0\|, \, \bar{\alpha}_n = w^{-1}\left(\|F_0'\left(x_n\right)\| - \underline{h}\right), \bar{\bar{\alpha}}_n = w_0^{-1}\left(\|F_0'\left(x_n\right)\| - \underline{h}_0\right),$$

$$\left(\text{or}, \bar{\bar{\alpha}}_n = w_0^{-1}\left(\|F_0'\left(x_n\right)\| - \underline{h}\right)\right) \qquad (2.5.9)$$

$$\bar{\varepsilon}_n = \left\|F_0'\left(x_n\right)^{-1} F_0\left(x_n\right)\right\| \quad (n \ge 0).$$

As in Theorem 4.3 in [98, pp. 831] but using w_0 (i.e., (2.5.4)) instead of w (i.e., (2.5.5)) for the computation of the upper bounds of the inverses $F_0'\left(x_n\right)^{-1}$ we show:

$$\bar{t}_n \le \bar{t}_n + \bar{\varepsilon}_n, \, \bar{\alpha}_{n+1} \ge \left(\bar{\alpha}_n - \bar{\varepsilon}_n\right)^+, \, \bar{\bar{\alpha}}_{n+1} \ge \left(\bar{\bar{\alpha}}_n - \bar{\varepsilon}_n\right)^+, \qquad (2.5.10)$$

$$\bar{\varepsilon}_{n+1} \le \frac{Q\left(\bar{\alpha}_n, \bar{\varepsilon}_n\right)}{1 - w_0\left(\bar{\bar{\alpha}}_{n+1} + \bar{t}_{n+1}\right) + w_0\left(\bar{\bar{\alpha}}_{n+1}\right)} \qquad (2.5.11)$$

where function Q is given by (2.5.7).

Consider the sequence $u_n = \left(t_n, \alpha_n, \alpha_n^0, \varepsilon_n\right)$ given by

$$t_{n+1} = t_n + \varepsilon_n, \, \alpha_{n+1} = \left(\alpha_n - \varepsilon_n\right)^+, \, \alpha_{n+1}^0 = \left(\alpha_{n+1}^0 - \varepsilon_n\right), \qquad (2.5.12)$$

$$\left(\text{or } \alpha_{n+1}^0 = \alpha_{n+1} \quad n \ge 0\right)$$

$$\varepsilon_{n+1} = \frac{Q(\alpha_n, \varepsilon_n)}{1 - w_0(\alpha_{n+1}^0 + t_{n+1}) + w_0(\alpha_{n+1}^0)} \qquad (2.5.13)$$

for

$$t_0 = 0, \ \alpha_0 = w^{-1}(1 - \underline{h}), \ \alpha_0^0 = w_0^{-1}(1 - \underline{h}_0) \text{ and } \varepsilon_0 \geq \bar{\varepsilon}_0 \text{ [98], [99]}. \quad (2.5.14)$$

The sequence $\{u_n\}$ is well defined and converges if for all $n \geq 0$

$$w_0\left(\alpha_{n+1}^0 + t_{n+1}\right) + w_0\left(\alpha_{n+1}^0\right) < 1, \qquad (2.5.15)$$

or, equivalently

$$t_n < w_0^{-1}(1) \qquad (2.5.16)$$

(as sequence u_n will then be increasing and bounded above by the number $w_0^{-1}(1)$).

Denote by s_n the sequence given by (2.5.12), (2.5.13) when $w_0 = w$. If strict inequality holds in (2.5.6) we get by induction on $n \geq 0$

$$t_n < s_n \qquad (2.5.17)$$
$$t_{n+1} - t_n < s_{n+1} - s_n \qquad (2.5.18)$$

and

$$t^* = \lim_{n \to \infty} \leq \lim_{n \to \infty} s_n = s^* \qquad (2.5.19)$$

We can now show the following semilocal convergence result for Newton's method under regular smoothness:

Theorem 2.5.1. *Assume:*
Operator F_0 is w_0-regularly smooth with respect to $x_0 \in D$, and w-regularly smooth on D;
condition (2.5.16) holds;
and for $t^* = \lim_{n \to \infty} t_n$

$$\bar{U}(x_0, t^*) \subseteq D. \qquad (2.5.20)$$

Then sequence $\{x_n\}$ ($n \geq 0$) generated by NK is well defined, remains in $U(x_0, t^*)$ for all $n \geq 0$, and converges to a solution $x^* \in \bar{U}(x_0, t^*)$ of equation $F(x) = 0$. Moreover the following bounds hold for all $n \geq 0$:

$$\left\| F_0'(x_n)^{-1} \right\| \leq \gamma_n^{-1} = \left[1 - w_0\left(\alpha_n^0 + t_n\right) + w_0\left(\alpha_n^0\right) \right]^{-1}, \qquad (2.5.21)$$

$$(\text{or } [1 - w_0(\alpha_n + t_n) + w_0(\alpha_n)]^{-1})$$

$$\|x_{n+1} - x_n\| \leq \bar{\varepsilon}_{n+1} \leq \varepsilon_{n+1}, \qquad (2.5.22)$$

and

$$\|x_n - x^*\| \leq t^* - t_n. \qquad (2.5.23)$$

Furthermore if ε_0 is such that

$$t^* \leq \alpha_0^0, \qquad (2.5.24)$$

then the solution x^* is unique in $\overline{U}\left(x_0, \mathcal{P}_{h,2}^{-1}(0)\right)$, where function $\mathcal{P}_{h,2}^{-1}$ is the inverse of the restriction of $\mathcal{P}_h$ to $\left[w_0^{-1}(1), \infty\right]$, and function $\mathcal{P}_h$ was defined in [98, pp. 830].

Proof. We state that the derivation of (2.5.21) requires only (2.5.4) and not the stronger (2.5.5) used in [98, pp. 831] (see also (2.5.6)). The rest follows exactly in the proof of Theorem 4.3 in [98, pp. 831].

That completes the proof of the theorem.

Remark 2.5.2. (a) If equality holds in (2.5.6) then our Theorem 2.5.1 reduces to Theorem 4.3 in [98]. However if strict inequality holds in (2.5.6) then our error bounds on the distances $\|x_{n+1} - x_n\|$ (see (2.5.18) and (2.5.22)) are finer (smaller) than the corresponding ones in [98], [99]. Moreover condition (2.5.16) is weaker than the corresponding one in [98] (see 4.4 there) given by

$$s_n < w^{-1}(1) \ (n \geq 0) . \tag{2.5.25}$$

Furthermore the information on the location of the solution x^* is at least as precise, as our majorizing sequence is finer (smaller) (see (2.5.19)).

All the above advantages hold even if we choose

$$\overline{\alpha}_n = \overline{\overline{\alpha}}_n \text{ and we set } \alpha_n = \alpha_n^0 \ (n \geq 0) . \tag{2.5.26}$$

Note also that the above results are obtained under the same computational cost since computing function w requires the computation of w_0.

Definition 2.5.3. *Given a continuous operator* $f: R^m \rightarrow U \subseteq D$, *the set*

$$U(p) = \left\{u_0 \in U \,\middle|\, f^n(u_0) \rightarrow p\right\} \tag{2.5.27}$$

is called the attraction basin of p *[81].*

This set is not empty if and only of p is a fixed point f as it can be seen from the equality

$$u_{n+1} = f(u_n) . \tag{2.5.28}$$

It follows that the convergence domain

$$U_c = \{u_0 \in U \,|\text{sequence } \{u_n\} \text{ converges}\}$$
$$= \bigcup_{p \in U} U(p) = \bigcup_{a \in U_f} U(p) \tag{2.5.29}$$

where

$$U_f = \{p \,|\, f(p) = p\} . \tag{2.5.30}$$

Hence, the convergence domain of f can be constructed as the union of the attraction basins of its fixed points.

Example 2.5.4. Iteration (2.5.12), (2.5.13) can be rewritten as

$$t_+ = t + \varepsilon, \ \varepsilon_+ = \frac{Q\big((\alpha_0 - t)^+, \varepsilon\big)}{1 - w_0\left(\alpha_0^0 + t_+\right) + w_0\left(\alpha_0^0\right)} \tag{2.5.31}$$

(see also [99, p. 789]).

Its fixed points constitute the segment $\left[0, w_0^{-1}(1)\right)$ of the t-axis. When F_0 is Lipschitz smooth at x_0 with $w_0(t) = L_0 t$ $(L_0 \geq 0)$ and Lipschitz smooth on D with $w(t) = Lt$ $(L \geq 0)$, (2.5.31) reduces to:

$$t_+ = t + \varepsilon, \ \varepsilon_+ = \frac{.5L\varepsilon^2}{1 - L_0 t_+}, \tag{2.5.32}$$

and in the special case $L_0 = L$

$$s_+ = t + \varepsilon, \ \varepsilon_+ = \frac{.5L\varepsilon^2}{1 - Ls_+}. \tag{2.5.33}$$

It was shown in [99, p. 789] (i.e., using (2.5.33)) that

$$U_G = U_c^{L=L_0} = \left\{(t, \varepsilon) \mid 0 \leq \varepsilon < .5\left(L^{-1} - t\right)\right\}. \tag{2.5.34}$$

Denote by U_A the convergence domain if

$$L_0 = L \tag{2.5.35}$$

or

$$L_0 < L. \tag{2.5.36}$$

Then we can show that our convergence domain U_A contains U_w:

Proposition 2.5.5. *Under hypotheses of Theorem 4.3 in [98] (i.e., (2.5.25))*

$$U_G \subset U_A \tag{2.5.37}$$

where $\subset$ denotes strict inequality if (2.5.36) holds.

Proof. Condition (2.5.16) follows from (2.5.25). Hence the conclusions of Theorem 2.5.1 also hold. The rest follows from (2.5.32), (2.5.33), (2.5.35) and the definitions of sets U_G and U_A.

Remark 2.5.6. It was shown in Section 2.2 of this chapter that for $\delta \in [0, 2)$ the set

$$U_\delta(L_0, L) = \tag{2.5.38}$$

$$= \left\{(t, \varepsilon) \,\Big|\, K_\delta = L(t + \varepsilon) + \delta L_0 \varepsilon \leq \delta, \ L_0\left(t + \tfrac{2\delta\varepsilon}{2-\delta}\right) \leq 1, \ \tfrac{L_0\delta^2}{2-\delta} \leq L\right\}$$

contains pairs (t, ε) such that method (2.5.32) converges.

Clearly we have:

$$U_\delta(L_0, L) \subset U_A. \tag{2.5.39}$$

Moreover, we have

$$U_G \subseteq U_\delta(L_0, L), \tag{2.5.40}$$

as

$$U_1(L_0, L) \subseteq U_\delta(L_0, L). \tag{2.5.41}$$

Furthermore if $t = 0$ we get from (2.5.34), and (2.5.38) (for say $\delta = 1$)

$$K = 2Ln \le 1, \tag{2.5.42}$$

and

$$K_1 = (L_0 + L) n \le 1, \tag{2.5.43}$$

respectively.

Remark 2.5.7. It follows from the above that we also managed to enlarge the convergence domain U_G found in [99], and under the same computational cost.

Note also that condition (2.5.37) holds obviously for all possible choices of functions w_0 and w satisfying (2.5.6) (not only the ones given in Example 2.5.4). Claims (a)–(c) made in the introduction have now been justified.

Finally note that our technique used here only for NK method has been also illustrative and can be used easily on other methods appearing in [99] or elsewhere [43].

2.6 Convergence of NK method and operators with values in a cone

In this section, we are concerned with the solution of problems of the form

$$\text{Find } x^* \text{ such that } F\left(x^*\right) \in C, \tag{2.6.1}$$

where C is a nonempty closed convex cone in a Banach space Y, and F is a reflexive and continuously Fréchet-differentiable operator from a subset D_0 of a Banach space X into Y.

We used an extension of Newton's method to solve (2.6.1). The usual Newton's method corresponds with the special case when C is the degenerate cone $\{0\} \subseteq Y$. We provide a semilocal convergence analysis for Newton's method that generalizes the Newton-Kantorovich theorem. It turns out that our sufficient convergence conditions are weaker than the ones given by Robinson in [178], and under the same computational cost.

Let $p \in D_0$ be fixed. Define set-valued operator $G(p)$ from X into Y and its inverse by

$$G(p)x = F'(p)x - C, \quad x \in X, \tag{2.6.2}$$

$$G^{-1}(p)y = \left\{z \in F'(p)z \in y + C\right\}, \quad y \in Y, \tag{2.6.3}$$

where $F'(x)$ denotes the Fréchet-decreative of F evaluated at x. It is well-known that operator $G(p)$ as well as its inverse are convex [178]. Assume there exists an initial guess $x_0 \in D_0$ such that

$$G^{-1}(x_0)[-F(x_0)] \ne \varnothing. \tag{2.6.4}$$

Introduce algorithm so that given x_n, we choose x_{n+1} to be any solution of

$$\underset{n \geq 0}{\text{minimize}} \left\{ \|x - x_n\| \mid F(x_n) + F'(x_n)(x - x_n) \in C \right\}. \qquad (2.6.5)$$

The similarity with the usual NK method is now clear. We expect that problem (2.6.2) will be easier to solve than problem (2.6.1) [178].

We need the following lemma on majorizing sequences. The proof can essentially be found in Lemma 2.4.7 (see also Section 2.3):

Lemma 2.6.1. *Assume there exist parameters* $b > 0$, $\ell \geq 0$, $\ell_0 \geq 0$, *with* $\ell_0 \leq \ell$, $\eta \geq 0$, *and*

$$h_\delta = b(\delta \ell_0 + \ell) n \leq \delta, \quad \delta \in [0, 1] \qquad (2.6.6)$$

or

$$h_\delta \leq \delta, \quad \frac{2b\ell_0 n}{2\delta} \leq 1, \quad \frac{\ell_0 \delta^2}{2 - \delta} \leq \ell, \quad \delta \in [0, 2) \qquad (2.6.7)$$

or

$$\ell_0 n \leq 1 - \tfrac{1}{2}\delta, \quad \delta \in [\delta_0, 2) \qquad (2.6.8)$$

where

$$\delta_0 = \frac{-b_0 + \sqrt{b_0^2 + 8b_0}}{2}, \quad b_0 = \frac{\ell}{\ell_0} \quad \text{for } \ell_0 \neq 0. \qquad (2.6.9)$$

Then, iteration $\{t_n\}$ $(n \geq 0)$ *given by*

$$t_0 = 0, \quad t_1 = \eta, \quad t_{n+2} = t_{n+1} + \frac{b\ell(t_{n+1} - t_n)^2}{2(1 - b\ell_0 t_{n+1})} \quad (n \geq 0) \qquad (2.6.10)$$

is nondecreasing, bounded by $t^{**} = \frac{2\eta}{2-\delta}$, *and converges to some* t^* *such that*

$$0 \leq t^* \leq t^{**}. \qquad (2.6.11)$$

Moreover, the following error bounds hold for all $n \geq 0$:

$$0 \leq t_{n+2} - t_{n+1} \leq \frac{\delta}{2}(t_{n+1} - t_n) \leq \left(\frac{\delta}{2}\right)^{n+1} \eta. \qquad (2.6.12)$$

We can show the following generalization of the Newton-Kantorovich theorem:

Theorem 2.6.2. *Let* D_0, X, Y, C, F, *and* G *be as above.*
Assume: there exists a point $x_0 \in D_0$ *and nonnegative numbers* b, ℓ_0, ℓ, η, δ *such that (2.6.6) or (2.6.7) or (2.6.8) hold;*

$$\left\| G^{-1}(x_0) \right\| \leq b, \qquad (2.6.13)$$

$$\left\| F'(x) - F'(x_0) \right\| \leq \ell_0 \|x - x_0\|, \qquad (2.6.14)$$

$$\left\| F'(x) - F'(y) \right\| \leq \ell \|x - y\|, \qquad (2.6.15)$$

$$U(x_0, t^*) \subseteq D_0, \qquad (2.6.16)$$

and

$$\|x_1 - x_0\| \le \eta, \tag{2.6.17}$$

where x_1 is any point obtained from (2.6.5) (given x_0 satisfying (2.6.4) and t^* is given in Lemma 2.6.1).

Then, algorithm (2.6.4)–(2.6.5) generates at least one Newton-iteration $\{x_n\}$ ($n \ge 0$), which is well defined, remains in $U(x_0, t^*)$ for all $n \ge 0$ and converges to some $x^* \in U(x_0, t^*)$ such that $F(x^*) \in C$. Moreover the following estimates hold for all $n \ge 0$

$$\|x_{n+1} - x_n\| \le t_{n+1} - t_n \tag{2.6.18}$$

and

$$\|x_n - x^*\| \le t^* - t_n. \tag{2.6.19}$$

Proof. We first show sequence $\{x_n\}$ ($n \ge 0$) exists, $x_n \in \overline{U}(x_0, t^*)$ and

$$\|x_{k+1} - x_k\| \le t_{k+1} - t_k \quad (n \ge 0). \tag{2.6.20}$$

Point x_1 exists as $G(x_0)$ is an onto operator, which solves (2.6.5) for $n = 0$, and (2.6.20) holds for $k = 0$ (by (2.6.17)). Moreover we get

$$x_1 \in \overline{U}(x_0, t^*).$$

If (2.6.5) is feasible it must be solvable. Indeed, because $F'(x_k)$ is continuous and C is closed and convex the feasible set of (2.6.5) is also closed and convex. The existence of a feasible point q implies that any solution of (2.6.5) lie in the intersection of the feasible set of (2.6.5) and $\overline{U}(|x_k, 1\|q - x_k\|)$. Moreover this intersection is a closed, convex and bounded set. Furthermore because X is reflexive and function $\|x - x_k\|$ is weakly lower semicontinuous a solution of (2.6.5) exists [178]. Finally, because (2.6.5) is a convex minimization problem, any solution will be a global solution.

Now assume $x_1, x_2, ..., x_{n+1}$ exists satisfying (2.6.20).

Then, we get

$$\|x_{k+1} - x_0\| \le \|x_{k+1} - x_k\| + \|x_n - x_{k-1}\| + \cdots + \|x_1 - x_0\| \tag{2.6.21}$$
$$\le (t_{k+1} - t_k) + (t_k - t_{k-1}) + \cdots + (t_1 - t_0) = t_{k+1} \le t^*.$$

Hence, $x_{k+1} \in \overline{U}(x_0, t^*)$.

By (2.6.14) we have:

$$\left\|G^{-1}(x_0)\right\| \left\|F'(x_{k+1}) - F'(x_0)\right\| \le b\ell_0 \|x_{k+1} - x_0\|$$

$$\le b\ell_0 t_{k+1} < 1. \tag{2.6.22}$$

Therefore the convexity of $G(x_{k+1})$ carries to

$$F'(x_{k+1})x - C = \left\{F'(x_0) + \left[F(x_{k+1}) - F'(x_0)\right]\right\}x - C, \tag{2.6.23}$$

and by the Banach Lemma

$$\left\| G^{-1}(x_{k+1}) \right\| \leq \frac{\left\| G^{-1}(x_0) \right\|}{1 - \left\| G^{-1}(x_0) \right\| \left\| F'(x_{k+1}) - F'(x_0) \right\|} \quad (2.6.24)$$

$$\leq \frac{b}{1 - b\ell_0 \left\| x_{k+1} - x_0 \right\|}.$$

It follows that (2.6.5) is feasible, and hence solvable for $n = k = 1$, so that x_{k+1} exists. We need to solve for x:

$$F(x_{k+1}) + F'(x_{k+1})(x - x_{k+1}) \in F(x_k) + F'(x_k)(x_{k+1} - x_k) + C. \quad (2.6.25)$$

But x_{k+1} solves (2.6.5) with $n = k$, so the right-hand side of (2.6.25) is contained in C. Hence any x satisfying (2.6.25) also satisfies (2.6.5) for $n = k + 1$. We can rewrite (2.6.25) as

$$x - x_{k+1} \in G^{-1}(x_{k+1}) \left[-F(x_{k+1}) + F(x_k) + F'(x_k)(x_{k+1} - x_k) \right]. \quad (2.6.26)$$

Using (2.6.15) we get

$$\left\| -F(x_{n+1}) + F(x_k) + F'(x_k)(x_{k+1} - x_k) \right\| \leq \tfrac{1}{2}\ell \left\| x_{k+1} - x_k \right\|^2. \quad (2.6.27)$$

Because the right-hand side of (2.6.26) contains an element of least norms, there exists some q satisfying (2.6.26) and consequently (2.6.25) so that

$$\| q - x_{k+1} \| \leq \left\| G^{-1}(x_{k+1}) \right\| \left\| -F(x_{k+1}) + F'(x_k)(x_{k+1} - x_k) \right\|$$

$$\leq \frac{\tfrac{1}{2}b\ell \left\| x_{k+1} - x_k \right\|^2}{1 - b\ell_0 \left\| x_{k+1} - x_0 \right\|} \leq \frac{\tfrac{1}{2}b\ell (t_{k+1} - t_k)^2}{1 - b\ell_0 t_{k+1}} \quad (2.6.28)$$

That is q is also feasible for (b) with $n = k$, we have

$$\| x_{k+2} - x_{k+1} \| \leq \| q - x_{k+1} \| \leq t_{k+2} - t_{k+1} \quad (2.6.29)$$

and $x_{k+2} \in \overline{U}(x_0, t^*)$ which completes the induction. Hence sequence $\{x_n\}$ is Cauchy in X and as such it converges to some $x^* \in U(x_0, t^*)$. Then for any k

$$\left[F(x_{k+1}) - F(x^*) \right] - \left[F(x_{k+1}) - F(x_k) - F'(x_k)(x_{k+1} - x_k) \right] \in C - F(x^*). \quad (2.6.30)$$

The left-hand side of (2.6.30) approaches zero by the continuity assumptions, and as $C - F(x^*)$ is closed we get $F(x^*) \in C$.

Finally (2.6.19) follows from (2.6.18) by standard majorization techniques.

Remark 2.6.3. Our Theorem 2.6.2 reduces to Theorem 2 in [178, pp. 343] if $\ell_0 = \ell$.

The advantages of this approach have already been explained in Section 2.2 of Chapter 2.

2.7 Convergence theorems involving center-Lipschitz conditions

In this section, we are concerned with the problem of approximating a locally unique solution x^* of equation (2.1.1).

Most authors have used a Lipschitz-type hypotheses of the form

$$\left\| F'(x_0)^{-1} \left[F'(x) - F'(y) \right] \right\| \leq \alpha(r) \tag{2.7.1}$$

for all $x, y, \in \overline{U}(x_0, r) \subseteq \overline{U}(x_0, R) \subseteq D$ for some $R > 0$ and a continuous nonnegative function α in connection with the NK method. The computation of function α is very difficult or impossible in general (see Example 2.2.18). That is why we use instead hypotheses (2.7.4) in which the corresponding function w_0 is easier to compute.

Based on this idea, we produce local and semilocal convergence theorems for the NK. Our results can be weaker than the corresponding ones using (2.7.1). In the local case, we show that a larger convergence radius can be obtained.

We provide the following semilocal convergence results involving center-Lipschitz conditions:

Theorem 2.7.1. *Let $F: D \subseteq X \to Y$ be a Fréchet-differentiable operator. Assume:*
there exist a point $x_0 \in D$, $\eta \geq 0$, $R > 0$, and nonnegative continuous functions w_0, w such that:

$$F'(x_0)^{-1} \in L(X, Y), \tag{2.7.2}$$

$$\left\| F'(x_0)^{-1} F(x_0) \right\| \leq \eta \tag{2.7.3}$$

$$\left\| F'(x_0)^{-1} \left[F'(x) - F'(x_0) \right] \right\| \leq w_0(\|x - x_0\|), \tag{2.7.4}$$

$$\left\| F'(x)^{-1} F'(x_0) \right\| \leq w(\|x - x_0\|) \tag{2.7.5}$$

for all $x \in \overline{D}(x_0, r) \subseteq \overline{U}(x_0, r)$;
equation

$$w(r) \left\{ \left[\int_0^1 w_0(tr) \, dt + w_0(r) \right] r + \eta \right\} = r \tag{2.7.6}$$

has solutions on $(0, R]$. Denote by r_0 the smallest positive solution of equation (2.7.6):

$$q = 2w_0(r_0) w(r_0) < 1; \tag{2.7.7}$$

and

$$\overline{U}(x_0, R) \subseteq D. \tag{2.7.8}$$

Then, sequence $\{x_n\}$ $(n \geq 0)$ generated by NK method is well defined and remains in $\overline{U}(x_0, r_0)$ for all $n \geq 0$ and converges to a unique solution x^* of equation $F(x) = 0$ in $\overline{U}(x_0, r_0)$. Moreover the following estimates hold for all $n \geq 0$:

$$\|x_{n+2} - x_{n+1}\| \leq q \|x_{n+1} - x_n\| \tag{2.7.9}$$

and

$$\|x_n - x^*\| \le \frac{\eta}{1-q} q^n. \tag{2.7.10}$$

Moreover if there exists $r_1 \in (r_0, R]$ such that

$$w(r_0) \left[\int_0^1 w_0 \left[(1-t) r_0 + t r_1 \right] dt + w_0(r_0) \right] \le 1 \tag{2.7.11}$$

the solution x^* is in $U(x_0, r_1)$.

Proof. By (2.7.1), (2.7.3), (2.7.6) and the definition of $r_0 \|x_1 - x_0\| \le \eta \le r_0$. Hence $x_1 \in \overline{U}(x_0, r_0)$. Assume $x_k \in \overline{U}(x_0, r_0)$, $k = 0, 1, ..., n$. Using (2.7.1) we obtain the approximation

$$x_{k+1} - x_0 =$$
$$= x_k - F'(x_k)^{-1} F(x_k) - x_0$$
$$= - \left[F'(x_k)^{-1} F'(x_0) \right] F'(x_0)^{-1} \left\{ \int_0^1 \left[F'(x_0 + t(x_k - x_0)) - F'(x_0) \right] \right.$$
$$\left. \cdot (x_k - x_0) dt + \left(F'(x_0) - F'(x_k) \right) (x_k - x_0) + F(x_0) \right\}. \tag{2.7.12}$$

By (2.7.3)–(2.7.6) and (2.7.12) we get in turn

$$\|x_{k+1} - x_0\| \le \left\| F'(x_k)^{-1} F'(x_0) \right\|$$
$$\cdot \left\{ \left\| F'(x_0)^{-1} \left[F'(x_0 + t(x_k - x_0)) - F'(x_0) \right] \right\| \|x_k - x_0\| dt \right.$$
$$+ \left\| F'(x_0)^{-1} \left[F'(x_0) - F'(x_k) \right] \right\|$$
$$\cdot \|x_k - x_0\| + \left\| F'(x_0)^{-1} F(x_0) \right\| \right\}$$
$$\le w(\|x_k - x_0\|) \left\{ \left[\int_0^1 w_0(t \|x_k - x_0\|) dt \right. \right.$$
$$\left. + w_0(\|x_k - x_0\|) \right] \|x_k - x_0\| + \eta \right\}$$
$$\le w(r) \left\{ \left[\int_0^1 w_0(tr) dt + w_0(r) \right] r + \eta \right\} = r. \tag{2.7.13}$$

That is $x_{k+1} \in \overline{U}(x_0, r_0)$. Moreover by (2.7.1) we obtain the approximation

$$F'(x_0)^{-1} F(x_{k+1}) = F'(x_0)^{-1} \left[F(x_{k+1}) - F(x_k) - F'(x_k)(x_{k+1} - x_k) \right]$$
$$= F'(x_0)^{-1} \left\{ \int_0^1 \left[F'(x_k + t(x_{k+1} - x_k)) - F'(x_0) \right] dt \right.$$
$$\left. + F'(x_0)^{-1} \left[F'(x_0) - F'(x_k) \right] \right\} (x_{k+1} - x_k). \tag{2.7.14}$$

By (2.7.4) and (2.7.14) we get in turn

$$\left\| F'(x_0)^{-1} F(x_{k+1}) \right\| \tag{2.7.15}$$

$$\leq \left\{ \left\| F'(x_0)^{-1} \int_0^{1-1} \left[F'(x_k + t(x_{k+1} - x_k)) - F'(x_0) \right] \right\| dt \right.$$

$$+ \left. \left\| F'(x_0)^{-1} \left[F'(x_0) - F'(x_k) \right] \right\| \right\} \| x_{k+1} - x_k \|$$

$$\leq \left\{ \int_0^1 w_0 \left[(1-t) \| x_k - x_0 \| + t \| x_{k+1} - x_0 \| \right] dt \right.$$

$$+ w_0 \left(\| x_k - x_0 \| \right) \right\} \| x_{k+1} - x_k \|$$

$$\leq \left[\int_0^1 w_0 \left[(1-t) r_0 + t r_0 \right] dt + w_0 (r_0) \right] \| x_{k+1} - x_k \|$$

$$= 2 w_0 (r_0) \| x_{k+1} - x_k \|.$$

Hence by (2.7.1), (2.7.5) and (2.7.15) we obtain

$$\| x_{k+2} - x_{k+1} \| = \left\| \left[F'(x_{k+1})^{-1} F'(x_0) \right] \left[F'(x_0)^{-1} F(x_{k+1}) \right] \right\| \tag{2.7.16}$$

$$\leq \left\| F'(x_{k+1})^{-1} F'(x_0) \right\| \cdot \left\| F'(x_0)^{-1} F(x_{k+1}) \right\|$$

$$\leq w(r_0) \, 2w(r_0) \| x_{k+1} - x_k \|$$

$$= q \| x_{k+1} - x_k \| \leq q^{k+1} \eta,$$

which shows (2.7.9) for all $n \geq 0$.

Let $m > 1$, then we get using (2.7.9),

$$x_{n+m} - x_{n+1} = (x_{n+m} - x_{n+m-1}) + (x_{n+m-1} - x_{n+m-2}) \tag{2.7.17}$$

$$+ \cdots + (x_{n+2} - x_{n+1}),$$

and

$$\| x_{n+m} - x_{n+1} \| \leq q \| x_{n+m-1} - x_{n+m-2} \| \leq \cdots \leq q^{m-1} \| x_{n+1} - x_n \| \tag{2.7.18}$$

that

$$\| x_{n+m} - x_{n+1} \| \leq \left(q + \cdots + q^{m-2} + q^{m-1} \right) \| x_{n+1} - x_n \| \tag{2.7.19}$$

$$\leq q \frac{1 - q^{m-1}}{1 - q} q^n \eta.$$

It follows from (2.7.19) and (2.7.7) that sequence $\{x_n\}$ $(n \geq 0)$ is Cauchy in a Banach space X and as such it converges to some $x^* \in \overline{U}(x_0, r_0)$. By letting $k \to \infty$, $m \to \infty$ in (2.7.16) and (2.7.19) we get (2.7.10) and (2.7.19) we get (2.7.10) and $F(x^*) = 0$ respectively.

To show uniqueness in $\overline{U}(x_0, r_0)$ let x_1^* be a solution in $\overline{U}(x_0, r_0)$. Using the approximation

$$x_{n+1} - x_1^* = \tag{2.7.20}$$

$$= x_n - F'(x_n)^{-1} F(x_n) - x_1^*$$

$$= F'(x_n)^{-1} \left[F(x^*) - F(x_n) - F'(x_n)(x_1^* - x_n) \right]$$

$$= \left[F'(x_n)^{-1} F'(x_0) \right] F'(x_0)^{-1} \left\{ \int_0^1 \left[F'(x_n + t(x_1^* - x_n)) - F'(x_0) \right] dt \right.$$

$$\left. + \left[F'(x_0) - F'(x_n) \right] \right\} (x_1^* - x_n)$$

as in (2.7.13) we get

$$\|x_{n+1} - x_1^*\| \le w(\|x_n - x_0\|) \left[\int_0^1 w_0 \left[(1 - t) \|x_n - x_0\| \right. \right. \tag{2.7.21}$$

$$\left. \left. + t \|x_1^* - x_0\| \right] + w_0(\|x_n - x_0\|) \right] \|x_n - x_1^*\|$$

$$\le q \|x_n - x_1^*\|. \tag{2.7.22}$$

By (2.7.7) and (2.7.22) $\lim_{n \to \infty} x_n = x_1^*$. But we already showed $\lim_{n \to \infty} x_n = x^*$. Hence, we conclude

$$x^* = x_1^*.$$

Finally to show uniqueness in $U(x_0, r_1)$, let x_1^* be a solution of $F(x) = 0$ in $U(x_0, r_1)$. As in (2.7.21) we get

$$\|x_{n+1} - x_1^*\| < w(r_0) \left[\int_0^1 w_0 [(1 - t) r_0 + t r_1] dt + w_0(r_0) \right] \|x_n - x_1^*\|$$

$$\le \|x_n - x_1^*\|. \tag{2.7.23}$$

By (2.7.23) we get

$$\lim_{n \to \infty} x_n = x_1^*.$$

Hence, again we deduce:

$$x^* = x_1^*.$$

Remark 2.7.2. In order for us to compare our results with earlier ones, consider the Lipschitz condition (2.2.36) and the Newton-Kantorovich hypothesis (2.2.37). Define

$$w_0(r) = \ell_0 r^\lambda, \tag{2.7.24}$$

$$w(r) = \ell_1 \tag{2.7.25}$$

for some $\lambda \ge 0$, $\ell_0 \ge 0$, $\ell_1 > 0$ and all $r \in [0, R]$. Assuming conditions (2.7.4), (2.7.5) and (2.7.7) hold with the above choices then (2.7.6) and (2.7.7) reduce to

$$\frac{(\lambda + 2)}{\lambda + 1}\ell_0 r^{\lambda+1} - \frac{r}{\ell_1} + \eta = 0, \tag{2.7.26}$$

and

$$2\ell_0\ell_1 r^\lambda < 1. \tag{2.7.27}$$

Set $\lambda = 1$ then (2.7.26) and (2.7.27) are satisfied if

$$h_0 = 6\ell_0\ell_1^2\eta \le 1 \tag{2.7.28}$$

with r_0 being the small solution of equation (2.7.26). By comparing (2.2.37) and (2.7.28) we see that (2.7.28) is weaker if (2.7.28) holds, and

$$3\ell_1^2 < \frac{\ell}{\ell_0} \quad (\ell_0 \ne 0). \tag{2.7.29}$$

This can happen in practice as $\frac{\ell}{\ell_0}$ can be arbitrarily large and hence larger than $3\ell_1^2$ (see Section 2.2).

This comparison can become even more favorable if $\lambda > 1$. Such a case is provided in Example 2.7.4.

Assume there exist a zero x^* of F, $R > 0$, and nonnegative continuous functions v_0, v such that

$$F'(x^*)^{-1} \in L(Y, X), \tag{2.7.30}$$

$$\left\| F'(x^*)^{-1}[F'(x) - F'(x^*)] \right\| \le v_0(\|x - x^*\|), \tag{2.7.31}$$

$$\left\| F'(x)^{-1}F'(x^*) \right\| \le v(\|x - x^*\|) \tag{2.7.32}$$

for all $x \in \overline{U}(x^*, r) \subseteq \overline{U}(x^*, R)$; equation

$$v(r)\left[\int_0^1 v_0[(1-t)r]\,dt + v_0(r)\right] = 1 \tag{2.7.33}$$

has solutions in $[0, R]$. Denote by r^* the smallest;

$$\overline{U}(x^*, R) \subseteq D. \tag{2.7.34}$$

Then the following local convergence result holds for NK method.

Theorem 2.7.3. *Under the above stated hypotheses sequence $\{x_n\}$ $(n \ge 0)$ generated by NK method is well defined, remains in $\overline{U}(x^*, r^*)$ for all $n \ge 0$, and converges to x^* provided that $x_0 \in U(x^*, r^*)$.*

Moreover the following estimates hold for all $n \ge 0$

$$\|x_{n+1} - x^*\| \le a_n\|x_n - x^*\| \le a\|x_n - x^*\|, \tag{2.7.35}$$

where

$$a_n = w\left(\|x_n - x^*\|\right)\left[\int_0^1 w_0\left((1-t)\,\|x_n - x^*\|\right)dt + w_0\left(\|x_n - x^*\|\right)\right] \quad (2.7.36)$$

and

$$a = w\left(r^*\right)\left[\int_0^1 w_0\left((1-t)\,r^*\right)dt + w_0\left(r^*\right)\right]. \quad (2.7.37)$$

Proof. It follows as in Theorem 2.7.1 by using (2.7.30)–(2.7.34), induction on n and the approximation

$$x_{n+1} - x^* = \quad (2.7.38)$$

$$= \left[F'\left(x_n\right)^{-1} F'\left(x^*\right)\right]\left\{\int_0^1 \left[F'\left(x_n + t\left(x^* - x_n\right)\right)\right.\right.$$

$$\left.\left. - F'\left(x^*\right)\right]dt + \left[F'\left(x^*\right) - F'\left(x_n\right)\right]\right\}\left(x^* - x_n\right).$$

We complete this section with a numerical example to show that we can obtain a larger convergence radius than in earlier results.

Example 2.7.4. Let $X = Y = \mathbf{R}$, $D = U\left(0, 1\right)$, and define function F on D by

$$F\left(x\right) = \tfrac{1}{5}e^{x^5} - x - \tfrac{1}{5}. \quad (2.7.39)$$

Choose $v_0\left(r\right) = \ell_0 r^\mu$, $v\left(r\right) = b$. Then it can easily be seen from (2.7.30)–(2.7.32), (2.7.39) that $\ell_0 = e$, $\mu = 4$, and

$$b = 1.581976707 = \left\|F'\left(-1\right)^{-1} F'\left(0\right)\right\|.$$

Equation (2.7.33) becomes

$$\left[\int_0^1 e\left[(1-t)\,r\right]^4 dt + er^4\right]b = 1 \quad (2.7.40)$$

or

$$r^* = \left[\frac{5}{6eb}\right]^{\frac{1}{4}} = .663484905. \quad (2.7.41)$$

We saw earlier in Example 2.4.16 that Rheinboldt radius [175] is given by

$$r_1^* = .245252961. \quad (2.7.42)$$

Hence, we conclude:

$$r_1^* < r^*. \quad (2.7.43)$$

Example 2.7.5. We refer the reader to Example 2.2.18.

2.8 The radius of convergence for the NK method

Let $F: D \subseteq X \to Y$ be an m-times continuously Fréchet-differentiable operator ($m \geq 2$ an integer) defined on an open convex subset D of a Banach space X with values in a Banach space Y. Suppose there exists $x^* \in D$ that is a solution of the equation

$$F(x) = 0. \tag{2.8.1}$$

The most popular method for approximating such a point x^* is Newton's method

$$x_{n+1} = G(x_n) \quad (n \geq 0), \quad (x_0 \in D), \tag{2.8.2}$$

where

$$G(x) \equiv x - F'(x)^{-1} F(x) \quad (x \in D). \tag{2.8.3}$$

In the elegant paper by Ypma [216], affine invariant results have been given concerning the radius of convergence of Newton's method. Ypma used Lipschitz conditions on the first Fréchet derivative as the basis for his analysis. In this study, we use Lipschitz-like conditions on the mth Fréchet derivative $F^{(m)}(x) \in L\left(X_1^m, Y_2\right)$ ($x \in D$) ($m \geq 2$) an integer. This way we manage to enlarge the radius of convergence for Newton's method (2.8.2). Finally we provide numerical examples to show that our results guarantee convergence, where earlier ones do not [216]. This is important in numerical computations [43], [216].

We give an affine invariant form of the Banach lemma on invertible operators.

Lemma 2.8.1. *Let $m \geq 2$ be an integer, $\alpha_i \geq 2m$ ($2 \leq i \leq m$), $\eta \geq 0$, X, Y Banach spaces, D a convex subset of X and $F: D \to Y$ an m-times Fréchet-differentiable operator. Assume there exist $z \in D$ so that $F'(z)^{-1}$ exists, and some convex neighborhood $N(z) \subseteq D$*

$$\left\| F'(z)^{-1} F^{(i)}(z) \right\| \leq \alpha_i, \, i = 2, \ldots, m \tag{2.8.4}$$

and

$$\left\| F'(z)^{-1} \left[F^{(m)}(x) - F^{(m)}(z) \right] \right\| \leq \varepsilon_0 \, \text{ for all } \, x \in N(z), \varepsilon_0 > 0. \tag{2.8.5}$$

If $x \in N(z) \cap U(z, \delta)$, where δ is the positive zero of the equation $f'(t) = 0$, where

$$f(t) = \frac{\alpha_m + \varepsilon_0}{m!} t^m + \cdots + \frac{\alpha^2}{2!} t^2 - t + d \tag{2.8.6}$$

then $F'(x)^{-1}$ exists and for $\|x - z\| < t \leq \delta$

$$\left\| F'(z)^{-1} F''(x) \right\| s < f''(t) \tag{2.8.7}$$

and

$$\left\| F''(x)^{-1} F'(z) \right\| \leq -f'(t)^{-1}. \tag{2.8.8}$$

Proof. It is convenient to define ε, b_1, b_i, $i = 2, ..., m$ by

$$\varepsilon = x - z_0,$$
$$b_1 = z + \theta\varepsilon,$$
$$b_i = z + \theta_i\,(b_{i-1} - z), \qquad \theta \in [0, 1].$$

We can have in turn

$F''(x)$

$$= F''(z) + \left[F''(x) - F''(z)\right]$$

$$= F''(z) + \int_0^1 F'''[z + \theta_1\,(x - z)]\,(x - z)\,d\theta_1$$

$$= F''(z) + \int_0^1 \left[F'''(z + \theta_1\,(x - z)) - F'''(z)\right](x - z)\,d\theta_1$$

$$\quad + \int_0^1 F'''(z)\,(x - z)\,d\theta_1$$

$$= F''(z) + \int_0^1 F'''(z)\,(x - z)\,d\theta_1 + \int_0^1 \int_0^1 F^{(4)}\,\{z + \theta_2$$

$$\qquad \cdot [z + \theta_1\,(x - z) - z]\}\,[z + \theta_1\,(x - z)\,z]\,(x - z)\,d\theta_2 d\theta_1$$

$$= F''(z) + \int_0^1 F'''(z)\,\varepsilon d\theta_1 + \int_0^1 \int_0^1 F^{(4)}\,(b_2)\,(b_1 - z_0)\,\varepsilon d\theta_2 d\theta_1$$

$$= \cdots$$

$$= F''(z) + \int_0^1 F'''(z)\,\varepsilon d\theta_1 + \cdots + \int_0^1 \cdots \int_0^1 F^{(m)}\,(b_{m-2})\,(b_{m-3} - z)$$

$$\qquad \cdots (b_1 - z)\,d\theta_{m-2} \cdots d\theta_1$$

$$= F''(z) + \int_0^1 F'''(z)\,\varepsilon d\theta_1 + \cdots + \int_0^1 \cdots \int_0^1 F^{(m)}\,(z)\,(b_{m-3} - z)$$

$$\qquad \cdots (b_1 - z)\,\varepsilon d\theta_{m-2} \cdots d\theta_1$$

$$\quad + \int_0^1 \cdots \int_0^1 \left[F^{(m)}\,(b_{m-2}) - F^{(m)}\,(z)\right](b_{m-3} - z)$$

$$\qquad \cdots (b_1 - z)\,\varepsilon d\theta_{m-2} \cdots d\theta_1. \tag{2.8.9}$$

Using the triangle inequality, (2.8.4), (2.8.5), (2.8.6) in (2.8.9) after composing by $F'(z)^{-1}$, we obtain (2.8.7).

We also get

$$-F'(z)^{-1}\left[F'(z) - F'(x)\right]$$

$$= F'(z)^{-1}\left[F'(x) - F'(z) - F''(z)(x-z) + F''(z)(x-z)\right]$$

$$= \int_0^1 F'(z)^{-1}\left\{F''[z + \theta_1\varepsilon] - F''(z)\right\}d\theta_1\varepsilon + F'(z)^{-1}\int_0^1 F''(z)\,\varepsilon d\theta_1$$

$$= \int_0^1\int_0^1 F'(z)\,F'''(b_2)\,(b_1 - z)\,\varepsilon d\theta_2 d\theta_1 + F'(z)^{-1}\int_0^1 F''(z)\,\varepsilon d\theta_1$$

$$= \cdots$$

$$= \int_0^1 \cdots \int_0^1 F^{(m)}(b_{m-1})(b_{m-2} - z)$$

$$\cdots(b_1 - z)\,\varepsilon d\theta_{m-1}\varepsilon d\theta_{m-2}\cdots d\theta_2 d\theta_1$$

$$+ \int_0^1 \cdots \int_0^1 F^{(m-1)}(b_{m-2})(b_{m-3} - z)\cdots(b_1 - z)\,\varepsilon d\theta_{m-2}\cdots d\theta_2 d\theta_1$$

$$+ \cdots + \int_0^1 F'(z)^{-1}F''(z)\,\varepsilon d\theta_1$$

$$= \int_0^1\int_0^1 F'(z)^{-1}\left[F^{(m)}(b_{m-1}) - F^{(m)}(z)\right](b_{m-2} - z)$$

$$\cdots(b_1 - z)\,\varepsilon d\theta_{m-2}\cdots d\theta_1$$

$$+ \int_0^1 \cdots \int_0^1 F'(z)^{-1}F^{(m)}(z)(b_{m-2} - z)\cdots(b_1 - z)\,\varepsilon d\theta_{m-1}\cdots d\theta_1$$

$$+ \int_0^1 \cdots \int_0^1 F'(z)^{-1}F^{(m-1)}(z)(b_{m-3} - z)\cdots(b_1 - z)\,\varepsilon d\theta_{m-2}\cdots d\theta_1$$

$$+ \cdots + \int_0^1 F'(z)^{-1}F''(z)\,\varepsilon d\theta_1. \tag{2.8.10}$$

Because $f'(t) < 0$ on $[0, \delta]$, using (2.8.4), (2.8.5), (2.8.6) in (2.8.10) we obtain for $\|x - z\| < t$

$$\left\|-F'(z)^{-1}\left[F'(z) - F'(x)\right]\right\| \le 1 + f'(\|x - z\|) < 1 + f'(t) < 1. \tag{2.8.11}$$

It follows from the Banach Lemma on invertible operators (2.8.11) $F'(x)^{-1}$ exists, and

$$\left\|F'(x)^{-1}F'(z)\right\| \le \left[1 - \left\|F'(z)^{-1}\left[F'(z) - F'(x)\right]\right\|\right]^{-1} \le -f'(t)^{-1}.$$

which shows (2.8.8).

We need the following affine invariant form of the mean value theorem for m-Fréchet-differentiable operators.

Lemma 2.8.2. *Let $m \geq 2$ be an integer, $\alpha_i \geq 0$ $(2 \leq i \leq m)$, X, Y Banach spaces, D a convex subset of X and $F; D \to Y$ an m-times Fréchet-differentiable operator. Assume there exist $z \in D$ so that $F'(z)^{-1}$ exists, and some convex neighborhood $N(z)$ of z such that $N(z) \subseteq D$,*

$$\left\| F'(z)^{-1} F^{(i)}(z) \right\| \leq \alpha_i, \quad i = 2, ..., m,$$

and

$$\left\| F'(z)^{-1} \left[F^{(m)}(x) - F^{(m)}(z) \right] \right\| \leq \varepsilon_0 \quad \text{for all} \quad x \in N(z), \ \varepsilon_0 > 0.$$

Then for all $x \in N(z)$

$$\left\| F'(z)^{-1} [F(z) - F(x)(z-x)] \right\| \tag{2.8.12}$$

$$\leq \frac{\alpha_m + \varepsilon}{m!} \|x - z\|^m + \frac{\alpha_{m-1}}{(m-1)!} \|x - z\|^{m-1} + \cdots + \frac{\alpha_2}{2!} \|x - z\|^2 .$$

Proof. We can write in turn:

$$F(z) - F(x) - F'(x)(z-x)$$

$$= \int_0^1 \left[F'(x + \theta_1(z-x)) - F'(x) \right](z-x) \, d\theta_1$$

$$= \int_0^1 \left[F''(z + \theta_1(x-z)) - F''(z) \right] \theta_1 d\theta_1 (x-z)^2 + \int_0^1 \theta_1 F''(z)(x-z)^2 \theta_1$$

$$= \int_0^1 \int_0^1 \left[F'''(z + \theta_2 \theta_1(x-z)) - F'''(z) \right] \theta_1 (x-z) \, d\theta_2 \theta_1 d\theta_1 (x-z)^2$$

$$+ \int_0^1 \int_0^1 F'''(z) \theta_1 (x-z) \, d\theta_2 \theta_1 d\theta_1 (x-z)^2 + \int_0^1 \theta_1 F''(z)(x-z)^2 \, d\theta_1$$

$$= \cdots$$

$$= \int_0^1 \int_0^1 \cdots \int_0^1 \left[F^{(m)}(z + \theta_{m-1}\theta_{m-2} \cdots \theta_1(x-z)) - F^{(m)}(z) \right] \theta^1_{m-2}$$

$$\cdots \theta_3^{m-4} \theta_2^{m-3} \theta_1^{m-1} (x-z)^m \, d\theta_{m-1} d\theta_{m-2} \cdots d\theta_3 d\theta_2 d\theta_1$$

$$+ \cdots + \int_0^1 \int_0^1 F'''(z) \theta_1^2 (x-z)^3 \, d\theta_2 d\theta_1 + \int_0^1 \theta_1 F''(z)(x-z)^2 \, d\theta_1.$$

$$\tag{2.8.13}$$

Composing both sides by $F'(z)^{-1}$, using the triangle inequality, (2.8.5) and (2.8.6) we obtain (2.8.12).

Based on the above lemmas, we derive affine invariant convergence results for the class $T \equiv T(\{\alpha_i\}, 2 \leq i \leq m, \alpha)$ $(\alpha > 0, \alpha_i \geq 0, 2 \leq i \leq m)$ of operators F defined by $T \equiv \{F | F: D \subseteq X \to Y;$ D open and convex set, F m-times continuously Fréchet-differentiable on D; there exists $x^* \in D$ such that $F(x^*) = 0$; $F'(x)^{-1}$ exists; $U(x^*, \alpha) \subseteq D$; x^* is the only solution of equation $F(x) = 0$ in $U(x^*, \alpha)$; and for all $x \in U(x^*, \alpha)$,

$$\left\| F'(x^*)^{-1} \left[F^{(m)}(x^*) - F^{(m)}(x) \right] \right\| s < \varepsilon_0, \ \varepsilon_0 > 0, \tag{2.8.14}$$

and

$$\left\| F'(x^*)^{-1} F^{(i)}(x^*) \right\| \leq \alpha_i, \quad i = 2, ..., m. \tag{2.8.15}$$

Let $F \in T$ and $x \in U(x^*, b)$ where $b \leq \min\{\alpha, \delta\}$. By Lemma 2.8.1, $F'(x)^{-1}$ exists. Define

$$\mu(F, x) \equiv \sup \left\{ \left\| F'(x)^{-1} \left[F^{(m)}(y) - F^{(m)} \right] \right\| \mid y \in U(x^*, b) \right\}, \tag{2.8.16}$$

$$q_i = q_i(F, x) \equiv \left\| F'(x)^{-1} F^{(i)}(x^*) \right\|, \quad 2 \leq i \leq m, \ x \in U(x^*, b). \tag{2.8.17}$$

It follows from (2.8.14)–(2.8.17) that

$$\mu(F, x^*) \leq \varepsilon_0 = \varepsilon(x^*), \quad q_i(F, x^*) \leq \alpha_i, \quad 2s < i \leq m, \tag{2.8.18}$$

$F \in T(\{q_i\}, 2 \leq i \leq m, \mu(F, x^*), \alpha)$, and by Lemma 2.8.1

$$\mu(F, x) \leq \frac{\mu(F, x^*)}{1 - q_2 \|x - x^*\| - \cdots - \frac{\mu(F, x^*) + \varepsilon_0}{(m-1)!} \|x - x^*\|^{m-1}} \equiv \overline{\mu}(x). \tag{2.8.19}$$

We also have the estimates

$$\left\| F'(x)^{-1} F^{(i)}(x^*) \right\| \leq \left\| F'(x)^{-1} F'(x^*) \right\| \left\| F'(x^*)^{-1} F^{(i)}(x^*) \right\|$$

$$\leq q_i \left\| F'(x)^{-1} F'(x^*) \right\| \tag{2.8.20}$$

$$\leq \frac{q_i}{1 - \alpha_2 \|x - x^*\| - \cdots - \frac{\alpha_m + \varepsilon_0}{(m-1)!} \|x - x^*\|^{m-1}} \equiv \overline{q}_i(x).$$

The following lemma on fixed points is important.

Lemma 2.8.3. *Let F, x be as above. Then, the Newton operator G defined in (2.8.3) satisfies:*

$$\left\| G(x) - x^* \right\| \leq \frac{\mu(F, x) + q_m}{m!} \|x - x^*\| + \frac{q_{m-1}}{(m-1)!} \|x - x^*\|^{m-1}$$

$$+ \cdots + \frac{q_2}{2!} \|x - x^*\|^2 \tag{2.8.21}$$

and

$$\left\| G(x) - x^* \right\| \leq \frac{\frac{\alpha_m + \varepsilon_0}{m!} \|x - x^*\| + \frac{\alpha_{m-1}}{(m-1)!} \|x - x^*\|^{m-1} + \cdots + \frac{\alpha_2}{2!} \|x - x^*\|^2}{1 - \alpha_2 \|x - x^*\| - \cdots - \frac{(\alpha_m + \varepsilon_0)}{(m-1)!} \|x - x^*\|^{m-1}}.$$

$$\tag{2.8.22}$$

Proof. Using (2.8.3), we can write

$G(x) - x^* =$

$$= x - F'(x)^{-1} F(x) - x^* = F'(x)^{-1} \left[F'(x)(x - x^*) - F(x) \right]$$

$$= F'(x)^{-1} \left[F(x^*) - F(x) - F'(x)(x^* - x) \right]$$

$$= \left[F'(x)^{-1} F'(x^*) \right] \left\{ F'(x^*)^{-1} \left[F(x^*) - F(x) - F'(x)(x^* - x) \right] \right\} \quad (2.8.23)$$

As in Lemma 2.8.1 by taking norms in (2.8.23) and using (2.8.14), (2.8.15) we obtain (2.8.21). Moreover using Lemma 2.8.2 and (2.8.12) we get (2.8.22).

Remark 2.8.4. Consider Newton method (2.8.2)–(2.8.3) for some $x_0 \in U(x^*, b)$. Define sequence $\{c_n\}$ $(n \geq 0)$ by

$$c_n \equiv \|x_n - x^*\| \quad (n \geq 0) \tag{2.8.24}$$

and function g on $[0, \delta)$ by

$$g(t) \equiv \frac{\frac{\alpha_m + \varepsilon_0}{m!} t^m + \frac{\alpha_{m-1}}{(m-1)!} t^{m-1} + \cdots + \frac{\alpha_2}{2!} t^2}{1 - \alpha_2 t - \cdots - \frac{\alpha_m + \varepsilon_0}{(m-1)!} t^{m-1}}. \tag{2.8.25}$$

Using (2.8.24) and (2.8.25), estimate (2.8.22) becomes

$$c_{n+1} \leq g(c_n) \quad (n \geq 0). \tag{2.8.26}$$

It is simple algebra to show that $g(t) < t$ iff $t < \delta_0$, where δ_0 is the positive zero of the equation

$$h(t) = 0, \tag{2.8.27}$$

where

$$h(t) = \frac{(\alpha_m + \varepsilon_0)(m+1)}{m!} t^{m-1} + \frac{m \alpha_{m-1}}{(m-1)!} t^{m-2} + \cdots + \frac{3}{2!} \alpha_2 t - 1. \tag{2.8.28}$$

Note that for $m = 2$, using (2.8.28) we obtain

$$\delta_0 = \frac{2}{3(\alpha_2 + \varepsilon_0)}. \tag{2.8.29}$$

Hence, we proved the following local convergence result for the NK method (2.8.2)–(2.8.3).

Theorem 2.8.5. *NK method $\{x_n\}$ $(n \geq 0)$ generated by (2.8.2)–(2.8.3) converges to the solution x^* of equation $F(x) = 0$, for all $F \in T$, iff the initial guess x_0 satisfies*

$$\|x_0 - x^*\| < \min\{\alpha, \delta_0\}. \tag{2.8.30}$$

We also have the following consequence of Theorem 2.8.5.

Theorem 2.8.6. *NK method* $\{x_n\}$ *($n \geq 0$) generated by (2.8.2)–(2.8.3) converges to the solution x^* of equation $F(x) = 0$, for all $F \in T$, if $F'(x_0)^{-1}$ exists at the initial guess x_0, and*

$$\|x_0 - x^*\| < \min\{\alpha, \bar{\delta}_0\}, \qquad (2.8.31)$$

where $\bar{\delta}_0$ is the positive zero of the equation resulting from (2.8.28) by replacing α_{m+1} by $\mu(F, x_0)$ (defined by (2.8.10)) and α_i, $2 \leq i \leq m$ by $q_i(F, x_0)$ (defined by (2.8.17)).

Proof. By Lemma 2.8.1, because $F'(x_0)^{-1}$ exists and $\|x_0 - x^*\| < \bar{\delta}_0$, we get

$$\mu(F, x^*) \leq m_0 \equiv \frac{\mu(F, x_0)}{1 - q_2(F, x_0)\|x_0 - x^*\| - \cdots - \frac{\mu(F, x_0) + \varepsilon_0}{(m-1)!}\|x - x_0\|^{m-1}}. \qquad (2.8.32)$$

Moreover, we have

$$q_i(F, x^*) =$$
$$= \left\| F'(x^*)^{-1} F^{(i)}(x^*) \right\| \leq \left\| F'(x^*)^{-1} F'(x_0) \right\| \left\| F'(x_0)^{-1} F^{(i)}(x^*) \right\|$$
$$\leq q_0^i \equiv \frac{q_i(F, x_0)}{1 - q_2(F, x_0)\|x_0 - x^*\| - \cdots - \frac{\mu(F, x_0) + \varepsilon_0}{(m-1)!}\|x_0 - x^*\|^{m-1}}. \qquad (2.8.33)$$

Denote by $\bar{\bar{\delta}}_0$ the positive zero of the equation resulting from (2.8.28) by replacing ε_0 by $\mu(F, x^*)$ (defined by (2.8.16)) and α_i, $2 \leq i \leq m$ by $q_i(F, x^*)$. Furthermore denote by $\bar{\bar{\delta}}_0$ the positive zero of the equation resulting from (2.8.28) by replacing ε_0 by m_0 and α_i, $2 \leq i \leq m$ by q_0^i.

Using the above definitions we get

$$\bar{\bar{\delta}}_0 \geq \bar{\bar{\delta}}_0 \geq \frac{q_m(F, x_0) + \mu(F, x_0)}{m!}\|x_0 - x^*\|^m + \frac{q_2(F, x_0)}{(m-1)!}\|x_0 - x^*\|^{m-1}$$
$$+ \cdots + \frac{q_2(F, x_0)}{2!}\|x_0 - x^*\| \geq \|G(x_0) - x^*\|. \qquad (2.8.34)$$

The result now follows from (2.8.34) and Theorem 2.8.5.

Remark 2.8.7. Let us assume equality in (2.8.26) and consider the iteration $c_{n+1} = g(c_n)$ ($n \geq 0$). Denote the numerator of function g by g_1 and the denominator by g_2. By Ostrowski's theorem for convex functions [155] iteration $\{c_n\}$ ($n \geq 0$) converges to 0 if $c_0 \in \left[0, \bar{\bar{\delta}}\right)$, $g'(c_0) < 1$. Define the real function h_0 by

$$h_0(t) = g_2(t)^2 - g_1'(t) g_2(t) + g_2'(t) g_1(t), \qquad (2.8.35)$$

where $\varepsilon_0(x^*) = \mu(F, x^*)$ and $\bar{\alpha}_i = q_i(F, x^*)$, $2 \leq i \leq m$ replace α_{m+1} and α_i in the definition of g respectively. Note that h is a polynomial of degree $2(m-1)$ and can be written in the form

$$h_0(t) = \frac{m^2 - m + 2}{(m-1)!}\varepsilon_0^2(x^*) t^{2(m-1)} + \text{(other lower order terms)} + 1. \qquad (2.8.36)$$

Because h_0 is continuous, and

$$h_0 (0) = 1 > 0, \tag{2.8.37}$$

we deduce that there exists $t_0 > 0$ such that $h_0 (t) > 0$ for all $t \in [0, t_0)$.

Set

$$\bar{c}_0 = \min \left\{ t_0, \bar{\bar{\delta}} \right\}. \tag{2.8.38}$$

It is simple algebra to show that $g' (c_0) < 1$ iff $h_0 (c_0) > 0$. Hence, NK method converges to x^* for all $F \in T$ if the initial guess x_0 satisfies

$$\|x_0 - x^*\| \le \min \{\alpha, \bar{c}_0\}. \tag{2.8.39}$$

Condition (2.8.39) is weaker than (2.8.31).

Although Theorem 2.8.5 gives an optimal domain of convergence for Newton's method, the rate of convergence may be slow for x_0 near the boundaries of that domain. However, it is known that if the conditions of the Newton-Kantorovich theorem are satisfied at x_0 then convergence is rapid. The proof of this theorem can be found in [27].

Theorem 2.8.8. *Let $m \ge 2$ be an integer, X, Y be Banach spaces, D an open convex subset of X, $F: D \to Y$, and an m-times Fréchet-differentiable operator. Let $x_0 \in D$ be such that $F' (x_0)^{-1}$ exists, and suppose the positive numbers δ^*, $d (F, x_0)$, $\alpha_i (F, x_0)$, $2 \le i \le m$ satisfy*

$$\left\| F' (x_0)^{-1} F (x_0) \right\| \le d (F, x_0), \tag{2.8.40}$$

$$\left\| F' (x_0)^{-1} F^{(i)} (x_0) \right\| \le \alpha_i (F, x_0), \quad i = 2, ..., m, \tag{2.8.41}$$

and

$$\left\| F' (x_0)^{-1} \left[F^{(m)} (x) - F^{(m)} (x_0) \right] \right\| \le \varepsilon_0, \quad \varepsilon_0 = \varepsilon_0 (F, x_0) \tag{2.8.42}$$

for all $x \in U (x_0, \delta^*) \subseteq D$.

Denote by s the positive zero of the scalar equation

$$p' (t) = 0, \tag{2.8.43}$$

where

$$p (t) = \frac{\alpha_m (F, x_0) + \varepsilon_0}{m!} t^m + \frac{\alpha_{m-1} (F, x_0)}{(m - 1)!} t^{m-1}$$

$$+ \cdots + \frac{\alpha_2 (F, x_0)}{2!} t^2 - t + d (F, x_0). \tag{2.8.44}$$

If

$$p (s) \le 0, \tag{2.8.45}$$

and

$$\delta^* \geq r_1, \tag{2.8.46}$$

where r_1 is the smallest nonnegative zero of equation

$$p(t) = 0$$

guaranteed to exist by (2.8.45), then NK method (2.8.2)–(2.8.3) starting from x_0 generates a sequence that converges quadratically to an isolated solution x^* of equation $F(x) = 0$.

Remark 2.8.9. Using this theorem we obtain two further sufficiency conditions for the convergence of NK method.

It is convenient for us to set $\varepsilon_0 = \mu(F, x_0)$, and $\alpha_i(F, x_0) = q_i$ (q_i evaluated at x_0) $2 \leq i \leq m$. Condition can be written as

$$d(F, x_0) \leq s_0, \tag{2.8.47}$$

where

$$s_0 = s - \left[\tfrac{q_2}{2!} s^2 + \cdots + \tfrac{\varepsilon_0 + q_m}{m!} s^m \right] > 0 \tag{2.8.48}$$

by the definition of s. Define functions h_1, h_2 by

$$h_1(t) = \tfrac{q_m + \varepsilon_0}{m!} t^m + \tfrac{q_{m-1}}{(m-1)!} t^{m-1} + \cdots + \tfrac{q_2}{2!} t^2 + t - s_0, \tag{2.8.49}$$

and

$$h_2(t) = \tfrac{\bar{q}_m(x_0) + \varepsilon_0}{m!} t^m + \tfrac{\bar{q}_{m-1}(x_0)}{(m-1)!} t^{m-1} + \cdots + \tfrac{\bar{q}_2(x_0)}{2!} t^2 + t - s_0. \tag{2.8.50}$$

Because $h_1(0) = h_2(0) = -s_0 < 0$, we deduce that there exist minimum $t_1 > 0$, $t_2 > 0$ such that

$$h_1(t) \leq 0 \quad \text{for all} \quad t \in [0, t_1] \tag{2.8.51}$$

and

$$h_2(t) \leq 0 \quad \text{for all} \quad t \in [0, t_2]. \tag{2.8.52}$$

Theorem 2.8.10. *Let $F \in T$, and $x_0 \in U(x^*, \alpha)$. Then condition (2.8.45) holds, if either*
(a) $F'(x_0)^{-1}$ exists and $\|x_0 - x^*\| \leq \min\{\alpha, t_1\}$;
(b) $F'(x_0)^{-1}$ exists and $\|x_0 - x^*\| \leq \min\{\alpha, t_2\}$,
where t_1 and t_2 are defined in (2.8.51), and (2.8.52), respectively.

Proof. Choose $\delta^* > 0$ such that $U(x_0, \delta^*) \subseteq U(x^*, \alpha)$. By (2.8.3), and (2.8.21), we get (for $\varepsilon_0(G, x_0) = \mu(F, x_0)$, and $\alpha_i(F, x_0) = q_i$ (q_i evaluated at x_0) $g \leq i \leq m$):

$$\left\| F'(x_0)^{-1} F(x_0) \right\| = \|G(x_0) - x_0\| \leq \|F(x_0) - x^*\| + \|x^* - x_0\|$$

$$\leq \tfrac{q_m + \varepsilon_0}{m!} \|x_0 - x^*\|^m + \tfrac{q_{m-1}}{(m-1)!} \|x_0 - x^*\|^{m-1}$$

$$+ \cdots + \tfrac{q_2}{2!} \|x_0 - x^*\|^2 + \|x_0 - x^*\|. \tag{2.8.53}$$

Using (2.8.53) to replace $d\,(F, x_0)$ in (2.8.44), and setting $\|x_0 - x^*\| \leq t$, we deduce that (2.8.45) holds if $h_1\,(t) \leq 0$, which is true by the choice of t_1, and (a). Moreover, by replacing $\mu\,(G, x_0)$ and $q_i,\ 2 \leq i \leq m$ using (2.8.19), and (2.8.20), respectively, condition (2.8.45) holds if $h_2\,(t) \leq 0$, which is true by the choice of t_2, and (b).

In order for us to cover the case $m = 1$, we start from the identity

$$
x_{n+1} - x^* =
$$
$$
= x_n - F'\,(x_n)^{-1}\,F\,(x_n)
$$
$$
= \left[F'\,(x_n)^{-1}\,F'\,(x^*) \right] F'\,(x^*)^{-1}\,1\left[F\,(x^*) - F\,(x_n) - F'\,(x_n)\,(x^* - x) \right]
$$
$$
= \left[F'\,(x_n)^{-1}\,F'\,(x^*) \right] F'\,(x^*)^{-1} \int_0^1 \left[F'\,(x_n + t\,(x^* - x_n)) - F'\,(x_n) \right]\,(x^* - x_n)\,dt
$$
$$
= \left[F'\,(x_n)^{-1}\,F'\,(x^*) \right] F'\,(x^*)^{-1} \int_0^1 \left[F'\,(x_n + t\,(x^* - x_n)) - F'\,(x^*) \right]\,(x^* - x_n)\,dt
$$
$$
+ \left[F'\,(x_n)^{-1}\,F'\,(x^*) \right] F'\,(x^*)^{-1}\left[F'\,(x^*) - F'\,(x_n) \right]\,(x^* - x_n)\,.
$$

to show as in Lemma 2.8.3.

Theorem 2.8.11. *Let $F: D \to Y$ be a Fréchet-differentiable operator. Assume there exists a simple zero x^* of $F\,(x) = 0$, and for $\varepsilon_1 > 0$ there exists $\ell > 0$ such that*

$$
\left\| F'\,(x^*)^{-1}\left[F'\,(x) - F'\,(x^*) \right] \right\| < \varepsilon_1
$$

for all $x \in U\,(x^*, \ell)$.

Then, NK method $\{x_n\}$ $(n \geq 0)$ generated by (2.8.2)–(2.8.3) is well defined, remains in $U\,(x^*, 1)$, and converges to x^* with

$$
\|x_{n+1} - x^*\| \leq \tfrac{2\varepsilon_1}{1-\varepsilon_1}\,\|x_n - x^*\| \quad (n \geq 0)
$$

provided that

$$
3\varepsilon_1 < 1
$$

and

$$
x_0 \in U\,(x^*, l_1)\,.
$$

Example 2.8.12. Returning back to Example 2.4.16, for $m = 3,\ \alpha_2 = \alpha_3 = 1$ and $\varepsilon_0 = e - 1$.
We get using (2.8.28)

$$
\delta_0^3 = .43649019. \tag{2.8.54}
$$

To compare our results with earlier ones, note that in Theorem 3.7 [216, p. 111] the condition is

$$
\|x_0 - x^*\| < \min\left\{ \sigma, \tfrac{2}{3\rho} \right\} = \rho_0, \tag{2.8.55}
$$

where σ, ρ are such that $U\,(x^*, \sigma) \subseteq D$, and

$$\left\| F'\left(x^*\right)^{-1} \left(F'\left(x\right) - F'\left(y\right)\right) \right\| \le \rho \left\| x - y \right\| \quad \text{for all} \quad x, y \in U\left(x^*, \sigma\right). \quad (2.8.56)$$

Letting $\sigma = \alpha = 1$, we get using (2.8.56) $\rho = e$, and condition (2.8.55) becomes

$$\left\| x_0 - x^* \right\| < \rho \equiv .245253. \quad (2.8.57)$$

Remark 2.8.13. For $m = 2$, (2.8.28) gives (2.8.29). In general $\alpha_2 < \rho$. Hence, there exists $\varepsilon_0 > 0$ such that $\alpha_2 + \varepsilon_0 < \rho$, which shows that

$$\rho_0 > \rho_0. \quad (2.8.58)$$

(See Example 2.8.15 for such a case.)

Remark 2.8.14. Our analysis can be simplified if instead of (2.8.22) we consider the following estimate: because $x \in U\left(x^*, \alpha\right)$, there exist γ_1, γ_2 such that

$$2\left[\tfrac{\alpha_m + \varepsilon_0}{m!} \left\| x_0 - x^* \right\|^{m-2} + \cdots + \tfrac{\alpha_2}{2!} \right] \le \gamma_1, \quad (2.8.59)$$

and

$$\tfrac{\alpha_m + \varepsilon_0}{(m-1)!} \left\| x_0 - x^* \right\|^{m-2} + \cdots + \alpha_2 \le \gamma_2. \quad (2.8.60)$$

Hence estimate (2.8.22) can be written

$$\left\| G\left(x\right) - x^* \right\| \le \tfrac{\gamma_1}{2\left(1 - \gamma_2 \| x - x^* \|\right)} \left\| x - x^* \right\|^2, \quad (2.8.61)$$

and for $\gamma^* = \max\left\{ \gamma_1, \gamma_2 \right\}$

$$\left\| G\left(x\right) - x^* \right\| \le \tfrac{\gamma^*}{2\left(1 - \gamma^* \| x - x^* \|\right)} \left\| x - x^* \right\|^2. \quad (2.8.62)$$

The convergence condition of Theorem 3.7 [216, p. 111] and (2.8.61), (2.8.62), becomes respectively

$$\left\| x_0 - x^* \right\| \le \min\left\{ \alpha, \gamma \right\}, \gamma = \tfrac{2}{\gamma_1 + 2\gamma_2}, \quad (2.8.63)$$

and

$$\left\| x_0 - x^* \right\| \le \min\left\{ \sigma, \tfrac{2}{3\gamma^*} \right\}. \quad (2.8.64)$$

In particular, estimate (2.8.64) is similar to (2.8.55), and if $\gamma < \rho$, then (2.8.63) allows a wider range for the initial guess x_0 than (2.8.55).

Furthermore, assuming (2.8.4), (2.8.5), and (2.8.55) hold, our analysis can be based on the following variations of (2.8.22):

$$\left\| G\left(x\right) - x^* \right\| \le \tfrac{\frac{q_m + \varepsilon_0}{m!} \| x - x^* \|^m + \cdots + \frac{q_2}{2!} \| x - x^* \|^2}{1 - \rho \| x - x^* \|}, \quad (2.8.65)$$

and

$$\left\| G\left(x\right) - x^* \right\| \le \tfrac{\rho}{2\left[1 - \alpha_2 \| x - x^* \| - \cdots - \frac{\alpha_m + \varepsilon_0}{(m-1)!} \| x - x^* \|^{m-1} \right]} \left\| x - x^* \right\|^2. \quad (2.8.66)$$

Example 2.8.15. Let us consider the system of equations

$$F(x, y) = 0,$$

where

$$F: \mathbf{R}^2 \to \mathbf{R}^2,$$

and

$$F(x, y) = (xy - 1, \, xy + x - 2y).$$

Then, we get

$$F'(x, y) = \begin{bmatrix} y & x \\ y+1 & x-2 \end{bmatrix},$$

and

$$F'(x, y)^{-1} = \frac{1}{x + 2y} \begin{bmatrix} 2-x & x \\ y+1 & -y \end{bmatrix},$$

provided that (x, y) does not belong on the straight line $x + 2y = 0$. The second derivative is a bilinear operator on $\mathbf{R}^2$ given by the following matrix

$$F''(x, y) = \begin{bmatrix} 0 & 1 \\ 1 & 0 \\ -- & -- \\ 0 & 1 \\ 1 & 0 \end{bmatrix}.$$

We consider the max-norm in $\mathbf{R}^2$. Moreover in $L(\mathbf{R}^2, \mathbf{R}^2)$ we use for

$$A = \begin{bmatrix} a_{11} & a_{12} \\ a_{21} & a_{22} \end{bmatrix}$$

the norm

$$\|A\| = \max \{|a_{11}| + |a_{12}|, \, |a_{21}| + |a_{22}|\}.$$

As in [7], we define the norm of a bilinear operator B on $\mathbf{R}^2$ by

$$\|B\| = \sup_{\|z\|=1} \max_i \sum_{j=1}^{2} \left| \sum_{k=1}^{2} b_i^{jk} z_k \right|,$$

where

$$z = (z_1, z_2) \quad \text{and} \quad B = \begin{bmatrix} b_1^{11} & b_1^{12} \\ b_1^{21} & b_1^{22} \\ -- & -- \\ b_2^{11} & b_2^{12} \\ b_2^{21} & b_2^{22} \end{bmatrix}.$$

Using (2.8.4), (2.8.5), (2.8.29), (2.8.55), (2.8.56), for $m = 2$ and $(x^*, y^*) = (1, 1)$, we get $\rho = \frac{4}{3}$, $\rho_0 = .5$, $\alpha_2 = 1$. We can set $\varepsilon_0 = .001$ to obtain $\delta_0^2 = .666444519$. Because $\rho_0 < \delta_0^2$, a remark similar to the one at the end of Example 2.8.12 can now follow.

2.9 On a weak NK method

R. Tapia in [188] showed that the weak Newton method (to be precised later) converges in cases NK cannot under the famous Newton-Kantorovich hypothesis (see (2.9.9)). Using the technique we recently developed in Section 2.2, we show that (2.9.9) can always be replaced by the weaker (2.9.6), which is obtained under the same computational cost. This way we can cover cases [39] that cannot be handled by the work in [188].

We need the following definitions and Lemma whose proof can be found in [188, p. 540]:

Definition 2.9.1. *Let* $D_0 \subseteq D$ *be a closed subset of* X, D_1 *and open subset of* D_0. *For* $x \in D_1$, $M(x)$ *is a left inverse for* $F'(x)$ *relative to* D_0 *if:*

(a) $M(x) \in L(Y_x, X)$, *where* Y_x *is a closed linear subspace of* Y *containing* $F(D_1)$;

(b) $M(x) F(D_1) \subseteq D_0$;

and

(c) $M(x) F'(x) = I$

where I *is the identity operator from* D_0 *into* D_0.

Lemma 2.9.2. *Hypotheses* (a) *and* (b) *imply that for all* $y \in D_1$;

(d) $F'(y)(D_0) \subseteq Y_x$,

and

(e) $M(x) F'(y)(D_0) \subseteq D_0$.

Definition 2.9.3. *If* $x_0 \in D_1$, *then*

$$x_{n+1} = x_n - M(x_n) F(x_n) \quad (n \geq 0) \tag{2.9.1}$$

is called the weak Newton method.

The following result is a version of Theorem 2 in [39] (see also Section 2.3):

Theorem 2.9.4. *If there exist* $M \in L(Y, X)$ *and constants* $\eta, \delta, \ell, \ell_0, t^*$ *such that:*

$$M^{-1} \text{ exists;}$$

$$\|M F(x_0)\| \leq \eta; \tag{2.9.2}$$

$$\|I - M F'(x_0)\| \leq \delta < 1; \tag{2.9.3}$$

$$\|M(F'(x) - F'(x_0))\| \leq \ell_0 \|x - x_0\|; \tag{2.9.4}$$

$$\|M(F'(x) - F'(y))\| \leq \ell \|x - y\| \tag{2.9.5}$$

for all $x, y \in D$;

$$h_0 = \frac{\bar{\ell}\eta}{(1-\delta)^2} \leq \frac{1}{2}, \quad \bar{\ell} = \frac{\ell_0 + \ell}{2} \tag{2.9.6}$$

and

$$\overline{U}(x_0, t^*) \subseteq \overline{U}(x_0, 2\eta) \subseteq D. \tag{2.9.7}$$

Then, sequence $\{x_n\}$ ($n \geq 0$) *generated by NK is well defined, remains in* $U(x_0, t^*)$ *for all* $n \geq 0$, *and converges to a unique solution* x^* *of equation* $F(x) = 0$ *in* $\overline{U}(x_0, t^*)$.

Proof. Simply use MF, D_0 instead of F, X respectively in the proof of Theorem 2 in [39] (see also Section 2.2).

Lemma 2.9.5. *[188] Newton sequences in Theorem 2.9.4 exists if and only if M is invertible.*

If M is not invertible, we can have the following Corollary of Theorem 2.9.4.

Corollary 2.9.6. *If there exists $M \in L(S, X)$, such that $MF(U(x_0, 2\eta)) \subseteq D_0$, where S is a closed linear subspace of Y containing $F(U(x_0, 2\eta))$, and (2.9.2)–(2.9.7) hold, then*

(a) sequence $\{x_n\}$ $(n \geq 0)$ generated by the weak Newton method (2.9.1) is well defined, remains in $U(x_0, t^*)$ for all $n \geq 0$, and converges to some point $x^* \in \overline{U}(x_0, t^*)$.

(b) If M is one-to-one, then $F(x^*) = 0$; or if $t^* < 2\eta$, and F has a solution in $\overline{U}(x_0, t^*)$, then again $F(x^*) = 0$.

Proof. It follows from Lemma 2.9.5 that for any $x \in D_1$, $F'(x): D_0 \to S$, and $MF'(x): D_1 \to D_0$ so that $\left[MF'(x) \right]^{-1} MF: D_1 \to D_0$ whenever it exists. The rest follows as in the proof of Theorem 2 in [39] with F, X replaced by MF, D_0 respectively.

Remark 2.9.7. If

$$\ell_0 = \ell, \tag{2.9.8}$$

then Theorem 2.9.4 and Corollary 2.9.6 reduce to Theorem 3.1 and Corollary 3.1 respectively in [188] (if F is twice Fréchet-differentiable on D). However $\ell_0 \leq \ell$, holds in general (see Section 2.2). It follows that the Newton-Kantorovich hypothesis

$$h = \frac{\ell\eta}{(1-\delta)^2} \leq \frac{1}{2} \tag{2.9.9}$$

used in the results in [188] mentioned above always implies (2.9.6) but not vice versa unless if (2.9.8) holds.

2.10 Bounds on manifolds

We recently showed the following weaker version of the Newton-Kantorovich theorem [39] (see Section 2.2):

Theorem 2.10.1. *Let $F: D = U(0, r) \subseteq X \to Y$ be a Fréchet-differentiable operator. Assume:*

$$F'(0)^{-1} \in L(Y, X), \tag{2.10.1}$$

and there exist positive constants a, ℓ, ℓ_0, η such that:

$$\| F'(0)^{-1} \| \leq a^{-1}, \tag{2.10.2}$$

$$\| F'(0)^{-1} F(0) \| \leq \eta_0 \leq \eta \tag{2.10.3}$$

$$\| F'(x) - F'(y) \| \leq \ell \| x - y \|, \tag{2.10.4}$$

$$\| F'(x) - F'(0) \| \leq \ell_0 \| x \| \tag{2.10.5}$$

for all x, y ∈ D,

$$h_0 = La^{-1}\eta < \tfrac{1}{2}, \quad L = \tfrac{\ell_0 + \ell}{2}, \tag{2.10.6}$$

$$M = \lim_{n \to \infty} t_n < r, \tag{2.10.7}$$

where,

$$t_0 = 0, \quad t_1 = \eta, \quad t_{n+2} = t_{n+1} - \frac{\ell(t_{n+1} - t_n)^2}{2(1 - \ell_0 t_{n+1})}. \tag{2.10.8}$$

Then equation F (x) = 0 has a solution x* ∈ D that is the unique zero of F in U (0, 2η).

Remark 2.10.2. The above result was also shown in affine invariant form and for any initial guess including 0. However, we want the result in the above form for simplicity, and in order to compare it with earlier ones [156].

Let us assume for X, Y being Hilbert spaces:
$A = F'(0) \in L(X, Y)$ is surjective, $A^+ \in L(Y, X)$ is a right inverse of A, and

$$\|A^+\| \leq a^{-1}, \tag{2.10.9}$$

$$\|A^+ F(0)\| = \bar{\eta} \leq \eta; \tag{2.10.10}$$

Conditions (2.10.3)–(2.10.6) hold (for $F'(0)^{-1}$ replaced by A^+).
It is convenient for us to introduce:

$$S = \{x \in D \mid F(x) = 0\}, \tag{2.10.11}$$

$$S_0 = \{x \in D \mid F(x) = F(0)\}, \tag{2.10.12}$$

$$N_1 = \text{Ker}(A), \tag{2.10.13}$$

and

$$N_2 \text{ the orthogonal complement of } N_1.$$

In Theorem 2.10.3 we provide an analysis in the normal space N_2 at 0 of S_0, which leads to an upper bound of $d(0, S)$.

Newton-Kantorovich-type condition (2.10.6) effects S to be locally in a convex cone. Theorem 2.10.8 gives the distance of 0 to that cone as a lower bound of $d(0, S)$.

This technique leads to a manageable way of determining for example sharp error bounds for an approximate solution of an undetermined system.

Finally we show that our approach provides better bounds than the ones given before in [156] (and the references there), and under the same computational cost.

The following results can be shown by simply using (2.10.4), (2.10.5) instead of (2.10.4) in the proofs of Theorem 2.10.3, Lemmas 2.10.4–2.10.7, Theorem 2.10.8, and Corollary 2.10.9, respectively.

Theorem 2.10.3. *Operator F: D ⊆ X → Y has a zero x* in U (0, M) ∩ N₂; x* is the unique zero of F/N₂ in U (0, 2η) ∩ N₂.*

Lemma 2.10.4. *The following bounds hold:*

$$r^* \leq \|x^*\| = b, \tag{2.10.14}$$
$$b < M < 2\eta, \tag{2.10.15}$$

and

$$\ell_0 b < 2\ell_0 \eta < a, \tag{2.10.16}$$

where,

$$r^* = \frac{a}{\ell_0}\left(-1 + \sqrt{1 + 2h_1}\right), \tag{2.10.17}$$

and

$$h_1 = \ell_0 \eta_0 a^{-1}. \tag{2.10.18}$$

It is convenient for us to introduce the notion:

$$V = \{x \in X | \|x\| < b\}, \quad P(x) = F'(x)/N_1, \tag{2.10.19}$$

$$Q(x) = F'(x)/N_2, \quad x \in V, \tag{2.10.20}$$

and

$$\alpha = \frac{\ell_0 b}{\lambda - \ell_0 b}. \tag{2.10.21}$$

Lemma 2.10.5. *The following hold*

$$Q(x) \text{ is regular for all } x \in V$$

and

$$\|Q(x)^{-1} P(x)\| \leq \alpha, \text{ for all } x \in V. \tag{2.10.22}$$

Let us define:

$$W = \{x = x_1 + x_2 \in X \mid \alpha \|x_1\| + \|x_2\| < b, \ x_i \in N_i, \ i = 1, 2, \} \tag{2.10.23}$$

and

$$K(w) = \{(1 - \theta) w_1 + x_2 \mid \theta \in [0, 1], \ x_2 \in N_2, \ \|x_2 - w_2\| \leq \alpha \|w_1\| \theta\} \tag{2.10.24}$$

where,

$$w = w_1 + w_2 \in X, \quad w_i \in N_i, \ i = 1, 2.$$

Lemma 2.10.6. *If $w = w_1 + w_2 \in W \cap V$, then $K(w) \subseteq V \cap W$.*

Lemma 2.10.7. *Operator F has no zeros in $W \cap V$.*

Theorem 2.10.8. *Let us define real function g by*

$$g(t) = t\left(\tfrac{a}{\ell_0} - t\right)\left(t^2 + \left(\tfrac{a}{\ell_0} - t\right)^2\right)^{-\frac{1}{2}}, \; for \, t \in \left(0, \tfrac{a}{\ell_0}\right).$$ (2.10.25)

Then, the following bounds hold:

$$d(0, S) \geq m = \begin{cases} g(M), & if \, \tfrac{1}{4}\sqrt{3} \leq h_0 < \tfrac{1}{2}, \\ & \sqrt{1 - 2h_0} + \tfrac{1}{2}(1 - 2h_0) \leq h_1 \leq h_0 \\ g(r^*), & if \, 0 < h_0 < \tfrac{1}{2}, \\ & h_1 \leq \min\left\{h_0, \sqrt{1 - 2h_0} + \tfrac{1}{2}(1 - 2h_0)\right\}, \end{cases}$$ (2.10.26)

where M and r^* are given by (2.10.7) and (2.10.14), respectively.

Corollary 2.10.9. *If $\eta_0 = \eta$ then the following bounds hold:*

$$m = d(0, S) \geq \begin{cases} g(M), & if \, \tfrac{\sqrt{3}}{4} \leq h_0 < \tfrac{1}{2} \\ g(r^*), & if \, h_0 < \tfrac{\sqrt{3}}{4}. \end{cases}$$ (2.10.27)

Remark 2.10.10. (a) Theorem 2.10.1 reduces to the corresponding one in [156] if

$$\ell_0 = \ell.$$ (2.10.28)

However, in general,

$$\ell_0 \leq \ell$$ (2.10.29)

holds. Let

$$h = \ell\eta < \tfrac{1}{2}.$$ (2.10.30)

Then note by (2.10.6) and (2.10.30)

$$h < \tfrac{1}{2} \Longrightarrow h_0 < \tfrac{1}{2}$$ (2.10.31)

but not necessarily vice versa unless if (2.10.28) holds.

(b) Our results reduce to the corresponding ones in [156] again if (2.10.28) holds. However, if strict inequality holds in (2.10.29), then our interval of bounds $[m, M]$ is always more precise than the corresponding one in [156] and are found under the same computational cost.

2.11 The radius of convergence and one-parameter operator embedding

In this section, we are concerned with the problem of approximating a locally unique solution of the nonlinear equation

$$F(x) = 0,$$ (2.11.1)

where F is a Fréchet-differentiable operator defined on closed convex subset of the mth Euclidean space X^m into X^m.

NK method when applied to (2.11.1) converges if the initial guess is "close" enough to the root of F. If no approximate roots are known, this and other iterative methods may be of little use. Here we are concerned with enlarging the region of convergence for Newton's method. Our technique applies to other iterative methods. That is, the use of Newton's method is only illustrative. The results obtained here extend immediately to hold when F is defined on a Banach spaces with values in a Banach space.

We assume operator F is differentiably embedded into a one-parameter family of operators $\{H(t, \cdot)\}$ such that $H(t_0, x_0) = 0$, and $H(t_1, x) = F(x)$ for some $x_0 \in D$ and two values t_0 and t_1 of the parameter.

We consider the commonly used embedding (homotopy)

$$H(t, x) = F(x) + (t - 1) F(x_0) \tag{2.11.2}$$

The solution of $F(x) = 0$ is then found by continuing the solution curve $x(t)$ of $H(t, x) = 0$ from t_0 until t_1.

Homotopies have been employed to prove existence results for linear and non-linear equations (see [139] and the references there).

In particular, the results obtained in [139] can be weakened, and the region of convergence for NK method can be enlarged if we simply replace the Lipschitz constant by the average between the Lipschitz constant and the center-Lipschitz constant. Moreover our results can be used in cases not covered in [139].

Motivated by advantages of the weaker version of the Newton-Kantorovich theorem that we provided in Section 2.2, we hope that the homotopy approach will be successful under the same hypotheses. In particular, we can show:

Theorem 2.11.1. *Let* $F: D \subseteq X^m \to X^m$ *be Fréchet-differentiable. Assume there exist* $x_0 \in D$, $\ell_0 \geq 0$, $\ell \geq 0$ *and* $\eta \geq 0$ *such that:*

$$F'(x_0)^{-1} \in L\left(X^m, X^m\right), \tag{2.11.3}$$

$$\left\| F'(x_0)^{-1} \left[F'(x) - F'(x_0) \right] \right\| \leq \ell_0 \|x - x_0\|, \tag{2.11.4}$$

$$\left\| F'(x_0)^{-1} \left[F'(x) - F'(y) \right] \right\| \leq \ell \|x - y\|, \tag{2.11.5}$$

$$\left\| F'(x_0)^{-1} F(x_0) \right\| \leq \eta, \tag{2.11.6}$$

$$h_0 = L_0 \eta \leq \tfrac{1}{2} \text{ for } \ell_0 < \ell, \tag{2.11.7}$$

or

$$h_0 < \tfrac{1}{2} \text{ for } \ell_0 = \ell, \tag{2.11.8}$$

where

$$L_0 = \tfrac{\ell_0 + \ell}{2}, \tag{2.11.9}$$

and

$$\overline{U}(x_0, 2\eta) \subseteq D. \tag{2.11.10}$$

Then the solution

$$x = -F'(x)^{-1} F(x_0), \quad x(0) = x_0 \tag{2.11.11}$$

exists, belongs in $\overline{U}(x_0, r_0)$ *for all* $t \in [0, 1]$, *and* $F(x(1)) = 0$.

Proof. Using (2.11.4), (2.11.6) for $x \in \overline{U} \in (x_0, r_0)$, we obtain in turn

$$\|F'(x)^{-1} F(x_0)\| \leq \|F'(x_0)^{-1} [F'(x_0) - F'(x)] F'(x)^{-1} F(x_0)\|$$
$$+ \|F'(x_0)^{-1} F(x_0)\|$$
$$\leq \ell_0 \|x - x_0\| \|F'(x)^{-1} F(x_0)\| + \eta$$

or

$$\|F'(x)^{-1} F(x_0)\| \leq \frac{\eta}{1-\ell_0\|x-x_0\|} \tag{2.11.12}$$

(as $\ell_0 \|x - x_0\| \leq \ell_0 2\eta < 1$ by (2.11.7)).

Define function h by

$$h(t, r) = \frac{\eta}{1-\ell_0 r}. \tag{2.11.13}$$

Then it is simple calculus to see that

$$r'(t) = h(t, r) \tag{2.11.14}$$

has a unique solution $r \leq r_0$, $r(0) = 0$ for $t \in [0, 1]$. Moreover by (2.11.7) we get $h(t, r) < \infty$ for $t \in [0, 1]$. Hence by Lemma 1.2 in [139], equation (2.11.2) has a solution $x(t)$ and $F(x(1)) = 0$.

Remark 2.11.2. (a) If F is twice Fréchet-differentiable, and

$$\ell_0 = \ell, \tag{2.11.15}$$

then Theorem 2.11.1 reduces to Corollary 2.1 in [139, p. 743]. However

$$\ell_0 \leq \ell \tag{2.11.16}$$

holds in general. Meyer in [139] used the famous Newton-Kantorovich hypothesis

$$h = \ell\eta < \tfrac{1}{2} \tag{2.11.17}$$

to show Corollary 2.1 in [139]. Note that

$$h \leq \tfrac{1}{2} \implies h_0 \leq \tfrac{1}{2}. \tag{2.11.18}$$

(b) The conclusion of the theorem holds if $r^* = 2\eta$ is replaced by

$$r_0 = \lim_{n\to\infty} t_n \leq 2\eta, \tag{2.11.19}$$

where,

$$t_0 = 0, \ t_1 = \eta, \ t_{n+2} = t_{n+1} + \frac{\ell(t_{n+1}-t_n)^2}{2(1-\ell_0 t_{n+1})} \quad (n \geq 0) \quad \text{(see Section 2.2), [39].} \tag{2.11.20}$$

Note also that

$$r_0 \leq \tfrac{\eta}{h}\left(1 - \sqrt{1 - 2h}\right) = r^* \tag{2.11.21}$$

in case (2.11.17) holds [39]. Note that r^* was used in Corollary 2.1 [139].

NK method corresponds with integrating

$$x'(\lambda) = -F'(x)^{-1} F(x), \quad x(0) = x_0, \quad \lambda \in [0, \infty), \tag{2.11.22}$$

with Euler's method of step size 1, or equivalently,

$$x'(t) = -F'(x)^{-1} F(x_0), \quad x(0) = x_0, \quad t \in [0, 1], \tag{2.11.23}$$

with Euler's method and variable step size

$$h_{k+1} = e^{-k}\left(1 - e^{-1}\right), \quad k \geq 0. \tag{2.11.24}$$

Hence the initial step is $h_1 \cong .63$, which is too large. This is a large step for approximating $x\left(1 - e^{-1}\right)$ unless if $F(x)$ is sufficiently controlled.

As in Meyer [139] we suggest an alternative: Integrate (2.11.23) with step size $h = \frac{1}{N}$. Choose the approximate solution

$$x(N_h) = x_N \tag{2.11.25}$$

as the initial guess for NK method.

This way we have the result:

Theorem 2.11.3. *Let* $F: X^m \to X^m$ *be Fréchet-differentiable and satisfying*

$$\| F'(x)^{-1} \| \leq a \|x\| + b \text{ for all } x \in X^m. \tag{2.11.26}$$

Let $x_0 \in X^m$ *be arbitrary and define ball* $U(x_0, \bar{r} + \delta)$ *for* $\delta > 0$ *by*

$$\bar{r} = \begin{cases} \left[\|x_0\| + \frac{b}{a}\right] \exp\left(a \|F(x_0)\|\right) - \left(\|x_0\| + \frac{b}{a}\right), & \text{if } a \neq 0, \\ b \|F(x_0)\|, & \text{if } a = 0. \end{cases} \tag{2.11.27}$$

Assume (2.11.23) is integrated from 0 to 1 with a numerical method of order h^p, *denoted by*

$$x_{k+1} = G(x_k, h) \tag{2.11.28}$$

and satisfying

$$\|x(1) - x_N\| \leq ch^p, \tag{2.11.29}$$

where c does not depend on h.

Moreover assume there exist constants d, ℓ_0, ℓ *such that:*

$$\| F'(x)^{-1} \| \leq d, \tag{2.11.30}$$

$$\| F'(x) - F'(x_0) \| \leq \ell_0 \|x - x_0\| \tag{2.11.31}$$

and

$$\| F'(x) - F'(y) \| \leq \ell \|x - y\| \tag{2.11.32}$$

for all $x, y \in \bar{U}(x_0, \bar{r} + \delta)$.

Then iteration

$$x_{k+1} = G(x_k, h) \quad k = 1, ..., N-1 \tag{2.11.33}$$

$$x_{k+1} = x_n - F'(x_k)^{-1} F(x_k) \quad k = N, ..., \tag{2.11.34}$$

converges to the unique solution of equation $F(x) = 0$ in $U(x_0, \bar{r} + \delta)$ provided that

$$h = \tfrac{1}{N} \leq \left(\tfrac{\sqrt{2}-1}{dL_0 c} \right)^{\frac{1}{p}}, \tag{2.11.35}$$

and

$$ch^{\mathbf{p}} < S. \tag{2.11.36}$$

Proof. Simply replace ℓ (Meyer denotes ℓ by L in [139]) by L_0 in the proof of Theorem 4.1 in [139, p. 750].

Remark 2.11.4. If $\ell_0 = \ell$ and F is twice Fréchet-differentiable then Theorem 2.11.3 reduces to Theorem 4.1 in [139]. However if strict inequality holds in (2.11.16), because the corresponding estimate (2.11.35) in [139] is given by

$$h_M = \tfrac{1}{N} \leq \left(\tfrac{\sqrt{2}-1}{d\ell c} \right)^{\frac{1}{p}} \tag{2.11.37}$$

we get

$$h_M < h. \tag{2.11.38}$$

Hence our technique allows a under step size h, and under the same computational cost, as the computations of ℓ require in practice the computation of ℓ_0.

2.12 NK method and Riemannian manifolds

In this section, we are concerned with the problem of approximating a locally unique solution x^* of equation

$$F(x) = 0, \tag{2.12.1}$$

where F is C^1 and defined on an open convex subset S of R^m (m a natural number) with values in R^m.

Newton-like methods are the most efficient iterative procedures for solving (2.12.1) when F is sufficiently many times continuously differentiable. In particular, Newton's method is given by

$$y_n = -F'(x_n)^{-1} F(x_n) \quad (x_0 \in S) \tag{2.12.2}$$

$$x_{n+1} = x_n + y_n \quad (n \geq 0). \tag{2.12.3}$$

We can extend this method to approximate a singularity of a vectorial field G defined on a Riemannian manifold M:

$$G(z) = 0, \quad z \in M. \tag{2.12.4}$$

Operator $F'(x_n)$ is replaced by the covariant derivative of G at z_n:

$$\nabla G_{z_n} : T_{z_n}(M) \to T_{z_n}(M) \tag{2.12.5}$$
$$y \to \nabla_y G$$

(we use $\nabla G_z y = \nabla_y G$). Therefore approximation (2.12.2) becomes:

$$y_n = -\nabla G_{z_n}^{-1} G(z_n), \tag{2.12.6}$$

and $y_n \in T_{z_n}(M)$ (if (2.12.6) is well defined for all $n \geq 0$).

In R^m, x_{n+1} is obtained from x_n using the secant line which passes through x_n with direction y_n, and at n distance $\|y_n\|$. In a Riemannian manifold, geodesics replace straight lines. Hence Newton's method in a Riemannian manifold becomes:

$$z_{n+1} = \exp_{z_n}(y_n) \quad (n \geq 0), \tag{2.12.7}$$

where y_n is given by (2.12.6) for all $n \geq 0$.

Ferreira and Svaiter in [94] extended the Newton-Kantorovich theorem to Riemannian manifolds. This elegant semilocal convergence theorem for Newton's method is based on the Newton-Kantorovich hypothesis (see (2.12.19)). Recently [39] they developed a new technique that on the one hand weakens (2.12.19) under the same computational cost, and on the other hand applies to cases not covered by the Newton-Kantorovich theorem (i.e., (2.12.19) is violated whereas (2.12.12) holds); fine error bounds on the distances involved are obtained and an at least as precise information on the location of the solution (if center-Lipschitz constant L_0 is smaller than Lipschitz constant L).

Here we extend our result from Banach spaces to Riemannian manifolds to gain the advantages stated above (in this new setting).

We refer the reader to [94] for fundamental properties and notations of Riemannian manifolds. Instead of working with Frobenius norm of rank-two tensors, we use "operator norm" of linear transformations on each tangent space.

We need the definitions:

Definition 2.12.1. *Let $S_z : T_z M \to T_z M$ be a linear operator. Define*

$$\|S_z\|_{op} = \sup\{\|S_z y\|, y \in T_z M, \|y\| = 1\} \tag{2.12.8}$$

Definition 2.12.2. *Let D be an open and convex subset of M, and let G be a C^1 vector field defined on D. We say: covariant derivative ∇G is Lipschitz if there exist a constant L for any geodesic γ, and $a, b \in R$ with $\gamma([a,b]) \subseteq D$ such that*

$$\left\| P(\gamma)_b^a \nabla G_{\gamma(b)} P(\gamma)_a^b - \nabla G_{\gamma(a)} \right\| \leq L \int_a^b \|\gamma'(t)\| \, dt, \tag{2.12.9}$$

where $P(\gamma)$ is the parallel transport along γ [94].

We use the notation $\nabla G \in \text{Lip}_L(D)$, and for the corresponding center-Lipschitz condition for $z_0 \in D$ fixed $\nabla G_{z_0} \in \text{Lip}_{L_0}(D)$.

Note that in general

$$L_0 \leq L \tag{2.12.10}$$

holds. Moreover $\frac{L}{L_0}$ can be arbitrarily large (see Section 2.2).

We can now show the following extension of the Newton-Kantorovich theorem on Riemannian manifolds using method (2.12.7):

Theorem 2.12.3. *Let D be an open and convex subset of a complete Riemannian manifold M. Let G be a continuous vector field defined on $\overline{D}$ that is C^1 on D with $\nabla G \in \text{Lip}_L(D)$, and for $z_0 \in D$ fixed $\nabla G_{z_0} \in \text{Lip}_{L_0}(D)$.*

Assume:

$$\nabla G_{z_0} \text{ is invertible;}$$

there exist constants c_0 and c_1 such that

$$\left\|\nabla G_{z_0}^{-1}\right\| \leq c_0, \quad \left\|\nabla G_{z_0}^{-1} G(z_0)\right\| \leq c_1 \tag{2.12.11}$$

$$h_0 = c_0 c_1 C \leq \tfrac{1}{2}, \quad c = \tfrac{L_0 + L}{2}, \tag{2.12.12}$$

and

$$U(z_0, t^*) \subseteq D \tag{2.12.13}$$

where,

$$t^* = \lim_{n \to \infty} t_n, \tag{2.12.14}$$

$$t_0 = 0, \quad t_1 = c_1, \quad t_{n+2} = t_{n+1} + \frac{L(t_{n+1} - t_n)^2}{2(1 - L_0 t_{n+1})} \quad (n \geq 0). \tag{2.12.15}$$

Then
(a) sequence $\{t_n\}$ $(n \geq 0)$ is monotonically increasing and converges to t^* with

$$t^* \leq 2n; \tag{2.12.16}$$

(b) sequence $\{z_n\}$ $(n \geq 0)$ generated by Newton's method (2.12.7) is well defined, remains in $U(z_0, t^*)$ for all $n \geq 0$, and converges to z^*, which is the unique singularity of G in $\overline{U}(z_0, t^*)$. Moreover if strict inequality holds in (2.12.10), z^* is the unique singularity of G in $U(z_0, 2n)$. Furthermore the following error bounds hold:

$$d(z_{n+1}, z_n) \leq t_{n+1} - t_n; \tag{2.12.17}$$

and

$$d(z_n, z^*) \leq t^* - t_n \quad (n \geq 0). \tag{2.12.18}$$

Proof. Simply use L_0 instead of L where the use of the center-Lipschitz (and not L) suffices in the proof of Theorem 3.1 in [94] (e.g., in the computation of an upper bound on $\left\|\nabla G_{z_n}^{-1}\right\|$).

Remark 2.12.4. If equality holds in (2.12.10) then Theorem 2.12.3 reduces to Theorem 3.1 in [94]. Denote the corresponding Newton-Kantorovich-type hypothesis there by:

$$h = c_0 c_1 \leq \tfrac{1}{2}. \tag{2.12.19}$$

By (2.12.10), (2.12.12), and (2.12.18) we see

$$h \leq \tfrac{1}{2} \Longrightarrow h_0 \leq \tfrac{1}{2} \tag{2.12.20}$$

but not vice versa unless if equality holds in (2.12.10).

The rest of the claims made at the introduction can now follow along the same lines of our work in Section 2.2 [39].

2.13 Computation of shadowing orbits

In this section, we are concerned with the problem of approximating shadowing orbits for dynamical systems. It is well-known in the theory of dynamical systems that actual computations of complicated orbits rarely produce good approximations to the trajectory. However under certain conditions, the computed section of an orbit lies in the shadow of a true orbit. Hence using product spaces and a recent result of ours (Section 2.2) [39], we show that the sufficient conditions for the convergence of Newton's method to a true orbit can be weakened under the same computational cost as in the elegant work by Hadeller in [108]. Moreover the information on the location of the solutions is more precise and the corresponding error bounds are finer.

Let f be a Fréchet-differentiable operator defined on an open convex subset D of a Banach space X with values in X.

The operator f defines a local dynamical system as follows:

$$x_{n+1} = f(x_k) \quad (x_0 \in D) \tag{2.13.1}$$

as long as $x_k \in D$.

A sequence $\{x_m\}_{i=0}^{N}$ in D with $x_{m+1} = f(x_m)$, $i = 0, ..., N-1$ is called an orbit. Any sequence $\{x_m\}_{i=0}^{N}$, $x_m \in D$, $m = 0, ..., N$ is called a pseudo-orbit of length N.

We can now pass to product spaces. Let $y = X^{N+1}$ equipped with maximum norm. The norm $\mathbf{x} = (x_0, ..., x_N) \in Y$ is given by

$$\|\mathbf{x}\| = \max_{0 \leq m \leq N} \|x_m\|.$$

Set $S = D^{N+1}$. Let $F: S \to Y$ be an operator associated with $\dot{f}$:

$$F(\mathbf{x}) = F \begin{bmatrix} x_0 \\ \vdots \\ x_N \end{bmatrix} = \begin{bmatrix} f(x_0) - x_1 \\ \vdots \\ f(x_N) \end{bmatrix}.$$

Assume there exist constants l_0, l, L_0, L such that:

$$\|f'(u) - f'(x_0)\| \le l_0 \|u - x_0\|$$
$$\|f'(u) - f'(v)\| \le l \|u - v\|$$
$$\|F'(\mathbf{u}) - F'(\mathbf{x_0})\| \le L_0 \|\mathbf{u} - \mathbf{x_0}\|$$
$$\|F'(\mathbf{u}) - F'(\mathbf{v})\| \le L \|\mathbf{u} - \mathbf{v}\|$$

for all $u, v \in D$, $\mathbf{u}, \mathbf{v} \in S$.

From now on we assume: $l_0 = L_0$ and $l = L$.

For $y \in X$ define an operator

$$F_y : S \to Y$$

by

$$F_y(\mathbf{x}) = \left\| \begin{matrix} f(x_0) - x_1 \\ \vdots \\ f(x_N) - y \end{matrix} \right\|.$$

It follows that $F_y'(\mathbf{x}) = F'(\mathbf{x})$.

As in [108], define the quantities

$$a(\mathbf{x}) = \max_{0 \le i \le N} \sum_{j=i}^{N} \|f'(x_i) \dots f'(x_j)\|^{-1},$$

$$b(\mathbf{x}) = \max_{0 \le i \le N-1} \left\| \sum_{j=i}^{N-1} f'(x_i) \dots f'(x_j) (f(x_j) - x_{j+1}) \right\|,$$

and

$$b_y(\mathbf{x}) = b(\mathbf{x}) + \|f'(x_N)(f(x_N) - y)\|$$

for $x \in Y$.

That is $a(x)$ is the operator norm of $F'(x)^{-1}$ and by (x) is the norm of the Newton convection $F'(\mathbf{x})^{-1} F_y(\mathbf{x})$.

Remark 2.13.1. The interpretation to the measures studied in [61], [108] is given by:

(a) The dilation measures $a(\mathbf{x})$ and by $(\mathbf{x})$ are the norm of $F'(x)^{-1}$ and the norm of the Newton-correction $F'(x)^{-1} F_y(x)$, respectively;

(b) the solution of equation

$$F_y(\mathbf{x}) = 0$$

yields a section $(x_0, \dots, x_N)$ of length $N + 1$ of a true orbit that meets the prescribed point y at the Nth iteration step [108].

Using a weak variant of the Newton-Kantorovich theorem, we recently showed in [39] (see Section 2.2) we obtain the following existence and uniqueness result for a true orbit:

Theorem 2.13.2. *Let* $\mathbf{x} \in Y$, $y \in X$ $a(\mathbf{x})$, *by* $(\mathbf{x})$ *be as above and* $\{x_i\}_{i=0}^{N}$ *be a pseudo-orbit.*
   Assume:

$$h_0 = \bar{L}, \quad a(\mathbf{x}) \, by(\mathbf{x}) \le \tfrac{1}{2}, \quad \bar{L} = \tfrac{L_0 + L}{2} \tag{2.13.2}$$

and

$$\bar{U}\left(\mathbf{x}_0 = \mathbf{x}, \ r^* = 2by(\mathbf{x}) = \left\{\mathbf{z} \in Y \mid \|\mathbf{x} - \mathbf{z}\| \le r^*\right\}\right) \subseteq S.$$

Then there is a unique true orbit $\mathbf{x}^* = \left(x_0^*, ..., x_N^*\right)$ *inside* $U(\mathbf{x}, r^*)$ *satisfying* $f\left(x_N^*\right) = y$.

We also have a more neutral form of Theorem 2.13.2:

Theorem 2.13.3. *Let* $\{x_i\}_{i=0}^{N}$ *be a pseudo-orbit of length* $N + 1$. *Assume:*

$$h_0^1 = \bar{L} \, a(\mathbf{x}) \, b(\mathbf{x}) \le \tfrac{1}{2} \tag{2.13.3}$$

and

$$\bar{U}\left(\mathbf{x}, r_1^* = 2b(x)\right) \subseteq S,$$

where $a(\mathbf{x})$, $b(\mathbf{x})$, r^*, $\bar{L}$ *are as defined above. Then there is a unique true orbit*

$$\mathbf{x}^* = \left(x_0^*, ..., x_N^*\right) \in U\left(\mathbf{x}, r_1^*\right)$$

satisfying $f\left(x_N^*\right) = f(x_N)$.

Remark 2.13.4. If

$$L_0 = L, \tag{2.13.4}$$

then Theorems 2.13.2 and 2.13.3 reduce to Theorems 1 and 2 in [108], respectively. However in general

$$L_0 \le L. \tag{2.13.5}$$

The conditions corresponding with (2.13.2) and (2.13.3), respectively, in Theorem 1 and 2 in [108] are given by

$$h = La(\mathbf{x}) \, by(\mathbf{x}) \le \tfrac{1}{2} \tag{2.13.6}$$

and

$$h^1 = L \, a(\mathbf{x}) \, b(\mathbf{x}) \le \tfrac{1}{2}. \tag{2.13.7}$$

It follows from (2.13.2), (2.13.3), (2.13.5), (2.13.6), and (2.13.7) that:

$$h \le \tfrac{1}{2} \Longrightarrow h_0 \le \tfrac{1}{2} \tag{2.13.8}$$

$$h^1 \le \tfrac{1}{2} \Longrightarrow h_0^1 \le \tfrac{1}{2} \tag{2.13.9}$$

but not vice versa unless if (2.13.4) holds. Hence we managed to weaken the sufficient convergence conditions given in [108], and under the same computational cost, as the evaluation of L requires in precise the evaluation of L_0.

Moreover the information on the location of the true orbit is more precise and the corresponding error bounds are finer [39] (see also Section 2.2).

2.14 Computation of continuation curves

In this study, we are concerned with approximating a locally unique solution x^* of the nonlinear equation

$$F(x) = 0, \tag{2.14.1}$$

where F is a continuously Fréchet-differentiable operator defined on an open convex subset D of R^m (a positive integer) in to R^m.

In recent years, a number of approaches have been proposed for the numerical computation of continuation curves, and with techniques for overcoming turning points [175], [205]. It turns out that all numerical continuation methods are of the predictor-corrector type. That is, information on the already computed portion of the curve is used to calculate an extrapolation approximating an additional curve portion. At the end, a point on the so constructed curve is chosen as the initial guess for the corrector method to converge to some point of the continuation curve.

Consider the system of n equations

$$F(x) = F\left(x^0\right), x \in \mathcal{R}(F), x_0 \in \mathcal{R}(F), \tag{2.14.2}$$

together with the (popular) choice [77], [176]

$$u^T x = z, \tag{2.14.3}$$

where u is derived from fixing the value of one of the variables. For example, set $u = e^i$, where e^i is the ith unit-basis vector of R^{n+1}.

System (2.14.2)–(2.14.3) can now be rewritten as

$$G(x) = 0, \tag{2.14.4}$$

where

$$G(x) = \begin{bmatrix} F(x) - F\left(x^0\right) \\ \left(e^i\right)^T x - z \end{bmatrix} \tag{2.14.5}$$

with z a not known yet constant.

Clearly, for $T(x) = u$:

$$\det G'(x) = \det \begin{bmatrix} F'(x) \\ \left(e^i\right)^T \end{bmatrix} = \begin{bmatrix} F'(x) \\ (T(x))^T \end{bmatrix} \left[I + T(x)\left(e^i - T(x)\right)^T \right]$$

$$= \begin{bmatrix} T(x)^T e^i \end{bmatrix} \det \begin{bmatrix} F'(x) \\ (T(x))^T \end{bmatrix}. \tag{2.14.6}$$

Therefore i should be chosen so that $\left| T(x)^T e^i \right|$ is as large as possible.

Here we address the length of the step-size. In particular, we show that under the same hypotheses and computational cost as before, we can enlarge the step size of the iteration process [39] (see Section 2.2). This observation is important in computational mathematics.

As in the elegant paper by Rheinboldt [176], we use NK method as the corrector method.

We need the following local convergence result of ours concerning the radius of convergence for NK method [39]:

Lemma 2.14.1. *Let* $G: D \subseteq R^m \rightarrow R^m$ *be a Fréchet-differentiable operator. Assume: there exists a solution* x^* *of equation* $G(x) = 0$ *such that* $G'(x^*)^{-1}$ *is irrevertible;*

$$\left\| G'\left(x^*\right)^{-1} \left[G'(x) - G'(y)\right] \right\| \leq \ell \|x - y\| \tag{2.14.7}$$

$$\left\| G'\left(x^*\right)^{-1} \left[G'(x) - G'\left(x^*\right)\right] \right\| \leq \ell_0 \|x - x^*\| \tag{2.14.8}$$

for all $x, y \in D$;
and

$$\overline{U}\left(x^*, r_A\right) \subseteq D, \tag{2.14.9}$$

where

$$r_A = \frac{2}{\ell_0 + \ell}. \tag{2.14.10}$$

Then NK method applied to G is well defined, remains in $U(x^*, r_A)$, *and converges to* x^* *provided that* $x_0 \in U(x^*, r_A)$.

Moreover the following error bounds hold for all $n \geq 0$:

$$\|x_{n+1} - x^*\| \leq \frac{\ell}{[1 - \ell_0 \|x_n - x^*\|]} \|x_n - x^*\|^2. \tag{2.14.11}$$

Remark 2.14.2. In general

$$\ell_0 \leq \ell \tag{2.14.12}$$

holds.

The corresponding radius r_R given by Rheinboldt [175]:

$$r_R = \frac{2}{3\ell} \tag{2.14.13}$$

is smaller than r_A if strict inequality holds in (2.14.12). Consequently, the step-size used with Newton's method as corrector can be increased (as it depends on r_A).

Indeed as in [176], the Lipschitz conditions (2.14.7) and (2.14.8) hold in compact subset C of R^{n+1}. We can have:

$$\left\| G'(x)^{-1} \right\| \leq \left(1 + \frac{2}{|T(x)^T e^i|}\right) \sqrt{1 + b(x)}, \tag{2.14.14}$$

(see (4.9) in [176]),
where,

$$b(x) = \| F'(x) (F'(x)^T)^{-1} \|_2. \tag{2.14.15}$$

If $x^* \in R(F)$ is n solution of equation (2.14.5), let

$$\tau(x^*) = \max_{j=1,\dots,n} \left| T(x^*) e^j \right| \delta(x^*, C) = \text{dist}(x^*, \delta C). \tag{2.14.16}$$

For $\theta \in (0, 1) \cup (x^*, r_A (x^*))$ with

$$r_A (x^*) = \min \left\{ \delta (x^*, C), \frac{2\theta\tau(x^*)}{(2+\tau(x^*))(2\ell_0+\ell)\sqrt{1+b(x^*)^2}} \right\} \geq r_R (x^*), \quad (2.14.17)$$

where ℓ_0, ℓ depend on C (not D) but we used the same symbol, and $r_R (x^*)$ is defined as r_A (with $\ell_0 = \ell$ in (2.14.17)) (clearly $U x^*, r_A (x^*) \subseteq \mathcal{R} (F)$). By Lemma 2.14.1, NK iteration for (2.14.5) converges to x^*.

Continuing as in [176], let

$$x: I \rightarrow R \rightarrow \mathcal{R} (F), \left\| x' (s) \right\|_2 = 1 \text{ for all } x \in I, x (s_0) = x^0, s_0 \in I \quad (2.14.18)$$

be the unique C^1 operator—parameterized in terms of the path length—that solves equation (2.14.2). We use the Euler-line

$$y_E (s) = x (s_0) + T (x) (s_0) (s - s_0) \quad s \in I \quad (2.14.19)$$

as predictor with Newton's method as corrector. Let $x^k \in \mathcal{R} (F)$ be n known approximation to $x (s_k), s_k \in I$; then one step of the process is:

1. Compute $T (x^k)$;
2. Determine i such that $\left| (e^i)^T (x^k) \right| = \max_{j=1,...,n+1} \left| (e^j)^T (T (x^*)) \right|$;
3. Choose the step-size $h_{k+1} > 0$;
4. Compute the predicted point $y = x^k + h_{k+1} (T (x^k))$;
5. Apply Newton's method to (2.14.5) with $z = (e^i)^T (y)$ with y as starting point;
6. If "satisfactory convergence," then $x^{k+1} = $ last iterate;

$$\text{else replace } h_{k+1} \text{ by } q h_k \quad (2.14.20)$$

for some $q \in (0, 1)$ and go to step 4.

7. $S_{k+1} = S_k + \left\| x^{k+1} - x^k \right\|_2$.

(A) Assume: We want to compute $x: \overline{I}_0 \rightarrow \mathcal{R} (F)$ of (2.14.18) for $\overline{I}_0 = [\underline{s}, \overline{s}] \subset I, \underline{s} < \overline{s}$. There exists $\delta > 0$ such that

$$C = \left\{ x \in R^{n+1} / \text{dist} (x, x (\overline{I}_0)) \leq \delta \right\} \subseteq \mathcal{R} (F). \quad (2.14.21)$$

We can have:

$$r_A (x) \geq r_A^0 = \min \left(\delta, \frac{\theta\tau_0}{(2+\tau_0)(2\ell_0+\ell)\sqrt{1+(\overline{b})^2}} \right)$$

$$\geq r_R^0 = \min \left(\delta, \frac{2\theta\tau_0}{3(2+\tau_0)\sqrt{1+(\overline{b})^2}} \right) > 0 \quad (2.14.22)$$

for all $s \in \overline{I}_0, \theta \in (0, 1)$,
where

$$\bar{b} = \sup \left\{ b\left(x\left(s \right) \right), s \in \bar{I}_0 \right\} < \infty,$$

$$\tau_0 = \inf \left\{ \max_{j=1,\dots,n+1} \left| T\left(x \right)^T e^j \right|, x \in C \right\} > 0. \tag{2.14.23}$$

(B) Moreover assume:

approximation x^k of $x\left(s_k \right)$, $s_k \in \bar{I}_0$ satisfies:

$$\left(e^i \right)^T \left(x^k - x\left(s_k \right) \right) = 0, \quad \left\| x^k - x\left(s_k \right) \right\|_2 \le \frac{\min\left(\delta, r_A^0 \right)}{2} \tag{2.14.24}$$

(C) Furthermore assume:

$$n_k = \min \left\{ \bar{s} - s_k, \tfrac{1}{2}\tau_0 \delta, \tfrac{\tau_0}{2\ell_1} \right\} > 0, \tag{2.14.25}$$

where ℓ_1 is the Lipschitz constant of T on C. For any point $x\left(s_k + \sigma \right)$ on (2.14.18) with $\sigma \in \bar{I}_k = [s_k, s_k + n_k]$ there exists $y = x^k + g\left(\sigma \right) T\left(x^k \right)$ on the Euler line with the same ith component, i.e., point y with

$$g\left(\sigma \right) = \frac{\left(e^i \right)^T \left(x(s_k+\sigma)-x^k \right)}{\left(e^i \right)^T T(x^k)}. \tag{2.14.26}$$

By Rheinboldt [176] we have:

$$\left| g\left(\sigma \right) \right| \le \tfrac{1}{2}\delta, \tag{2.14.27}$$

$$x^k + g\left(\sigma \right) T\left(x^k \right) \in C \text{ for all } \sigma \in \bar{I}_k \tag{2.14.28}$$

and

$$y = x^k + h_{k+1} T\left(x^k \right) \in U\left(x\left(s_k + \sigma_k \right), r_0 \right) \tag{2.14.29}$$

with

$$h_{k+1}^A = g\left(\sigma_k^A \right) \ge g\left(\sigma_k^R \right) \tag{2.14.30}$$

$$0 < \sigma_k^A = \min \left(n_k, \left[\frac{\tau_0 r_A^0}{\ell_1(1+\tau_0)} \right]^{1/2} \right) \ge \sigma_k^R = \min \left(n_k, \left[\frac{\tau_0 r_R^0}{\ell_1(1+\tau_0)} \right]^{1/2} \right).$$
$$\tag{2.14.31}$$

Hence the convergence of NK method for (2.14.5) from y to $x\left(s_k + \sigma_k^A \right)$ is ensured.

Define

$$\sigma_*^A = \min \left(\tfrac{1}{2}\tau_0 \delta, \tfrac{\tau_0}{2\ell_1}, \left[\frac{\tau_0 r_A^0}{\ell_1(1+\tau_0)} \right]^{1/2} \right)$$

$$\ge \min \left(\tfrac{1}{2}\tau_0 \delta, \tfrac{\tau_0}{2\ell_1}, \left[\frac{\tau_0 r_R^0}{\ell_1(1+\tau_0)} \right]^{1/2} \right) = \sigma_*^R. \tag{2.14.32}$$

Then we get

$$\sigma_*^A = \sigma_*^A \text{ for } 0 \le s_k \le \bar{s} - \sigma_* \tag{2.14.33}$$

and for $s_k \in \left[\bar{s} - s^*, \bar{s} \right]$ we reach $\bar{s}$ in one step, whereas interval $\bar{I}_0$ is traversed in finitely many steps.

Hence we showed as in Theorem 4.2 in [176]:

Theorem 2.14.3. *Under hypotheses (A)–(C) there exists $s_k \in \bar{I}_0$, a step $h_{k+1} > 0$ along the euler line such that Newton's method of step 5 is well defined and converges to some $x \left(s_k + \sigma_k^A \right)$, $\sigma_k^A > 0$. Starting from $s_0 = \underline{s}$, we can choose h_k^A $k = 0, 1, \ldots$ such that $s_k = \underline{\sigma} + k\sigma_*^A$, $k = 0, 1, \ldots, M^A$, $s_{m+1} = \bar{s}$ with a constant $\sigma_*^A > 0$ for which*

$$M^A \sigma_*^A \leq \bar{\sigma} - \underline{\sigma} \leq \left(M^A + 1 \right) \sigma_*^A. \tag{2.14.34}$$

Remark 2.14.4. Under hypotheses of Theorem 2.14.3 and Theorem 4.2 in [176], because of (2.14.17), (2.14.22), (2.14.31), and (2.14.32) (if strict inequality holds in (2.14.12) for C instead of D), we conclude:

$$h_k^R \leq h_k^A \tag{2.14.35}$$

$$\sigma_*^R \leq \sigma_*^A \tag{2.14.36}$$

$$\sigma_k^R \leq \sigma_k^A \tag{2.14.37}$$

and

$$M^A \leq M^R. \tag{2.14.38}$$

Estimates (2.14.35)–(2.14.38) justify the claims made in the introduction about the improvements on the step-size. Note also that strict inequalities will hold in (2.14.35)–(2.14.38) if the "minimum" is expressed in terms of r_0^A in the definition of the above quantities (see (2.14.22)).

Some comments on a posteriori, asymptotic estimates are given next:

Remark 2.14.5. Rheinboldt also showed [176, p. 233] that if the solution (2.14.18) of equation (2.14.2) is three times continuously Fréchet-differentiable on the open interval I, then σ should be chosen by

$$\sigma^R = \theta \sqrt{\frac{\rho_R}{\left\| w^k - \gamma_k T \left(x^k \right) \right\|_2}} \tag{2.14.39}$$

where w^k, γ_k are given (4.27) and (4.29) in [176, p. 233], $\theta \in (0, 1)$ and ρ_R is a "safe" radius of convergence of NK method at $x \left(s_k + \sigma^R \right)$. Because again our corresponding radius of convergence ρ_A is such that

$$\rho_R < \rho_A \tag{2.14.40}$$

we deduce (if strict inequality holds in (2.14.12)):

$$\sigma^R < \sigma^A, \tag{2.14.41}$$

where σ^A is given by (2.14.39) for ρ_R replaced by ρ_A.

2.15 Gauss-Newton method

In this section, we are concerned with the problem of approximating a point x^* minimizing the objective operator

$$Q(x) = \tfrac{1}{2}\|F(x)\|_2^2 = \tfrac{1}{2}F^T(x)F(x) \qquad (2.15.1)$$

where F is a Fréchet-differentiable regression operator defined on an open subset D of R^j with values in R^m ($j \leq m$).

It is well-known that for x^* to be a local minimum, it is necessary to be a zero of the gradient ∇Q of Q, too.

That is why Ben-Israel [46] suggested the so called Gauss-Newton method:

$$x_{n+1} = x_n - J^+(x_n)F(x_n) \quad (n \geq 0), \qquad (2.15.2)$$

where, $J(x) = F'(x)$, the Fréchet derivative of F. Here M^+ denotes the pseudo inverse of a matrix M satisfying:

$$\left(M^+M\right)^T = M^+M,\ \left(MM^+\right)^T = MM^+,\ M^+MM^+ = M^+,\ MM^+M = M. \tag{2.15.3}$$

Moreover, if rank-(m, j) matrix M is of full rank, then its pseudo inverse becomes

$$M^+ = \left(M^T M\right)^{-1} M^T. \qquad (2.15.4)$$

A semilocal convergence analysis for method (2.15.2) has already been given in the elegant paper in [110]. However, we noticed that under weaker hypotheses, we can provide a similar analysis with the following advantages over the ones in [110], and under the same computational cost:

(a) our results apply whenever the ones in [110] do but not vice versa;
(b) error bounds $\|x_{n+1} - x_n\|$, $\|x_n - x^*\|$ ($n \geq 0$) are finer;
(c) the information on the location of the solution x^* is more precise.

The results obtained here can be naturally extended to hold in arbitrary Banach spaces using outer or generalized inverses [59] (see also Chapter 8).

We need the following result on majorizing sequences for method (2.15.2).

Lemma 2.15.1. *Let $a \geq 0$, $b > 0$, $c \geq 0$, $L_0 \geq 0$, $L \geq 0$ be given parameters. Assume there exists $d \in [0, 1)$ with $c \leq d$ such that for all $k \geq 0$*

$$\left[\tfrac{1}{2}bL(1-d)d^k + dbL_0\left(1 - d^{k+1}\right)\right]a + (c - d)(1 - d) \leq 0, \qquad (2.15.5)$$

and

$$\frac{bL_0a}{1-d}\left(1 - d^k\right) < 1. \qquad (2.15.6)$$

Then, iteration $\{s_n\}$ ($n \geq 0$) given by

$$s_0 = 0,\ s_1 = a,\ s_{n+2} = s_{n+1} + \frac{\tfrac{1}{2}bL(s_{n+1}-s_n)+c}{1-bL_0 s_{n+1}} \cdot (s_{n+1} - s_n) \qquad (2.15.7)$$

is nondecreasing, bounded above by $s^{**} = \frac{a}{1-d}$, *and converges to some* s^* *such that*

$$0 \leq s^* \leq s^{**}. \tag{2.15.8}$$

Moreover, the following estimates hold for all $n \geq 0$:

$$0 \leq s_{n+2} - s_{n+1} \leq d\,(s_{n+1} - s_n) \leq d^{n+1}a. \tag{2.15.9}$$

Proof. We shall show using induction that for all $k \geq 0$

$$\tfrac{1}{2}bL\,(s_{k+1} - s_k) + dbL_0 s_{k+1} + c \leq d, \tag{2.15.10}$$
$$s_{k+1} - s_k \geq 0, \tag{2.15.11}$$

and

$$1 - bL_0 s_{k+1} > 0. \tag{2.15.12}$$

Using (2.15.5)–(2.15.7), estimates (2.15.10)–(2.15.12) hold. But then (2.15.7) gives

$$0 \leq s_2 - s_1 \leq d\,(s_1 - s_0).$$

Let us assume (2.15.9)–(2.15.12) hold for all $k \leq n + 1$.
We can have in turn

$$\tfrac{1}{2}bL\,(s_{k+2} - s_{k+1}) + dbL_0 s_{k+2} + c \leq \tag{2.15.13}$$
$$\leq \tfrac{1}{2}bLd^{k+1} + dbL_0 \left[s_1 + d\,(s_1 - s_0) + d^2\,(s_1 - s_0) + \cdots + d^{k+1}\,(s_1 - s_0) \right] + c$$
$$\leq \tfrac{1}{2}bLd^{k+1} + dbL_0 \tfrac{1-d^{k+2}}{1-d}a + c \leq d \quad \text{(by (2.15.5))}.$$

Moreover we show:

$$s_k \leq s^{**}. \tag{2.15.14}$$

For $k = 0, 1, 2$, $s_0 = 0 \leq s^{**}$, $s_1 = a \leq s^{**}$, $s_2 \leq a + da = (1 + d)\,a \leq s^{**}$.
It follows from (2.15.9) that for all $k \leq n + 1$

$$s_{k+2} \leq s_{k+1} + d\,(s_{k+1} - s_k) \leq \cdots \leq s_1 + d\,(s_1 - s_0) + \cdots + d\,(s_{k+1} - s_k) \tag{2.15.15}$$

$$\leq \left[1 + d + d^2 + \cdots + d^{k+1} \right] a = \tfrac{1-d^{k+2}}{1-d}a \leq s^{**}.$$

Furthermore, we get

$$bL_0 s_{k+1} \leq bL_0 \tfrac{1-d^{k+1}}{1-d}a < 1. \tag{2.15.16}$$

Finally (2.15.9), (2.15.11) hold by (2.15.7), (2.15.13)–(2.15.16).
The induction is now complete.
Hence, sequence $\{s_n\}$ $(n \geq 0)$ is nondecreasing and bounded above by s^{**}, and as such it converges to some s^* satisfying (2.15.8).
That completes the proof of the Lemma.

We can show the following main semilocal convergence result for method (2.15.2).

Theorem 2.15.2. *Let $F: D_0 \subseteq D \subseteq R^j \to R^m$ be a Fréchet-differentiable operator, where D_0 is a convex set. Assume:*
there exists $x_0 \in D_0$ with rank $(J(x_0)) = r \leq m, r \geq 1$ *and* rank $(J(x)) \leq r$ *for all $x \in D_0$;*

$$\|J^+(x_0) F(x_0)\| \leq a, \tag{2.15.17}$$
$$\|J(x) - J(x_0)\| \leq L_0 \|x - x_0\|, \tag{2.15.18}$$
$$\|J(x) - J(y)\| \leq L \|x - y\|, \tag{2.15.19}$$
$$\|J^+(x_0)\| \leq b, \tag{2.15.20}$$
$$\|J^+(y) q(x)\| \leq c(x) \|x - y\| \tag{2.15.21}$$

with $q(x) = (I - J(x) J^+(x)) F(x)$, and $q(x) \leq c < 1$, for all $x, y \in D_0$; conditions (2.15.5) and (2.15.6) hold; and

$$\overline{U}(x_0, s^*) \subseteq D_0, \tag{2.15.22}$$

where s^* is defined in Lemma 2.15.1.
    Then,

(a) *sequence $\{x_n\}$ $(n \geq 0)$ generated by method (2.15.2) is well defined, remains in $U(x_0, s^*)$ for all $n \geq 0$, and converges to a solution $x^* \in \overline{U}(x_0, s^*)$ of equation $J^+(x) F(x) = 0$;*
(b) rank $(J(x)) = r$ *for all $x \in U(x_0, s^*)$;*
(c) rank $(J(x^0)) = r$ *if strict inequality holds in (2.15.5) or equality and $c > 0$.*

    Moreover the following estimates hold for all $n \geq 0$

$$\|x_{n+1} - x_n\| \leq s_{n+1} - s_n, \tag{2.15.23}$$

and

$$\|x_n - x^*\| \leq s^* - s_n. \tag{2.15.24}$$

Furthermore, if

$$\text{rank}(J(x_0)) = m, \quad \text{and } F(x^*) = 0, \tag{2.15.25}$$

then x^* is the unique solution of equation $F(x) = 0$ in $U(x_0, s^{**})$, and the unique zero of equation $J^+(x) F(x) = 0$ in $U(x_0, s^*)$.

Proof. We shall show $\{s_n\}$ $(n \geq 0)$ is a majorizing sequence for $\{x_n\}$ so that estimate (2.15.23) holds, and iterates $s_n \in U(x_0, s^*)$ $(n \geq 0)$.
    It follows from the Banach Lemma, and the estimate

$$\|J(x) - J(x_0)\| \leq L_0 \|x - x_0\| \leq L_0 s^* < 1 \quad \text{(by (2.15.6))}$$

for all $x \in U(x_0, s^*)$ that (b) and (c) above hold, with

$$\left\| J^+ (x) \right\| \leq \frac{b}{1-bL\|x-x_0\|} \text{ for all } x \in U\left(x_0, s^*\right).$$ (2.15.26)

Consequently, operator

$$P(x) = x - J^+ (x) F(x)$$ (2.15.27)

is well defined on $U(x_0, s^*)$. If $x, P(x) \in U(x_0, s^*)$ using (2.15.2), (2.15.17)–(2.15.21) we can obtain in turn:

$$\| P(P(x)) - P(x) \| =$$

$$= \left\| J^+ (P(x)) \int_0^1 \{ J(x + t(P(x) - x)) - J(x) \} (P(x) - x) \, dt \right.$$ (2.15.28)

$$\left. + J^+ \left(P(x) \left(I - J(x) J^+ (x) \right) \right) F(x) \right\|$$

$$\leq \frac{1}{1-bL_0\|P(x)-x_0\|} \left(\tfrac{1}{2} bL \| P(x) - x \| + c \right) \| P(x) - x \|.$$

Estimate (2.15.23) holds for $n = 0$ by the initial conditions. Assuming by induction: $\|x_i - x_{i-1}\| \leq s_i - s_{i-1}$ $(i = 1, 2, ..., k)$ it follows

$$\|x_i - x_0\| \leq s_k - s_0 \text{ for } i = 1, 2, ..., k.$$ (2.15.29)

Hence, we get $\{x_n\} \subset (x_0, s^*)$.

It follows from (2.15.7) and (2.15.29) that (2.15.23) holds for all $n \geq 0$.

That is $\{x_n\}$ $(n \geq 0)$ is a Cauchy sequence in R^m and as such it converges to some $x^* \in \overline{U}(x_0, s^*)$ (because $\overline{U}(x_0, s^*)$ is a closed set).

Using the continuity of $J(x)$, $F(x)$, and the estimate

$$\left\| J^+ F(x_k) \right\| \leq \left\| J^+ (x^*) \left(I - J(x_k) J^+ (x_k) F(x_k) \right) \right\|$$

$$+ \left\| J^+ (x^*) \right\| \cdot \left\| J(x_k) J^+ (x_k) F(x_k) \right\|$$

$$\leq c \left\| x_k - x^* \right\| + \left\| J^+ (x^*) \right\| \left\| J(x_k) \right\| \left\| x_{k+1} - x_k \right\|$$ (2.15.30)

we conclude $J^+ (x^*) F(x^*) = 0$.

The uniqueness part follows exactly as in Theorem 2.4 in [110] (see also [39] or Section 2.2, or Theorem 12.5.5 in [154]).

Remark 2.15.3. Conditions (2.15.5), (2.15.6) are always present in the study of Newton-type methods. We wanted to leave conditions (2.15.5) and (2.15.6) as uncluttered as possible. We may replace (2.15.5) and (2.15.6) by the stronger

$$\left[\tfrac{1}{2} bL (1 - d) + dbL_0 \right] a + (c - d)(1 - d) \leq 0$$ (2.15.31)

and

$$\frac{bL_0 a}{1-d} < 1,$$ (2.15.32)

respectively. Clearly conditions (2.15.5) and (2.15.6) are weaker than the Newton-Kantorovich-type hypothesis

$$h = \frac{abL}{(1-c)^2} \leq \tfrac{1}{2} \qquad (2.15.33)$$

used in Theorem 2.4 in [110, p. 120].

Indeed first of all

$$L_0 \leq L \qquad (2.15.34)$$

holds in general. If equality holds in (2.15.35), then iteration $\{s_n\}$ reduces to $\{t_n\}$ $(n \geq 0)$ in [110] (simply set $L_0 = L$ in (2.15.7)), and Theorem 2.15.2 reduces to Theorem 2.4 in [110]. However, if strict inequality holds in (2.15.34), then our estimates on the distances $\|x_{n+1} - x_n\|$, $\|x_n - x^*\|$ are more precise than the ones in [110]. Indeed we immediately get

$$s_{n+1} - s_n < t_{n+1} - t_n \qquad (n \geq 1), \qquad (2.15.35)$$
$$s^* - s_n \leq t^* - t_n \qquad (n \geq 0) \qquad (2.15.36)$$

and

$$s^* \leq t^*. \qquad (2.15.37)$$

For $c = 0$ and $d = \tfrac{1}{2}$, conditions (2.15.5) and (2.15.6) hold provided that

$$h_1 = abL_1 \leq \tfrac{1}{2} \qquad (2.15.38)$$

where,

$$L_1 = \frac{L_0 + L_1}{2}. \qquad (2.15.39)$$

Corresponding condition (e) in Theorem 2.4 in [110] becomes the famous Newton-Kantorovich hypothesis

$$h_2 = abL \leq \tfrac{1}{2}. \qquad (2.15.40)$$

Note that (2.15.39) is weaker than (2.15.41) if strict inequality holds in (2.15.35). Hence, we have

$$h_2 \leq \tfrac{1}{2} \implies h_1 \leq \tfrac{1}{2} \qquad (2.15.41)$$

but not necessarily vice versa unless if $L_0 = L$.

Remark 2.15.4. Along the lines of our comments above, the corresponding results in [110, pp. 122–124] can now be improved (see also Section 2.2).

2.16 Exercises

2.16.1. Show that f defined by $f(x, y) = |\sin y| + x$ satisfies a Lipschitz condition with respect to the second variable (on the whole xy-plane).

2.16.2. Does f defined by $f(t, x) = |x|^{1/2}$ satisfy a Lipschitz condition?

2.16.3.

(a) Let $F: D \subseteq X \to X$ be an analytic operator. Assume:

- there exists $\alpha \in [0, 1)$ such that

$$\|F'(x)\| \leq \alpha \quad (x \in D); \qquad (2.16.1)$$

- $$\gamma = \sup_{\substack{k>1 \\ x \in D}} \left\| \tfrac{1}{k!} F^{(k)}(x) \right\|^{\frac{1}{k-1}} \qquad \text{is finite;}$$

- there exists $x_0 \in D$ such that

$$\|x_0 - F(x_0)\| \le \eta \le \tfrac{3-\alpha-2\sqrt{2-\alpha}}{\gamma}, \quad \gamma \ne 0;$$

- $\bar{U}(x_0, r_1) \subseteq D$, where, r_1, r_2 with $0 \le r_1 \le r_2$ are the two zeros of function f, given by

$$f(r) = \gamma(2 - \alpha)r^2 - (1 + \eta\gamma - \alpha)r + \eta.$$

Show: method of successive substitutions is well defined, remains in $\bar{U}(x_0, r_1)$ for all $n \ge 0$ and converges to a fixed point $x^* \in \bar{U}(x_0, r_1)$ of operator F. Moreover, x^* is the unique fixed point of F in $\bar{U}(x_0, r_2)$. Furthermore, the following estimates hold for all $n \ge 0$:

$$\|x_{n+2} - x_{n+1}\| \le \beta \|x_{n+1} - x_n\|$$

and

$$\|x_n - x^*\| \le \tfrac{\beta^n}{1-\beta}\eta,$$

where

$$\beta = \tfrac{\gamma\eta}{1-\gamma\eta} + \alpha.$$

The above result is based on the assumption that the sequence

$$\gamma_k = \left\| \tfrac{1}{k!} F^{(k)}(x) \right\|^{\frac{1}{k-1}} \quad (x \in D), \quad (k > 1)$$

is bounded above by γ. This kind of assumption does not always hold. Let us then not assume sequence $\{\gamma_k\}$ $(k > 1)$ is bounded and define "function" f_1 by

$$f_1(r) = \eta - (1 - \alpha)r + \sum_{k=2}^{\infty} \gamma_k^{k-1} r^k.$$

(b) Let $F: D \subseteq X \to X$ be an analytic operator. Assume (2.16.1) holds and for $x_0 \in D$ function f_1 has a minimum positive zero r_3 such that

$$\bar{U}(x_0, r_3) \subseteq D.$$

Show: method of successive substitutions is well defined, remains in $\bar{U}(x_0, r_3)$ for all $n \ge 0$ and converges to a unique fixed point $x^* \in \bar{U}(x_0, r_3)$ of operator F. Moreover the following estimates hold for all $n \ge 0$

$$\|x_{n+2} - x_{n+1}\| \le \beta_1 \|x_{n+1} - x_n\|$$

and

$$\|x_n - x^*\| \le \frac{\beta_1^n}{1-\beta_1}\eta,$$

where,

$$\beta_1 = \sum_{k=2}^{\infty} \gamma_k^{k-1}\eta^{k-1} + \alpha.$$

2.16.4.

(a) It is convenient to define:

$$\gamma = \sup_{k>1} \left\|\tfrac{1}{k!}F^{(k)}(x^*)\right\|^{\frac{1}{k-1}}$$

with $\gamma = \infty$, if the supremum does not exist. Let $F: D \subseteq X \to X$ be an analytic operator and $x^* \in D$ be a fixed point of F. Moreover, assume that there exists α such that

$$\|F'(x^*)\| \le \alpha, \qquad (2.16.2)$$

and

$$\bar{U}(x^*, r^*) \subseteq D,$$

where,

$$r^* = \begin{cases} \infty, & \text{if } \gamma = 0 \\ \frac{1}{\gamma}\cdot\frac{1-\alpha}{2-\alpha}, & \text{if } \gamma \ne 0. \end{cases}$$

Then, if

$$\beta = \alpha + \frac{\gamma r^*}{1-\gamma r^*} < 1,$$

show: the method of successive substitutions remains in $\bar{U}(x^*, r^*)$ for all $n \ge 0$ and converges to x^* for any $x_0 \in U(x^*, r^*)$. Moreover, the following estimates hold for all $n \ge 0$:

$$\|x_{n+1} - x^*\| \le \beta_n\|x_n - x^*\| \le \beta\|x_n - x^*\|,$$

where,

$$\beta_0 = 1, \quad \beta_{n+1} = \alpha + \frac{\gamma r^*\beta_n}{1-\gamma r^*\beta_n} \quad (n \ge 0).$$

The above result was based on the assumption that the sequence

$$\gamma_k = \left\|\tfrac{1}{k!}F^{(k)}(x^*)\right\|^{\frac{1}{k-1}} \quad (k \ge 2)$$

is bounded by γ. In the case where the assumption of boundedness does not necessarily hold, we have the following local alternative.

(b) Let $F: D \subseteq X \to X$ be an analytic operator and $x^* \in D$ be a fixed point of F. Moreover, assume: $\max_{r>0} \sum_{k=2}^{\infty}(\gamma_k r)^{k-1}$ exists and is attained at some $r_0 > 0$. Set

$$p = \sum_{k=2}^{\infty}(\gamma_k r_0)^{k-1};$$

there exist α, δ with $\alpha \in [0, 1), \delta \in (\alpha, 1)$ such that (2.16.2) holds,

$$p + \alpha - \delta \le 0$$

and

$$\bar{U}(x^*, r_0) \subseteq D.$$

Show: the method of successive substitutions $\{x_n\}$ $(n \ge 0)$ remains in $\bar{U}(x^*, r_0)$ for all $n \ge 0$ and converges to x^* for any $x_0 \in \bar{U}(x^*, r_0)$. Moveover the following error bounds hold for all $n \ge 0$:

$$\|x_{n+1} - x^*\| \le \alpha \|x_n - x^*\| + \sum_{k=2}^{\infty} \gamma_k^{k-1} \|x_n - x^*\|^k \le \delta \|x_n - x^*\|.$$

2.16.5. Let x^* be a solution of Equation (2.1.1). If the linear operator $F'(x^*)$ has a bounded inverse, and $\lim_{\|x-x^*\| \to 0} \|F'(x) - F'(x^*)\| = 0$, then show NK method converges to x^* if x_0 is sufficiently close to x^* and

$$\|x_n - x^*\| \le d\varepsilon^n \quad (n \le 0),$$

where ε is any positive number; d is a constant depending on x_0 and ε.

2.16.6. The above result cannot be strengthened, in the sense that for every sequence of positive numbers c_n such that: $\lim_{n \to \infty} \frac{c_{n+1}}{c_n} = 0$, there is an equation for which NK converges less rapidly than c_n. Define

$$s_n = \begin{cases} c_{n/2}, & \text{if } n \text{ is even} \\ \sqrt{c_{(n-1)/2} c_{(n+1)/2}}, & \text{if } n \text{ is odd.} \end{cases}$$

Show: $s_n \to 0$, $\frac{s_{n+1}}{s_n} \to 0$, and $\lim_{n \to \infty} \frac{c_n}{s_{n+k}} = 0$, $(k \ge 1)$.

2.16.7. Assume operator $F'(x)$ satisfies a Hölder condition

$$\|F'(x) - F'(y)\| \le a \|x - y\|^b,$$

with $0 < b < 1$ and $U(x_0, R)$. Define $h_0 = b_0 a \eta_0^b \le c_0$, where c_0 is a root of

$$\left(\frac{c}{1+b}\right)^b = (1 - c)^{1+b} \quad (0 \le c \le 1)$$

and let $R \ge \frac{\eta_0}{1-d_0} = r_0$, where $d_0 = \frac{h_0}{(1+b)(1-h_0)}$. Show that NK method converges to a solution x^* of Equation $F(x) = 0$ in $U(x_0, r_0)$.

2.16.8. Let K, B_0, η_0 be as in Theorem 2.2.4. If $h_0 = b_0 \eta_0 K < \frac{1}{2}$, and

$$r_0 = \frac{1 - \sqrt{1 - 2h_0}}{h_0} \eta_0 \le r.$$

Then show: modified Newton's method (2.1.5) converges to a solution $x^* \in U(x_0, r_0)$ of Equation (2.1.1). Moreover, if

$$r_0 \le r < \frac{1 + \sqrt{1 - 2h_0}}{h_0} \eta_0,$$

then show: Equation (2.1.1) has a unique solution x^* in $U(x_0, r)$. Furthermore show: $\bar{x}_{n+1} = \bar{x}_n - F'(x_0)^{-1} F(\bar{x}_n)$ $(n \ge 0)$ converges to a solution x^* of Equation (2.1.1) for any initial guess $\bar{x}_0 \in U(x_0, r)$.

2.16.9. Under the hypotheses of Theorem 2.2.4, let us introduce $\bar{U} = \bar{U}(x_1, r_0 - \eta)$, sequence $\{t_n\}$ $(n \geq 0)$, $t_0 = 0$, $t_{n+1} = t_n - \frac{f(t_n)}{f'(t_n)}$, $f(t) = \frac{1}{2}Kt^2 - t + \eta$, $\Delta = r^* - r_0$, $\theta = \frac{r_0}{r^*}$, $\nabla t_{n+1} = t_{n+1} - t_n$, $d_n = \|x_{n+1} - x_n\|$, $\Delta_n = \|x_n - x_0\|$, $\bar{U}_0 = \bar{U}$, $\bar{U}_n = \bar{U}(x_n, r_0 - t_n)$ $(n \geq 1)$, $K_0 = L_0 = K$,

$$K_n = \sup_{\substack{x,y \in \bar{U}_n \\ x \neq y}} \frac{\|F'(x_n)^{-1}(F'(x)-F'(y))\|}{\|x-y\|} \quad (n \geq 1),$$

$$L_n = \sup_{\substack{x,y \in \bar{U} \\ x \neq y}} \frac{\|F'(x_n)^{-1}(F'(x)-F'(y))\|}{\|x-y\|} \quad (n \geq 1),$$

$$\underline{\lambda}_n = \frac{2d_n}{1+\sqrt{1+2L_n d_n}} \quad (n \geq 0),$$

$$\lambda_n = \frac{2d_n}{1+\sqrt{1-2L_n d_n}} \quad (n \geq 0), \quad \underline{\kappa}_n = \frac{2d_n}{1+\sqrt{1+2K_n d_n}},$$

$$k_n = \frac{2d_n}{1+\sqrt{1-2K_n d_n}} \quad (n \geq 0),$$

$$s_0 = 1, \quad s_n = \frac{s_{n-1}^2}{2^{n-1}\sqrt{1-2h}+s_{n-1}(1-\sqrt{1-2h})^{2^n-1}} \quad (n \geq 0).$$

With the notation introduced above show (Yamamoto [206]):

$$\|x^* - x_n\| \leq K_n \, (n \geq 0) \leq \lambda_n \, (n \geq 0)$$

$$\leq \frac{2d_n}{1 + \sqrt{1 - 2K(1 - K\Delta_n)^{-1}d_n}} \quad (n \geq 0)$$

$$\leq \frac{2d_n}{1 + \sqrt{1 - 2K(1 - Kt_n)^{-1}d_n}} \quad (n \geq 0)$$

$$= \frac{2d_n}{1 + \sqrt{1 - 2KB_n d_n}} \quad (n \geq 0)$$

$$= \begin{cases} \dfrac{2d_n}{1+\sqrt{1-\frac{4}{\Delta} \cdot \frac{1-\theta^{2^n}}{1+\theta^{2^n}} d_n}} & (2h < 1) \\[3mm] \dfrac{2d_n}{1+\sqrt{1-\frac{2^n}{\eta} d_n}} & (2h = 1) \end{cases} \quad (n \geq 0)$$

$$\leq \frac{r_0 - t_n}{\nabla t_{n+1}} d_n \quad (n \geq 0)$$

$$= \frac{2d_n}{1 + \sqrt{1 - 2h_n}} \quad (n \geq 0)$$

$$\leq \frac{KB_n d_{n-1}^2}{1+\sqrt{1-2h_n}} \quad (n \geq 0)$$

$$= \frac{r_0 - t_n}{(\nabla t_n)^2} d_{n-1}^2 \quad (n \geq 0)$$

$$= \begin{cases} \dfrac{1-\theta^{2^n}}{\Delta} d_{n-1}^2 & (2h < 1) \\[3mm] \dfrac{2^{n-1}}{\eta} d_{n-1}^2 & (2h = 1) \, (n \geq 1) \end{cases}$$

$$\leq \frac{K d_{n-1}^2}{\sqrt{1-2h}+\sqrt{1-2h+(Kd_{n-1})^2}} \quad (n \geq 1)$$

$$\leq \frac{K \eta_{n-1} d_{n-1}}{\sqrt{1-2h}+\sqrt{1-2h+(K\eta_{n-1})^2}} \quad (n \geq 1)$$

$$= e^{-2^{n-1}\varphi} d_{n-1} \quad (n \geq 1)$$

$$= \theta^{2^{n-1}} d_{n-1} \quad (n \geq 1)$$

$$= \frac{r_0 - t_n}{\nabla t_n} d_{n-1} \quad (n \geq 1)$$

$$\leq r_0 - t_n \quad (n \geq 0)$$

$$= \frac{2\eta_n}{1 + \sqrt{1-2h_n}} \quad (n \geq 0)$$

$$= \begin{cases} e^{-2^{n-1}\varphi \frac{\sinh \varphi}{\sinh 2^{n-1}\varphi} \eta} & (2h < 1) \\ 2^{1-n}\eta & (2h < 1) \end{cases}$$

$$= \begin{cases} \dfrac{\Delta \theta^{2^n}}{1 - \theta^{2^n}} & (2h < 1) \\ 2^{1-n}\eta & (2h = 1)\,(n \geq 0) \end{cases}$$

$$= \frac{s_n}{2^n K} \left(\frac{2h}{1 + \sqrt{1-2h}} \right)^{2^n} \quad (n \geq 0)$$

$$\leq \frac{1}{2^n} K \left(\frac{2h}{1 + \sqrt{1-2h}} \right)^{2^n} \quad (n \geq 0)$$

$$\leq \frac{1}{2^{n-1}} (2h)^{2n-1} \eta \quad (n \geq 0),$$

$$\left\| x^* - x_n \right\| \leq \lambda_n \quad (n \geq 0)$$

$$\leq \frac{L_n d_{n-1}^2}{1 + \sqrt{1-(L_n d_{n-1})^2}} \quad (n \geq 1)$$

$$\leq \frac{L_{n-1} d_{n-1}^2}{1 - L_{n-1} d_{n-1} + \sqrt{1 - 2L_{n-1} d_{n-1}}} \quad (n \geq 1)$$

$$\leq \frac{L_{n-1} d_{n-1}^2}{1 - L_{n-1} d_{n-1}} \quad (n \geq 1),$$

$$\left\| x^* - x_n \right\| \leq \lambda_n$$

$$\leq \frac{2 d_n}{1 + \sqrt{1 - 2L_0 (1 - L_0 \Delta_n)^{-1} d_n}}$$

$$\leq \frac{2\|F'(x_0)^{-1} F(x_n)\|}{1 - L_0\Delta_n + \sqrt{(1 - L_0\Delta_n)^2 - 2L_0\|F'(x_0)^{-1} F(x_n)\|}}$$

$$\leq \frac{L_0 d_{n-1}^2}{1 - L_0\Delta_n + \sqrt{(1 - L_0\Delta_n)^2 - (L_0 d_{n-1})^2}},$$

$$\|x^* - x_n\| \geq \underline{\kappa}_n \ (n \geq 0) \geq \underline{\lambda}_n \ (n \geq 0)$$

$$\geq \frac{2d_n}{1 + \sqrt{1 + 2K(1 - K\Delta_n)^{-1} d_n}} \quad (n \geq 0)$$

$$\geq \frac{2d_n}{1 + \sqrt{1 + 2K(1 - Kt_n)^{-1} d_n}} \quad (n \geq 0)$$

$$= \frac{2d_n}{1 + \sqrt{1 + 2K B_n d_n}} \quad (n \geq 0)$$

$$= \frac{2d_n}{1 + \sqrt{1 + 4 \cdot \frac{r_0 - t_{n+1}}{(r_0 - t_n)^2} d_n}} \quad (n \geq 0)$$

$$= \frac{2d_n}{1 + \sqrt{1 + 4 \cdot \frac{\nabla t_{n+1}}{(\nabla t_n)^2} d_n}} \quad (n \geq 0)$$

$$= \frac{2d_n}{1 + \sqrt{1 + \frac{2K d_n}{\sqrt{1 - 2h + (K\eta_{n-1})^2}}}} \quad (n \geq 0)$$

$$\geq \frac{2d_n}{1 + \sqrt{1 + \frac{2K d_n}{\sqrt{1 - 2h + (K d_{n-1})^2}}}} \quad (n \geq 0)$$

$$= \frac{2d_n}{1 + \sqrt{1 + \frac{2d_n}{\sqrt{a^2 + d_{n-1}^2}}}} \quad \left(a = \sqrt{1 - 2h}/K, n \geq 1\right)$$

$$\geq \frac{2d_n}{1 + \sqrt{1 + \frac{2d_n}{d_n + \sqrt{a^2 + d_n^2}}}} \quad (n \geq 0)$$

$$\geq \frac{2d_n}{1 + \sqrt{1 + 2h_n}} \quad (n \geq 0)$$

$$= \frac{2d_n}{1 + \sqrt{1 + \frac{4\theta^{2^n}}{(1 + \theta^{2^n})^2}}} \quad (n \geq 0),$$

$$\|x^* - x_{n+1}\| \leq \kappa_{n+1} \leq \kappa_n - d_n \leq \frac{r_0 - t_{n+1}}{\nabla t_{n+1}} d_n,$$

$$\|x^* - x_{n+1}\| \leq \lambda_{n+1} \leq \lambda_n - d_n \leq \frac{r_0 - t_{n+1}}{\nabla t_{n+1}} d_n,$$

$$d_n \leq \tfrac{1}{2} K_n d_{n-1}^2$$
$$\leq \tfrac{1}{2} L_n d_{n-1}^2$$
$$\leq \tfrac{1}{2} K \left(1 - K \Delta_n\right)^{-1} d_{n-1}^2$$
$$\leq \tfrac{1}{2} K \left(1 - K \Delta_{n-1} - K d_{n-1}\right)^{-1} d_{n-1}^2$$
$$\leq \tfrac{1}{2} K \left(1 - K t_n\right)^{-1} d_{n-1}^2$$
$$= \tfrac{1}{2} K B_n d_{n-1}^2$$
$$= \frac{r_0 - t_{n+1}}{(r_0 - t_n)^2} d_{n-1}^2$$
$$= \frac{\nabla t_{n+1}}{(\nabla t_n)^2} d_{n-1}^2$$
$$= \frac{d_{n-1}^2}{2\sqrt{a^2 + \eta_{n-1}^2}}$$
$$\leq \frac{d_{n-1}^2}{2\sqrt{a^2 + d_{n-1}^2}}$$
$$\leq \frac{\nabla t_{n+1}}{\nabla t_n} d_{n-1}$$
$$= \frac{\eta_n}{\eta_{n-1}} d_{n-1}$$
$$= \frac{1}{2 \cosh 2^{n-1}\varphi} d_{n-1}$$
$$\leq \tfrac{1}{2} d_{n-1}$$
$$\leq \tfrac{1}{2} \eta_{n-1} = \tfrac{1}{2} \nabla t_n,$$

and

$$d_n \leq \eta_n$$
$$= \nabla t_{n+1}$$
$$= (r_0 - t_{n+1}) \theta^{-2^n}$$
$$= (r_0 - t_{n+1}) e^{2^n} \varphi$$
$$= \begin{cases} \dfrac{\sinh \varphi}{\sinh 2^n \varphi} \eta & (2h < 1) \\ 2^{-n} \eta & (2h = 1) \end{cases}$$
$$= \frac{\Delta \theta^{2^n}}{1 - \theta^{2^{n+1}}} \quad (2h < 1).$$

3

Applications of the Weaker Version of the NK Theorem

There is a very extensive literature on popular results that have used the NK Theorem 2.2.4 on specific real-life problems. Here we provide an incomplete list of the most popular of them and replace the NK Theorem by its weaker version introduced in Section 2.2. The advantages of this approach have already been explained in detail in Section 2.2.

3.1 Comparison of Kantorovich and Moore theorems

In this section, we are concerned with the problem of approximating a locally unique solution x^* of equation

$$F(x) = 0 \qquad (3.1.1)$$

where $F: D \subseteq \mathbb{R}^k \to \mathbb{R}^k$ is continuously differentiable on an open convex set D, and k is a positive integer.

Rall in [171] compared the theorems of Kantorovich and Moore. This comparison showed that the Kantorovich theorem has only a slight advantage over the Moore theorem with regard to sensitivity and precision, whereas the latter requires less computational cost. Later Neumaier and Shen [146] showed that when the derivative in the Krawczyk operator is replaced with a suitable slope, then the corresponding existence theorem is at least as effective as the Kantorovich theorem with respect to sensitivity and precision. At the same time, Zuhe and Wolfe [148] showed that the hypotheses in the affine invariant form of the Moore theorem are always implied by the Kantorovich theorem but not necessarily vice versa.

Here we show that this implication is not true in general for a weaker version of the Kantorovich theorem shown in Section 2.2.

We will need the following semilocal convergence theorem for NK method due to Deuflhard and Heindl [78]:

Theorem 3.1.1. *Let F be a Fréchet-differentiable operator defined on an open convex subset D of a Banach space X with values in a Banach space Y. Suppose that $x_0 \in D$ is such that:*

I.K. Argyros, *Convergence and Applications of Newton-type Iterations*,
DOI: 10.1007/978-0-387-72743-1_3, © Springer Science+Business Media, LLC 2008

$$F'(x_0)^{-1} \in L(Y, X), \tag{3.1.2}$$

$$\|F'(x_0)^{-1}F(x_0)\| \leq \eta, \tag{3.1.3}$$

$$\|F'(x_0)^{-1}[F'(x) - F'(y)]\| \leq \ell\|x - y\| \quad \text{for all } x, y \in D, \tag{3.1.4}$$

$$h = 2\ell\eta \leq 1, \tag{3.1.5}$$

and

$$\bar{U}(x_0, r_1) \subseteq D \tag{3.1.6}$$

where,

$$r_1 = \frac{1 - \sqrt{1 - h}}{\ell} \quad (\ell \neq 0). \tag{3.1.7}$$

Then sequence $\{x_n\}$ generated by Newton's method

$$x_{n+1} = x_n - F'(x_n)^{-1}F(x_n) \quad (x_0 \in D) \quad (n \geq 0) \tag{3.1.8}$$

is well defined, remains in $U(x_0, r_1)$ for all $n \geq 0$, and converges to a solution x^* of equation $F(x) = 0$ that is unique in $\bar{U}(x_0, r_1) \cup (D \cap U(x_0, r_2))$, where,

$$r_2 = \frac{1 + \sqrt{1 - h}}{\ell}. \tag{3.1.9}$$

Moreover the following error bounds hold for all $n \geq 0$:

$$\|x_{n+1} - x_n\| \leq s_{n+1} - s_n \tag{3.1.10}$$

and

$$\|x_n - x^*\| \leq r_1 - s_n, \tag{3.1.11}$$

where

$$s_0 = 0, \quad s_1 = \eta, \quad s_{n+2} = s_{n+1} + \frac{\ell(s_{n+1} - s_n)^2}{2(1 - \ell - s_{n+1})} \quad (n \geq 0). \tag{3.1.12}$$

Remark 3.1.2. In the case of Moore's theorem [143] suppose that hypotheses of Theorem 3.1.1 are valid with $X = Y = \mathbb{R}^k$, $x_0 = z = \text{mid}(\underline{x}_\gamma)$, where $\underline{x}_\gamma \in I(D)$ is given by

$$\underline{x}_\gamma = U_\infty[z, \gamma] = [z - \gamma e, z + \gamma e], \ e = (1, \ldots, 1)^T \in \mathbb{R}^n, \ \gamma > 0, \tag{3.1.13}$$

and define the Krawczyk operator

$$\underline{K}(\underline{x}_\gamma) = w - \gamma\{F'(z)^{-1}\underline{F}[z, \underline{x}_\gamma] - I\}[-e, e], \tag{3.1.14}$$

where

$$w = z - F'(z)^{-1}F(z), \tag{3.1.15}$$

and

$$\underline{F}[z, \underline{x}_\gamma] = \int_0^1 \underline{F'}(z + t(\underline{x}_\gamma - z))dt \tag{3.1.16}$$

in which integration is defined as in [143].

The following result relates a weaker version of Theorem 3.1.1 with Moore's theorem.

Theorem 3.1.3. *Assume for $x_0 = z = \text{mid}(\underline{x}_\gamma)$: hypotheses (3.1.2) and (3.1.3) of Theorem 3.1.1 hold;*

$$\| F'(x_0)^{-1} \left[F'(x) - F'(x_0) \right] \| \leq \ell_0 \| x - x_0 \| \quad \text{for all } x \in D; \tag{3.1.17}$$

$$\bar{h} = 2\ell_0 \eta \leq 1; \tag{3.1.18}$$

and

$$\bar{U}(x_0, r_3) \subseteq D, \tag{3.1.19}$$

where,

$$r_3 = \frac{1 - \sqrt{1 - \bar{h}}}{\ell_0} \quad (\ell_0 \neq 0). \tag{3.1.20}$$

Then

$$\underline{K}\left(\underline{x}_\gamma\right) \subseteq \underline{x}_\gamma, \tag{3.1.21}$$

where $\underline{x}_\gamma$, $\underline{K}$ are given by (3.1.13) and (3.1.14)–(3.1.16), respectively.

Proof. If $u \in \underline{K}(\underline{x}_\gamma)$, then $u = w + v$

$$v = \gamma \left\{ \left[F'(z)^{-1} \underline{F}\left[z, \underline{x}_\gamma \right] - I \right\} [-e, e] \tag{3.1.22}$$

(by (3.1.15) and (3.1.16)).
We have in turn

$$\| z - u \|_\infty \leq \| z - w \|_\infty + \| v \|_\infty$$

$$\leq \eta + \gamma \| F'(z)^{-1} \{ \underline{F}\left[z, \underline{x}_\gamma \right] - F'(z) \} \|_\infty$$

$$\leq \eta + \tfrac{1}{2} \ell_0 \gamma^2. \tag{3.1.23}$$

Hence (3.1.21) holds if

$$\eta + \tfrac{1}{2} \ell_0 \gamma^2 \leq \gamma, \tag{3.1.24}$$

which is true by (3.1.18)–(3.1.20).
That completes the proof of the Theorem.

Remark 3.1.4. (a) Note that Moore's theorem [143] guarantees the existence of a solution x^* of equation (3.1.1) if (3.1.21) holds.
(b) Theorem 3.1.3 reduces to Theorem 2 in [146] if $\ell_0 = \ell$. However in general

$$\ell_0 \leq \ell \tag{3.1.25}$$

holds. Note also that

$$h \leq 1 \Rightarrow \bar{h} \leq 1 \tag{3.1.26}$$

but not vice versa unless if $\ell_0 = \ell$. Hence our Theorem 3.1.3 weakens Theorem 2 in [146] and can be used in cases not covered by the latter.

An example was given in [171] and [146] to show that the hypotheses of the affine invariant form of the Moore theorem may be satisfied even when those of Theorem 3.1.1 are not.

Example 3.1.5. Let $k = 1$, $a \in \left[0, \frac{1}{2}\right)$, $x \in \underline{x}_\gamma = [a, 2 - a]$ and define function $F : \underline{x}_\gamma \to \mathbb{R}$ by

$$F(x) = x^3 - a. \tag{3.1.27}$$

Choose $z = \text{mid}(\underline{x}_\gamma)$ then (3.1.5) becomes

$$h = \frac{4}{3}(1 - a)(2 - a) > 1 \quad \text{for all } a \in \left[0, \frac{1}{2}\right). \tag{3.1.28}$$

That is, there is no guarantee that NK method generated by (3.1.8) converges to a solution of $F(x) = 0$, as the Newton-Kantorovich hypothesis (3.1.5) is violated. However using (3.1.13)–(3.1.16) and (3.1.27) we get

$$\underline{K}(\underline{x}_\gamma) = \frac{1}{3}\left[a^3 - 6a^2 + 10a + 2, -a^3 + 3a^2 - 8a + 6\right] \tag{3.1.29}$$

and if

$$a \in \left[.44, \frac{1}{2}\right) \tag{3.1.30}$$

then (3.1.21) holds. That is Moore's theorem guarantees convergence of NK method (3.1.8) to a solution x^* of equation $F(x) = 0$ provided that (3.1.30) holds.

However we can do better. Indeed, because $\ell_0 = 3 - a$

$$\bar{h} = \frac{2}{3}(1 - a)(3 - a) \leq 1 \tag{3.1.31}$$

if

$$a \in \left[\frac{4 - \sqrt{10}}{2}, \frac{1}{2}\right), \tag{3.1.32}$$

which improves (3.1.30).

Remark 3.1.6. If (3.1.5) holds as a strict inequality NK method (3.1.8) converges quadratically to x^*. However (3.1.18) guarantees only the linear convergence to x^* of the modified MNK method.

$$y_{n+1} = y_n - F'(y_0)^{-1} F(y_n) \quad (y_0 = x_0), \ (n \geq 0). \tag{3.1.33}$$

In practice, the quadratic convergence of NK method is desirable. So we wonder if it is possible to find a condition weaker than (3.1.5) (but probably stronger than (3.1.18)) so that the quadratic convergence of Newton's method is guaranteed.

3.2 Comparison of Kantorovich and Miranda theorems

In this section, we are concerned with the problem of approximating a locally unique solution x^* of an equation

$$F(x) = 0, \tag{3.2.1}$$

where F is defined on an open convex subset S of $\mathbb{R}^n$ (n a positive integer) with values in $\mathbb{R}^n$.

NK method

$$x_{m+1} = x_m - F'(x_m)^{-1} F(x_m) \quad (x_0 \in S) \ (m \geq 0) \tag{3.2.2}$$

has been used to generate a sequence approximating x^*.

Here we first weaken the generalization of Miranda's theorem (Theorem 4.3 in [136]). Then we show that operators satisfying the weakened Newton-Kantorovich conditions satisfy those of the weakened Miranda's theorem.

In order for us to compare our results with earlier ones, we need to list the following theorems guaranteeing the existence of solution x^* of equation (3.2.1) (see Chapter 2).

Theorem 3.2.1. *Let $F: S \to R^n$ be a Fréchet-differentiable operator. Assume:*
there exists $x_0 \in S$ such that $F'(x_0)^{-1} \in L(Y, X)$, and set

$$G(x) = F'(x_0)^{-1} F(x) \quad (x \in S); \tag{3.2.3}$$

there exists an $\eta \geq 0$ such that

$$\|G(x_0)\| \leq \eta; \tag{3.2.4}$$

there exists an $\ell > 0$ such that

$$\|G'(x) - G'(y)\| \leq \ell \|x - y\| \quad \text{for all } x, y \in S; \tag{3.2.5}$$
$$h = 2\ell\eta \leq 1, \tag{3.2.6}$$

and

$$\bar{U}(x_0, r^*) \subseteq S, \tag{3.2.7}$$

where,

$$r^* = \frac{1 - \sqrt{1 - h}}{\ell}. \tag{3.2.8}$$

Then there exists a unique solution $x^* \in \bar{U}(x_0, r^*)$ of equation $F(x) = 0$.

Remark 3.2.2. Theorem 3.2.1 is the portion of the famous Newton-Kantorovich theorem (see Chapter 2). The following theorem is due to Miranda [142], which is a generalization of the intermediate value theorem:

Theorem 3.2.3. *Let $b > 0$, $x_0 \in R^n$, and define*

$$Q = \{x \in \mathbb{R}^n \mid \|x - x_0\|_\infty \le b\}, \tag{3.2.9}$$

$$Q_k^+ = \{x \in Q : x_k = x_{0,k} + b\}, \tag{3.2.10}$$

and

$$Q_k^- = \{x \in Q : x_k = x_{0,k} - d\} \quad \text{for } k = 1, \dots, n. \tag{3.2.11}$$

Let $F = (F_k)$ $(1 \le k \le n): Q \to R^n$ be a continuous operator satisfying for all $k = 1, 2, \dots, n$,

$$F_k(x) F_k(y) \le 0 \quad \text{for all } x \in Q_k^+ \text{ and } y \in Q_k^-. \tag{3.2.12}$$

Then there exists $x^* \in Q$ such that $F(x^*) = 0$.

The following result connected Theorems 3.2.1 and 3.2.3.

Theorem 3.2.4. *Suppose $F: S \to R^n$ satisfies all hypotheses of Theorem 3.2.1 in the maximum norm, then G satisfies the conditions of Theorem 3.2.4 on $\bar{U}(x_0, r^*)$.*

In the elegant study [136], a generalization of Theorem 3.2.4 was given (Theorem 4.3). We first weaken this generalization.

Let $\mathbb{R}^n$ be equipped with a norm denoted by $\| \cdot \|$ and $\mathbb{R}^{n \times n}$ with a norm $\| \cdot \|$ such that $\|M \cdot x\| \le \|M\| \cdot \|x\|$ for all $M \in \mathbb{R}^{n \times n}$ and $x \in \mathbb{R}^n$. Choose constants $c_0, c_1 > 0$ such that for all $x \in \mathbb{R}^n$

$$c_0 \|x\|_\infty \le \|x\| \le c_1 \|x\|_\infty, \tag{3.2.13}$$

since all norms on finite-dimensional spaces are equivalent.

Set

$$c = \frac{c_0}{c_1} \le 1. \tag{3.2.14}$$

Definition 3.2.5. *Let $S \subseteq \mathbb{R}^n$ be an open convex set, and let $G: S \to \mathbb{R}^n$ be a differentiable operator on S. Let $x_0 \in S$, and assume:*

$$G'(x_0) = I \quad \text{(the identity matrix)} \tag{3.2.15}$$

there exists $\eta \ge 0$ such that

$$\|G(x_0)\| \le \eta; \tag{3.2.16}$$

there exists an $\ell_0 \ge 0$ such that

$$\|G'(x) - G'(x_0)\| \le \ell_0 \|x - x_0\| \quad \text{for all } x \in S. \tag{3.2.17}$$

Define:

$$h_0 = 2\ell_0 \eta. \tag{3.2.18}$$

We say that G satisfies the weak center-Kantorovich conditions in x_0 if

$$h_0 \le 1. \tag{3.2.19}$$

We also say that G satisfies the strong center-Kantorovich conditions in x_0 if

$$h_0 \le \frac{c^2}{2}.$$ (3.2.20)

Moreover define

$$r_1 = \frac{c - \sqrt{c^2 - h_0}}{\ell_0},$$ (3.2.21)

$$r_2 = \frac{c + \sqrt{c^2 - h_0}}{\ell_0}$$ (3.2.22)

and

$$R = [r_1, r_2] \quad \text{for } \ell_0 \neq 0.$$ (3.2.23)

Furthermore if $\ell_0 = 0$, define

$$r_1 = \frac{\eta}{c}$$ (3.2.24)

and

$$R = [r_1, \infty).$$ (3.2.25)

Remark 3.2.6. The weak and strong center-Kantorovich conditions are equivalent only for the maximum norm.

As in [136], we need to define certain concepts. Let $r > 0$, $x_0 \in \mathbb{R}^n$, and define

$$U(r) = \{z \in \mathbb{R}^n \mid \|z\| \le r\},$$ (3.2.26)

$$U(x_0, r) = \{x = x_0 + z \in \mathbb{R}^n \mid z \in U(r)\},$$ (3.2.27)

$$U_k^+(r) = \{z \in \mathbb{R}^n \mid \|z\| = r, z_k = \|z\|_\infty\},$$ (3.2.28)

$$U_k^-(r) = \{z \in \mathbb{R}^n \mid \|z\| = r, z_k = -\|z\|_\infty\},$$ (3.2.29)

$$U_k^+(x_0, r) = \{x = x_0 + z \in \mathbb{R}^n \mid z \in U_k^+(r)\},$$ (3.2.30)

$$U_k^-(x_0, r) = \{x = x_0 + z \in \mathbb{R}^n \mid z \in U_k^-(r)\} \quad \text{for all } k = 1, 2, \ldots, n.$$ (3.2.31)

We show the main result:

Theorem 3.2.7. *Let $G: S \to R^n$ be a differentiable operator defined on an open convex subset of R^n. Assume G satisfies the strong center-Kantorovich conditions. Then, for any $r \in R$ with $U(x_0, r) \subseteq S$ the following hold:*

(a) $U = U(r) = U(x_0, r)$ *is a Miranda domain,*   (3.2.32)

and

$$U_1 = U_1(r) = \{U_1^+(x_0, r), \ U_1^-(x_0, r), \ldots, U_n^+(x_0, r), \ U_n^-(x_0, r)\}$$ (3.2.33)

is a Miranda partition [136] of the boundary ∂U. It is a canonical Miranda partition [136] for $r > 0$ and a trivial Miranda domain and partition for $r = 0$;

(b) $G_k(x) \geq 0$ for all $x \in U_k^+(x_0, r)$, $k = 1, \ldots, n$ (3.2.34)

and

$$G_k(x) \leq 0 \quad \text{for all } x \in U_k^-(x_0, r), \quad k = 1, \ldots, n; \quad (3.2.35)$$

(c) G satisfies the Miranda conditions on (U, U_1);

(d) if $G(x_0) = 0$ and $\ell_0 > 0$, then G satisfies the Miranda conditions on (U, U_1) for any $r \in \left[0, \frac{2c}{\ell_0}\right]$ such that $U(x_0, r) \subseteq S$.

Proof. The first point of the theorem follows exactly as in Theorem 4.3 in [136]. For the rest, we follow the proof of Theorem 4.3 in [136] (which is essentially the reasoning of Theorem 3.2.4) but with some differences stretching the use of center-Lipschitz condition (3.2.17) instead of the stronger Lipschitz condition (3.2.5), which is not really needed in the proof. However, it was used in both proofs mentioned above.

Using the intermediate value theorem for integration we first obtain the identity

$$G(x_0 + rz) - G(x_0)$$

$$= \int_0^1 G'(x_0 + rtz) rz \, dt$$

$$= \int_0^1 \left[G'(x_0 + rtz) - G'(x_0) \right] rz \, dt + \int_0^1 G'(x_0) rz \, dt$$

$$= \int_0^1 \left[G'(x_0 + rtz) - G'(x_0) \right] rz \, dt + rz \int_0^1 dt \quad (G'(x_0) = I). \quad (3.2.36)$$

Let e_k denote the kth unit vector. Then we can have:

$$G_k(x_0 + rz) = G_k(x_0) + \int_0^1 e_k^T \left[G'(x_0 + rtz) - G'(x_0) \right] rz \, dt + rz_k, \quad (3.2.37)$$

and by (3.2.17)

$$\left| \int_0^1 e_k^T \left[G'(x_0 + rtz) - G'(x_0) \right] rz \, dt \right| \leq \int_0^1 \left| e_k^T \left[G'(x_0 + rtz) - G'(x_0) \right] rz \, dt \right|$$

$$\leq \int_0^1 \frac{1}{c_1} \|(G'(x_0 + rtz) - G'(x_0)) rz\| \, dt$$

$$\leq \int_0^1 \frac{1}{c_1} \|G'(x_0 + rtz) - G'(x_0)\| \, \|rz\| \, dt$$

$$\leq \frac{1}{c_1} \int_0^1 \ell_0 \|rtz\| \, \|rz\| \, dt$$

$$= \frac{\ell_0 r^2}{c_1} \int_0^1 t \, dt = \frac{\ell_0 r^2}{2c_1}. \quad (3.2.38)$$

Let $z \in U_k^+(1)$. Using (3.2.38), and

$$|G_k(x_0)| \leq \frac{1}{c_1}\|G(x_0)\| \leq \frac{\eta}{c_1}, \tag{3.2.39}$$

$$z_k \geq \frac{1}{c_2} \quad \text{for } u \in U_k^+(1), \tag{3.2.40}$$

$$z_k \leq -\frac{1}{c_2} \quad \text{for } u \in U_k^-(1), \tag{3.2.41}$$

we get from (3.2.37)

$$G_k(x_0 + rz) \geq -|G(x_0)| - \frac{\ell_0 r^2}{2c_1} + rz_k \geq -\frac{\eta}{c_1} - \frac{\ell_0 r^2}{2c_1} + \frac{r}{c_2} \geq 0 \tag{3.2.42}$$

for $r \in R$ (by (3.2.20) and (3.2.23)). Similarly,

$$G_k(x_0 + rz) \leq |G(x_0)| + \frac{\ell_0 r^2}{2c_1} + rz_k \leq \frac{\eta}{c_1} + \frac{\ell_0 r^2}{2c_1} - \frac{r}{c_2} \leq 0 \tag{3.2.43}$$

for $r \in R$. If $G(x_0) = 0$, let $\eta = 0$, which implies $h_0 = 0$.

Remark 3.2.8. If $\ell = \ell_0$, then our Theorem 3.2.7 becomes Theorem 4.3 in [136]. Moreover if $\|\cdot\|$ is the maximum norm, then Theorem 3.2.7 becomes Theorem 3.2.4. However in general

$$\ell_0 \leq \ell. \tag{3.2.44}$$

Hence the strong Kantorovich condition is such that

$$h_1 = 2\ell\eta \leq \frac{c^2}{2}, \tag{3.2.45}$$

$$h_1 \leq \frac{c^2}{2} \Rightarrow h_0 \leq \frac{c^2}{2}. \tag{3.2.46}$$

Similarly, the Kantorovich condition (3.2.6) is such that

$$h \leq 1 \Rightarrow h_2 = 2\ell_0\eta \leq 1, \tag{3.2.47}$$

but not vice versa unless if $\ell_0 = \ell$. If strict inequality holds in (3.2.44) and conditions (3.2.45) or (3.2.6) are not satisfied, then the conclusions of Theorem 4.3 in [136] respectively do not necessarily hold. However if (3.2.9) holds, the conclusions of our Theorem 3.2.7 hold.

Remark 3.2.9. Condition (3.2.6) guarantees the quadratic convergence of NK method to x^*. However this is not the case for condition (3.2.19). To rectify this and still use a condition weaker than (3.2.6) (or (3.2.45)), define

$$p(\delta) = p_\delta = (\ell + \delta\ell_0)\eta, \quad \delta \in [0, 2). \tag{3.2.48}$$

We showed in Section 2.2 that if

$$p_\delta \le \delta, \tag{3.2.49}$$

$$\frac{2\ell_0\eta}{2-\delta} \le 1, \tag{3.2.50}$$

and

$$\frac{\ell_0\delta^2}{2-\delta} \le \ell \tag{3.2.51}$$

replace (3.2.6) then NK method converges to $x^* \in \bar{U}(x_0, r_3)$ with $r_3 \le r^*$. Moreover finer error bounds on the distances between iterates or between iterates and x^* are obtained. If we restrict $\delta \in [0, 1]$, then (3.2.50) and (3.2.51) hold if only (3.2.49) is satisfied. Choose $\delta = 1$ for simplicity in (3.2.49). Then again

$$h \le 1 \Rightarrow p_1 \le 1, \tag{3.2.52}$$

$$h_1 \le \frac{c^2}{2} \Rightarrow p_1 \le \frac{c^2}{2} \tag{3.2.53}$$

but not vice versa unless if $\ell = \ell_0$. Hence if (3.2.19) is replaced by

$$p_1 \le \frac{c^2}{2} \tag{3.2.54}$$

then all conclusions of Theorem 3.2.7 hold.

3.3 The secant method and nonsmooth equations

In this section, we are concerned with the problem of approximating a locally unique solution x^* of equation (2.1.1). Here we take D to be a closed convex subset of X.

The most popular iterative procedures for approximations x^* are the so-called Newton-like methods. The essence of these methods is to replace F by an approximate operator (linearization) that can be solved more easily.

When operator F is nonsmooth, the linearization is no longer available. In [180], a replacement was introduced through the notion of a point-based approximation (to be precised later). The properties of this approximation are similar to those of linearization and were successfully used for the NK method. However, we noticed (see the numerical example at the end of the section) that such an approximation may not exist. Therefore in order to solve a wider range of problems, we introduce a more flexible and precise point-based approximation that is more suitable for Newton-like methods and in particular for secant-type iterative procedures.

A local as well as a semilocal convergence analysis for the secant method is provided, and our approach is justified through numerical examples.

We need a definition of a point-based approximation (PBA) for operator F that is suitable for the secant method.

Definition 3.3.1. *Let F be an operator from a closed subset D of a metric space (X, d) into a normed linear space Y. Operator F has a (PBA) on D at the point*

$x_0 \in D$ *if there exists an operator* $A: D \times D \times D \to Y$ *and scalars* ℓ_0, ℓ *such that* u, v, w, x, y *and* z *in* D,

$$\|F(w) - A(u, v, w)\| \le \ell d(u, w) d(v, w), \tag{3.3.1}$$

$$\| [A(x, y, z) - A(x_0, x_0, z)] - [A(x, y, w) - A(x_0, x_0, w)] \| \tag{3.3.2}$$
$$\le \ell_0 [d(x, x_0) + d(y, x_0)] d(z, w),$$

and

$$\| [A(x, y, z) - A(u, v, z)] - [A(x, y, w) - A(u, v, w)] \|$$
$$\le \ell [d(x, u) + d(u, v)] d(z, w), \tag{3.3.3}$$

where x_0 *is a given point in* D.
   We then say A *is a (PBA) for* F.

   This definition is suitable for the application of the secant method. Indeed let X be also a normed linear space, D a convex set and F having a divided difference of order one on $D \times D$ denoted by $[x, y; F]$ and satisfying the standard condition (see Section 1.2):

$$\| [u, v; F] - [w, x; F] \| \le \ell(\|u - w\| + \|v - x\|) \tag{3.3.4}$$

for all u, v, w and x in D. If we set

$$A(u, v, w) = F(v) + [u, v; F] (w - v) \tag{3.3.5}$$

then (3.3.1) becomes

$$\|F(w) - F(v) - [u, v; F] (w - v)\| \le \ell \|u - w\| \|v - w\|, \tag{3.3.6}$$

whereas (3.3.2) and (3.3.3) are equivalent to property (3.3.4) of linear operator $[\cdot, \cdot; F]$. Note that a (PBA) does not imply differentiability.
   It follows by (3.3.1) that one way of finding a solution x^* of equation (2.1.1) is to solve for w the equation

$$A(x, y, w) = 0 \tag{3.3.7}$$

provided that x and y are given.
   We now need a definition also used in [179], [180], which amounts to the reciprocal of a Lipschitz constant for the inverse operator.

Definition 3.3.2. *Let* X, D, Y *and* F *be as in Definition 3.3.1, and let* $F: D \to Y$. *Then*

$$\delta(F, D) = \inf \left\{ \frac{\|F(u) - F(v)\|}{d(u, v)}, \ u \ne v, \ u, v \in D \right\}. \tag{3.3.8}$$

Clearly, if $\delta(F, D) \ne 0$, *then* F *is* $1 - 1$ *on* D. *We also define*

$$\delta_0(F, D) = \inf \left\{ \frac{\|F(u) - F(x_0)\|}{d(u, x_0)}, \ u \ne x_0, \ u, x_0 \in D \right\}.$$

Set $d = \delta(F, D)$ *and* $d_1 = \delta_0(F, D)$.

We state and prove the following generalization of the classic Banach Lemma on invertible operators:

Lemma 3.3.3. *Let X, D and Y be as in Definition 3.3.1. Assume further X is a Banach space. Let F and G be operators from D into Y with G being Lipschitzian with modulus ℓ and center-Lipschitzian with modulus ℓ_0. Let $x_0 \in D$ with $F(x_0) = y_0$. Assume that:*

$$\bar{U}(y_0, \alpha) \subseteq F(D); \tag{3.3.9}$$

$$0 \leq \ell < d; \tag{3.3.10}$$

$$\bar{U}(x_0, d_1^{-1}\alpha) \subseteq D, \tag{3.3.11}$$

and

$$\theta_0 = (1 - \ell_0 d_1^{-1})\alpha - \|G(x_0)\| \geq 0. \tag{3.3.12}$$

Then the following hold:

$$\bar{U}(y_0, \theta_0) \subseteq (F + G)(\bar{U}(x_0, d_1^{-1}\alpha)) \tag{3.3.13}$$

and

$$\delta(F + G, D) \geq d - \ell > 0. \tag{3.3.14}$$

Proof. Define operator $Ty(x) = F^{-1}(y - G(x))$, for each fixed $y \in \bar{U}(y_0, \theta_0)$, and $x \in \bar{U}(x_0, d_1^{-1}\alpha)$. We can get:

$$\|y - G(x) - y_0\| \leq \|y - y_0\| + \|G(x) - G(x_0)\| + \|G(x_0)\|$$

$$\leq \theta_0 + \ell_0 d_1^{-1}\alpha + \|G(x_0)\| = \alpha.$$

Therefore $Ty(x)$ is a singleton set as $d_1 \geq d > 0$. That is, Ty is an operator on $\bar{U}(x_0, d_1^{-1}\alpha)$. This operator maps $\bar{U}(x_0, d_1^{-1}\alpha)$ into itself. Indeed for $x \in \bar{U}(x_0, d_1^{-1}\alpha)$:

$$d(Ty(x), x_0) = d(F^{-1}(y - G(x)), F^{-1}(y_0)) \leq d_1^{-1}\alpha.$$

Moreover let u, v be in $\bar{U}(x_0, d_1^{-1}\alpha)$, then

$$d(Ty(u), Ty(v)) \leq d(F^{-1}(y - G(u)), F^{-1}(y - G(v)))$$

$$\leq d_1^{-1}\ell d(u, v). \tag{3.3.15}$$

It follows by the contraction mapping principle (see Section 1.3) and (3.3.11) that operator Ty is a strong contraction, and as such it has a fixed point $x(y)$ in $U(x_0, d_1^{-1}\alpha)$ with $(F + G)(x(y)) = y$. Such a point $x(y)$ in D is unique in D because

$$\delta(F + G, D) = \inf\left\{ \frac{\|[F(u) - F(v)] + [G(u) - G(v)]\|}{d(u, v)}, u \neq v, u, v \in D \right\}$$

$$\geq \delta(F, D) - \sup\left\{ \frac{\|G(u) - G(v)\|}{d(u, v)}, u \neq v, u, v \in D \right\} \geq d - \ell > 0.$$

That is $F + G$ is one-to-one on D.

Remark 3.3.4. In general

$$\ell_0 \le \ell, \quad \text{and} \quad d \le d_1 \tag{3.3.16}$$

hold and $\frac{\ell}{\ell_0}$, $\frac{d_1}{d}$ can be arbitrarily large (see Section 2.2). If equality holds in both inequalities in (3.3.16) then our Lemma 3.3.3 reduces to the corresponding Lemma 3.1 in [179, p. 298]. Otherwise our Lemma 3.3.3 improves (enlarges) the range for θ given in [179, p. 298], and under the same computational cost because in practice the computation of ℓ (or d) requires that of ℓ_0 (or d_1). This observation is important in computational mathematics.

The following lemma is used to show uniqueness of the solution in the semilocal case and convergence of secant method in the local case.

Lemma 3.3.5. *Let X and Y be normed linear spaces, and let D be a closed subset of X. Let $F: D \to Y$, and let A be a (PBA) for operator F on D at the point $x_0 \in D$. Denote by d the quantity $\delta(A(x_0, x_0, \cdot), D)$. If $U(x_0, \rho) \subseteq D$, then*

$$\delta(F, U(x_0, \rho)) \ge d - (2\ell_0 + \ell)\rho. \tag{3.3.17}$$

In particular, if $d - (2\ell_0 + \ell)\rho > 0$, then F is one-to-one on $U(x_0, \rho)$.

Proof. Let w, z be points in $U(x_0, \rho)$. We can write

$$F(w) - F(z) = [F(w) - A(x, y, w)] + [A(x, y, w) - A(x, y, z)] \\ + [A(x, y, z) - F(z)] \tag{3.3.18}$$

By (3.3.1) we can have

$$\|F(w) - A(x, y, w)\| \le \ell \|x - w\| \, \|y - w\|$$

and

$$\|F(z) - A(x, y, z)\| \le \ell \|x - z\| \, \|y - z\|.$$

Moreover we can find

$$\|A(x, y, u) - A(x, y, v)\| \ge \|A(x_0, x_0, u) - A(x_0, x_0, v)\| \\ - \|[A(x, y, u) - A(x_0, x_0, u)] - [A(x, y, v) - A(x_0, x_0, v)]\|$$

and therefore

$$\delta(A(x, y, \cdot), D) \\ \ge \delta(A(x_0, x_0, \cdot), D) \\ - \sup \left\{ \frac{\|[A(x, y, u) - A(x_0, x_0, u)] - [A(x, y, v) - A(x_0, x_0, v)]\|}{\|u - v\|}, \right. \\ \left. u \ne v, \; u, v \in D \right\} \\ \ge d - \ell_0(\|x - x_0\| + \|y - x_0\|) \ge d - 2\ell_0\rho.$$

Furthermore, we can now have

$$\|F(w) - F(z)\| \geq (d - 2\ell_0\rho)\|w - z\| - \ell\left[\|x - w\|\,\|y - w\| + \|x - z\|\,\|y - z\|\right]$$

$$\geq (d - 2\ell_0\rho)\|w - z\| - \frac{\ell}{2}\|w - z\|^2$$

and for $w \neq z$,

$$\frac{\|F(w) - F(z)\|}{\|w - z\|} \geq d - (2\ell_0 + \ell)\rho.$$

That completes the proof of our Lemma.

Remark 3.3.6. In order for us to compare our result with the corresponding Lemma 2.4 in [180, p. 294], first note that if:

(a) equality holds in both inequalities in (3.3.16), $u = v$ and $x = y$ in (3.3.1)–(3.3.3), then our result reduces to Lemma 2.4 by setting $\frac{k}{2} = \ell = \ell_0$.

(b) Strict inequality holds in any of the inequalities in (3.3.16), $u = v$ and $x = y$ then our Lemma 3.3.5 improves (enlarges) the range for ρ, and under the same computational cost. The implications of that are twofold (see Theorems 3.3.8 and 3.3.10 that follow): in the semilocal case the uniqueness ball is more precise, and in the local case the radius of convergence is enlarged.

We will need our result on majorizing sequences for the secant method. The proof using conditions (C_1)–(C_3) can be found in Section 2.3, whereas for well-known condition (C_4) see, e.g., [162]. Detailed comparisons between conditions (C_1)–(C_4) were given in Section 2.3. In particular if strict inequality holds in (3.3.16) (first inequality) the error bounds under (C_1)–(C_3) are more precise and the limit of majorizing sequence more accurate than under condition (C_4).

Lemma 3.3.7. *Assume for $\eta \geq 0$, $c \geq 0$, $d_0 > 0$:*

(C_1) *for all $n \geq 0$ there exists $\delta \in [0, 1)$ such that:*

$$\ell\delta^n(1 + \delta)\eta + \frac{\delta\ell_0}{1 - \delta}(2 - \delta^{n+2} - \delta^{n+1})\eta + \delta\ell_0 c \leq \delta d_0,$$

and

$$\frac{\delta\ell_0}{1 - \delta}(2 - \delta^{n+2} - \delta^{n+1})\eta + \delta\ell_0 c < d_0,$$

or

(C_2) *there exists $\delta \in [0, 1)$ such that:*

$$\ell(1 + \delta)\eta + \frac{2\delta\ell_0\eta}{1 - \delta} + \delta\ell_0 c \leq \delta d_0,$$

and

$$\frac{2\delta\ell_0\eta}{1 - \delta} + \delta\ell_0 c < d_0,$$

or

(C_3) *there exists $a \in [0, 1]$ and*

$$\delta \in \begin{cases} \left[0, \dfrac{-1+\sqrt{1+4a}}{2a}\right], & a \neq 0 \\ [0, 1), & a = 0 \end{cases}$$

such that

$$(\ell + \delta \ell_0)(c + \eta) \leq d_0 \delta, \quad \eta \leq \delta c \text{ and } \ell_0 \leq a\ell,$$

or

(C4) $d_0^{-1}\ell c + 2\sqrt{d_0^{-1}\ell\eta} \leq 1 \text{ for } \ell_0 = \ell.$

Then,

(a) iteration $\{t_n\}$ $(n \geq -1)$ given by

$$t_{-1} = 0, \quad t_0 = c, \quad t_1 = c + \eta,$$

$$t_{n+2} = t_{n+1} + \frac{d_0^{-1}\ell(t_{n+1} - t_{n-1})}{1 - d_0^{-1}\ell_0 \left[t_{n+1} - t_0 + t_n\right]}(t_{n+1} - t_n), \tag{3.3.19}$$

is nondecreasing, bounded above by r

$$r = \frac{\eta}{1 - \delta} + c,$$

and converges to some t^* such that

$$0 \leq t^* \leq r.$$

Moreover, the following estimates hold for all $n \geq 0$:

$$0 \leq t_{n+2} - t_{n+1} \leq \delta(t_{n+1} - t_n) \leq \delta^{n+1}\eta.$$

(b) Iteration $\{s_n\}$ $(n \geq -1)$ given by

$$s_{-1} - s_0 = c, \quad s_0 - s_1 = \eta,$$

$$s_{n+1} = s_{n+2} + \frac{d_0^{-1}\ell(s_{n-1} - s_{n+1})}{1 - d_0^{-1}\ell_0 \left[(s_0 + s_{-1}) - (s_n + s_{n+1})\right]}(s_n - s_{n+1}), \tag{3.3.20}$$

provided that $s_{-1} \geq 0$, $s_0 \geq 0$, $s_1 \geq 0$ is nonincreasing, bounded below by s given by

$$s = s_0 - \frac{\eta}{1 - \delta},$$

and converges to some s^* such that

$$0 \leq s^* \leq s.$$

Moreover, the following estimates hold for all $n \geq 0$:

$$0 \leq s_{n+1} - s_n \leq \delta(s_n - s_{n+1}) \leq \delta^{n+1}\eta.$$

We denote by (C), Conditions (C_1) or (C_2) or (C_3) or (C_4).

We can state and prove the main semilocal convergence result for the secant method involving a (PBA).

Theorem 3.3.8. *Let X and Y be Banach spaces, D a closed convex subset of X, x_{-1}, and $x_0 \in D$ with $\|x_{-1} - x_0\| \le c$, and F a continuous operator from D into Y. Suppose operator F has a (PBA) on D at the point x_0. Moreover assume:*

$$\delta(A(x_{-1}, x_0, \cdot), D) \ge d_0 > 0;$$

Condition (C) holds;
for each $y \in U(x_0, d_0(t^* - t_1)$ the equation $A(x_{-1}, x_0, x) = y$ has a solution x; the solution $S(x_{-1}, x_0)$ of $A(x_{-1}, x_0, S(x_{-1}, x_0)) = 0$ satisfies

$$\|S(x_{-1}, x_0) - x_0\| \le \eta;$$

and

$$U(x_0, t^*) \subseteq D.$$

Then the secant iteration defining x_{n+1} by

$$A(x_{n-1}, x_n, x_{n+1}) = 0 \qquad (3.3.21)$$

remains in $U(x_0, t^*)$, and converges to a solution $x^* \in U(x_0, t^*)$ of equation $F(x) = 0$.

Moreover the following estimates hold for all $n \ge 0$:

$$\|x_{n+1} - x_n\| \le t_{n+1} - t_n \qquad (3.3.22)$$

and

$$\|x_n - x^*\| \le t^* - t_n, \qquad (3.3.23)$$

where sequence $\{t_n\}$ is defined by (3.3.19) and $t^* = \lim_{n \to \infty} t_n$.

Proof. We use Lemma 3.3.3 with quantities F, G, x_0 and y_0 replaced by $A(w, x, \cdot)$, $A(x_0, x_1, \cdot) - A(v, x, \cdot)$, $x_1 = S(x_{-1}, x_0)$, and 0 respectively. Hypothesis (3.3.9) of the Lemma follows from the fact that $A(x_{-1}, x_0, x) = y$ has a unique solution x_1.

For hypothesis (3.3.10) we have using (3.3.2)

$$\delta(A(x_0, x_1, \cdot), D) \ge \delta(A(x_{-1}, x_0, \cdot), D) - \ell_0(\|x_0 - x_{-1}\| + \|x_1 - x_0\|)$$
$$\ge d_0 - \ell_0 t_1 > 0 \quad \text{by } (C)). \qquad (3.3.24)$$

To show (3.3.11), we must have $U(x_1, t^* - t_1) \subseteq D$. Instead by hypothesis $U(x_0, t^*) \subseteq D$, it suffices to show:

$$U(x_1, t^* - t_1) \subset U(x_0, t^*)$$

which is true because

$$\|x_1 - x_0\| + t^* - t_1 \le t^*.$$

We also have by (3.3.1)

$$\|A(x_{-1}, x_0, x_1) - A(x_1, x_1, x_1)\| \leq \ell(t_1 - t_{-1})(t_1 - t_0). \tag{3.3.25}$$

It follows by (3.3.12) and (3.3.25) that $\theta_0 \geq 0$ if

$$\theta_0 \geq \left[1 - \ell_0 d_1^{-1}(c + \eta)\right] d_0 \eta - \ell(t_1 - t_{-1})(t_1 - t_0) \geq 0, \tag{3.3.26}$$

which is true by (3.3.19) and because $d_0 \leq d_1 = \delta_0(A(x_{-1}, x_0, \cdot), D)$.

It follows from Lemma 3.3.3 that for each $y \in U(0, r - \|x_1 - x_0\|)$, the equation $A(x_0, x_1, z) = y$ has a unique solution because $\delta(A(x_0, x_1, \cdot), D) > 0$. We also have

$$A(x_0, x_1, x_2) = A(x_{-1}, x_0, x_1) = 0.$$

By Definition 3.3.2, the induction hypothesis, (3.3.1) and (3.3.19) we get in turn:

$$\begin{aligned} \|x_2 - x_1\| &\leq \delta(A(x_0, x_1, \cdot), D)^{-1} \|A(x_{-1}, x_0, x_1) - F(x_1)\| \\ &\leq \frac{\ell \|x_{-1} - x_0\| \, \|x_0 - x_1\|}{d_0 - \ell_0(\|x_0 - x_{-1}\| + \|x_1 - x_0\|)} \\ &\leq \frac{d_0^{-1}\ell(t_1 - t_{-1})(t_1 - t_0)}{1 - d_0^{-1}\ell_0 t_1} = t_2 - t_1. \end{aligned} \tag{3.3.27}$$

Hence we showed:

$$\|x_{n+1} - x_n\| \leq t_{n+1} - t_n, \tag{3.3.28}$$

and

$$U(x_{n+1}, t^* - t_{n+1}) \subset U(x_n, t^* - t_n) \tag{3.3.29}$$

hold for $n = 0, 1$. Moreover for every $v \in U(x_1, t^* - t_1)$

$$\|v - x_0\| \leq \|v - x_1\| + \|x_1 - x_0\| \leq t^* - t_1 + t_1 - t_0,$$

implies $v \in U(x_0, t^* - t_0)$. Given they hold for $n = 0, 1, \ldots, j$, then

$$\|x_{j+1} - x_0\| \leq \sum_{i=1}^{j+1} \|x_i - x_{i-1}\| \leq \sum_{i=1}^{j+1}(t_i - t_{i-1}) = t_{j+1} - t_0.$$

The induction for (3.3.28) and (3.3.29) can easily be completed by simply replacing x_{-1}, x_0, x_1, by x_{n-1}, x_n, x_{n+1}, respectively. Indeed, corresponding with (3.3.26), we have:

$$\theta_0 \geq \left[1 - \ell_0 d_1^{-1}(t_{n+1} - t_0 + t_n)\right] d_0(t_{n+2} - t_{n+1}) - \ell(t_{n+1} - t_{n-1})(t_{n+1} - t_n) = 0, \tag{3.3.30}$$

which is true by (3.3.19).

Scalar sequence $\{t_n\}$ is Cauchy. From (3.3.28) and (3.3.29) it follows $\{x_n\}$ is Cauchy, too, in a Banach space X, and as such it converges to some $x^* \in U(x_0, t^*)$.

Moreover we have by (3.3.1) and (3.3.28)

$$\|F(x_{n+1})\| = \|F(x_{n+1}) - A(x_{n-1}, x_n, x_{n+1})\|$$
$$\leq \ell\|x_n - x_{n-1}\| \|x_n - x_{n+1}\|$$
$$\leq \ell(t_n - t_{n-1})(t_{n+1} - t_n) \longrightarrow 0 \text{ as } n \to \infty.$$

By the continuity of F we deduce $F(x^*) = 0$. Finally estimate (3.3.23) follows from (3.3.22) by using standard majorization techniques.

That completes the proof of the Theorem.

Remark 3.3.9. The uniqueness of the solution x^* was not considered in Theorem 3.3.8. Indeed, we do not know if under the conditions stated above the solution x^* is unique, say in $U(x_0, t^*)$. However using Lemma 3.3.5 we can obtain a uniqueness result, so that if ρ satisfies

$$t^* < \rho < \frac{d_0}{2\ell_0 + \ell}, \tag{3.3.31}$$

then operator F is one-to-one in a neighborhood of x^*, as $x^* \in U(x_0, t^*)$. That is, x^* is an isolated zero of F in this case.

The corresponding local convergence result for the secant method is given by:

Theorem 3.3.10. *Assume:*

$x^* \in D$ is an isolated zero if F on D;

operator F has a (PBA) on D at the point x^* of modulus (L, L_0).

Moreover assume that the following hold:

$$\delta(A(x^*, x^*, \cdot), D) \geq d^* > 0;$$

$$0 \leq r^* < \frac{d^*}{2L_0 + 3L};$$

for each $y \in U(0, d^*r^*)$ the equation $A(x_{-1}, x_0, x) = y$ has a solution x satisfying

$$\|x - x^*\| \leq r^*;$$

and

$$U(x^*, r^*) \subseteq D.$$

Then secant method $\{x_n\}$ generated by (3.3.21) is well defined, remains in $U(x^*, r^*)$ for all $n \geq 0$, and converges to x^* provided $x_{-1}, x_0 \in U^0(x^*, r^*)$.

Moreover the following estimates hold for all $n \geq 0$:

$$\|x_{n+1} - x^*\| \leq \frac{(d^*)^{-1}L(\|x_{n-1} - x_n\| + \|x_n - x^*\|)}{1 - (d^*)^{-1}L_0(\|x_{n-1} - x^*\| + \|x_n - x^*\|)}\|x_n - x^*\|.$$

Proof. The proof is omitted as it is similar to Theorem 3.3.8. Note that local results were not given in [180].

We now show how to choose operator A in cases not covered in [180].

Example 3.3.11. Let $X = Y = (\mathbf{R}^2, \|\cdot\|_\infty)$. Consider the system

$$3x^2y + y^2 - 1 + |x - 1| = 0$$
$$x^4 + xy^3 - 1 + |y| = 0. \tag{3.3.32}$$

Set for $v = (v_1, v_2)$, $\|v\|_\infty = \|(v_1, v_2)\|_\infty = \max\{|v_1|, |v_2|\}$, $F(v) = P(v) + Q(v)$, $P = (P_1, P_2)$, $Q = (Q_1, Q_2)$. Define

$$P_1(v) = 3v_1^2 v_2 + v_2^2 - 1, \quad P_2(v) = v_1^4 + v_1 v_2^3 - 1,$$

$$Q_1(v) = |v_1 - 1|, \quad Q_2(v) = |v_2|.$$

We shall take divided differences of order one $[x, y; P]$, $[x, y; Q] \in M_{2\times2}(\mathbf{R})$ to be for $w = (w_1, w_2)$:

$$[v, w, P]_{i,1} = \frac{P_i(w_1, w_2) - P_i(v_1, w_2)}{w_1 - v_1},$$

$$[v, w, P]_{i,2} = \frac{P_i(v_1, w_2) - P_i(v_1, v_2)}{w_2 - v_2}$$

provided that $w_1 \neq v_1$ and $w_2 \neq v_2$. If $w_1 = v_1$ or $w_2 = v_2$ replace $[x, y; P]$ by P'. Similarly we define

$$[v, w; Q]_{i,1} = \frac{Q_i(w_1, w_2) - Q_i(v_1, w_2)}{w_1 - v_1},$$

$$[v, w; Q]_{i,2} = \frac{Q_i(v_1, w_2) - Q_i(v_1, v_2)}{w_2 - v_2}$$

for $w_1 \neq v_1$ and $w_2 \neq v_2$. If $w_1 = v_1$ or $w_2 = v_2$ replace $[x, y; Q]$ by the zero 2×2 matrix in $M_{2\times2}(\mathbf{R})$.

We consider three interesting choices for operator A:

$$A(v, v, w) = P(v) + Q(v) + P'(v)(w - v), \tag{3.3.33}$$

$$A(u, v, w) = P(v) + Q(v) + ([u, v; P] + [u, v; Q])(w - v) \tag{3.3.34}$$

and

$$A(u, v, w) = P(v) + Q(v) + (P'(v) + [u, v; Q])(w - v). \tag{3.3.35}$$

Using method (3.3.33) for $x_0 = (1, 0)$, and both methods (3.3.34) and (3.3.35) for $x_{-1} = (5, 5)$, $x_0 = (1, 0)$ we obtain the following three tables respectively.

We did not verify the hypotheses of Theorem 3.3.8 for the above starting points. However, it is clear that the hypotheses of Theorem 3.3.8 are satisfied for all three methods for starting points closer to the solution

$$x^* = (.894655373334687, .327826521746298),$$

chosen from the lists of the tables displayed in Tables 3.3.1–3.3.3.

Table 3.3.1. Newton's method (3.3.33)

n	$x_n^{(1)}$	$x_n^{(2)}$	$\|x_n - x_{n-1}\|$
0	1	0	
1	1	0.333333333333333	3.333E–1
2	0.906550218340611	0.354002911208151	9.344E–2
3	0.885328400663412	0.338027276361322	2.122E–2
4	0.891329556832800	0.326613976593566	1.141E–2
5	0.895238815463844	0.326406852843625	3.909E–3
6	0.895154671372635	0.327730334045043	1.323E–3
7	0.894673743471137	0.327979154372032	4.809E–4
8	0.894598908977448	0.327865059348755	1.140E–4
9	0.894643228355865	0.327815039208286	5.002E–5
10	0.894659993615645	0.327819889264891	1.676E–5
11	0.894657640195329	0.327826728208560	6.838E–6
12	0.894655219565091	0.327827351826856	2.420E–6
13	0.894655074977661	0.327826643198819	7.086E–7
...			
39	0.894655373334687	0.327826521746298	5.149E–19

Table 3.3.2. Secant method (3.3.34)

n	$x_n^{(1)}$	$x_n^{(2)}$	$\|x_n - x_{n-1}\|$
–1	5	5	
0	1	0	5.000E+00
1	0.989800874210782	0.012627489072365	1.262E–02
2	0.921814765493287	0.307939916152262	2.953E–01
3	0.900073765669214	0.325927010697792	2.174E–02
4	0.894939851625105	0.327725437396226	5.133E–03
5	0.894658420586013	0.327825363500783	2.814E–04
6	0.894655375077418	0.327826521051833	3.045E–04
7	0.894655373334698	0.327826521746293	1.742E–09
8	0.894655373334687	0.327826521746298	1.076E–14
9	0.894655373334687	0.327826521746298	5.421E–20

Table 3.3.3. Newton's method (3.3.35)

n	$x_n^{(1)}$	$x_n^{(2)}$	$\|x_n - x_{n-1}\|$
–1	5	5	
0	1	0	5
1	0.909090909090909	0.363636363636364	3.636E–01
2	0.894886945874111	0.329098638203090	3.453E–02
3	0.894655531991499	0.327827544745569	1.271E–03
4	0.894655373334793	0.327826521746906	1.022E–06
5	0.894655373334687	0.327826521746298	6.089E–13
6	0.894655373334687	0.327826521746298	2.710E–20

Note that the results in [18] cannot apply here because operator A no matter how it is chosen cannot satisfy the Lipschitz conditions (a) or (b) in Definition 2.1 in [180, p. 293] needed for the application of Theorem 3.2 in the same paper.

Other possible applications of operators equations with a (PBA) are already noted in [43], [180, p. 293] and the references there.

Hence method (3.3.1) (i.e., method (3.3.35) in this case) converges faster than (3.3.33) suggested in Chen and Yamamoto [58], Zabrejko and Nguen [146] in this case, and the method of chord (3.3.34) (see also Section 5.3).

Application 3.3.12. *In the case of the NK method, the proof of Robinson's theorem 32 in [180] was based on the crucial Newton-Kantorovich-type hypothesis*

$$h_K = d_0^{-1} L r_0 \leq \frac{1}{2} \tag{3.3.36}$$

which is the sufficient condition for the monotone convergence of majorizing sequence $\{v_n\}$ $(n \geq 0)$ given by

$$v_{n+2} = v_{n+1} + \frac{d_0^{-1} L \left(v_{n+1} - v_n \right)^2}{2 \left(1 - d_0^{-1} L v_{n+1} \right)}, \quad v_0 = 0, \ v_1 = r_0,$$

where L is the Lipschitzian constant appearing in the definition of a (PBA) approximation for F on D, i.e.,

$$\| F(v) - A(u, v) \| \leq \frac{1}{2} L d(u, v)^2.$$

Moreover, by assuming operator $A(u, \cdot) - A(x_0, \cdot)$ is Lipschitzian on D with modulus $L_0 d(u, x_0)$, we can show by simply repeating the proof of Theorem 3.2 in [180] or our Theorem 4.3.1 (or Theorem 2.2.11) that hypothesis (3.3.36) can be replaced by

$$h_A = d_0^{-1} \bar{L} r_0 \leq \frac{1}{2}, \quad \bar{L} = \frac{L + L_0}{2},$$

and $\{v_n\}$ by the finer majorizing sequence $\{w_n\}$ given by

$$w_{n+2} = w_{n+1} + \frac{d_0^{-1} L \left(w_{n+1} - w_n \right)^2}{2 \left(1 - d_0^{-1} L_0 w_{n+1} \right)}, \quad w_0 = 0, \ w_1 = r_0.$$

Note that if $L_0 = L$, our hypotheses reduce to the ones in Robinson's Theorem 3.2 [180]. Otherwise, i.e., if $L_0 < L$, then our results are weaker. The rest of the advantages of our approach have already been explained in Section 2.2.

3.4 Improvements on curve tracing of the homotopy method

The local convergence of the NK method for the tracing of an implicitly defined smooth curve is analyzed. The domain of attraction is shown to be larger than before [62]. Moreover, finer error bounds on the distances involved are obtained and quadratic instead of geometrical order is established.

Finally, a numerical example is provided to justify our theoretical results.

Local convergence for the curve tracing of the homotopy method

We are concerned with the following problem: Suppose that a smooth curve $\Gamma \subset \mathbf{R}^{n+1}$ is implicitly defined by

$$F(x, t) = 0 \tag{3.4.1}$$

where $F: \mathbf{R}^n \times \mathbf{R} \to \mathbf{R}^n$ is a C^2 function. We intend to numerically trace curve Γ from the point (x_0, t_0) to the point (x^*, t^*). We assume the $n \times (n+1)$ Jacobian matrix $DF(x, t)$ has full rank at every point in Γ.

A survey of such tehniques can be found in [2], [176] and the references there.

We will use the following algorithmic form:

(a) Let $y_i = (x_i, t_i) \in \mathbf{R}^{n+1}$ be an approximation for Γ. Use the predictor

$$z_0 = y_i + h_i \tau_i \tag{3.4.2}$$

for the next approximating point, where h_i is an appropriate step length and τ_i is the tangent vector of Γ at y_i;

(b) Starting from z_0, take a sequence of Newton iterations by requiring z_k to lie on the hyperplane normal to a certain vector (usually the tangent vector τ_i).

(c) Set $y_{i+1} = z$ where z is the point of convergence for the sequence $\{z_k\}$.

We need some preliminaries:

A point (x, t) in $\mathbf{R}^{n+1}$ will be denoted by y. Let σ be the arc length, along the curve Γ, then an initial value problem is implicitly defined by

$$DF(y) \cdot \dot{y} = 0; \quad y(0) = y_0, \tag{3.4.3}$$

where $\cdot = \frac{d}{d\sigma}$. It is known that vector field $\dot{y}$ is locally Lipschitzian [176].

We assume $DF(y)$ is full rank along the solution curve, then equation

$$DF(y) y' = -F(y) \tag{3.4.4}$$

can be reduced to

$$y' = -DF^+(y) F(y) \tag{3.4.5}$$

where $DF^+(y) = DF^T(y) \left[DF(y) DF^T(y)\right]^{-1}$ is the Moore-Penrose generalized inverse of $DF(y)$. By the result

$$\text{Rang}\left(DF^+\right) = \text{Rang}\left(DF^T\right) = \text{Kernel}(DF)^\perp \tag{3.4.6}$$

and equation

$$F(y(\tau)) = e^{-\tau} F(y(0)) \tag{3.4.7}$$

we conclude a solution $y(\tau)$ of (3.4.5) is such that the magnitude of $F(y)$ is reduced and also remains perpendicular to the 1-dimensional kernel space of $F(y)$.

Consider the Euler step of (3.4.5). This corresponds with the Newton method in the form

$$y_{k+1} = y_k - DF^+(y_k) F(y_k). \tag{3.4.8}$$

In the next section, we analyze the local convergence of method (3.4.8).

We state a result whose proof can be found in [62, p. 327]:

Theorem 3.4.1. *Let* $F: D \subseteq \mathbf{R}^{n+1} \to \mathbf{R}^n$ *be a* C^2 *function such that*

$$\|DF(x) - DF(y)\| \leq \ell \|x - y\|, \quad for\ all\ x, y \in D. \tag{3.4.9}$$

Suppose that $F(x^*)$ *and* $DF(x^*)$ *is full rank. Let* $\delta \in (0, \frac{3-\sqrt{5}}{2})$ *and define*

$$M = \min\left\{\frac{2}{3\|DF^+(x^*)\|\ell}, \operatorname{dist}(x^*, \partial D)\right\}. \tag{3.4.10}$$

If $r \in (0, \delta M = r_0)$ *is such that for every* $x \in U(x^*, r)$ *we have*

$$\|F(x)\| \leq \frac{\delta\ell M^2}{2}, \tag{3.4.11}$$

then for any $x_0 \in U(x^*, r) \subseteq D$, *method (3.4.8) is well defined and converges geometrically to a point in* $\Gamma \cap U(x^*, M)$.

Remark 3.4.2. Under the hypotheses of Theorem 3.4.1, method (3.4.8) converges only geometrically and condition (3.4.1) should hold. To do so we first introduce the center-Lipschitz condition

$$\|DF(x) - DF(x^*)\| \leq \ell_0 \|x - x^*\|, \quad for\ all\ x \in D. \tag{3.4.12}$$

We note that in general

$$\ell_0 \leq \ell \tag{3.4.13}$$

holds and $\frac{\ell}{\ell_0}$ can be arbitrarily large. In practice the computation of ℓ requires that of ℓ_0.

Then we can show the following improvement over Theorem 3.4.1.

Theorem 3.4.3. *Suppose hypotheses of Theorem 3.4.1 and (3.4.12) hold but* M *is defined as*

$$M_0 = \min\left\{\frac{2}{(2\ell_0 + \ell)\|DF^+(x^*)\|}, \operatorname{dist}(x^*, \partial D)\right\} \tag{3.4.14}$$

then the conclusions of Theorem 3.4.1 hold with M_0 *replacing* M.

Proof. For any $x \in U(x^*, M_0)$, we get using Lemma 3.1 in [62, p. 326] and (3.4.12)

$$\|DF(x) - DF(x^*)\|\|DF^+(x^*)\| \leq \ell_0\|x - x^*\|\|DF^+(x^*)\|$$
$$< \frac{2}{3} < 1. \tag{3.4.15}$$

The rest of the proof follows exactly as in Theorem 1 in [62, p. 326] (with M_0 replacing M).

That completes the proof of the theorem.

Remark 3.4.4. If equality holds in (3.4.13), then Theorem 3.4.3 reduces to Theorem 3.4.1. Otherwise

$$M < M_0 \tag{3.4.16}$$

holds and the bounds on the distances $\|y_{n+1} - y_n\|$, $\|y_{n+1} - x^*\|$ ($n \geq 0$) are finer in Theorem 3.4.3. This improvement allows a wider choice of initial guesses x_0. Such an observation is important in computational mathematics. By comparing (3.4.10) and (3.4.14), we see that M_0 can be (at most) three times larger than M (if $\ell_0 = \ell$).

In order to show that it is possible to achieve quadratic convergence and drop strong condition (3.4.11) we use a modification of our Theorem 2 in [40] (where we have replaced $F'(x)^{-1}$ by $DF(x)^+$) and use Lemma 3.1 in [62] instead of Banach Lemma on invertible operators in the proof of Theorem 2 in [40] to obtain the proof of Theorem 3.4.5 that follows:

Theorem 3.4.5. *Assume conditions of Theorem 3.4.3 hold excluding (3.4.11). If*

$$U_1\left(x^*, r_1\right) \subseteq D, \tag{3.4.17}$$

where

$$r_1 = \frac{1}{\ell_0 \left\| DF\left(x^*\right)^+ \right\|}, \tag{3.4.18}$$

then for all $x_0 \in U_2\left(x^*, r_2\right)$, where

$$r_2 = \frac{2 + \gamma - \sqrt{\gamma^2 + 2\gamma}}{(2 + \gamma)\,\ell_0 \left\| DF\left(x^*\right)^+ \right\|}, \quad \text{for } \gamma \geq 2, \ell = \frac{\gamma}{2}\ell_0 \tag{3.4.19}$$

the following hold:

Newton-Kantorovich hypothesis

$$h = 2\ell \left\| DF\left(x_0\right)^+ \right\| \left\| DF\left(x_0\right)^+ F\left(x_0\right) \right\| \leq 1 \tag{3.4.20}$$

holds as strict inequality, and consequently the Newton-Kantorovich theorem guarantees method (3.4.8) is well-defined and converges quadratically to a point in $\Gamma \cap U\left(x^*, r_1\right)$.

Remark 3.4.6. Even if equality holds in (3.4.13) we can set $\gamma = 2$ and r_2 can be written as

$$r_2 = \frac{2 - \sqrt{2}}{2\ell_0 \left\| DF\left(x^*\right)^+ \right\|} \tag{3.4.21}$$

which is larger than r_0 as

$$\delta < \frac{2 - \sqrt{2}}{2}. \tag{3.4.22}$$

If strict inequality holds in (3.4.13), then r_2 is enlarged even further (see also Example 3.4.7 as follows).

Convergence radius r_2 can be extended even further by using Theorem 3 in [40] based on an even weaker hypothesis than (3.4.20) found by us in Section 2.2:

$$h_0 = (\ell + \ell_0) \left\| DF\left(x_0\right)^+ \right\| \left\| DF\left(x_0\right)^+ F\left(x_0\right) \right\| \leq 1. \tag{3.4.23}$$

However we do not pursue this here, leaving it for the motivated reader.

Instead we provide an example where strict inequality holds in (3.4.13).

Example 3.4.7. Let $D = U(0, 1)$ and define function F on the real line by

$$F(x) = e^x - 1. \tag{3.4.24}$$

For simplicity we take $x_0 = x^*$. We obtain

$$\ell = e,$$
$$\ell_0 = e - 1,$$
$$\left\| DF(x^*)^+ \right\| = 1,$$
$$\gamma = 3.163953415,$$
$$\delta = .381966011,$$
$$M = .245252961,$$
$$M_0 = .324947231,$$
$$r_0 = \delta M = .093678295,$$
$$\bar{r}_0 = \delta M_0 = .124118798,$$
$$r_1 = .581976707,$$
$$r_2 = .126433594.$$

Therefore we conclude

$$M < M_0 < r_1$$

and

$$r_0 < \bar{r}_0 < r_2,$$

which demonstrate the superiority of our results over the ones in [62].

3.5 Nonlinear finite element analysis

We provide a discretization result to find finite element solutions of elliptic bound-ary value problems. Our analysis is based on the weaker version of the Newton-Kantorovich theorem established in Section 2.2 (see Theorem 2.2.11). The advan-tages of this approach over Newton-Kantorovich theorem 2.2.4 have already been explained in Section 2.2.

Finally we provide examples of elliptic boundary value problems where our re-sults apply.

We state the version of our main Theorem 2.2.11 needed in this study.

Theorem 3.5.1. *Let* $F: D \subseteq A \to B$ *be a nonlinear Fréchet-differentiable operator. Assume:*

there exists a point $x_0 \in D$ *such that the Fréchet derivative* $F'(x_0) \in L(A, B)$ *is an isomorphism and* $F(x_0) \neq 0$;

there exists positive constants ℓ_0 and ℓ such that the following center-Lipschitz and Lipschitz conditions are satisfied:

$$\left\| F'(x_0)^{-1} \left[F'(x) - F'(x_0) \right] \right\| \leq \ell_0 \|x - x_0\| \qquad (3.5.1)$$

$$\left\| F'(x_0)^{-1} \left[F'(x) - F'(y) \right] \right\| \leq \ell \|x - y\| \qquad (3.5.2)$$

for all $x, y \in D$;
Setting $\eta = \left\| F'(x_0)^{-1} F(x_0) \right\|$ and $h_1 = (\ell_0 + \ell)\,\eta$, we further assume

$$h_1 \leq 1; \qquad (3.5.3)$$

$$\bar{U}\left(x_1, t^* - \eta\right) \subseteq D, \qquad (3.5.4)$$

where, $x_1 = x_0 - F'(x_0)^{-1} F(x_0)$, and t^* a well defined point in $[\eta, 2\eta]$.
Then equation $F(x) = 0$ has a solution $x^* \in \bar{U}(x_1, t^* - \eta)$ and this solution is unique in $U(x_0, t^*) \cap D$, if $\ell_0 = \ell$ and $h_1 < 1$, and $\bar{U}(x_0, t^*) \cap D$, if $\ell_0 = \ell$ and $h_1 = 1$. If $\ell_0 \neq \ell$ the solution x^* is unique in $U(x_0, R)$ provided that $\frac{1}{2}(t^* + R)\ell_0 \leq 1$ and $U(x_0, R) \subseteq D$.
Moreover, we have the estimate

$$\left\| x^* - x_0 \right\| \leq t^*. \qquad (3.5.5)$$

We will simply use $\|\cdot\|$ if the norm of the element involved is well understood. Otherwise we will use $\|\cdot\|_X$ for the norm on a particular set X.

We assume the following:

(A_1) there exist Banach spaces $Z \subseteq X$ and $U \subseteq Y$ such that the inclusions are continuous, and the restriction of F to Z, denoted again by F, is a Fréchet-differentiable operator from Z to U.

(A_2) For any $v \in Z$ the derivative $F'(v) \in L(Z.U)$ can be extended to $F'(v) \in L(X, Y)$ and it is:

—Locally Lipschitz continuous on Z, i.e., for any bounded convex set $T \in Z$ there exists a positive constant c_1 depending on T such that

$$\left\| F'(v) - F'(w) \right\| \leq c_1 \|v - w\|, \quad \text{for all } v.w \in T. \qquad (3.5.6)$$

—center locally Lipschitz continuous at a given $u_0 \in Z$, i.e., for any bounded convex set $T \in Z$ with $u_0 \in T$ there exists a positive constant c_0 depending on u_0 and T such that

$$\left\| F'(v) - F'(u_0) \right\| \leq c_0 \|v - u_0\|, \quad \text{for all } v \in T. \qquad (3.5.7)$$

(A_3) There are Banach spaces $V \subseteq Z$ and $W \subseteq U$ such that the inclusions are continuous. We suppose that there exists a subset $S \subseteq V$ for which the following holds: "if $F'(u) \in L(V, W)$ is an isomorphism between V and W at $u \in S$, then $F'(u) \in L(X, Y)$ is an isomporhism between X and Y as well."

To define discretized solutions of $F(u) = 0$, we introduce the finite-dimensional subspaces $S_d \subseteq Z$ and $S_d \subseteq U$ parameterized by d, $0 < d < 1$ with the following properties:

(A$_4$) There exists $r \geq 0$ and a positive constant c_2 independent of d such that

$$\|V_h\|_Z \leq \frac{c_2}{d^r} \|V_h\|_X, \quad \text{for all } v_d \in S_d. \tag{3.5.8}$$

(A$_5$) There exists projection $\Pi_d: X \to S_d$ for each S_d such that, if $u_0 \in S$ is a solution of $F(u) = 0$, then

$$\lim_{d \to 0} d^{-r} \|u_0 - \Pi_d u_0\|_X = 0 \tag{3.5.9}$$

and

$$\lim_{d \to 0} d^{-r} \|u_0 - \Pi_d u_0\|_Z = 0. \tag{3.5.10}$$

We can show the following result concerning the existence of locally unique solutions of discretized equations.

Theorem 3.5.2. *Assume that conditions (A$_1$)–(A$_5$) hold. Suppose $F'(u_0) \in L(V, W)$ is an isomorphism, and $u_0 \in S$. Moreover, assume $F'(u_0)$ can be decomposed into $F'(u_0) = Q + R$, where $Q \in L(X, Y)$ and $R \in L(X, Y)$ is compact. The discretized nonlinear operator $F_d: Z \to U$ is defined by*

$$F_d(u) = (I - P_d) Q(u) + P F_d(u) \tag{3.5.11}$$

where I is the identity of Y, and $P_d: Y \to S_d$ is a projection such that

$$\lim_{d \to 0} \|v - P_d v\|_Y = 0, \quad \text{for all } v \in Y, \tag{3.5.12}$$

and

$$(I - P_d) Q(v_d) = 0, \quad \text{for all } v_d \in S_d. \tag{3.5.13}$$

Then, for sufficiently small $d > 0$, there exists $u_d \in S_d$ such that $F_d(u_d) = 0$, and u_d is locally unique.

Moreover the following estimate holds

$$\|u_d - \Pi_d(u_0)\| \leq \ell_1 \|u_0 - \Pi_d(u_0)\| \tag{3.5.14}$$

where ℓ_1 is a positive constant independent of h.

Proof. The proof is similar to the corresponding one in [136, Th. 2.1, p. 126]. However, there are some crucial differences where weaker (3.5.7) is used (needed) instead of stronger condition (3.5.6).

Step 1. We claim that there exists a positive constant c_3, independent of d, such that, for sufficiently small $h > 0$,

$$\left\| F'_d(\Pi_d(u_0)) v_d \right\|_Y \geq c_3 \|v_d\|_X, \quad \text{for all } v_d \in S_d. \tag{3.5.15}$$

From (A3) and $u_0 \in S$, $F'(u_0) \in L(X, Y)$ is an isomorphism. Set $B_0 = \left\| F'(u_0)^{-1} \right\|$.

We can have in turn

$$F_d'(\Pi_d(u_0)) v_d = F'(u_0) v_d + P_d \left(F'(\Pi_d(u_0)) - F'(v_0) \right) v_d \tag{3.5.16}$$
$$- (I - P_d) \left(-Q + F'(u_0) \right) v_d.$$

Because $-Q + F'(u_0) \in L(X, Y)$ is compact we get by (3.5.12) that

$$\lim_{d \to 0} \left\| (I - P_d) \left(-Q + F'(u_0) \right) \right\| = 0. \tag{3.5.17}$$

By (3.5.12) there exists a positive constant c_4 such that

$$\sup_{d > 0} \| P_d \| \leq c_4. \tag{3.5.18}$$

That is, using (3.5.7) we get

$$\left\| P_d \left(F'(\Pi_d(u_0)) - F'(u_0) \right) \right\| \leq c_0 c_4 \| \Pi_d(u_0) - u_0 \|. \tag{3.5.19}$$

Hence, by (3.5.10) we can have

$$\left\| F_d'(\Pi_d(u_0)) v_d \right\| \geq \left(\tfrac{1}{B_0} - \delta(d) \right) \| v_d \|, \tag{3.5.20}$$

where $\lim_{d \to 0} \delta(d) = 0$, and (3.5.15) holds with $c_3 = \frac{B_0^{-1}}{2}$.

Step 2. We shall show:

$$\lim_{d \to 0} d^{-r} \left\| F_d'(\Pi_d(u_0))^{-1} F_d(\Pi_d(u_0)) \right\| = 0. \tag{3.5.21}$$

Note that

$$\| F_d(\Pi_d(u_0)) \| \leq c_4 \| F_d(\Pi_d(u_0)) - F_d(u_0) \|$$
$$\leq c_4 \int_0^1 \| G_t \| \, dt \, \| \Pi_d(u_0) - u_0 \|$$
$$\leq c_4 c_5 \| \Pi_d(u_0) - u_0 \|, \tag{3.5.22}$$

where

$$G_t = F'((1 - t) u_0 + t \Pi_d(u_0)) \tag{3.5.23}$$

and we used

$$\| G_t \| \leq \left\| G_t - F'(u_0) \right\| + \left\| F'(u_0) \right\|$$
$$\leq c_0 t \| \Pi_d(u_0) - u_0 \| + \left\| F'(u_0) \right\| \leq c_5 \tag{3.5.24}$$

where c_5 is independent of d.

The claim is proved.

Step 3. We use our modification of the Newton-Kantorovich theorem with the following choices:

$A = S_d \subseteq Z$, with norm $d^{-r} \|w_d\|_X$,

$B = S_d \subseteq U$ with norm $d^{-r} \|w_d\|_Y$,

$x_0 = \Pi_d (u_0)$,

$F = F_d$.

Notice that $\|S\|_{L(A,B)} = \|S\|_{L(X,Y)}$ for any linear operator $S \in L(S_d, S_d)$.

By Step 1, we know $F_d' (\Pi_d (u_0)) \in L(S_d, S_d)$ is an isomorphism.

It follows from (3.5.6) and (A_4) that for any $w_d, v_d \in S_d$,

$$\left\| F_d' (w_d) - F_d' (v_d) \right\| \leq c_1 c_4 \|w_d - v_d\|_Z$$

$$\leq c_1 c_2 c_4 d^{-r} \|w_d - v_d\|_X \qquad (3.5.25)$$

Similarly, we get using (3.5.7) and (A_4) that

$$\left\| F_d' (w_d) - F_d' (\Pi_d (u_0)) \right\| \leq c_1 c_2 c_4 d^{-r} \|w_d - x_0\|_X .$$

Hence assumptions are satisfied with

$$\ell = c_1 c_2 c_3^{-1} c_4 \quad \text{and} \quad \ell_0 = c_0 c_2 c_3^{-1} c_4. \qquad (3.5.26)$$

From Step 2, we may take sufficiently small $d > 0$ such that $(\ell_0 + \ell) \eta \leq 1$, where

$$\eta = d^{-r} \left\| F_d' (\Pi_d (u_0))^{-1} F_d (\Pi_d (u_0)) \right\|_X .$$

That is, assumption $h_1 \leq 1$ is satisfied.

Hence for sufficiently small $d > 0$ there exists a locally unique $u_d \in S_d$ such that $F_d (u_d) = 0$ and

$$\|u_d - \Pi_d (u_0)\|_X \leq 2d^r \eta \leq 2c_3^{-1} \|F_d (\Pi_d (u_0))\|_Y$$

$$\leq 2c_3^{-1} c_4 c_5 \|u_0 - \Pi_d (u_0)\|_X .$$

It follows (3.5.14) holds with $\ell_1 = 2c_3^{-1} c_4 c_5$.

That completes the proof of the Theorem.

Remark 3.5.3. In general

$$c_0 \leq c_1 \quad \text{(i.e., } \ell_0 \leq \ell \text{)} \qquad (3.5.27)$$

holds and $\frac{\ell}{\ell_0}$ can be arbitrarily large, where ℓ and ℓ_0 are given by (3.5.26).

If $\ell = \ell_0$ our Theorem 3.5.2 reduces to the corresponding Theorem 2.1 in [194, p. 126].

Otherwise our condition $h_1 \leq 1$ is weaker than the corresponding one in [194] using the Newton-Kantorovich hypothesis $h = 2\ell\eta \leq 1$.

Note also that our parameter d will be smaller than the corresponding one in [194], which in turn implies fewer computations and smaller dimension subspaces S_d are used to approximate u_d. This observation is very important in computational mathematics.

The above observations suggest that all results obtained in [194] can be improved if rewritten with weaker $h_1 \leq 1$ instead of stronger $h \leq 1$.

However, we do not attempt this here (leaving this task to the motivated reader). Instead we provide examples of nonlinear problems already reported in [194] where finite element methods apply along the lines of our theorem above.

Example 3.5.4. Find $u \in H_0^1(J)$, $J = (b, c) \subseteq \mathbf{R}$ such that

$$\langle F(u), v \rangle = \int_J \left[g_0\left(x, u, u'\right) v' + g\left(x, u, u'\right) v \right] dx$$

$$= 0, \qquad \text{for all } v \in H_0^1(J) \tag{3.5.28}$$

where g_0 and g_1 are sufficiently smooth functions from $J \times \mathbf{R} \times \mathbf{R}$ to $\mathbf{R}$.

Example 3.5.5. For the N-dimensional case ($N = 2, 3$) let $D \subseteq \mathbf{R}^N$ be a bounded domain with a Lipschitz boundary. Then consider the problem:
find $u \in H_0^1(D)$ such that

$$\langle F(u), v \rangle = \int_D [q_0(x, u, \nabla u) \cdot \nabla v + q(x, u, \nabla u) \cdot v] dx$$

$$= 0, \qquad \text{for all } v \in H_0^1(D), \tag{3.5.29}$$

where $q_0 \in D \times \mathbf{R} \times \mathbf{R}^N$ to $\mathbf{R}$ are sufficiently smooth functions.

Example 3.5.6. Because equations (3.5.28) and (3.5.29) are defined in divergence form, their finite element solutions are defined in a natural way. Finite element methods applied to nonlinear elliptic boundary value problems have also been considered by other authors [93], [168].

3.6 Convergence of the structured PSB update in Hilbert space

A finer semilocal convergence analysis for the structured PSB update in Hilbert space is provided here based on Theorem 2.2.11 instead of the NK Theorem 2.2.4 used in [134]. Our results extend the applicability of the update algorithm. The advantages of our approach have already been explained in Section 2.2.

The motivation and the definition of the quantities introduced in the algorithm as well as applications to optimal shape design can be found in the elegant paper by Laumen [134] (see also the references therein). Laumen used Theorem 3.2 given by Dennis in [74, p. 438] to provide his Newton-Kantorovich-type Theorem 2.2 upon which the semilocal convergence of the algorithm was based. In particular, he justified the choice of the PSB Update (Powell symmetric Broyden update),

$$B_+ = B + [(q - Bw) \otimes w + w \otimes (q - Bw)] / \langle v, w \rangle$$
$$- [\langle q - Bw, w \rangle] w \otimes w / \langle w, w \rangle^2.$$

We are concerned with the problem of approximating a locally unique solution u^* of the minimization problem

$$\min_{u \in H} F(u) \tag{3.6.1}$$

using the algorithm [134]:

Structured Quasi-Newton method in Hilbert Space H.

Step 1. Given $u \in H$, $E \in L(H)$, $B = C(u) + E \in L(H)$.

Step 2. Compute w as the solution of

$$\langle Bw, v \rangle = \langle -F'(u), v \rangle, \quad \forall v \in H.$$

Step 3. Set $u_+ = u + w$.

Step 4. Choose $q^{\#}$ approximately.

Step 5. Set $q = C(u_+) + q^{\#}$.

Step 6. Update the quasi-Newton operator

$$E_+ = B\left(E, q^{\#}, w\right),$$

and set

$$B_+ = C(u_+) + E_+.$$

We state and prove the main semilocal convergence result for the structured PSB Update.

Theorem 3.6.1. *Let H be a Hilbert space, and let $F'(\cdot): U \subseteq H \to L(H)$ be Fréchet-differentiable. Suppose there exist $u_0 \in U$ and parameters $\delta \in [0, 2)$, $\gamma, \rho, C_0, L_{F''}^0, L_{F''}, L_c$, such that*

$$B_0^{-1} = [C(u_0) + E_0]^{-1} \in L(H),$$

$$\left\| B_0^{-1}(B_0 - F''(u_0)) \right\| \le \gamma,$$

$$\left\| B_0^{-1} F(u_0) \right\| \le \rho,$$

$$\left\| F''(u) - F''(u_0) \right\| \le L_{F''}^0 \|u - u_0\|, \quad \forall u \in U, \tag{3.6.2}$$

$$\left\| F''(u) - F''(w) \right\| \le L_{F''} \|u - w\|, \quad \forall u, w \in U,$$

$$\|C(u) - C(w)\| \le L_C \|u - w\|, \quad \forall u, w \in U,$$

$$\left\| q^{\#} - D(u_+) w \right\| \le C_0 \|w\|^2, \quad \forall u, w \in U,$$

$$\left\| B_\eta - F''(u_\eta) \right\| \le \left\| B_0 - F''(u_0) \right\| + (2C_0 + L_C + L_{F''}) \sum_{j=1}^{\eta} \|u_j - u_{j-1}\|,$$

$$h_\delta = \left(3L_{F''} + 4C_0 + 2L_C + L_{F''}^0\right) \rho \le \delta - [2\gamma + (\gamma_0 + \gamma)\delta] \tag{3.6.3}$$

$$2\gamma + (\gamma_0 + \gamma)\delta \le \delta, \tag{3.6.4}$$

and

$$U\left(u_0, \frac{2\rho}{2-\delta}\right) \subseteq U.$$

Then, the quasi-Newton method with structured PSB Update is well defined and converges to $u^* \in U\left(u_0, \frac{2\rho}{2-\delta}\right)$, where u^* is the unique solution of $F'(u) = 0$ in $U(u_0, t^*)$, where

$$t^* = \lim_{n \to \infty} t_n \leq \frac{2\rho}{2-\delta},$$

$$t_0 = 0, t_1 = \rho,$$

$$t_{n+2} = t_{n+1} + \frac{1}{2a_n}\left[L_{F''}(t_{n+1} - t_n) + 2\gamma + 2(2C_0 + L_C + L_{F''})t_n\right](t_{n+1} - t_n)$$

and

$$a_n = 1 - \left[\gamma_0 + \gamma + \frac{2\rho}{2-\delta}\left(1 - \left(\frac{\delta}{2}\right)^{n+1}\right)\left(2C_0 + L_C + L_{F''} + L_{F''}^0\right)\right].$$

Moreover, the solution u^* is unique in $U^0\left(u_0, t_1^*\right)$, provided that

$$U^0\left(u_0, t_1^*\right) \subseteq U,$$

and

$$\frac{L_{F''}^0}{2}\left(t^* + t_1^*\right) \leq 1.$$

Furthermore, the following estimates hold for all $n \geq 0$:

$$\|u_{n+1} - u_n\| \leq t_{n+1} - t_n,$$

and

$$\|u_{n+1} - u^*\| \leq t^* - t_n.$$

Proof. It follows immediately from Lemma 5.1.1 and Theorem 5.1.2 in Section 5.1 by simply replacing $b_n - \Delta$, $c_n - a_n$, h_δ^n, K_0, K_1, K, d, q_n given in Section 5.1 (see also [41]) by γ, a, $h_\delta = \frac{(2L_{F''} + 2C_0 + L_C)\rho + 2\gamma}{a}$, $L_{F''}^0$, $2C_0 + L_C + L_{F''}$, γ, $1 - a$ defined above respectively.

Remark 3.6.2. Lemma 5.1.1 and Theorem 5.1.2 in Section 5.1 (see also [41]) were shown under even weaker hypotheses. However in order for us to compare with Theorem 2.2 [134, p. 404] given below it is preferred to provide only the above stated results.

Although the results in [134] were not given in affine invariant form, we modify and present them here in such a way that they will be comparable with the corresponding ones in our Theorem 3.6.1 above, so that an equitable comparison can be made.

Theorem 3.6.3. *[134, p. 404]. Assume conditions of Theorem 3.6.1 but replace (3.6.2), (3.6.3), t^*, t_1^* by (3.6.5), (3.6.6) r^*, r_1^**

$$h = \frac{(4C_0 + 2L_C + 3L_{F''})\rho}{(1 - 3\gamma)^2} \leq \frac{1}{2}, \qquad (3.6.5)$$

$$3\gamma < 1 \qquad (3.6.6)$$

$$r^* = \frac{(1 - \sqrt{1 - 2h})(1 - 3\gamma)}{4C_0 + 2L_C + 3L_{F''}}$$

and

$$r_1^* = \frac{\left(1 - \sqrt{1 - 2h^1}\right)(1 - \gamma)}{L_{F''}},$$

where

$$h^1 = \frac{CL_{F''}}{(1 - \gamma)^2} \leq \frac{1}{2},$$

respectively.

Then the conclusions of Theorem 3.6.1 hold in this setting.

Note that condition (3.6.2) is not used in Theorem 3.6.3. This allows a greater flexibility. On one hand, Theorem 3.6.3 can be reduced to Theorem 3.6.1 if $L_{F''}^0 = L_{F''}$.

However, in general

$$L_{F''}^0 \leq L_{F''}$$

holds and $\frac{L_{F''}}{L_{F''}^0}$ can be arbitrarily large. Moreover, it can easily be seen (simply compare (3.6.5) with (3.6.3)) that condition (3.6.5) $\Longrightarrow$ (3.6.3), provided that (3.6.4) holds together with

$$\delta \in [\delta_0, 2),$$

and

$$4(\gamma_0 + 2\gamma) + (1 - 3\gamma)^2 < 4,$$

where

$$\delta_0 = \frac{4\gamma + (1 - 3\gamma)^2}{2[1 - (\gamma_0 + \gamma)]},$$

and ρ is sufficiently small.

Note also that in an even more general setting (see Theorem 2, Remark 1 in [41] and Theorem 3.2 in [74]), it was shown in [41] that $t^* \leq r^*$ and upper bounds on the distances $\|u_n - u_{n-1}\|$, $\|u_n - u^*\|$ are finer.

Finally note that all the above advantages are obtained under the same computational cost because in practice the computation of $L_{F''}$ requires that of $L_{F''}^0$.

Hence their usefulness in optimizing the convergence of the structured PSB Update has been established.

3.7 On the shadowing lemma for operators with chaotic behavior

It is well-known that complicated behavior of dynamical systems can easily be detected via numerical experiments. However, it is very difficult to prove mathematically in general that a given system behaves chaotically.

Several authors have worked on various aspects of this problem, see, e.g., [157], [186], and the references therein. In particular, the shadowing lemma [157, p. 1684] proved via the celebrated Newton-Kantorovich Theorem 2.2.4 was used in [157] to present a computer-assisted method that allows us to prove that a discrete dynamical system admits the shift operator as a subsystem. Motivated by this work and using a weaker version of the Newton-Kantorovich Theorem 2.2.4 reported by us in Theorem 2.2.11 (see Theorem 3.7.1 that follows), we show that it is possible to weaken the shadowing Lemma on on which the work in [157] is based. In particular, we show that under weaker hypotheses and the same computational cost, a larger upper bound on the crucial norm of operator L^{-1} (see (3.7.7)) is found and the information on location of the shadowing orbit is more precise. Other advantages have already been reported in Section 2.2. Clearly this approach widens the applicability of the shadowing lemma.

We need the definitions: Let $D \subseteq \mathbf{R}^k$ be an open subset of $\mathbf{R}^k$ (k a natural number), and let $f: D \to D$ be an injective operator. Then the pair (D, f) is a discrete dynamical system. Denote by $S = l^\infty (\mathbf{Z}, \mathbf{R}^k)$ the space of $\mathbf{R}^k$ valued bounded sequences $x = \{x_n\}$ with norm $\|x\| = \sup_{n \in \mathbf{Z}} |x_n|_2$. Here we use the Euclidean norm in $\mathbf{R}^k$ and denote it by $|\cdot|$, ommitting the index 2. A δ_0-pseudo-orbit is a sequence $y = \{y_n\} \in D^{\mathbf{Z}}$ with $|y_{n+1} - f(y_n)| \le \delta_0$ ($n \in \mathbf{Z}$). A r-shadowing orbit $x = \{x_n\}$ of a δ_0-pseudo-orbit y is an orbit of (D, f) with $|y_n - x_n| \le 2$ ($n \in \mathbf{Z}$).

We need the following version for Theorem 2.2.11.

Theorem 3.7.1. *Let $F: D \subseteq X \to Y$ be a Fréchet-differentiable operator. Assume there exist $x_0 \in D$ and positive constant η, β, L_0 and L such that $F'(x_0)^{-1} \in L(Y, X)$,*

$$\left\| F'(x_0)^{-1} \right\| \le \beta, \tag{3.7.1}$$

$$\left\| F'(x_0)^{-1} F(x_0) \right\| \le \eta, \tag{3.7.2}$$

$$\left\| F'(x) - F'(y) \right\| \le L \|x - y\|, \quad \textit{for all } x, y \in D, \tag{3.7.3}$$

$$\left\| F'(x) - F'(x_0) \right\| \le L_0 \|x - x_0\|, \quad \textit{for all } x \in D, \tag{3.7.4}$$

$$h_A = \beta (L_0 + L) \eta \le 1 \tag{3.7.5}$$

and

$$\bar{U}(x_0, s^*) \subseteq D,$$

where $s^* = \lim_{n \to \infty} s_n$,

$$s_0 = 0, s_1 = \eta, s_{n+2} = s_{n+1} + \frac{L(s_{n+1} - s_n)}{2(1 - L_0 s_{n+1})} \quad (n \ge 0).$$

Then sequence $\{y_n\}$ $(n \geq 0)$ generated by NK method

$$y_{n+1} = y_n - F'(y_n)^{-1} F(y_n) \quad (n \geq 0)$$

is well defined, remains in $\bar{U}(x_0, s^*)$ for all $n \geq 0$ and converges to a unique solution $y^* \in \bar{U}(x_0, s^*)$, so that estimates

$$\|y_{n+1} - y_n\| \leq s_{n+1} - s_n$$

and

$$\|y_n - y^*\| \leq s^* - s_n \leq 2\eta - s_n$$

hold for all $n \geq 0$.

Moreover y^* is the unique solution of equation $F(y) = 0$ in $U(x_0, R)$ provided that

$$L_0 (s^* + R) \leq 2$$

and

$$U(x_0, R) \subseteq D.$$

The advantages of Theorem 3.7.1 over the Newton-Kantorovich Theorem 2.2.4 have been explained in detail in Section 2.2.

From now on we set $X = Y = \mathbf{R}^k$.

Sufficient conditions for a δ_0-pseudo-orbit y to admit a unique r-shadowing orbit are given in the following main result.

Theorem 3.7.2. *(Weak version of the shadowing lemma) Let $D \subseteq \mathbf{R}^k$ be open, $f \in C^{1,\text{Lip}}(D, D)$ be injective, $y = \{y_n\} \in D^{\mathbf{Z}}$ be a given sequence, $\{A_n\}$ be a bounded sequence of $k \times k$ matrices and let $\delta_0, \delta, \ell_0, \ell$ be positive constants. Assume that for the operator*

$$L: S \to S \text{ with } \{Lz\}_n = z_{n+1} - A z_n \tag{3.7.6}$$

is invertible and

$$\left\|L^{-1}\right\| \leq a = \frac{1}{\delta + \sqrt{(\ell + \ell_0)\delta_0}}. \tag{3.7.7}$$

Then the numbers t^*, R given by

$$t^* = \lim_{n \to \infty} t_n \tag{3.7.8}$$

and

$$R = \frac{2}{\ell_0} - t^* \tag{3.7.9}$$

satisfy $0 < t^* \leq R$, where sequence $\{t_n\}$ is given by

$$t_0 = 0, t_1 = \eta, t_{n+2} = t_{n+1} + \frac{\ell (t_{n+1} - t_n)^2}{2 (1 - \ell_0 t_{n+1})} \quad (n \geq 0) \tag{3.7.10}$$

and

$$\eta = \frac{\delta_0}{\frac{1}{\|L^{-1}\|} - \delta}. \tag{3.7.11}$$

Let $r \in [t^*, R]$. Moreover, assume that

$$\overline{\bigcup_{n \in \mathbf{Z}} U(y_n, r)} \subseteq D \tag{3.7.12}$$

and for every $n \in \mathbf{Z}$

$$|y_{n+1} - f(y_n)| \leq \delta_0, \tag{3.7.13}$$

$$|A_n - Df(y_n)| \leq \delta, \tag{3.7.14}$$

$$\left| F'(u) - F'(0) \right| \leq \ell_0 |u| \tag{3.7.15}$$

and

$$\left| F'(u) - F'(v) \right| \leq \ell |u - v|, \tag{3.7.16}$$

for all $u, v \in U(y_n, r)$.

Then there is a unique t^*-shadowing orbit $x^* = \{x_n\}$ of y. Moreover, there is no orbit $\bar{x}$ other than x^* such that

$$\|\bar{x} - y\| \leq r. \tag{3.7.17}$$

Proof. We shall solve the difference equation

$$x_{n+1} = f(x_n) \quad (n \geq 0) \tag{3.7.18}$$

provided that x_n is close to y_n. Setting

$$x_n = y_n + z_n \tag{3.7.19}$$

and

$$g_n(z_n) = f(z_n + y_n) - A_n z_n - y_{n+1} \tag{3.7.20}$$

we can have

$$z_{n+1} = A_n z_n + g_n(z_n). \tag{3.7.21}$$

Define $D_0 = \{z = \{z_n\} : \|z\| \leq 2\}$ and nonlinear operator $G : D_0 \to S$, by

$$(G(z))_n = g_n(z_n). \tag{3.7.22}$$

Operator G can naturally be extended to a neighborhood of D_0. Equation (3.7.21) can be rewritten as

$$F(x) = Lx - G(x) = 0, \tag{3.7.23}$$

where F is an operator from D_0 into S.

We will show the existence and uniqueness of a solution $x^* = \{x_n\}$ $(n \geq 0)$ of equation (3.7.23) with $\|x^*\| \leq r$ using Theorem 3.7.1. Clearly we need to express η, L_0, L and β in terms of $\|L^{-1}\|$, δ_0, δ, ℓ_0 and ℓ.

(i) $\left\| F'(0)^{-1} F(0) \right\| \leq \eta$.

Using (3.7.13), (3.7.14) and (3.7.20), we get $\| F(0) \| \leq \delta_0$ and $\left\| G'(0) \right\| \leq \delta$, as $\left[G'(0)(w) \right]_n = \left(F'(y_n) - A_n \right) w_n$.

By (3.7.7) and the Banach lemma on invertible operators we get $F'(0)^{-1}$ exists and

$$\left\| F'(0)^{-1} \right\| \leq \left(\frac{1}{\| L^{-1} \|} - \delta \right)^{-1}. \tag{3.7.24}$$

That is, η can be given by (3.7.11).

(ii) $\left\| F'(0)^{-1} \right\| \leq \beta$

By (3.7.24) we can set

$$\beta = \left(\frac{1}{\| L^{-1} \|} - \delta \right)^{-1}. \tag{3.7.25}$$

(iii) $\left\| F'(u) - F'(v) \right\| \leq L \| u - v \|$

We can have using (3.7.16)

$$\left| \left(F'(u) - F'(v) \right)(w)_n \right| = \left| \left(F'(y_n + u_n) - F'(y_n + v_n) \right) w_n \right|$$
$$\leq \ell \left| u_n - v_n \right| | w_n |. \tag{3.7.26}$$

Hence we can set $L = \ell$.

(iv) $\left\| F'(u) - F'(0) \right\| \leq L_0 \| u \|$.

By (3.7.17) we get

$$\left| \left(F'(u) - F'(0) \right)(w)_n \right| = \left| \left(F'(y_n + u_n) - F'(y_n + 0) \right) w_n \right|$$
$$\leq \ell_0 | u_n | | w_n |. \tag{3.7.27}$$

That is, we can take $L_0 = \ell_0$.

Crucial condition (3.7.5) is satisfied by (3.7.7) and with the above choices of η, β, L and L_0.

Therefore the claims of Theorem 3.7.2 follow immediately from the conclusions of Theorem 3.7.1.

That completes the proof of the theorem.

Remark 3.7.3. In general

$$\ell_0 \leq \ell \tag{3.7.28}$$

holds and $\frac{\ell}{\ell_0}$ can be arbitrarily large. If $\ell_0 = \ell$, Theorem 3.7.2 reduces to Theorem 1 in [157, p. 1684]. Otherwise our Theorem 3.7.2 improves Theorem 1 in [157]. Indeed, the upper bound in [157, p. 1684] is given by

$$\left\| L^{-1} \right\| \leq b = \frac{1}{\delta + \sqrt{2\ell\delta_0}}. \tag{3.7.29}$$

By comparing (3.7.7) with (3.7.29) we deduce

$$b < a$$

(if $\ell_0 < \ell$).

3.8 The mesh independence principle and optimal shape design problems

Shape optimization is described by finding the geometry of a structure that is optimal in the sense of a given minimization cost function with respect to certain constraints. A Newton's mesh independence principle was very efficiently used to solve optimal design problems in [133]. Here motivated by optimization considerations, we show that under the same computational cost, an even finer mesh independence principle can be given.

We are concerned with the problem

$$\min_{u \in U} F(u) \tag{3.8.1}$$

where $F(u) = J(u, S(u), z(u)) + \frac{\varepsilon}{2} \|u - u_T\|^2$, $\varepsilon \in \mathbf{R}$, functions u_T, S, z, and J are defined on a function space (Banach or Hilbert) U with values in another function space V. Many optimal shape design problems can be formulated as in (3.8.1) [133]. In the excellent paper by W. Laumen [133], the mesh independence principle (see also [2]) was transferred to the minimization problem by the necessary first-order condition

$$F'(u) = 0 \text{ in } U. \tag{3.8.2}$$

The most popular method for solving (3.8.2) is given for $n \in \mathbf{N}$ by Newton's method

$$F''(u_{n-1})(w)(v) = -F'(u_{n-1})(v)$$

$$u_n = u_{n-1} + w,$$

where F, F', and F'' also depend on functions defined on the infinite-dimensional Hilbert space V. The discretization of this method is obtained by replacing the infinite-dimensional space V and U with the finite-dimensional subspaces V^M, U^M and the discretized NK method

$$F_N''\left(u_{n-1}^M\right)\left(w^M\right)\left(v^M\right) = -F_N'\left(u_{n-1}^M\right)\left(v^M\right),$$

$$u_n^M = u_{n-1}^M + w^M.$$

Here we show that under the same hypotheses and computational cost, a finer mesh independence principle can be given.

Let u_0 be chosen in the closed ball

$$U_* = U(u_*, r_*)$$

in order to guarantee convergence to the solution u_*. The assumptions concerning the cost function F_N, which are assumed to hold on a possible smaller ball $\hat{U}_* = U(u_*, \hat{r}_*)$ with $\hat{r}_* \leq r_*$ are stated below.

Assumption C1. There exist positive constants L_0, L and δ such that for all $u, v \in U_*$

$$\left\| F''(u) - F''(u_*) \right\| \leq L_0 \left\| u - u_* \right\|$$
$$\left\| F''(u) - F''(v) \right\| \leq L \left\| u - v \right\|$$
$$\left\| F''(u_*)^{-1} \right\| \leq \delta.$$

Assumption C2. There exist uniformly bounded Lipschitz constants $L_N^{(i)}$, $i = 1, 2$, such that

$$\left\| F_N'(u) - F_N'(v) \right\| \leq L_N^{(1)} \left\| u - v \right\|, \quad \text{for all } u, v \in \hat{U}_*, N \in \mathbf{N},$$
$$\left\| F_N''(u) - F_N''(v) \right\| \leq L_N^{(2)} \left\| u - v \right\|, \quad \text{for all } u, v \in \hat{U}_*, N \in \mathbf{N}.$$

Without loss of generality, we assume $L_N^{(i)} \leq L$, $i = 1, 2$, for all N.

Assumption C3. There exist a sequence $z_N^{(1)}$ with $z_N^{(1)} \to 0$ as $N \to \infty$, such that

$$\left\| F_N'(u) - F'(u) \right\| \leq z_N^{(1)}, \quad \text{for all } u \in \hat{U}_*, N \in \mathbf{N},$$
$$\left\| F_N''(u) - F''(u) \right\| \leq z_N^{(2)}, \quad \text{for all } u \in \hat{U}_*, N \in \mathbf{N}.$$

Assumption C4. There exists a sequence $z_N^{(2)}$ with $z_N^{(2)} \to 0$ as $N \to \infty$ such that for all $N \in \mathbf{N}$ there exists a $\hat{u}^N \in U^N \times \hat{U}_*$ such that

$$\left\| \hat{u}^N - u_* \right\| \leq z_N^{(2)}.$$

Assumption C5. F_N' and F_N'' correspond with the derivatives of F_N.

The cost function F is assumed to be twice continuously Fréchet-differentiable. Therefore, its first derivative is also Lipschitz continuous:

$$\left\| F'(u) - F'(v) \right\| \leq \hat{L} \left\| u - v \right\|, \quad \text{for all } u, v \in U_*.$$

Without loss of generality we assume $\hat{L} \leq L$.

Remark 3.8.1. In general

$$L_0 \leq L \tag{3.8.3}$$

holds and $\frac{L}{L_0}$ can be arbitrarily large. If $L_0 = L$, our Assumptions C1–C5 coincide with the ones in [133, p. 1074].

Otherwise our assumptions are finer and under the same computational cost as in practice the evaluation of L requires the evaluation of L_0. This modification of the assumptions in [133] will result in larger convergence balls U_* and $\hat{U}_*$, which in turn implies a wider choice of initial guesses for Newton's method and finer bounds on the distances involved. This observation is important in computational mathematics.

We now justify the claims made in the previous remark, as follows:

$$\delta \left\| F''(u_*) - F_N''\left(\hat{u}^M\right)\right\| \leq \delta \left[\left\| F''(u_*) - F''\left(\hat{u}^M\right)\right\| + \left\| F''\left(\hat{u}^M\right) - F_N''\left(\hat{u}^M\right)\right\|\right]$$

$$\leq \delta \left[L_0 z_M^{(2)} + z_N^{(1)}\right]$$

$$\leq \delta \hat{z} < 1$$

hold for a constant $\hat{z}$ if M and N are sufficiently large. It also follows by the Banach Lemma on invertible operators that $F''\left(\hat{u}^M\right)^{-1}$ exists and

$$\left\| F''\left(\hat{u}^M\right)^{-1}\right\| \leq \frac{\delta}{1 - \delta \hat{z}} = \hat{\delta}.$$

We showed in Section 2.4 that if

$$r_* \leq \frac{2}{(2L_0 + L)\,\delta} < \frac{1}{\delta L_0}, \tag{3.8.4}$$

then the estimates

$$\delta \left\| F''(u_i) - F''(u_*)\right\| \leq \delta L_0 \left\| u_i - u_*\right\| \leq \delta L_0 r_* < 1$$

hold, which again also imply the existence of $F''(u_i)$ with

$$\left\| F''(u_i)^{-1}\right\| \leq \frac{\delta}{1 - \delta L_0 r_*} = \hat{\delta}. \tag{3.8.5}$$

Hence by redefining δ by $\hat{\delta}$ if necessary, we assume that

$$\left\| F''(u_i)^{-1}\right\| \leq \delta, \quad \text{for all } i \in \mathbf{N} \tag{3.8.6}$$

$$\left\| F_N''\left(\hat{u}^M\right)^{-1}\right\| \leq \delta, \quad \text{for all } \hat{u}^M \in U^M, N \in \mathbf{N}, \tag{3.8.7}$$

for M and N satisfying

$$\delta \left[L_0 z_M^{(2)} + z_N^{(1)}\right] \leq \delta \hat{z} < 1. \tag{3.8.8}$$

The next result is a refinement of Theorem 2.1 in [133, p. 1075], which also presents sufficient conditions for the existence of a solution of the problem

$$\min_{u \in U^M} F_N\left(u^M\right) \tag{3.8.9}$$

and shows the convergence of Newton's method for $M, N \to \infty$.

Theorem 3.8.2. *Assume C1–C5 hold and parameters M, N satisfy*

$$Z_{MN} = 2\delta \left[\max\{1, L_0\} + \frac{1}{2\delta}\right]\left(z_N^{(1)} + z_M^{(2)}\right) \leq \min\left\{r_*, \frac{1}{\delta L}\right\}. \tag{3.8.10}$$

Then the discretized Newton's method has a local solution $u_*^M \in \hat{U}_*$ satisfying

$$\left\| u_*^M - u_*\right\| < z_{MN}. \tag{3.8.11}$$

Proof. We apply the Newton-Kantorovich Theorem 2.2.4 to Newton's method starting at $u_0^M = \hat{u}^M$ to obtain the existence of a solution u_*^M of the infinite-dimensional minimization problem. Using C2–C4, (3.8.7), and (3.8.10), we obtain in turn

$$
\begin{aligned}
2h = 2\delta L \left\| F_N'' \left(\hat{u}^M \right)^{-1} F_N' \left(\hat{u}^M \right) \right\| \\
\leq 2\delta L \left\| F_N'' \left(\hat{u}^M \right)^{-1} \right\| \left\| F_N' \left(\hat{u}^M \right) \right\| \\
\leq 2\delta^2 L \left(\left\| F_N' \left(\hat{u}^M \right) - F' \left(\hat{u}^M \right) \right\| + \left\| F' \left(\hat{u}^M \right) - F' \left(u_* \right) \right\| \right) \\
\leq 2\delta^2 L \left(z_N^{(1)} + L_0 \left\| \hat{u}^M - u_* \right\| \right) \\
\leq 2\delta^2 L \left(\max \{1, L_0\} + \frac{1}{2\delta} \right) \left(z_N^{(1)} + z_M^{(2)} \right) \\
\leq \delta L z_{MN} \leq 1
\end{aligned}
\tag{3.8.12}
$$

which imply the required assumption $2h < 1$ (for the quadratic convergence).

We also need to show $U \left(\hat{u}^M, r(h) \right) \subset U \left(u_*, \hat{r}_* \right)$. By C4 is suffices to show

$$
r(h) = \frac{1}{\delta L} \left(1 - \sqrt{1 - 2h} \right) \leq \hat{r}_* - z_M^{(2)}.
\tag{3.8.13}
$$

But by (3.8.10) and the definition of $r(h)$, we get

$$
\begin{aligned}
r(h) = 2\delta \max \{1, L_0\} \left(z_N^{(1)} + z_M^{(2)} \right) \\
< 2\delta \left(\max \{1, L_0\} + \frac{1}{2\delta} \right) \left(z_N^{(1)} + z_M^{(2)} \right) - z_M^{(2)} \\
\leq z_{MN} - z_M^{(2)} \\
\leq \hat{r}_* - z_M^{(2)},
\end{aligned}
\tag{3.8.14}
$$

which shows estimate (3.8.13).

Hence, there exists a solution $u_*^M \in U \left(\hat{u}^M, r(h) \right)$ such that

$$
\left\| u_*^M - u_* \right\| \leq \left\| u_*^M - \hat{u}^M \right\| + \left\| \hat{u}^M - u_* \right\| < z_{MN} - z_M^{(2)} + z_M^{(2)} = z_{MN}.
\tag{3.8.15}
$$

That completes the proof of the Theorem.

Remark 3.8.3. If equalities hold in (3.8.7), then our Theorem 3.8.2 reduces to Theorem 2.1 in [133]. Otherwise it is an improvement (and under the same computational cost) as $\hat{z}, M, N, z_{MN}$ are smaller and $\hat{\delta}$ (i.e., δ), $r_*, \hat{r}_*$ are larger than the corresponding ones in [133, p. 1075] and our condition (3.8.12) is weaker than the corresponding (2.5) in [133] (i.e., set $L_0 = L$ in (3.8.12)).

That is, the claim made in Remark 3.8.1 is justified, and our Theorem extends the applicability of the mesh independence principle.

Remark 3.8.4. In practice, we want $\min\left\{r_*, \frac{1}{\delta L}\right\}$ to be as large as possible. It then immediately follows from (3.8.4), (3.8.12), and (3.8.14) that the conclusions of Theorem 3.8.2 hold if z_{MN} given in (3.8.10) is replaced by

$$z_{MN}^1 = \frac{2\delta L}{L_0}\left[\max\{1, L_0\} + \frac{1}{2\delta}\right]\left(z_N^{(1)} + z_M^{(2)}\right) \le \min\left\{r_*, \frac{1}{\delta L_0}\right\}. \qquad (3.8.16)$$

Another way is to rely on our Theorem 2.2.11 using the weaker (than (3.8.12)) Newton-Kantorovich-type hypothesis

$$h_0 = (L_0 + L)\,\delta\left\|F_N''\left(\hat{u}^M\right)^{-1} F_N'\left(\hat{u}^M\right)\right\| \le 1 \qquad (3.8.17)$$

or as in (3.8.16) for (3.8.17) to hold we must have

$$h_0 < L_0\left(1 + \frac{L}{L_0}\right)\delta^2\left[\max\{1, L_0\} + \frac{1}{2\delta}\right]\left(z_N^{(1)} + z_M^{(2)}\right)$$

$$\le \delta L_0 z_{MN}^0 \le 1,$$

provided that

$$z_{MN}^0 = \left(1 + \frac{L}{L_0}\right)\delta\left[\max\{1, L_0\} + \frac{1}{2\delta}\right]\left(z_N^{(1)} + z_M^{(2)}\right)$$

$$\le \min\left\{r_*, \frac{1}{\delta L_0}\right\}. \qquad (3.8.18)$$

The other hypothesis for the application of our Theorem 2.2.11: $U\left(\hat{u}^M, r^1\left(h\right)\right) \subseteq U\left(u_*, \hat{r}_*\right)$, where

$$r^1\left(h\right) = 2\delta \max\{1, L_0\}\left(z_N^{(1)} + z_M^{(2)}\right).$$

Hence we arrived at:

Theorem 3.8.5. *Under the hypotheses of Theorem 3.8.2 with z_{MN} replaced by z_{MN}^0 (given in (3.8.15)) the conclusions of this theorem hold.*

So far we showed that a solution

$$u_*^M \in U\left(\hat{u}^M, r\left(h\right)\right) \ \left(\text{or } \left(\hat{u}^M, r^1\left(h\right)\right)\right) \subset U\left(u_*, \hat{r}_*\right)$$

of the discretized minimization problem exists.

Next, in the main results of this section we show two different ways of improving the corresponding Theorem 2.2 in [133, p. 1076], where it was shown that the discretized Newton's method converges to the solution u_*^M for any $u_0^M \in U\left(u_*, r_1\right)$ for sufficiently small r_1.

In order to further motivate the reader let us provide a simple numerical example.

Example 3.8.6. Let $U = \mathbf{R}$, $U_* = U(0, 1)$ and define the real function F on U_* by

$$F(x) = e^x - 1. \tag{3.8.19}$$

Then we obtain using (3.8.19) that $L = e$, $L_0 = e - 1$ and $\delta = 1$. We let $z_N^{(1)} = 0$, $z_M^{(2)} = \frac{1}{M}$. Set
$r^* = \hat{r}_*$ and $N = 0$.

The convergence radius given in [133, p. 1075] is

$$r_*^L = \hat{r}_*^L = \frac{2}{3\delta L} = .24525246,$$

whereas by (3.8.4) ours is given by

$$r^* = \hat{r}_* = \frac{2}{(2L_0 + L)\delta} = .324947231. \tag{3.8.20}$$

That is, (3.8.3) holds as a strict inequality and

$$r_*^L = \hat{r}_*^L < r^* = \hat{r}_*.$$

The condition (2.4) used in [133, p. 1075] corresponding to ours (3.8.10) is given by

$$z_{MN}^L = 2\delta \left[\max\{1, L_0\} + \frac{1}{2\delta} \right] \left(z_N^{(1)} + z_M^{(2)} \right) \leq \min\left\{ \hat{r}_*^L, \frac{1}{\delta L} \right\}. \tag{3.8.21}$$

We can tabulate the following results containing the minimum M for which conditions (3.8.8), (3.8.10), (3.8.16), and (3.8.15) are satisfied.

M	z_{MN} (3.8.21)	r_*^L (3.8.18)	z_{MN}(3.8.10)	z_{MN}(3.8.16)	z_{MN}(3.8.15)	r_*(3.8.20)
27	.238391246	.24525296	.164317172	.259945939	.212131555	.324947231
22				.319024562		
19			.23350335			
18					.318197333	

Table 3.8.1. Comparison table.

The above table indicates the superiority of our results over the ones in [133, p. 1075].

We can now present a finer version than Theorem 2.2 in [133] of the mesh independence principle.

Theorem 3.8.7. *Suppose:*

Assumptions C1–C5 are satisfied and there exist discretization parameters M and N such that

$$z_{MN} \leq \frac{1}{6} \min \left\{ \frac{\hat{r}_*}{4}, \frac{1}{(2L_0 + 3L)\,\delta + 1} \right\}. \qquad (3.8.22)$$

Then the discretized Newton's method converges to u_*^M for all starting points $u_0^M \in U(u_*, r_1)$, where

$$r_1 = \frac{3}{4} \min \left\{ \frac{1}{(2L_0 + L)\,\delta}, \frac{\hat{r}_*}{2} \right\}. \qquad (3.8.23)$$

Moreover, if

$$\left\| u_0^M - u_0 \right\| \leq \tau,$$

where

$$\tau = \frac{2 \left(\frac{1}{2} + \| u_0 - u_* \| \right) z_{MN}}{b^2 + \sqrt{b^2 - 6L\delta \left(\frac{1}{2} + \| u_0 - u_* \| \right) z_{MN}}} \qquad (3.8.24)$$

and

$$b = 1 + \frac{1}{2} z_{MN} - 2\delta L \left\| u_0^M - u_* \right\|$$

the following estimates hold for $c_i \in \mathbf{R}$, $i = 1, 2, 3, 4$, $n \in \mathbf{N}$:

$$\left\| u_{n+1}^M - u_*^M \right\| \leq c_1 \left\| u_n^M - u_*^M \right\|^2, \qquad (3.8.25)$$

$$\left\| u_n^M - u_n \right\| \leq c_2 z_{MN}$$

$$\left\| F_N' \left(u_n^M \right) - F'(u_n) \right\| \leq c_3 z_{MN}$$

and

$$\left\| u_n^M - u_*^M \right\| \leq \| u_n - u_* \| + c_4 z_{MN}.$$

Proof. We first show the convergence of the discretized Newton's method for all u_0^M in a suitable ball around u_*. Because the assumptions of Theorem 3.8.2 are satisfied, the existence of a solution $u_*^M \in \hat{U}_*$ is guaranteed. We shall show that the discretized Newton method converges to u_*^M if $u_0^M \in U(u_*, r_2)$, where

$$r_2 = \min \left\{ \frac{1}{(2L_0 + L)\,\delta}, \frac{\hat{r}_*}{2} \right\}.$$

The estimates

$$\left\| u_*^M - u_* \right\| + \left\| u_0^M - u_*^M \right\| \leq 2 \left\| u_*^M - u_* \right\| + \left\| u_0^M - u_* \right\|$$

$$\leq 2 z_{MN} + r_2 \leq \hat{r}_*$$

imply

$$U \left(u_*^M, \left\| u_0^M - u_*^M \right\| \right) \subset \hat{U}_*.$$

Hence, Assumptions C1–C5 hold in $U \left(u_*^M, \left\| u_0^M - u_*^M \right\| \right)$.

We can also have

$$\left\| F''(u_*)^{-1} \right\| \left\| F_N''\left(u_*^M\right) - F''(u_*) \right\| \leq$$

$$\leq \delta \left[\left\| F_N''\left(u_*^M\right) - F_N''(u_*) \right\| + \left\| F_N''(u_*) - F''(u_*) \right\| \right]$$

$$\leq \delta \left[L \left\| u_*^M - u_* \right\| + z_{MN} \right]$$

$$\leq (\delta L + 1) z_{MN}$$

$$\leq \frac{\delta L + \frac{1}{2}}{(2_0 + 3L)\delta + 1} < 1, \tag{3.8.26}$$

where we used

$$z_n^{(1)} \leq \frac{1}{2\delta} z_{MN} \text{ (by (3.8.10))}.$$

It follows by (3.8.26) and the Banach Lemma on invertible operators that $F_N''\left(u_*^M\right)^{-1}$ exists and

$$\left\| F_N''\left(u_*^M\right)^{-1} \right\| \leq \frac{\delta}{1 - \left(\delta L + \frac{1}{2}\right) z_{MN}}.$$

By the theorem on quadratic convergence of Newton's method and since all assumptions hold, the convergence to u_*^M has been established.

Using a refined formulation of this theorem given by us in Section 2.4, the convergence is guaranteed for all $u_0^M \in U\left(u_*^M, r_3\right)$, where

$$r_3 = \frac{2}{(2L_0 + L) \left\| F_N''\left(u_*^M\right)^{-1} \right\|}.$$

Therefore we should show

$$U(u_*, r_2) \subset U\left(u_*^M, r_3\right)$$

or equivalently

$$\left\| u_0^M - u_*^M \right\| \leq \left\| u_0^M - u_* \right\| + \left\| u_* - u_*^M \right\|$$

$$\leq r_2 + z_{MN}$$

$$\leq \frac{1}{(2L_0 + L)\delta} + z_{MN}$$

$$= \frac{1 + (2L_0 + L)\delta}{(2L_0 + L)\delta}$$

$$\leq \frac{2\left(1 - L\delta z_{MN} - \frac{1}{2} z_{MN}\right)}{(2L_0 + L)\delta}$$

$$\leq \frac{2}{(2L_0 + L) \left\| F_N''\left(u_*^M\right)^{-1} \right\|} = r_3.$$

Hence, the discretized Newton's method converges to u_*^M for all $u_0^M \in U(u_*, r_2)$ such that (3.8.25) holds for $c_1 = \delta L$.

Next a proof by induction is used to show

$$\left\| u_n^M - u_n \right\| \leq \tau \leq c_2 z_{MN} \tag{3.8.27}$$

for all $u_0^M \in U(u_*, r_1)$, $r_1 = \frac{3}{4} r_2$, where τ is given by

$$\tau = \frac{2\left(\frac{1}{2} + \|u_0 - u_*\|\right) z_{MN}}{b^2 + \sqrt{b^2 - 6L\delta \left(\frac{1}{2} + \|u_0 - u_*\|\right) z_{MN}}}$$

$$\leq \frac{\left(\frac{1}{2} + \|u_0 - u_*\|\right) z_{MN}}{b^2} =: c_2 z_{MN}$$

with $b = 1 + \frac{1}{2} z_{MN} - 2\delta L \left\| u_0^M - u_* \right\|$. The constant τ is well defined, as the inequalities

$$6L\delta \left(\frac{1}{2} + \|u_0 - u_*\|\right) z_{MN} \leq \frac{2L\delta + 1}{4\left((2L_0 + 3L)\delta + 1\right)} < \frac{1}{4} \text{ and } b \geq 1 - 2\delta L \rho_1 \geq \frac{1}{2}$$

imply $b^2 \geq \frac{1}{4} \geq 6L\delta \left(\frac{1}{2} + \|u_0 - u_*\|\right) z_{MN}$.

While the assertion (3.8.27) is fulfilled by assumption for $n = 0$, the induction step is based on the simple decomposition

$$u_{i+1}^M - u_{i+1} = F_N''\left(u_i^M\right)^{-1} \left\{ F_N''\left(u_i^M\right)\left(u_i^M - u_i\right) - F_N'\left(u_i^M\right) + F_N'(u_i) \right.$$

$$+ \left(F_N''\left(u_i^M\right) - F_N''(u_i)\right) F''(u_i)^{-1} F'(u_i)$$

$$+ F_N''(u_i) F''(u_i)^{-1} F'(u_i) - F'(u_i)$$

$$\left. + F'(u_i) - F_N'(u_i) \right\}. \tag{3.8.28}$$

Assumptions C1–C4, equation (3.8.27), and the definition of z_{MN} imply

$$\delta \left\| F_N''\left(u_i^M\right) - F''(u_i) \right\| \leq \delta \left\| F_N''\left(u_i^M\right) - F''(u_i) \right\| + \delta \left\| F_N''(u_i) - F''(u_i) \right\|$$

$$\leq \delta \left(L\tau + z_N^{(1)}\right)$$

$$\leq \delta L\tau + \frac{1}{2} z_{MN}$$

$$\leq \frac{\delta L z_{MN} + 2\|u_0 - u_*\| \delta L z_{MN}}{1 - 2\delta L \|u_0 - u_*\|} + \frac{1}{2} z_{MN}$$

$$\leq \frac{\frac{1}{3}\delta L + \frac{1}{4}}{(2L_0 + 3L)\delta + 1} < 1$$

resulting in the inequality $\left\| F_N'' \left(u_i^M \right)^{-1} \right\| \leq \frac{\delta}{1 - \left(L\delta\tau + \frac{1}{2} z_{MN} \right)}$. We obtain

$$\left\| F_N'' \left(u_i^M \right) \left(u_i^M - u_i \right) - F_N' \left(u_i^M \right) + F_N' \left(u_i \right) \right\| \leq \frac{1}{2} L \left\| u_i^M - u_i \right\|^2 \leq \frac{1}{2} L\tau^2,$$

and the convergence assertion $\| u_i - u_* \| \leq \| u_0 - u_* \|$ yields

$$\left\| \left(F_N'' \left(u_i^M \right) - F_N'' \left(u_i \right) \right) F'' \left(u_i \right)^{-1} F' \left(u_i \right) \right\| \leq L \left\| u_i^M - u_i \right\| \| u_i - u_{i+1} \|$$

$$\leq 2L\tau \| u_0 - u_* \|.$$

The assumptions of the Theorem lead to

$$\left\| F_N'' \left(u_i \right) F'' \left(u_i \right)^{-1} F' \left(u_i \right) - F' \left(u_i \right) \right\| \leq$$

$$\leq \left\| -F_N'' \left(u_i \right) \left(u_{i+1} - u_i \right) + F'' \left(u_i \right) \left(u_{i+1} - u_i \right) \right\|$$

$$\leq \left\| F_N'' \left(u_i \right) - F'' \left(u_i \right) \right\| \| u_{i+1} - u_i \|$$

$$\leq z_n^{(1)} 2 \| u_0 - u_* \|$$

$$\leq \frac{1}{\delta} z_{MN} \| u_0 - u_* \|$$

and $\left\| F' \left(u_i \right) - F_N' \left(u_i \right) \right\| \leq z_n^{(1)} \leq \frac{1}{2\delta} z_{MN}$. Using the decomposition (3.8.28), the last inequalities complete the induction proof by

$$\left\| u_{i+1}^M - u_{i+1} \right\| \leq$$

$$\leq \frac{\delta}{1 - \left(L\delta\tau + \frac{1}{2} z_{MN} \right)} \left\{ \frac{1}{2} L\tau^2 + 2L \| u_0 - u_* \| \tau + \left(\frac{1}{2} + \| u_0 - u_* \| \right) \frac{z_{MN}}{\delta} \right\}$$

$$= \tau.$$

The last equality is based on the fact that τ is equal to the smallest solution of the quadratic equation $3L\delta\tau^2 - 2b\tau + 2z_{MN} \left(\frac{1}{2} + \| u_0 - u_* \| \right) = 0$.

Finally, inequality (3.8.27) is shown by

$$\left\| F_N' \left(u_n^M \right) - F' \left(u_n \right) \right\| \leq \left\| F_N' \left(u_n^M \right) - F_N' \left(u_n \right) \right\| + \left\| F_N' \left(u_n \right) - F' \left(u_n \right) \right\|$$

$$\leq L \left\| u_n^N - u_n \right\| + z_{MN}$$

$$\leq (Lc_2 + 1) z_{MN} =: c_3 z_{MN}$$

and inequality (3.8.24) results from

$$\left\| \left(u_n^M - u_*^M \right) - \left(u_n - u_* \right) \right\| \leq \left\| u_n^M - u_n \right\| + \left\| u_*^M - u_* \right\|$$

$$\leq c_2 z_{MN} + z_{MN}$$

$$\leq (c_2 + 1) z_{MN} =: c_4 z_{MN}$$

Remark 3.8.8. The upper bounds on z_{MN} and r_1 were defined in [133] by

$$\frac{1}{6} \min \left\{ \frac{\hat{r}_*}{4}, \frac{1}{6L\delta + 1} \right\} \qquad (3.8.29)$$

and

$$\frac{3}{4} \min \left\{ \frac{1}{3L\delta}, \frac{\hat{r}_*}{2} \right\} \qquad (3.8.30)$$

respectively. By comparing (3.8.22) and (3.8.23) with (3.8.29) and (3.8.30), respectively, we conclude that our choices of z_{MN} and r_1 (or r_2 or r_3) are finer than the ones in [133]. However, we leave the details to the motivated reader.

3.9 The conditioning of semidefinite programs

In this section, we are motivated by the elegant work in [144] concerning the conditioning of semidefinite programs (SDP). In particular, we show how to refine their results by using a weaker version of the Newton-Kantorovich Theorem 2.2.4 given by us in 2.2.11.

Let S^n be the space of real, symmetric $n \times n$ matrices. As in [144], we consider the semidefinite program in the form

$$\min C \bullet X \text{ such that } A_k \bullet X = b_k, \ k = 1, 2, \ldots, m, \ X \geq 0, \qquad (3.9.1)$$

where C, A_k, and X belong to S^n, b_k are scalars, $\bullet$ denotes inner product, and by $X \geq 0$ we mean that X lies in the closed, convex cone of positive semidefinite matrices. The dual of (3.9.1) is

$$\max b^T y \text{ such that } \sum_{k=1}^{m} y_k A_k + Z = C; \ Z \geq 0, \qquad (3.9.2)$$

where $Z \in S^n$ is a positive semidefinite dual slack variable.

The following assumptions are used thorought the section:

Assumption 1. The matrices A_k are linearly independent.

Assumption 2. There exists a primal feasible X and a dual feasible (y, Z) with X and Z strictly positive definite (slater condition).

Assumption 3. The primal (3.9.1) and the dual (3.9.2) programs have solutions X_0 and (y_0, Z_0) satisfying strict complementarity, primal nondegeneracy and dual nondegeneracy.

We will mapping $n \times n$ symmetric matrices onto vectors of length $\frac{n(n+1)}{2}$, so let $\text{vec}: S^n \to \mathbf{R}^{\frac{n(n+1)}{2}}$ be an isometry, then

$$A \bullet B = (\text{vec } A)^T (\text{vec } B) \text{ for all } A, B \in S^n.$$

The primal and dual equality constraints become

$$A \text{ vec } X = b; \quad A^T y + \text{vec } Z = \text{vec } C,$$

where $A \in \mathbf{R}^{m \times n(n+1)/2}$ is a matrix whose $k + h$ row is $(\text{vec } A_k)^T$, and $b = [b_1, \ldots, b_m]^T \in \mathbf{R}^m$.

The optimality conditions become:

$$A \text{ vec } X = b; \quad X \geq 0 \tag{3.9.3}$$

$$A^T y + \text{vec } Z = \text{vec } C; \quad Z \geq 0, \tag{3.9.4}$$

$$XZ = 0. \tag{3.9.5}$$

Solving (3.9.3)–(3.9.5) reduces to finding a root of the function

$$F(X, y, Z) = \begin{pmatrix} A \text{ vec } X - b \\ A^T y + \text{vec}(Z - C) \\ \frac{1}{2} \text{vec}(XZ + ZX) \end{pmatrix} \tag{3.9.6}$$

such that $X \geq 0$, $Y \geq 0$.

Let I be the identity matrix, and let $\text{mat}: \mathbf{R}^{\frac{n(n+1)}{2}} \to S^n$ be the inverse of vec. We use $\circledast$ to denote the symmetrized Kronecker product given by

$$(A_1 \circledast B_1) v = \frac{1}{2} \text{vec}(A_1 (\text{mat } v) B_1 + B_{11} (\text{mat } v)) A_1, \tag{3.9.7}$$

where $A_1, B_1 \in S^n$, $v \in \mathbf{R}^{\frac{n(n+1)}{2}}$.

Because F is a map from $\mathbf{R}^{m+n(n+1)}$ to itself, the Jacobian of F is given by

$$J(X, y, Z) = \begin{pmatrix} A & 0 & 0 \\ 0 & A^T & I \circledast I \\ Z \circledast I & 0 & X \circledast I \end{pmatrix}. \tag{3.9.8}$$

We will now define a certain type of norm already used in [144]. However we note that all our results here can be reintroduced with different norms.

For any two vectors $x = [x^1, \ldots, x^n]^T$ and $y = [y^1, \ldots, y^m]^T$, the pair (x, y) is used to denote the vector $[x^1, \ldots, x^n, y^1, \ldots, y^m]^T$.

We use the Euclidean norm $\|\cdot\|$ for vectors, and the induced 2-norm for matrices. The Frobenius norm of a matrix is denoted by $\|\cdot\|_F$. We have

$$\|A\|_F = \|\text{vec } A\| = \sqrt{A \bullet A} \tag{3.9.9}$$

for any real and symmetric matrix A. Then for $u = (X, y, Z) \in S^n \times \mathbf{R}^m \times S^n = D$ we use the norm

$$\|u\| = \|(\text{vec } X, y, \text{vec } Z)\| = \left[\|X\|_F^2 + \|y\|^2 + \|Z\|_F^2 \right]. \tag{3.9.10}$$

We denote by

$$U(u, r) = \{u_1 \in D : \|u - u_1\| < r\|\}$$

and by $\bar{U}(u, r)$ the corresponding ball.

By $\text{Lip}_\gamma (U(u, r))$, we mean the class of all functions that are Lipschitz continuous in $U(u, r)$, γ being the Lipschitz constant using the 2-norm. We also use the compact notation $[A, b, C]$ to denote the SQP's in (3.9.1) and (3.9.2).

Consider a perturbation of the problem parameters A_k, b and C in (3.9.1) as follows:

$$\bar{A} = A + \Delta A, \ \bar{b} = b + \Delta b, \ \bar{C} = C + \Delta C, \qquad (3.9.11)$$

where ΔC is symmetric, and ΔA is a matrix whose kth row is $(\text{vec} \, \Delta A_k)^T$, with ΔA_k symmetric.

Therefore (3.9.6) and (3.9.8) become for the perturbed system respectively

$$\bar{F}(u) = \bar{F}(X, y, Z) = \begin{pmatrix} \bar{A} \, \text{vec} \, X - \bar{b} \\ \bar{A}^T y + \text{vec}(Z - \bar{C}) \\ \frac{1}{2} \text{vec}(XZ + ZX) \end{pmatrix} \qquad (3.9.12)$$

and

$$\bar{J}(X, y, Z) = \begin{pmatrix} \bar{A} & 0 & 0 \\ 0 & \bar{A}^T & I \circledast I \\ Z \circledast I & 0 & X \circledast I \end{pmatrix}. \qquad (3.9.13)$$

We shall denote the solution of the original problem by $u_0 = (X_0, y_0, Z_0)$ and the solution of the perturbed problem by $\bar{u}_0 = (\bar{X}_0, \bar{y}_0, \bar{Z}_0)$.

We state a version of our main Theorem 2.2.11 suitable for our purposes here:

Theorem 3.9.1. *Let $r_0 > 0$, $u_0 \in \mathbf{R}^P$, $G: \mathbf{R}^P \to \mathbf{R}^P$, and that G is continuously differentiable in $U(u_0, r_0)$. Assume for a vector norm and the induced norm that the Jacobian $G'(u) \in \text{Lip}_\gamma (U(u_0, r_0))$ for $u \neq u_0$ and $G'(u) \in \text{Lip}_{\gamma_0}(U(u_0, r_0))$ if $u = u_0$, with $G'(u_0)$ nonsingular. Set*

$$\beta \geq \left\| G'(u_0)^{-1} \right\|, \ \eta \geq \left\| G'(u_0)^{-1} G(u_0) \right\|, \ h_0 = \beta \bar{\gamma} \eta, \ \bar{\gamma} = \frac{\gamma + \gamma_0}{2}, \quad (3.9.14)$$

$$r_1 = \lim_{t \to \infty} t_n, \ r_2 = 2\eta, \qquad (3.9.15)$$

where scalar sequences $\{t_k\}$ $(k \geq 0)$ is given by

$$t_0 = 0, t_1 = \eta, t_{i+2} = t_{i+1} + \frac{\gamma (t_{i+1} - t_i)^2}{2(1 - \gamma_0 t_{i+1})}. \qquad (3.9.16)$$

If

$$(a) \ h_0 \leq \frac{1}{2}, \qquad (3.9.17)$$

and

$$(b) \ r_1 \leq r_0 (\text{or} \, r_2 \leq r_0), \qquad (3.9.18)$$

then

(i) G has a unique zero $\bar{u}_0$ in $\bar{U}(u_0, r)$, $r_1 \leq r_2$, and

(ii) Newton's method with unit steps, starting at u_0 converges to the unique zero $\bar{u}_0$.

Remark 3.9.2. If $\gamma_0 = \gamma$, our Theorem 3.9.1 coincides with Theorem 1 in [144, p. 529]. Set $h = \beta\gamma\eta$. However

$$\gamma_0 \leq \gamma, \tag{3.9.19}$$

holds in general and $\frac{\gamma}{\gamma_0}$ can be arbitrarily large.

Then

$$h \leq \frac{1}{2} \implies h_0 \leq \frac{1}{2}, \tag{3.9.20}$$

but not vice versa unless if $\gamma = \gamma_0$. Moreover finer bounds on the Newton distances are obtained and at least as precise information on the location of the solution, as

$$r_1 \leq \frac{1 - \sqrt{1 - 2h}}{\beta\gamma} = r_3.$$

We assume from now on that condition (3.9.17) holds.

Motivated by these advantages, we improve the rest of the results in [144] as follows:

Corollary 3.9.3. *Under the hypotheses of Theorem 3.9.1 further assume* $h_0 < \frac{1}{2}$, *then* $G'(\bar{u}_0)$ *is nonsingular, where* $\bar{u}_0$ *is the zero of G guaranteed to exist by Theorem 3.9.1.*

Proof. We have in turn

$$\left\| G'(\bar{u}_0) - G'(u_0) \right\| \leq \gamma_0 \|\bar{u}_0 - u_0\| \leq 2\gamma_0\eta$$

$$\leq (\gamma_0 + \gamma)\eta < \frac{1}{\beta} = \frac{1}{\left\| G'(u_0)^{-1} \right\|}. \tag{3.9.21}$$

It follows by the Banach Lemma on invertible operators and (3.9.21) that $G'(\bar{u}_0)^{-1}$ exists and

$$\left\| G'(\bar{u}_0)^{-1} \right\| \leq \frac{\beta}{1 - 2\beta\gamma_0\eta}.$$

We need two lemmas for the semilocal convergence analysis of Newton's method:

Lemma 3.9.4. *[1, Th. 1]. Let* $[A, b, C]$ *define an SDP satisfying the Assumptions. Then, the Jacobian at the solution,* $J(u_0)$ *is nonsingular. Conversely, if an SDP has a solution* u_0 *such that* $J(u_0)$ *is nonsingular, then strict complementarity and nondegeneracy hold at* u_0 *[109].*

Lemma 3.9.5. *[144, Th. 1]. Let* $[A, b, C]$ *define an SDP, not necessarily satisfying the Assumptions. Then the Jacobian* $J(u)$ *satisfies*

$$\|J(u_2) - J(u_1)\| \leq \gamma \|u_2 - u_1\|, \tag{3.9.22}$$

for some fixed $u_1 = (X_1, y_1, Z_1) \in S^n \times \mathbf{R}^m \times S^n$ *and for any* $u_2 = (X_2, y_2, Z_2) \in S^n \times \mathbf{R}^m \times S^n$ *and* $\gamma = 1$.

From now on, we assume $u_0 = (X_0, y_0, Z_0)$ is a solution for an SDP $[A, b, C]$ satisfying the assumptions, and

Assumption 4. There exists γ_0 such that

$$\|J(u_1) - J(u_0)\| \le \gamma_0 \|u_1 - u_0\|, \tag{3.9.23}$$

for all $u_1 = (X_1, y_1, Z_1) \in S^n \times \mathbf{R}^m \times S^n$.

It follows by (3.9.22) and (3.9.23) that

$$\gamma_0 \le 1$$

holds in general and $\frac{1}{\gamma_0}$ can be arbitrarily large.

It is convenient to define the following quantities that will be used for the semilocal convergence that follows:

$$\beta_0 = \left\| J(u_0)^{-1} \right\|,$$
$$\beta_1 = \|M\|,$$

where M consists of the first $m + \frac{n(n+1)}{2}$ columns of $J(u_0)^{-1}$, and

$$\delta_0 = \min\left(\min_{1 \le i \le n} \left\{ \lambda_0^i : \lambda_0^i > 0 \right\}, \min_{1 \le i \le n} \left\{ w_0^i : w_0^i > 0 \right\} \right).$$

We can state the main result:

Theorem 3.9.6. *Let u_0 be the primal-dual solution of the SDP$[A, b, C]$. Suppose the Assumptions 1–4 hold, and let*

$$[\bar{A}, \bar{b}, \bar{C}] = [A + \Delta A, b + \Delta b, C + \Delta C].$$

Set

$$\varepsilon_0 = \|\Delta A\| \|(\text{vec } X_0, y_0)\| + \|(\Delta b, \text{vec } \Delta C)\|,$$
$$\beta = \frac{\beta_0}{1 - \beta_0 \|\Delta A\|},$$
$$\eta = \frac{\beta_0 \beta_1 \varepsilon_0}{1 - \beta_0 \|\Delta A\|}.$$

If

$$\|\Delta A\| \le a = \frac{1}{\beta_0}\left[1 - \beta_0 \sqrt{\beta_1 \varepsilon_0 (1 + \gamma_0)} \right] \tag{3.9.24}$$

and either

$$\varepsilon_0 < \varepsilon_1 = \frac{1}{\beta_0^2 \beta_1 (1 + \gamma_0)} \tag{3.9.25}$$

and

$$r_1 < \delta_0, \tag{3.9.26}$$

or

$$\varepsilon_0 < \varepsilon_2 = \frac{1}{\beta_0 \beta_1} \min \left\{ \frac{1}{\beta_0 (1 + \gamma_0)}, \frac{\delta_0 (1 - \beta_0 \|\Delta A\|)}{2} \right\}, \tag{3.9.27}$$

then

   (i) the SDP defined by $[\bar{A}, \bar{b}, \bar{C}]$ has a unique solution u_0 in $\bar{U}(u_0, r_1)$ provided that (3.9.24)–(3.9.26) hold or in $\bar{U}(u_0, r_2)$ if (3.9.24) and (3.9.27) hold.

   (ii) the solution to $[\bar{A}, \bar{b}, \bar{C}]$ is unique.

   (iii) Newton's method with unit steps applied to $\tilde{F}$ and starting from u_0 converges quadratically to $\bar{u}_0$.

Proof. (i) In order to use Theorem 3.9.1, we first note that $J(u_0)^{-1}$ exists and $\gamma = 1$.
   Then we can have

$$\Delta J = \tilde{J}(u_0) - J(u_0) = \begin{bmatrix} \Delta A & 0 & 0 \\ 0 & \Delta A & 0 \\ 0 & 0 & 0 \end{bmatrix},$$

and

$$\left\| J(u_0)^{-1} \Delta J \right\| \leq \beta_0 \|\Delta A\| < 1, \text{ by (3.9.24)}. \tag{3.9.28}$$

   It follows by (3.9.28) and the Banach Lemma on invertible operators that $J(u_0)$ is nonsingular with

$$\left\| J(u_0)^{-1} \right\| \leq \beta.$$

We can write

$$\tilde{F}(u_0) = \begin{bmatrix} (A + \Delta A) \text{ vec } X_0 - (b + \Delta b) \\ (A + \Delta A)^T y_0 + \text{vec } Z_0 - \text{vec } (C + \Delta C) \\ \frac{1}{2} \text{vec } (X_0 Z_0 + Z_0 X_0) \end{bmatrix}$$

$$= \begin{bmatrix} (\Delta A) \text{ vec } X_0 - \Delta b \\ (\Delta A)^T y_0 - \text{vec } (\Delta C) \\ 0 \end{bmatrix}$$

and

$$\left\| J(u_0)^{-1} \tilde{F}(u_0) \right\| \leq \beta_1 (\|\Delta A\| \|(\text{vec } X_0, y_0)\| + \|(\Delta b, \text{vec } (\Delta C))\|)$$

$$= \beta_1 \varepsilon_0.$$

Therefore, we have

$$\left\| \tilde{J}(u_0)^{-1} \tilde{F}(u_0) \right\| \leq \left\| \tilde{J}(u_0)^{-1} \right\| \left\| \tilde{F}(u_0) \right\|$$

$$\leq \eta.$$

Hence, we get

$$\beta \eta \bar{\gamma} = \beta \eta \frac{1 + \gamma_0}{2} \leq \frac{1}{2}$$

by the choices of ε_0 and $\|\Delta A\|$.

By Theorem 3.9.1, we conclude that $\tilde{F}$ has a unique zero $\tilde{u}$ in $\bar{U}(x_0, r_1)$ if (3.9.24)–(3.9.26) hold or in $\bar{U}(x_0, r_2)$ if (3.9.24) and (3.9.27) hold.

To show that this root is a solution of the SDP, we shall show $\bar{X} \geq 0$ and $\bar{Z}_0 \geq 0$. First note that if either (3.9.24), (3.9.25), (3.9.26) or (3.9.24) and (3.9.27) hold, then

$$\left(\|\bar{X}_0 - X_0\|_F^2 + \|\bar{y}_0 - y_0\| + \|\bar{Z}_0 - Z_0\|_F^2 \right)^{1/2} = \|\bar{u}_0 - u_0\| < \delta_0.$$

Let $\lambda_0(w_0)$ be the vector of eigenvalues of $\bar{X}_0$ (of $\bar{Z}_0$), arranged in nonincreasing (nondecreasing) order. For $1 \leq j \leq n$

$$\lambda_0^j > 0 \implies \bar{\lambda}_0^j > 0$$

and

$$\lambda_0^j = 0 \implies w_0^j > 0$$
$$\implies \bar{w}_0^j > 0 \implies \bar{\lambda}_0^j = 0.$$

That is, $\bar{X}_0 \geq 0$. Similarly we show $\bar{Z}_0 \geq 0$. The proof of part (i) is now completed.

The proof of (ii) follows from Corollary 3.9.3 and (iii) follows from (b) of Theorem 3.9.1. and the existence of $\bar{J}(\bar{u}_0)^{-1}$.

That completes the proof of the Theorem.

Remark 3.9.7. If $\gamma_0 = 1$, our Theorem 3.9.6 reduces to Theorem 2 in [144]. Otherwise it is finer as a, ε_1 (or ε_2) are more flexible than $a_1 = \frac{1}{2\beta_1}$, $\varepsilon_3 = \min\left\{ \frac{\sigma-1}{2\sigma^2\beta_0\beta_1}, \frac{\delta_0}{2\sigma\beta_1} \right\}$ for some $1 < \sigma \leq 2$ given in [144] and as $r_1 \leq r_3$.

The rest of the results given in [144] can be improved along the same lines. However, we leave the details to the motivated reader.

3.10 Exercises

3.10.1. Consider an equation

$$F(z) = 0 \tag{3.10.1}$$

where F is a nonlinear operator between the Banach spaces E, $\widehat{E}$. Under certain conditions, Newton's method

$$z_{n+1} = z_n - F'(z_n)^{-1} F(z_n), \quad n = 0, 1, \ldots, \tag{3.10.2}$$

produces a sequence that converges quadratically to a solution z^* of (3.10.1). Because the formal procedure (3.10.2) can rarely be executed in infinite-dimensional spaces, (3.10.1) is replaced in practice by a family of discretized equations

$$\phi_h(\zeta) = 0 \tag{3.10.3}$$

—indexed by some real numbers $h > 0$—where now ϕ_h is a nonlinear operator between finite-dimensional spaces E_h, $\hat{E}_h$. Let the discretization on E be defined by the bounded linear operators $\Delta_h \colon E \to E_h$. Then, under appropriate assumptions, the equations (3.10.3) have solutions

$$\zeta_h^* = \Delta_h z^* + (h^p)$$

which are the limit of the Newton sequence applied to (3.10.3) and started at $\Delta_h z_0$; that is,

$$\zeta_0^h = \Delta_h z_0, \quad \zeta_{n+1}^h = \zeta_n^h - \phi_h'(\zeta_n^h)^{-1} \phi_h(\zeta_n^h), \quad n = 0, 1, \ldots \quad (3.10.4)$$

In many applications, it turns out that the solution z^* of (3.10.1) as well as the Newton iterates $\{z_n\}$ have "better smoothness" properties than the elements of E. This is a motivation for considering a subset $W^* \subset E$ such that

$$z^* \in W^*, \ z_n \in W^*, \ z_n - z^* \in W^*, \ z_{n+1} - z_n \in W^*, n = 0, 1, \ldots . \quad (3.10.5)$$

The discretization methods are described by a family of triplets.

$$\left\{ \phi_h, \Delta_h, \hat{\Delta}_h \right\}, \ h > 0 \quad (3.10.6)$$

where

$$\phi_h \colon D_h \subset E_h \to \hat{E}_h, \ h > 0$$

are nonlinear operators and

$$\Delta_h \colon E \to E_h, \ \hat{\Delta}_h \colon \hat{E} \to \hat{E}_h, \ h > 0,$$

are bounded linear (discretization) operators such that

$$\Delta_h \left(W^* \cap B^* \right) \subset D_h, \ h > 0. \quad (3.10.7)$$

The discretization (3.10.6) is called *Lipschitz uniform* if there exist scalars $\rho > 0$, $L > 0$ such that

$$\bar{B} \left(\Delta_h, z^*, \rho \right) \subset D_h, \ h > 0, \quad (3.10.8)$$

and

$$\left\| \phi_h'(\eta) - \phi_h'(\xi) \right\| \le L \left\| \eta - \xi \right\|, \ h > 0, \ \eta, \xi \in \bar{U} \left(\Delta_h z^*, \rho \right). \quad (3.10.9)$$

Moreover, the discretization family (3.10.6) is called: *bounded* if there is a constant $q > 0$ such that

$$\left\| \Delta_h u \right\| \le q \left\| u \right\|, \ u \in W^*, \ h > 0, \quad (3.10.10)$$

stable if there is a constant $\sigma > 0$ such that

$$\left\| \phi_h'(\Delta_h u)^{-1} \right\| \le \sigma, \ u \in W^* \cap B^*, \ h > 0, \quad (3.10.11)$$

consistent of order p if there are two constants $c_0 > 0$, $c_1 > 0$ such that

$$\left\| \hat{\Delta}_h F(z) - \phi_h(\Delta_h z) \right\| \leq c_0 h^p, \qquad z \in W^* \cap B^*, \ h > 0, \quad (3.10.12)$$

$$\left\| \hat{\Delta}_h \left(F'(u) v - \phi'_h(\Delta_h u) \Delta_h v \right) \right\| \leq c_1 h^p, \quad u \in W^* \cap B^*, \ v \in W^*,$$
$$h > 0. \qquad (3.10.13)$$

Let $F: D \subset E \to \hat{E}$ be a nonlinear operator such that F' is γ Lipschitz continuous on $U(z^*, r^*) \subseteq D$ with z^* such that $F(z^*) = 0$, $\|F'(z^*)^{-1}\| = \beta$ and $r^* = \frac{2}{3\beta\gamma}$, and consider a Lipschitz uniform discretization (3.10.6) that is bounded, stable, and consistent of order p. Then
Show:
(a) (3.10.3) has a locally unique solution

$$\zeta_h^* = \Delta_h z^* + (h^p) \qquad (3.10.14)$$

for all $h > 0$ satisfying

$$0 < h \leq h_0 = \left[\frac{1}{2\sigma c_0} \min\left(\rho, (\sigma L)^{-1} \right) \right]^{1/p} \qquad (3.10.15)$$

(b) there exist constants $h_1 \in (0, h_0]$, $r_1 \in (0, r^*]$ such that the discrete process (3.10.4) converges to ζ_h^*, and that

$$\zeta_n^h = \Delta_h z_n + (h^p), \ n = 0, 1, \ldots, \qquad (3.10.16)$$

$$\phi_h(\zeta_n^h) = \hat{\Delta}_h F(z_n) + (h^p), \ n = 0, 1, \ldots, \qquad (3.10.17)$$

$$\zeta_n^h - \zeta_h^* = \Delta_h (z_n - z^*) + (h^p), \ n = 0, 1, \ldots, \qquad (3.10.18)$$

for all $h \in (0, h_1]$, and all starting points $z_0 \in B(z^*, r_1)$.

3.10.2. Suppose that the hypotheses of Exercise 3.10.1 hold and that there is a constant $\delta > 0$ for such

$$\liminf_{h>0} \|\Delta_h u\| \geq 2\delta \|u\| \quad \text{for each} \ u \in W^*. \qquad (3.10.19)$$

Then show that for some $\bar{r} \in (0, r_1]$ and for any fixed $\varepsilon > 0$ and $z_0 \in U(z^*, \bar{r})$ there exists a constant $\bar{h} = \bar{h}(\varepsilon, z_0) \in (0, h_1]$ such that

$$\left| \min\left\{ n \geq 0, \|z_n - z^*\| < \varepsilon \right\} - \min\left\{ n \geq 0, \|\zeta_n^h - \zeta_h^*\| < \varepsilon \right\} \right| \leq 1 \quad (3.10.20)$$

for all $h \in (0, \bar{h}]$.

3.10.3. Suppose that the hypothesis of Exercise 3.10.1 is satisfied and that

$$\lim_{h \to 0} \|\Delta_h u\| = \|u\|$$

holds uniformly for $u \in W^*$. Then show there exists a constant $\bar{r}_1 \in (0, r_1]$ and, for any fixed $\varepsilon > 0$, some $\bar{h}_1 = \bar{h}(\varepsilon) \in (0, h_1]$ such that (3.10.20) holds for all $h \in (0, \bar{h}_1]$ and all starting points $z^0 \in U(z^*, \bar{r}_1)$.

3.10.4. Consider the operator $F: D \subset C^2 [0, 1] \to C [0, 1] \times \mathbb{R}^2$,

$$F(y) = \{y'' - f(x, y, y'); 0 \le x \le 1, y(0) - \alpha, y(1) - \beta\},$$

where D and f are assumed to be such that (3.10.1) has a unique solution $z^* \in D$, and

$$f \in C^3 (U(z^*, \rho)),$$
$$U(z^*, \rho) =$$
$$= \{(x_1, x_2, x_3) \in \mathbb{R}^3; 0 \le x_1 \le 1, |x_2 - x^*(x_1)| \le \rho, |x_3 - z^{*\prime}(x_1)| \le \rho\}.$$

Under these assumptions it follows that $z^* \in C^5 [0, 1]$. Indeed, from $z^{*\prime\prime} = f(x, z^*, z^{*\prime})$ we deduce that $z^{*\prime\prime\prime}$ exists and

$$z^{*\prime\prime\prime} = f^{(1,0,0)}(x, z^*, z^{*\prime}) + f^{(0,1,0)}(x, z^*, z^{*\prime}) + f^{(0,0,1)}(x, z^*, z^{*\prime}) z^{*\prime\prime}$$

which, in turn, gives the existence of $z^{*(iv)}$, etc. Here $f^{(1,0,0)}$, etc., denotes the partial derivatives of f.
As usual, we equip $C^k [0, 1]$, $k \ge 0$, with the norm

$$\|u\| = \left\{ \left(\max |u^i(x)|, 0 \le x \le 1 \right), i = 0, \ldots, k \right\}.$$

The Fréchet derivative of F is

$$F'(y) u =$$
$$= \left\{ u'' - f^{(0,1,0)}(x, y, y') u - f^{(0,0,1)}(x, y, y') u', \ 0 \le x \le 1, u(0), u(1) \right\}$$

and hence, for given $z_n \in D$, Newton's method specifies z_{n+1} as the solution of the linear equation

$$z''_{n+1} = f(x, z_n, z'_n) - f^{(0,1,0)}(x, z_n, z'_n)(z_n - z_{n+1})$$
$$- f^{(0,0,1)}(x, z_n, z'_n)(z'_n - z'_{n+1}) \qquad (3.10.21)$$

subject to the boundary conditions $z_{n+1}(0) = \alpha$, $z_{n+1}(1) = \beta$.
From (3.10.21) it follows easily that if $z_0 \in C^3 [0, 1]$ then $z_{n+1} \in C^4 [0, 1]$, $n = 0, 1, 2, \ldots$. We shall assume also that $z_0 \in C^4 [0, 1]$. Moreover, (3.10.21) and the fact that z_n converges to z^* in the norm of $C^2 [0, 1]$ imply that there exists a constant $K > 0$ such that

$$z_n \in W_K = \left\{ z \in C^4 [0, 1]; \sup_x |z^{(i)}(x)| \le K, i = 0, 1, 2, 3, 4 \right\},$$

$n = 0, 1, \ldots$. By choosing, if necessary, a larger K, it is not restrictive to assume that $z^* \in W_K$, $z_n - z^* \in W_K$ and $z_n - z_{n+1} \in W_K$, $n = 0, 1, \ldots$, which is (3.10.5).
The discretization method $\{\phi_h, \Delta_h, \bar{\Delta}_h\}$ is specified as follows

$$h = 1/n, \quad n = 1, 2, \ldots,$$

$$G_h = \{x_i = ih, \ i = 0, 1, \ldots, n\}, \quad \mathring{G}_h = G_h \setminus \{0, 1\},$$

$$E_h = \{\eta : G_h \to \mathbb{R}\}, \quad \eta_i = \eta(x_i), \quad i = 0, 1, \ldots, n,$$

$$\hat{E}_h = \left\{ (\eta, a, b) ; \eta : \mathring{G}_h \to \mathbb{R}, \ a, b \in \mathbb{R} \right\},$$

$$\Delta_h y = y|_{G_h}, \quad \hat{\Delta}_h (y, a, b) = \left(y|_{\mathring{G}_h}, a, b \right),$$

$$\phi_h (\eta) = \left\{ \left[\frac{\eta_{i+1} - 2\eta_i + \eta_{i-1}}{h^2} - f\left(x_i, \eta_i, \frac{\eta_{i+1} - \eta_{i-1}}{2h} \right) \right] ; \right.$$

$$\left. i = 1, 2, \ldots, n-1 ; (\eta_0 - \alpha), (\eta_n - \beta) \right\}.$$

We use the following norms

$$\|y\| = \max \left\{ |y^{(i)} (x)|, 0 \le x \le 1, i = 0, 1, 2 \right\}, \quad y \in C^2 [0, 1],$$

$$\|v\| = \max \{ |u (x)|, a, b; 0 \le x \le 1 \}, \quad v = (u, a, b) \in C [0, 1] \times \mathbb{R}^2,$$

$$\|\eta\| = \left\{ |\eta_0|, |\eta_n|, |\eta_i|, \left| \frac{\eta_{i+1} - \eta_{i-1}}{2h} \right|, \left| \frac{\eta_{i+1} - 2\eta_i + \eta_{i-1}}{h^2} \right|, \right.$$

$$\left. i = 1, \ldots, n-1 \right\}, \quad \eta \in E_h.$$

It is easily seen that for $y \in W_K$, we have

$$\left| \frac{y_{i+1} - y_{i-1}}{2h} \right| \le \tfrac{1}{6} K h^2, \quad \left| \frac{y_{i+1} - 2y_i + y_{i-1}}{h^2} - y_i'' \right| \le \tfrac{1}{12} K g^2,$$

where $y_i = y(x_i)$, $y_i' = y'(x_i)$, $y_i'' = y''(x_i)$, $i = 1, 2, \ldots, n-1$. It is not difficult to prove that, with the above norms, (3.10.10) holds with $q = 1$ and (3.10.12), (3.10.13) are satisfied with $p = 2$. It is also easily seen that

$$\|\Delta_h u\| \le \|u\| \le \|\Delta_h u\| + K \left(\tfrac{1}{6} (h + 1) \right) h$$

for $u \in W_K$ and hence that $\lim_{h \to 0} \|D_h u\| = \|u\|$.

Thus the conclusions of Exercises 3.10.1 and 3.10.2.

3.10.5. (a) Let F be a Fréchet-differentiable operator defined on a convex subset D of a Banach space X with values in a Banach space Y. Assume that the equations $F(x) = 0$ has a simple zero $x^* \in D$ in the sense that $F'(x^*)$ has an inverse $F'(x^*)^{-1} \in L(Y, X)$. Moreover assume

$$\left\| F'(x^*)^{-1} \left[F'(x) - F'(x^*) \right] \right\| \le \ell_1 \|x - x^*\| \quad \text{for all } x \in D. \quad (3.10.22)$$

Then, show: sequence $\{x_n\}$ $(n \ge 0)$ generated by Newton's method is well defined, remains in $U(x^*, r^*)$ for all $n \ge 0$, and converges to x^* with

$$\|x_{n+1} - x^*\| \le \frac{3\ell_1}{2[1 - \ell_1 \|x_n - x^*\|]} \|x_n - x^*\|^2, \quad r^* = \frac{2}{5\ell_1} \quad (n \ge 0) \quad (3.10.23)$$

provided that $x_0 \in U(x^*, r^*)$ and $U = U(x^*, r^*) \subseteq D$.

(b) Let F be as in (a).
Assume:

(1) there exists $x_0 \in D$ such that $F'(x_0)^{-1} \in L(Y, X)$;

(2) $\|F'(x_0)^{-1}\left[F'(x) - F'(x_0)\right]\| \leq \ell_0\|x - x_0\|$ for all $x \in D$; (3.10.24)

(3) $\|F'(x_0)^{-1}F(x_0)\| \leq \eta$ some $\eta \geq 0$, (3.10.25)

$(5 + 2\sqrt{6})\ell_0\eta \leq 1$, (3.10.26)

(4) $U(x_0, r_2) \subseteq D$, where, r_1, r_2 are the real zeros ($r_1 \leq r_2$) of equations

$$f(t) = 3\ell_0 r^2 - (1 + \ell_0\eta)r + \eta = 0.$$ (3.10.27)

Then, show: sequence $\{x_n\}$ ($n \geq 0$) generated by Newton's method is well defined, remains in $U(x_0, r_1)$, and converges to a unique solution x^* of equation $F(x) = 0$ in $\bar{U}(x_0, r_1)$. Moreover, the following estimates hold for all $n \geq 0$. The solution x^* is unique in $U(x_0, r_2)$.

$$\|x_{n+2} - x_{n+1}\| \leq \tfrac{2\ell_0 r_1}{1 - \ell_0 r_1}\|x_{n+1} - x_n\|$$ (3.10.28)

and

$$\|x_{n+1} - x^*\| \leq \tfrac{c^n}{1-c}\|x_n - x^*\|,$$ (3.10.29)

where,

$$c = \tfrac{2\ell_0 r_1}{1 - \ell_0 r_1}.$$ (3.10.30)

(c) Let $F: D \subseteq X \to Y$ be a nonlinear operator satisfying the hypotheses of (a), and consider a Lipschitz uniform discretization that is bounded, stable, and consistent of order p. Then equation $T_h(v) = 0$ has a locally unique solution

$$y_h^* = L_h(x^*) + (h^p)$$

for all $h > 0$ satisfying

$$0 < h \leq h_0 = \left[\tfrac{1}{c_0\sigma} \min\left\{\tfrac{\rho}{2}, \tfrac{5 - 2\sqrt{6}}{\ell\sigma}\right\}\right]^{1/p}.$$

Moreover, there exist constants $h_1 \in (0, h_0]$ and $r_3 \in (0, r^*]$ such that the discrete process converges to y_h^* for all $h \in (0, h_1]$ and all starting points $x_0 \in U(x^*, r_1)$.

3.10.6. (a) Let F be a Fréchet-differentiable operator defined on a convex subset D of a Banach space X with values in a Banach space Y. Assume that the equation $F(x) = 0$ has a simple zero $x^* \in D$ in the sense that $F'(x^*)$ has an inverse $F'(x^*)^{-1} \in L(Y, X)$. Then

(1) for all $\varepsilon_1 > 0$ there exists $\ell_1 > 0$ such that

$$\|F'(x^*)^{-1}\left[F'(x) - F'(y)\right]\| < \varepsilon_1$$

for all $x, y \in U(x^*, \ell_1) \subseteq D$.

(2) If $\varepsilon_1 \in \left[0, \frac{1}{2}\right)$ and $x_0 \in U(x^*, \ell_1)$ then show: sequence $\{x_n\}$ $(n \geq 0)$ generated by Newton's method is well defined, remains in $U(x^*, \ell_1)$ for all $n \geq 0$, and converges to x^* with

$$\|x_{n+1} - x^*\| \leq \tfrac{\varepsilon_1}{1-\varepsilon_1}\|x_n - x^*\| \quad (n \geq 0).$$

(b) Let F be as in (a). Assume:

(1) there exist $\eta \geq 0$, $x_0 \in D$ such that $F'(x_0)^{-1} \in L(Y, X)$,

$$\|F'(x_0)^{-1}F(x_0)\| \leq \eta;$$

Then, for all $\varepsilon_0 > 0$ there exists $\ell_0 > 0$ such that

$$\|F'(x_0)^{-1}\left[F'(x) - F'(y)\right]\| < \varepsilon_0$$

for all $x, y \in U(x_0, \ell_0) \subseteq D$;

(2)

$$\tfrac{\eta(1-\varepsilon_0)}{1-2\varepsilon_0} \leq \ell_0$$

and

$$\varepsilon_0 < \tfrac{1}{2}.$$

Then, show: sequence $\{x_n\}$ $(n \geq 0)$ generated by Newton's method is well defined, remains in $U(x_0, \ell_0)$, and converges to a unique solution x^* of equation $F(x) = 0$ in $\bar{U}(x_0, \ell_0)$. Moreover, the following estimates hold for all $n \geq 0$

$$\|x_{n+2} - x_{n+1}\| \leq \tfrac{\varepsilon_0}{1-\varepsilon_0}\|x_{n+1} - x_n\|$$

and

$$\|x_{n+1} - x^*\| \leq \tfrac{c^n}{1-c}\|x_n - x^*\|,$$

where

$$c = \tfrac{\varepsilon_0}{1-\varepsilon_0}.$$

(c) Let $F: D \subseteq X \to Y$ be a nonlinear operator satisfying a Fréchet uniform discretization $\{T_h, L_h, \hat{L}_h\}$, $h > 0$ that is bounded, stable, and consistent of order p. Then show: equation $T_h(v) = 0$ has a locally unique solution

$$y_h^* = L_h(x^*) + (h^p)$$

for all $h > 0$ satisfying

$$0 < h \leq h_0 = \left[\tfrac{\rho(1-2\ell\sigma)}{(1-\ell\sigma)\sigma c_0}\right]^{1/p}.$$

Moreover, there exist constants $h_1 \in (0, h_0]$ and $r_1 \in (0, r^*]$ such that the discrete process

$$y_0^h = L_h(x_0), \quad y_{n+1}^h = y_n^h - T_h'(y_n^h)^{-1}T_h(y_n^h) \quad (n \geq 0)$$

converges to y_h^* for all $h \in (0, h_1]$ and all starting points $x_0 \in U(x^*, r_1)$, where $r^* = \min \rho \left\{\tfrac{1}{2q}, 1\right\}$.

4

Special Methods

Efficient and special iterative methods other than NKs are studied under weaker conditions than before.

4.1 Broyden's method

In this section, we are concerned with the problem of approximating a locally unique solution x^* of the nonlinear equation

$$F(x) = 0, \tag{4.1.1}$$

where F is a Fréchet-differentiable operator defined on an open subset D of a Banach space X with values in a Banach space Y.

C.G. Broyden suggested the method

$$x_{n+1} = x_n - H_n F(x_n) \quad (n \geq 0) \quad (x_0 \in D) \tag{4.1.2}$$

to generate a sequence approximating x^*, [52], [75]. Here $H_n \in L(Y, X)$ $(n \geq 0)$. Operators H_n are required to satisfy the equation

$$H_{n+1}(y_n) = H_{n+1}(F(x_{n+1}) - F(x_n)) = x_{n+1} - x_n \tag{4.1.3}$$

or equivalently

$$\int_{x_n}^{x_{n+1}} (H_{n+1} F'(x) - I) \, dx = 0, \tag{4.1.4}$$

where $F'(x)$ denotes the Fréchet derivative of operator F.

It seems that H_{n+1} is a reasonable approximation to the inverse of $F'(x)$ (Jacobian) in the neighborhood between x_n and x_{n+1} in the direction of $x_{n+1} - x_n$.

In the case of $X = Y = \mathbf{R}^j$ for example, and for single rank methods, we choose H_{n+1} from the class of $j \times j$ matrices satisfying (4.1.3) that are given by

$$H_{n+1} = H_n - (H_n(y_n) + H_n F(x_n)) \, d_n^T / d_n^T(y_n) \tag{4.1.5}$$

I.K. Argyros, *Convergence and Applications of Newton-type Iterations*,
DOI: 10.1007/978-0-387-72743-1_4, © Springer Science+Business Media, LLC 2008

where $d_n \in \mathbf{R}^j$, chosen so that $d_n^T(y_n) \neq 0$.

If $A(x_n) = A_n = H_n^{-1}$, then

$$A_{n+1} = A_n - (y_n + F(x_n)) \, d_n^T A_n / d_n^T F(x_n) \quad (n \geq 0). \tag{4.1.6}$$

J.E. Dennis in [75] provided a local and a semilocal convergence analysis for method (4.1.2) using a Newton-Kantorovich-type approach (see also Section 2.2).

Here we show by using more precise majorizing sequences first in the semilocal case that under the same hypotheses and computational cost, we can find weaker sufficient convergence conditions for method, finer error bounds on the distances involved, and provide a more precise information on the location of the solution. Moreover in the local case, we provide a larger radius of convergence.

We need the following result on majorizing sequences in order to study the semilocal convergence of method (4.1.1).

Lemma 4.1.1. *Assume there exist nonnegative numbers* K, M, L, μ, η *and* $\delta \in [0, 2)$ *such that for all* $n \geq 0$

$$h_\delta^n = \left\{ K \left(\tfrac{\delta}{2} \right)^n + \tfrac{2\delta(L+M)}{2-\delta} \left[1 - \left(\tfrac{\delta}{2} \right)^{n+1} \right] + \tfrac{4M}{2-\delta} \left[1 - \left(\tfrac{\delta}{2} \right)^n \right] \right\} \eta + 2\delta\mu \leq \delta, \tag{4.1.7}$$

and

$$(L+m) \frac{1 - \left(\tfrac{\delta}{2} \right)^n}{1 - \tfrac{\delta}{2}} \eta < 1 - 2\mu. \tag{4.1.8}$$

Then, iteration $\{t_n\}$ $(n \geq 0)$, *given by*

$$t_0 = 0, \ t_1 = \eta, \ t_{n+2} = t_{n+1} + \tfrac{1}{2} \frac{K(t_{n+1}-t_n) + 2(\mu + Mt_n)}{1 - [2\mu + (L+M)t_{n+1}]} (t_{n+1} - t_n) \tag{4.1.9}$$

is nondecreasing, bounded above by $t^{**} = \tfrac{2\eta}{2-\delta}$, *and converges to some* t^* *such that*

$$0 \leq t^* \leq t^{**}. \tag{4.1.10}$$

Moreover, the following estimates hold for all $n \geq 0$:

$$0 \leq t_{n+2} - t_{n+1} \leq \tfrac{\delta}{2} (t_{n+1} - t_n) \leq \left(\tfrac{\delta}{2} \right)^{n+1} \eta. \tag{4.1.11}$$

Proof. The result clearly holds if $\eta = 0$ or $K = 0$ or $\delta = 0$. Let us assume $K \neq 0, n \neq 0$ and $\delta \neq 0$.

We shall show using induction on $i \geq 0$:

$$K(t_{i+1} - t_i) + 2(\mu + Mt_i) + 2\delta\mu + \delta(L+M)t_{i+1} \leq \delta \tag{4.1.12}$$

$$1 - 2\mu - (L+M)t_{i+1} > 0 \tag{4.1.13}$$

and

$$t_{i+1} - t_i \geq 0. \tag{4.1.14}$$

Estimate (4.1.11) can then follow immediately from (4.1.9) and (4.1.12)–(4.1.14). For $i = 0$, (4.1.9), (4.1.12)–(4.1.14) hold by (4.1.7) and (4.1.8).

We also get:

$$0 \le t_2 - t_1 \le \tfrac{\delta}{2} (t_1 - t_0). \tag{4.1.15}$$

Let us assume (4.1.11)–(4.1.14) hold for all $i \le n+1$. We can have in turn:

$$K (t_{i+2} - t_{i+1}) + 2 (Mt_{i+1} + \mu) + 2\delta\mu + \delta (L + M) t_{i+2} \le$$

$$\le K\eta \left(\tfrac{\delta}{2}\right)^{i+1}$$

$$+ 2 \left\{ M \left[t_1 + \tfrac{\delta}{2} (t_1 - t_0) + \left(\tfrac{\delta}{2}\right)^2 (t_1 - t_0) + \cdots + \left(\tfrac{\delta}{2}\right)^i (t_1 - t_0) \right] + \mu \right\}$$

$$+ \delta (L + M) \left[t_1 + \tfrac{\delta}{2} (t_1 - t_0) + \cdots + \left(\tfrac{\delta}{2}\right)^{i+1} (t_1 - t_0) \right] + 2\delta\mu$$

$$= h_\delta^{i+1} \le \delta \qquad \text{(by (4.1.7))}. \tag{4.1.16}$$

Moreover, we shall show:

$$t_i \le t^{**} \quad (i \ge 0). \tag{4.1.17}$$

Inequality (4.1.17) holds for $i = 0, 1, 2$ by the initial conditions. Assume (4.1.17) holds for all $i \le n$. It then follows from (4.1.11)

$$t_{i+2} \le t_{i+1} + \tfrac{\delta}{2} (t_{i+1} - t_i) \le \cdots \le \eta + \tfrac{\delta}{2}\eta + \cdots + \left(\tfrac{\delta}{2}\right)^{i+1} \eta$$

$$= \frac{\left[1 - \left(\tfrac{\delta}{2}\right)^{i+2} \right]}{1 - \tfrac{\delta}{2}} \eta < \frac{2\eta}{2-\delta} = t^{**} \tag{4.1.18}$$

Furthermore,

$$(L + M) t_{i+2} \le (L + M) \frac{1 - \left(\tfrac{\delta}{2}\right)^{i+2}}{1 - \tfrac{\delta}{2}} \eta < 1 - 2\mu, \quad \text{(by (4.1.8))}, \tag{4.1.19}$$

which shows (4.1.13) for all $i \ge 0$.

Then induction for (4.1.12)–(4.1.14) is now complete.

Hence, sequence $\{t_n\}$ $(n \ge 0)$ is: bounded above by t^{**}; nondecreasing and as such it converges some t^* satisfying (4.1.10).

That completes the proof of the Lemma.

Remark 4.1.2. We wanted to leave the conditions (4.1.7) and (4.1.8) as uncluttered as possible. However if verification of (4.1.7) and (4.1.8) is difficult, we can use instead respectively the pairs:

$$\overline{h}_\delta = \left[K + \tfrac{2\delta(L+M)}{2-\delta} + \tfrac{4M}{2-\delta} \right] \eta + 2\delta\mu \le \delta, \tag{4.1.20}$$

and

$$\frac{2(L+M)}{2-\delta} \le 1 - 2\mu, \tag{4.1.21}$$

or (4.1.21),

$$h_\delta = \left[K + (L + M) \delta + \tfrac{4M}{2-\delta} \right] \eta + 2\delta\mu \le \delta, \tag{4.1.22}$$

and

$$\tfrac{(L+M)\delta^2}{2-\delta} \le K. \tag{4.1.23}$$

Note that (4.1.20) and (4.1.21) follow immediately from (4.1.7), and (4.1.8), respectively, whereas for h_δ^n to be bounded above by δ it suffices to have:

$$\delta\,(L+M)\left[\tfrac{2}{2-\delta}\left(1-\left(\tfrac{\delta}{2}\right)^{i+2}\right)-1\right] \le K\left[1-\left(\tfrac{\delta}{2}\right)^{i+1}\right], \tag{4.1.24}$$

and

$$\tfrac{4M}{2-\delta}\left[1-\left(\tfrac{\delta}{2}\right)^{i+1}\right] \le \tfrac{4M}{2-\delta}. \tag{4.1.25}$$

Inequality (4.1.24) can be rewritten

$$\tfrac{(L+M)\delta^2}{2-\delta}\left[1-\left(\tfrac{\delta}{2}\right)^{\delta+1}\right] \le K\left[1-\left(\tfrac{\delta}{2}\right)^{\delta+1}\right], \tag{4.1.26}$$

which holds for all $i \ge 0$ by (4.1.23).

Moreover, (4.1.25) also holds for all $i \ge 0$.

We can show the following semilocal convergence result for method (4.1.2):

Theorem 4.1.3. *Let $F: D \subseteq X \to Y$ be a Fréchet-differentiable operator. Assume: there exists an approximation $A\,(x) \in L\,(X, Y)$ of operator $F'\,(x)$, an open convex subset D_0 of D, $x_0 \in D_0$, nonnegative parameters η, K, L, M, μ and $\delta \in [0, 2)$ such that:*

$$\|A\,(x_0)^{-1}\,F\,(x_0)\| \le \eta, \tag{4.1.27}$$

$$\|A\,(x_0)^{-1}\left[F'\,(x) - F'\,(y)\right]\| \le K\,\|x - y\|, \tag{4.1.28}$$

$$\|A\,(x_0)^{-1}\left[F'\,(x) - F'\,(x_0)\right]\| \le L\,\|x - x_0\|, \tag{4.1.29}$$

for all $x, y \in D_0$,

$$\|A\,(x_0)^{-1}\left(A\,(x_{n+1}) - F'\,(x_{n+1})\right)\| \le$$

$$\le \|A\,(x_0)^{-1}\left(A\,(x_0) - F'\,(x_0)\right)\| + M\sum_{i=0}^{n} \|x_{i+1} - x_i\|, \tag{4.1.30}$$

$$\|A\,(x_0)^{-1}\left(A\,(x_0) - F'\,(x_0)\right)\| \le \mu; \tag{4.1.31}$$

conditions (4.1.7), (4.1.8) hold,
and

$$\overline{U}\,(x_0, t^*) \subseteq D_0, \tag{4.1.32}$$

where t^* is defined in Lemma 4.1.1.

Then, sequence $\{x_n\}$ $(n \ge 0)$ generated by Broyden's method (4.1.2) is well defined, remains in $U\,(x_0, t^*)$ for all $n \ge 0$, and converges to a solution $x^* \in \overline{U}\,(x_0, t^*)$ of equation $F\,(x) = 0$.

Moreover, the following estimates hold for all $n \ge 0$:

$$\|x_{n+1} - x_n\| \le t_{n+1} - t_n, \tag{4.1.33}$$

and

$$\|x_n - x^*\| \leq t^* - t_n. \tag{4.1.34}$$

Furthermore, the solution x^* is unique in $\overline{U}(x_0, t^*)$ if $M \neq 0$ or $\mu \neq 0$. Finally, if there exists $t_1^* > t^*$ such that

$$U(x_0, t_1^*) \subseteq D_0, \tag{4.1.35}$$

and

$$\frac{L}{2}(t^* + t_1^*) \leq 1, \tag{4.1.36}$$

then the solution x^* is unique in $U(x_0, t_1^*)$.

Proof. Using induction on the whole integer k, we shall show

$$\|x_{k+1} - x_k\| \leq t_{k+1} - t_k, \tag{4.1.37}$$

and

$$x_{k+1} \in \overline{U}(x_0, t^*), \quad \sum_{i=0}^{k} \|x_{i+1} - x_i\| \leq t^*, \tag{4.1.38}$$

for all $k \geq 0$.

Estimates (4.1.37) and (4.1.38) hold for $k = 0$. Assume they hold for all $k \leq n$. Then, we have:

$$\begin{aligned}
\|x_{k+1} - x_0\| &\leq \|x_{k+1} - x_k\| + \|x_k - x_{k-1}\| + \cdots + \|x_1 - x_0\| \\
&\leq (t_{k+1} - t_k) + (t_k - t_{k-1}) + \cdots + (t_1 - t_0) \\
&= t_{k+1} - t_0 = t_{k+1} \leq t^*,
\end{aligned} \tag{4.1.39}$$

and

$$\sum_{i=0}^{k} \|x_{i+1} - x_i\| \leq t^*. \tag{4.1.40}$$

By (4.1.8), (4.1.28)–(4.1.31), we obtain in turn:

$$\begin{aligned}
\|A(x_0)^{-1} &\left[A(x_{k+1}) - A(x_0)\right]\| \leq \\
&\leq \|A(x_0)^{-1}\left(A(x_{k+1}) - F'(x_{k+1})\right)\| + \|\left(F'(x_{k+1}) - F'(x_0)\right)\| \\
&\quad + \|A(x_0)^{-1} A(x_0) - F'(x_0)\| \\
&\leq \mu + M \sum_{i=0}^{k} \|x_{i+1} - x_i\| + L \|x_{k+1} - x_0\| \\
&\leq \mu + (L + M) t_{k+1} < 1.
\end{aligned} \tag{4.1.41}$$

It follows from (4.1.41) and the Banach Lemma on invertible operators that $A(x_{k+1})^{-1}$ exists and

$$\|A(x_{k+1})^{-1} A(x_0)\| \leq \left[1 - \mu - (L + M) t_{k+1}\right]. \tag{4.1.42}$$

Moreover, using (4.1.2), (4.1.28), (4.1.30), and (4.1.31), we obtain:

$$A(x_0)^{-1} F(x_{k+1}) = \tag{4.1.43}$$

$$= A(x_0)^{-1} \left[F(x_{k+1}) - F(x_k) - A(x_k)(x_{k+1} - x_k) \right]$$

$$= A(x_0)^{-1} \left\{ \int_0^1 \left[F'(x_{k+1} + \theta(x_k - x_{k+1})) - F'(x_k) \right](x_{k+1} - x_k)\, d\theta \right.$$

$$\left. + \left[F'(x_k) - A(x_k) \right](x_{k+1} - x_k) \right\},$$

and

$$\| A(x_0)^{-1} F(x_{k+1}) \| \le$$

$$\le \tfrac{1}{2} K \|x_{k+1} - x_k\|^2 + \left(\mu + M \sum_{i=0}^{k-1} \|x_{i+1} - x_i\| \right) \|x_{k+1} - x_k\|$$

$$\le \tfrac{1}{2} K (t_{k+1} - t_k)^2 + (\mu + M t_k)(t_{k+1} - t_k). \tag{4.1.44}$$

Furthermore, by (4.1.2), (4.1.9), (4.1.42), and (4.1.44), we have:

$$\|x_{k+2} - x_{k+1}\| = \left\| \left[A(x_{k+1})^{-1} A(x_0) \right] \left[A(x_0)^{-1} F(x_{k+1}) \right] \right\| \tag{4.1.45}$$

$$\le \left\| A(x_{k+1})^{-1} A(x_0) \right\| \left\| A(x_0)^{-1} F(x_{k+1}) \right\|$$

$$\le \frac{\tfrac{1}{2} K (t_{k+1} - t_k)^2 + (\mu + M t_k)(t_{k+1} - t_k)}{1 - \mu - (L + M) t_{k+1}} = t_{k+2} - t_{k+1},$$

which completes the induction for (4.1.37).

The induction is now completed, as

$$\sum_{i=0}^{k+1} \|x_{i+1} - x_i\| \le t_{k+1} \le r_0 \text{ and } \|x_{k+2} - x_0\| \le t_{k+2} \le r_0. \tag{4.1.46}$$

It follows from (4.1.37) and (4.1.38) that sequence $\{x_n\}$ $(n \ge 0)$ is Cauchy in a Banach space X, and as such it converges to same $x^* \in \overline{U}(x_0, t^*)$ (as $\overline{U}(x_0, t^*)$ is a closed set). By letting $k \to \infty$ in (4.1.45), we obtain $F(x^*) = 0$.

Estimate (4.1.34) follows from (4.1.33). Indeed we have:

$$\|x_{k+i} - x_k\| \le \|x_{k+i} - x_{k+i-1}\| + \|x_{k+i-1} - x_{k+i-2}\| + \cdots + \|x_{k+1} - x_k\|$$

$$\le (t_{k+i} - t_{k+i-1}) + (t_{k+i-1} - t_{k+i-2}) + \cdots + (t_{k+1} - t_k)$$

$$= t_{k+i} - t_k. \tag{4.1.47}$$

By letting $i \to \infty$ in (4.1.47), we obtain (4.1.34).

To show uniqueness in $\overline{U}(x_0, t^*)$, let y^* be a solution of equation $F(x) = 0$. By (4.1.29) and (4.1.8) we can have:

$$\left\| A\left(x_0\right)^{-1} \int_0^1 \left[F'\left(y^* + \theta\left(x^* - y^*\right)\right) - F'\left(x_0\right)\right] d\theta \right\|$$

$$\leq L \int_0^1 \left\| \left[y^* + \theta\left(x^* - y^*\right) - x_0\right] \right\| d\theta$$

$$\leq L \int_0^1 \left[\theta \left\| x^* - x_0 \right\| + (1 - \theta) \left\| y^* - x_0 \right\|\right] d\theta \qquad (4.1.48)$$

$$\leq L t^* < 1. \qquad (4.1.49)$$

It follows from (4.1.49), and the Banach Lemma on invertible operators, that linear operator

$$L = \int_0^1 F'\left(y^* + \theta\left(x^* - y^*\right)\right) d\theta \qquad (4.1.50)$$

is invertible. Using the identity

$$0 = F\left(x^*\right) - F\left(y^*\right) = L\left(x^* - y^*\right) \qquad (4.1.51)$$

we deduce

$$x^* = y^*. \qquad (4.1.52)$$

Similarly if $y^* \in u\left(x_0, t_1^*\right)$, we obtain again that linear operator L is invertible, as by (4.1.48)

$$\left\| A\left(x_0\right)^{-1} \left[L - F'\left(x_0\right)\right] \right\| < \tfrac{L}{2}\left(t^* + t_1^*\right) \leq 1. \qquad (4.1.53)$$

Hence, again we get (4.1.52).

That completes the proof of the theorem.

Remark 4.1.4. Dennis in [75, Theorem 3, p. 562] has provided a similar semilocal convergence result. He is not using condition (4.1.29), which is what is really needed, but the stronger (4.1.28) to find upper bounds on the norms $\left\| A\left(x_n\right)^{-1} A\left(x_0\right) \right\|$ ($n \geq 0$) (see Chapter 2). However in general, $L \leq K$ holds.

Finally note that the derivation of condition (4.1.30) and its significance has been explained in [75, see Theorem 1] (for $M = \frac{3K}{2}$).

Finally we can show the following local result for method (4.1.2)

Theorem 4.1.5. *Let $F: D \subseteq X \to Y$ be a Fréchet-differentiable operator. Assume: there exist an approximation $A(x) \in L(X, Y)$ of operator $F'(x)$, an open convex subset D_0 of D, a solution x^* of equation (4.1.1) such that $A(x^*) \in L(Y, X)$ and nonnegative parameters α_i, $i = 0, 1, ..., 4$ such that the following conditions hold for all $x_n, x \in D_0$*

$$\left\| A\left(x^*\right)^{-1}\left(F'(x) - F'\left(x^*\right)\right) \right\| \leq \alpha_0 \left\| x - x^* \right\|, \qquad (4.1.54)$$

$$\left\| A\left(x^*\right)^{-1}\left(A(x) - A\left(x^*\right)\right) \right\| \leq \alpha_1 + \alpha_2 \left\| x - x^* \right\|, \qquad (4.1.55)$$

and

$$\left\| A\left(x^*\right)^{-1}\left(A\left(x_n\right)-F'\left(x^*\right)\right)\right\| \leq \alpha_3+\alpha_4\left\|x_n-x^*\right\|; \tag{4.1.56}$$

equation

$$\left(\tfrac{1}{2}\alpha_0+\alpha_q+\alpha_4\right)r+\alpha_1+\alpha_3-1=0 \tag{4.1.57}$$

has a minimal nonnegative zero r^* satisfying

$$\alpha_2 r+\alpha_1<1, \tag{4.1.58}$$

and

$$U\left(x^*,r^*\right)\subseteq D_0. \tag{4.1.59}$$

Then, sequence $\{x_n\}$ $(n\geq 0)$ generated by Broyden's method (4.1.2) is well defined, remains in $U\left(x^*,r^*\right)$ for all $n\geq 0$ and converges to x^* provided that $x_0\in U\left(x^*,r^*\right)$.

Moreover the following estimates hold for all $n\geq 0$

$$\left\|x_{n+1}-x^*\right\|\leq\frac{\left(\tfrac{1}{2}\alpha_0+\alpha_4\right)\|x_n-x^*\|+\alpha_3}{1-(\alpha_1+\alpha_2\|x_n-x^*\|)}\left\|x_n-x^*\right\|. \tag{4.1.60}$$

Proof. By hypothesis $x_0\in U\left(x^*,r^*\right)$. Let $x\in U\left(x^*,r^*\right)$. Then by (4.1.55) and (4.1.58) we get:

$$\left\|A\left(x^*\right)^{-1}\left(A\left(x\right)-A\left(x^*\right)\right)\right\|\leq\alpha_1+\alpha_2\left\|x-x^*\right\|\leq\alpha_1+\alpha_2 r^*<1. \tag{4.1.61}$$

It follows from (4.1.61) and the Banach Lemma on invertible operators that $A\left(x\right)^{-1}$ exists so that:

$$\left\|A\left(x\right)^{-1}A\left(x^*\right)\right\|\frac{1}{1-(\alpha_1+\alpha_2\|x-x^*\|)}\leq\frac{1}{1-(\alpha_1+\alpha_2 r^*)}. \tag{4.1.62}$$

Assume $x_k\in U\left(x^*,r^*\right)$ for all $k\leq n$. Then using (4.1.2), (4.1.54)–(4.1.59), we obtain in turn

$$\left\|x^*-x_{k+1}\right\|=$$

$$=\left\|x^*-x_k+A\left(x_n\right)^{-1}F\left(x_k\right)-A\left(x_k\right)^{-1}F\left(x^*\right)\right\|$$

$$\leq\left\|A\left(x_k\right)^{-1}A\left(x^*\right)\right\|\left\{\left\|\int_0^1 A\left(x^*\right)^{-1}\left[F'\left(x^*+t\left(x_k-x^*\right)\right)\right]-F'\left(x^*\right)\right\|\right.$$

$$+\left.\left\|A\left(x^*\right)^{-1}\left(F'\left(x^*\right)-A\left(x_k\right)\right)\right\|\right\}\left\|x^*-x_k\right\|$$

$$\leq\frac{1}{1-(\alpha_1+\alpha_2\|x^*-x_k\|)}\left[\tfrac{1}{2}\alpha_0\left\|x^*-x_k\right\|+\alpha_3+\alpha_4\left\|x^*-x_k\right\|\right]$$

$$<\frac{\left(\tfrac{1}{2}\alpha_0+\alpha_4\right)r^*-\alpha_3}{1-(\alpha_1+\alpha_2 r^*)}\left\|x^*-x_k\right\|,$$

which shows (4.1.60), and $x_{k+1}\in U\left(x^*,r^*\right)$.

Moreover, by (4.1.62) we have:

$$\left\| x^* - x_{k+1} \right\| \leq \left\| x^* - x_k \right\| < \left\| x^* - x_0 \right\|. \tag{4.1.63}$$

Hence we deduce from (4.1.63) that $\lim_{n \to \infty} x_n = x^*$.

That completes the proof of the theorem.

Remark 4.1.6. Dennis in [75, Theorem 5, p. 564] provided a local result for Broyden's method (4.1.2) in the special case when $X = Y = \mathbf{R}^j$. In particular, he stated that if

$$\left\| F'(x) - F'(x^*) \right\| \leq \alpha_5 \left\| x - x^* \right\|, \tag{4.1.64}$$

$$\left\| F'(x^*)^{-1} \right\| \leq \alpha_6 \tag{4.1.65}$$

then there exist real positive numbers ε and ε_0 such that if $A(x_0)$ is a real $j \times j$ matrix, $\left\| A(x_0) - F'(x^*) \right\| \leq \varepsilon_0$ and $\|x_0 - x^*\| = r_D < \varepsilon$, Broyden's method (4.1.2) converges to x^* from this starting point.

The proof was given for

$$\varepsilon_0 \leq \tfrac{1}{6\alpha_6}, \tag{4.1.66}$$

and

$$\varepsilon \leq \tfrac{2\varepsilon_0}{5\alpha_5}. \tag{4.1.67}$$

We now apply this result on a certain numerical example in order to compare it with our Theorem 4.1.5.

Example 4.1.7. Let $D = X = Y = \mathbf{R}$, $D_0 = U(0, 1)$, $A(x) = F'(x)$, and define function F on D_0 by

$$F(x) = e^x - 1. \tag{4.1.68}$$

Using (4.1.64)–(4.1.68) we see

$$\alpha_5 = e - 1, \text{ and } \alpha_6 = 1. \tag{4.1.69}$$

Consequently, the maximum possible convergence radius r_D given by Dennis is

$$r_D = \tfrac{1}{15\alpha_5} = .038798447. \tag{4.1.70}$$

By (4.1.54)–(4.1.57) we get

$$\alpha_0 = e - 1, \alpha_1 = e - 1, \ \alpha_3 = 0, \ \alpha_4 = e - 1, \tag{4.1.71}$$

and

$$r^* = \tfrac{2}{7(e-1)} = .0232790683. \tag{4.1.72}$$

Hence by (4.1.70) and (4.1.72), we deduce

$$r_D < r^*.$$

That is we provide a six times larger convergence radius than Dennis' no matter how x_0 is chosen in D_0.

4.2 Stirling's method

In this section, we approximate a locally unique fixed point x^* of the nonlinear equation

$$F(x) = x, \tag{4.2.1}$$

were F is a nonlinear operator defined on a closed convex subset D of a Banach space E with values on itself.

We propose Stirling's method

$$x_{n+1} = x_n - \left[I - F'(P(x_n))\right]^{-1}(x_n - F(x_n)) \quad (n \geq 0). \tag{4.2.2}$$

Here $P: D \subseteq E \to E$ is a continuous operator and $F'(x)$ denotes the Fréchet derivative of operator F. Special cases of (4.2.2), namely NK method ($P(x_n) = x_n$ ($n \geq 0$)), the modified form of Newton's method ($P(x_n) = x_0$ ($n \geq 0$)), and the ordinary Stirling's method ($P(x_n) = F(x_n)$ ($n \geq 0$)), have been studied extensively [43], [185]. Stirling's method can be viewed as a combination of the method of successive substitutions and Newton's method. In terms of the computational effort, Stirling's and Newton's methods require the same computational cost.

In this section, we provide sufficient conditions for the convergence of method (4.2.2) to x^*. Moreover, we find a ball centered at a certain point $x_0 \in D$ including same center convergence balls found in earlier works (see [43], [185], and the references there). Consequently, we find a ring containing infinitely many new starting points from which x^* can be accessed via method (4.2.2).

To achieve this goal, we define the operator $G: D \to E$ by

$$G(x) = x - \left[I - F'(P(x))\right]^{-1}(x - F(x)). \tag{4.2.3}$$

We then use the degree of logarithmic convexity of G, which is defined to be the Fréchet derivative G' of G.

Finally, we complete our study with an example where our results compare favorably with earlier ones.

Let $a \in [0, 1)$, $b \geq 0$, and $x_0 \in D$ be given. Define the real function g on $[0, +\infty)$,

$$g(r) = b(1+a)r^2 - \left[(1-a)^2 - b\|x_0 - F(x_0)\|\right]r + (1-a)\|x_0 - F(x_0)\|. \tag{4.2.4}$$

Set:

$$c = b\|x_0 - F(x_0)\|. \tag{4.2.5}$$

It can easily be seen, that if

$$c < \left(\sqrt{a^2 + (1-a)^2} - a\right)^2 = d, \tag{4.2.6}$$

then equation $g(r) = 0$ has two nonnegative zeros denoted by r_1 and r_2, with $r_1 \leq r_2$.

Define also:

$$r_3 = \frac{(1-a)^2 - b\|x_0 - F(x_0)\|}{b(1+a)}.$$ (4.2.7)

Finally, set:

$$I = [r_1, r_3].$$ (4.2.8)

We now state and prove the main semilocal convergence theorem for method (4.2.2).

Theorem 4.2.1. *Let F, P be continuous operators defined on a closed convex subset D of a Banach space E with values on itself. For $a \in [0, 1)$, $b \geq 0$ and $x_0 \in D$ fixed, assume:*

(a) F is twice continuously Fréchet-differentiable on D, and

$$\|F'(x) - F'(y)\| \leq b\|x - y\|,$$ (4.2.9)
$$\|F'(x)\| \leq a < 1,$$ (4.2.10)

for all $x, y \in D$;
(b) $U(x_0, r) \subseteq D$ for any $r \in I$, where I is given by (4.2.8).
(c) $c < d$, where c, d are given by (4.2.5), and (4.2.6), respectively;
(d) P is continuously Fréchet-differentiable on D,

$$\|P'(x)\| \leq a,$$ (4.2.11)
$$P(x) \in U(x_0, r),$$ (4.2.12)

and

$$\|x - P(x)\| \leq \|x - F(x)\|,$$ (4.2.13)

for all $x \in U(x_0, r)$.
Then, the following hold:

(i)

$$\|G'(x)\| \leq \frac{b}{(1-a)^2}\|x - F(x)\| \leq h(r) < 1,$$ (4.2.14)

where

$$h(r) = [(1+a)r + \|x_0 - F(x_0)\|]\frac{b}{(1-a)^2},$$ (4.2.15)

for all $r \in I$.
(ii) Iteration $\{x_n\}$ $(n \geq 0)$, generated by (4.2.2) is well defined, remains in $U(x_0 r)$ $(r \in I)$ for all $n \geq 0$ and converges to a fixed point x^* of G in $U(x_0, r_1)$ which is unique in $U(x_0, r_4)$, where $r_4 \in [r_1, r_5)$ and $r_5 = \min\{r_2, r_3\}$.

Moreover, the following error estimates hold for all $n \geq 0$:

$$\|x_n - x^*\| \leq h^n(r)r, \quad r \in I,$$ (4.2.16)

and

$$\|x_{n+1} - x^*\| \leq \frac{b}{1-a}\left[\|x_n - P(x_n)\| + \|P(x_n) - x^*\|\right]\|x_n - x^*\|$$
$$\leq \frac{b(1+2a)}{2(1-a)}\|x_n - x^*\|^2.$$ (4.2.17)

Proof. (i) By differentiating (4.2.3), we obtain in turn for $x \in D$ (4.2.18)

$$G'(x) = \tag{4.2.18}$$

$$= I - \left(\left[I - F'(P(x)) \right]^{-1} \right)' (x - F(x)) - \left[I - F'(P(x)) \right]^{-1} (x - F(x))'$$

$$= I + \left[I - F'(P(x)) \right]^{-1} F''(P(x)) P'(x) \left[I - F'(P(x)) \right]^{-1} (x - F(x))$$

$$\quad - \left[I - F'(P(x)) \right]^{-1} \left(I - F'(x) \right)$$

$$= \left[I - F'(P(x)) \right]^{-1} \left[I - F'(P(x)) \right.$$

$$\quad + F''(P(x)) P'(x) \left(I - F'(P(x))^{-1} (x - F(x)) \right) - I + F'(x) \left. \right]$$

$$= \left[I - F'(P(x)) \right]^{-1} \left[F'(x) - F'(P(x)) \right.$$

$$\quad + F''(P(x)) P'(x) \left(I - F'(P(x)) \right)^{-1} (x - F(x)) \left. \right].$$

Using (4.2.9)–(4.2.13), and the Banach lemma on invertible operators we obtain from (4.2.18)

$$\left\| G'(x) \right\| \le \tfrac{b}{(1-a)^2} \left\| x - F(x) \right\|. \tag{4.2.19}$$

In particular for $x \in U(x_0, r)$, (4.2.19), the choice of $r \in I$, and the estimate

$$\left\| x - F(x) \right\| = \left\| (x - x_0) + (x_0 - F(x_0)) + (F(x_0) - F(x)) \right\|$$

$$\le r + \left\| x_0 - F(x_0) \right\| + ar,$$

we obtain (4.2.14).

(ii) It follows from (4.2.4) that

$$r \ge \tfrac{\|x_0 - G(x_0)\|}{1 - h(r)}, \quad r \in I. \tag{4.2.20}$$

Hence, we can get

$$\left\| x_1 - x_0 \right\| = (1 - h(r)) r \le r, \quad r \in I$$

which shows $x_1 \in U(x_0, x)$ and (4.2.16) for $n = 1$. Assume that

$$x_k \in U(x_0, r), \quad \text{and} \quad \left\| x_k - x_0 \right\| \le \left(1 - h^k(r) \right) r \le r, r \in I \tag{4.2.21}$$

for $k = 1, 2, ..., n$.

Using (4.2.2) and part (i), we obtain in turn

$$\left\| x_{n+1} - x_n \right\| = \left\| G(x_n) - G(x_{n-1}) \right\| \le \sup_{y \in [x_{n-1}, x_n]} \left\| G'(y) \right\| \left\| x_n - x_{n-1} \right\|$$

$$\tag{4.2.22}$$

$$\le h(r) \left\| x_n - x_{n-1} \right\|,$$

$$\left\| x_{n+1} - x_n \right\| \le h(r) \left\| x_n - x_{n-1} \right\| \le \cdots \le h^n(r) \left\| x_1 - x_0 \right\| = (1 - h(r)) h^n(r) r,$$

and

$$\|x_{n+1} - x_0\| \le \|x_{n+1} - x_n\| + \|x_n - x_0\|$$
$$\le (1 - h(r)) h^n(r) r + \left(1 - h^n(r)\right) r$$
$$= \left(1 - h^{n+1}(r)\right) r \le r, \ r \in I.$$

That is, we showed (4.2.16) for all $k \in N$. Moreover by (4.2.22), we have for $n, m \in N$

$$\|x_{n+m} - x_n\| \le \left(1 - h^m(r)\right) h^n(r) r. \tag{4.2.23}$$

Estimate (4.2.23) shows that $\{x_n\}$ $(n \ge 0)$ is a Cauchy sequence in a Banach space E, and as such it converges to some $x^* \in U(x_0, r)$. Because of the continuity of F, F', P, and (4.2.22), we obtain $P(x^*) = x^*$, $G(x^*) = x^*$, $F(x^*) = x^*$.

To show uniqueness, let y^* be a fixed point of G in $U(x_0, r_4)$. Then using (4.2.14), we get

$$\|x^* - y^*\| = \|G(x^*) - G(y^*)\|$$
$$\le \sup_{y \in [x^*, y^*]} \|G'(y)\| \|x^* - y^*\|$$
$$\le h(r) \|x^* - y^*\|,$$

which shows $x^* = y^*$.

Furthermore by letting $m \to \infty$ in (4.2.23), we obtain (4.2.16). Finally by (4.2.2), we obtain for all $n \ge 0$

$$x_{n+1} - x^* =$$
$$= x_n - x^* - \left[I - F'(P(x_n))\right]^{-1} (x_n - F(x_n)) \tag{4.2.24}$$
$$= \left[I - F'(P(x_n))\right]^{-1} \left[(I - F'(p(x_n))) (x_n - x^*) - (x_n - F(x_n))\right]$$
$$= \left[I - F'(P(x_n))\right]^{-1} \left[F(x_n) - F(x^*) - F'(P(x_n)) (x_n - x^*)\right].$$

But we can also have by (4.2.11) and (4.2.13) that for all $n \ge 0$

$$\|x_n - P(x_n)\| = \|x_n - x^* + P(x^*) - P(x_n)\| \le (1 + a) \|x_n - x^*\|, \tag{4.2.25}$$

and

$$\|P(x_n) - x^*\| = \|P(x_n) - P(x^*)\| \le a \|x_n - x^*\|. \tag{4.2.26}$$

Estimate (4.2.17) now follows from (4.2.24)–(4.2.26) and the approximation

$$F(x_n) - F(x^*) - F'(P(x_n)) (x_n - x^*) =$$
$$= \int_0^1 \left[F'(tx_n + (1 - t) x^*) - F'(tP(x_n) + (1 - t) P(x_n))\right] (x_n - x^*) \, dt. \tag{4.2.27}$$

We now state the following theorem for comparison (see [169] and the references there for a proof).

Theorem 4.2.2. *Let F be a Fréchet-differentiable on $D \subseteq E$. Assume:*

(a_1) Condition (a) holds;
(b_1)

$$P(x) = F(x) \, (x \in D);\tag{4.2.28}$$

(c_1) $c < d_0$, where

$$d_0 = \frac{2(1-a)}{1+2a};\tag{4.2.29}$$

(d_1) $U(x_0, r_0) \subseteq D$, where

$$r_0 = \frac{2c}{b(1-a)}, \ \text{for } b \neq 0.\tag{4.2.30}$$

Then, Stirling's iteration $\{x_n\}$ ($n \geq 0$) converges to the unique fixed point x^* of F in $U(x_0, r_0)$ at the rate given by (4.2.17).

Remark 4.2.3. Favorable comparisons of Stirling's over NK method have been made in [169] and the references there.

Proposition 4.2.4. *Under the hypotheses of Theorem 4.2.1 and 4.2.2, assume:*

$$c < \frac{(1-a)^3}{3+a} = d_1.\tag{4.2.31}$$

Then the following hold:

$$r_1 < r_0 < r_3,\tag{4.2.32}$$

and

$$U(x_0, r_0) \subseteq U(x_0, r_3).\tag{4.2.33}$$

Proof. Estimates (4.2.32) and (4.2.33) follow immediately by the definition of r_1, r_0, r_3 and (4.2.31).

Remark 4.2.5. Let $d_2 = \min\{d_1, d, d_0\}$, under the hypotheses of Theorem 4.2.1 and 4.2.2. Then the conclusion of the proposition hold. This observation justifies the claim made at the introduction.

We complete this study with an example.

Example 4.2.6. Let $E = R$, $D = \left[-\frac{\pi}{4}, \frac{\pi}{4}\right]$, $P(x) = F(x)$ and

$$F(x) = \tfrac{1}{2} \sin x.$$

For $x_0 = .1396263 = 8^0$, we obtain $d = \frac{3-\sqrt{2}}{4} = .428932$, $d_0 = \frac{1}{2} = .5$, $d_1 = \frac{1}{28} = .0357143$, $a = b = \frac{1}{2}$, $\|x_0 - F(x_0)\| = .0700397$, $c = 0350199$, $r_0 = .2801592$ and $r_3 = .2866401$. With the above values, the hypotheses of Theorems 4.2.1, 4.2.2, and the Proposition 4.2.4 are satisfied. Hence we get

$$0 = x^* \in U(x_0, r_0) = [-.1405329, .4197855]$$
$$\subseteq (-.1470138, .4262664) = U^0(x_0, r_3).$$

That is there are infinitely many new starting points $U^0(x_0, r_3) - U(x_0, r_0)$ for which iteration (4.2.2) converges to x^* but Theorem 4.2.2 does not guarantee that, whereas Theorem 4.2.1 does.

4.3 Steffensen's method

Let E, Λ be Banach spaces and denote by $U(x_0, R)$ the closed ball with center $x_0 \in E$ and of radius $R \geq 0$. We will use the same symbol for the norm $\| \; \|$ in both spaces. Let P be a projection operator $(P = P^2)$ that projects E on its subspace E_p and set $Q = I - P$. Suppose that the nonlinear operators $F(x, \lambda)$ and $G(x, \lambda)$ with values in E are defined for $x \in D$, where D is some open convex subset of E containing $U(x_0, R)$, and $\lambda \in U(\lambda_0, S)$ for some $\lambda_0 \in \Lambda$, $S \geq 0$. For each fixed $\lambda \in U(\lambda_0, S)$, the operator $PF(\omega, \lambda)$ will be assumed to be Fréchet derivative of the operator $PF(\omega, \lambda)$ with respect to the argument $\omega = x$. Moreover for each fixed $\lambda \in U(\lambda_0, S)$, the operator $PG(\omega, \lambda)$ will be assumed to be continuous for all $\omega \in D$.

In this study, we are concerned with the problem of approximating a solution $x^* := x^*(\lambda)$ of the equation

$$F(x, \lambda) + G(x, \lambda) = 0. \tag{4.3.1}$$

We introduce the inexact Steffensen-Aitken-type method

$$x_{n+1}(\lambda) = \tag{4.3.2}$$
$$= x_n(\lambda) - A(x_n(\lambda), \lambda)^{-1} (F(x_n(\lambda), \lambda) + G(x_n(\lambda), \lambda)) - z(x_n(\lambda), \lambda),$$
$$(n \geq 0)$$

where by x_0 we mean $x_0(\lambda)$. That is, x_0 depends on the λ used in (4.3.2). $A(x, \lambda) \in L(E \times \Lambda, E)$ and is given by

$$A(x_n(\lambda), \lambda) = P\left[g^1(x_n(\lambda), \lambda), g^2(x_n(\lambda), \lambda); F \right] \tag{4.3.3}$$
$$+ P\left[g^3(x_n(\lambda), \lambda), g^4(x_n(\lambda), \lambda); G \right] \quad (n \geq 0)$$

where $[x(\lambda), y(\lambda); F]$ (or $[x(\lambda), y(\lambda); G]$) denotes divided difference of order one on F (or G) at the points $x(\lambda), y(\lambda) \in D$, satisfying

$$[x(\lambda), y(\lambda); F](y(\lambda) - x(\lambda)) = F(y(\lambda), \lambda) - F(x(\lambda), \lambda) \tag{4.3.4}$$

for all $x(\lambda) \neq y(\lambda)$, $\lambda \in U(\lambda_0, S)$ and

$$[x(\lambda), x(\lambda); F] = F'(x(\lambda), \lambda), \quad \lambda \in U(\lambda_0, S) \tag{4.3.5}$$

if $F(x(\lambda), \lambda)$ is Fréchet-differentiable at $x(\lambda)$ for all $\lambda \in U(\lambda_0, S)$. The operator $z: D \times U(\lambda_0, S) \to E$ is chosen so that iteration $\{x_n(\lambda)\}$ $(n \geq 0)$ generated by (4.3.2) converges to x^*. The operators $g^1, g^2, g^3, g^4: D \times U(\lambda_0, S) \to E$ are continuous.

The importance of studying inexact Steffensen-Aitken methods comes from the fact that many commonly used variants can be considered procedures of this type. Indeed, approximation (4.3.2) characterizes any iterative process in which corrections are taken as approximate solutions of Steffensen-Aitken equations. Moreover, we note that if for example an equation on the real line is solved $F(x_n(\lambda), \lambda) + G(x_n(\lambda), \lambda) \geq 0$ and $A(x_n(\lambda), \lambda)$ overestimates the derivative

$$x_n - A(x_n(\lambda), \lambda)^{-1} (F(x_n(\lambda), \lambda) + G(x_n(\lambda), \lambda))$$

is always "larger" than the corresponding Steffensen-Aitken iterate. In such cases, a positive $z(x_n(\lambda), \lambda)$ $(n \geq 0)$ correction term is appropriate.

It can easily be shown by induction on n that under the above hypotheses $F(x_n(\lambda), \lambda) + G(x_n(\lambda), \lambda)$ belong to the domain of $A(x_n(\lambda), \lambda)^{-1}$ for all $n \geq 0$.

Therefore, if the inverses exist (as it will be shown later in the theorem), then the iterates $\{x_n(\lambda)\}$ can be computed for all $n \geq 0$. The iterates generated when $P = I$ (identity operator on E) cannot easily be computed in infinite-dimensional spaces as the inverses may be too difficult or impossible to find. It is easy to see, however, that the solution of equation (4.3.2) reduces to solving certain operator equations in the space EP. If, moreover, EP is a finite-dimensional space of dimension N, we obtain a system of linear algebraic equations of at most order N. Special choices of the operators introduced above reduce our iteration (4.3.2) to earlier considered methods. Indeed we can have: for $g^1(x(\lambda), \lambda) = g^2(x(\lambda), \lambda) = x(\lambda)$, $g^3(x(\lambda), \lambda) = g^4(x(\lambda), \lambda) = 0$, $z = 0$ we obtain Newton methods, for $P = I$, no λ, $g^1(x) = g^2(x) = x$ $(x \in D)$, $g^3(x_n) = x_{n-1}$ $(n \geq 1)$, $g^4(x_n) = x_n$ $(n \geq 0)$ we obtain Cătinaş method [54]; for $P = I$, no λ, $G(x) = 0$ $(x \in D) m$, $z_n = 0$ $(n \geq 0)$, $g^3(x) = g^4(x) = 0$, $g^2(x) = g^1(F(x))$ $(x \in D)$, we obtain methods considered by Păvăloiu in [158], [159]. Our choices of the operators because they include all previous methods allow us to consider a wider class of problems.

We provide sufficient conditions for the convergence of iteration (4.3.2) to a locally unique solution $x^*(\lambda)$ of equation (4.3.1) as well as several error bounds on the distances $\|x_{n+1}(\lambda) - x_n(\lambda)\|$ and $\|x_n(\lambda) - x^*(\lambda)\|$ $(n \geq 0)$.

We can now state and prove the following semilocal convergence result:

Theorem 4.3.1. *Let F, G, P, Q be as in the introduction. Assume:*

(a) *there exist $x_0(\lambda) \in D$, $\lambda_0 \in \Lambda$ such that $C := C(\lambda) = A(x_0(\lambda), \lambda_0)$ is invertible. Set $B = C^{-1}$;*

(b) *there exist nonnegative numbers a_i, R, S, $i = 1, 2, ..., 15$ such that:*

$$\|BP([x, y; F] - [v, w; F])\| \leq a_1 (\|x - v\| + \|y - \omega\|), \tag{4.3.6}$$

$$\left\|x - g^1(x, \lambda)\right\| \leq a_2 \left\|A(x, \lambda)^{-1} (F(x, \lambda) + G(x, \lambda)) - z(x, \lambda)\right\| \tag{4.3.7}$$

$$\left\| x - g^2(x, \lambda) \right\| \le a_3 \left\| A(x, \lambda)^{-1} (F(x, \lambda) + G(x, \lambda)) - z(x, \lambda) \right\|, \quad (4.3.8)$$

$$\left\| g^1(x, \lambda) - g^1(y, \lambda) \right\| \le a_4 \|x - y\| \quad a_4 \in [0, 1), \quad (4.3.9)$$

$$\left\| g^2(x, \lambda) - g^2(y, \lambda) \right\| \le a_5 \|x - y\| \quad a_5 \in [0, 1), \quad (4.3.10)$$

$$\| B(QF(x, \lambda) - QF(y, \lambda)) \| \le a_6 \|x - y\|, \quad (4.3.11)$$

$$\| B(A(x_{n+1}, \lambda)) (z(x_{n+1}, \lambda)) - A(x_n, \lambda)(z(x_n, \lambda)) \| \le \quad (4.3.12)$$
$$\le a_7 \|x_{n+1} - x_n\| \quad (n \ge 0),$$

$$\left\| BP \left([x, y; G] - \left[g^3(x, \lambda), g^4(x, \lambda); G \right] \right) \right\| \le \quad (4.3.13)$$
$$\le a_8 \left(\left\| x - g^3(x, \lambda) \right\| + \left\| y - g^4(x, \lambda) \right\| \right),$$

$$\left\| x - g^3(x, \lambda) \right\| \le a_9 \left\| A(x, \lambda)^{-1} (F(x, \lambda) + G(x, \lambda)) - z(x, \lambda) \right\|, \quad (4.3.14)$$

$$\left\| x - g^4(x, \lambda) \right\| \le a_{10} \left\| A(x, \lambda)^{-1} (F(x, \lambda) + G(x, \lambda)) - z(x, \lambda) \right\| \quad (4.3.15)$$

$$\left\| g^3(x, \lambda) - g^3(y, \lambda) \right\| \le a_{11} \|x - y\| \quad a_{11} \in [0, 1), \quad (4.3.16)$$

$$\left\| g^4(x, \lambda) - g^4(y, \lambda) \right\| \le a_{12} \|x - y\| \quad a_{12} \in [0, 1), \quad (4.3.17)$$

$$\| B(QG(x, \lambda) - QG(y, \lambda)) \| \le a_{13} \|x - y\|, \quad (4.3.18)$$

$$\left\| B \left(\left[g^1(x_0, \lambda), g^2(x_0, \lambda); F \right] - \left[g^1(x_0, \lambda_0), g^2(x_0, \lambda_0); F \right] \right) \right\| \le$$
$$\le a_{14} \|\lambda - \lambda_0\|, \quad (4.3.19)$$

and

$$\left\| B \left(\left[g^3(x_0, \lambda), g^4(x_0, \lambda); G \right] - \left[g^3(x_0, \lambda_0), g^4(x_0, \lambda_0); G \right] \right) \right\| \le$$
$$\le a_{15} \|\lambda - \lambda_0\|, \quad (4.3.20)$$

for all $v, w, x, y \in U(x_0, R)$, $\lambda \in U(\lambda, S)$;

(c) the sequence $\{z(x_n(\lambda), \lambda)\}$ $(n \ge 0)$ *is null for all* $\lambda \in U(\lambda_0, S)$;

(d) for each fixed $\lambda \in U(\lambda_0, S)$ *there exists a minimum nonnegative number* $r^* := r_\lambda^*$
satisfying

$$T_\lambda(r^*) \le r^* \quad \text{and} \quad r^* \le R \quad (4.3.21)$$

with $r := r(\lambda)$,

$$T_\lambda(r) = n + \frac{b_1 r + b_2}{b(r) - b_3 r} r, \quad (4.3.22)$$

where

$$n := n(\lambda) \ge \|x_0(\lambda) - x_1(\lambda)\|, \quad (4.3.23)$$
$$b_1 = a_1(1 + a_4 + a_5) + a_8(1 + a_{11} + a_{12}),$$
$$b_2 = a_6 + a_{13} + a_7,$$
$$b_3 = a_1(a_2 + a_3) + a_8(a_9 + a_{10}),$$

$$b(r) := b(r, s) = 1 - (a_1 (a_4 + a_5) + a_8 (a_{11} + a_{12})) r - (a_{14} + a_{15}) S; \quad (4.3.24)$$

(e) r^*, R, S also satisfy:

$$b(r^*) - b_3 r^* > 0, \quad (4.3.25)$$

$$r^* \geq \max \left\{ \frac{\left\| g^1(x_0(\lambda), \lambda) - x_0(\lambda) \right\|}{1 - a_4}, \frac{\left\| g^2(x_0(\lambda), \lambda) - x_0(\lambda) \right\|}{1 - a_5}, \right.$$

$$\left. \frac{\left\| g^3(x_0(\lambda), \lambda) - x_0(\lambda) \right\|}{1 - a_{11}}, \frac{\left\| g^4(x_0(\lambda), \lambda) - x_0(\lambda) \right\|}{1 - a_{12}} \right\}, \quad (4.3.26)$$

$$c := c(\lambda) = d(r, R) < 1, \quad (4.3.27)$$

where

$$d(e_1, e_2) = \frac{b_1(e_1 + e_2) + b_4}{b(e_1) - b_3(e_1 + e_2)} \quad (4.3.28)$$

and

$$b_4 = a_6 + a_{13}. \quad (4.3.29)$$

Then

(i) For each fixed $\lambda \in U(\lambda_0, S)$ the scalar sequence $\{t_n(\lambda)\}$ $(n \geq 0)$ generated by

$$t_0(\lambda) = 0, \quad t_1(\lambda) = n, \quad (4.3.30)$$

$$t_{n+1}(\lambda) = t_n(\lambda) + \frac{b_1(t_n(\lambda) - t_{n-1}(\lambda)) + b_2}{\alpha_n \beta_n} (t_n(1) - t_{n-1}(\lambda))$$

$$(n \geq 1), \quad (4.3.31)$$

$$\alpha_n := \alpha_n(\lambda) = 1 - b_3 \gamma_n \quad (n \geq 0), \quad (4.3.32)$$

$$\beta_n := \beta_n(\lambda) = 1 - (a_{14} + a_{15}) S$$

$$- [a_1(a_4 + a_5) + a_8(a_{11} + a_{12})] t_n(\lambda) \quad (n \geq 0), \quad (4.3.33)$$

and

$$\gamma_n := \gamma_n(\lambda) = (t_n(\lambda) - t_{n-1}(\lambda)) \beta_n^{-1} \quad (n \geq 1), \quad (4.3.34)$$

is monotonically increasing, bounded above by r^* and $\lim_{n \to \infty} t_n(\lambda) = r^*$;

(ii) the inexact Steffensen-Aitken method generated by (4.3.2) is well defined, remains in $U(x_0(\lambda), r^*)$ for all $n \geq 0$, and converges to a solution $x^*(\lambda) \in U(x_0(\lambda), r^*)$ of equation (4.3.1). Moreover if $z = 0$ then $x^*(\lambda)$ is unique in $U(x_0(\lambda), R)$. Furthermore, the following estimates are true:

$$\| x_{n+1}(\lambda) - x_n(\lambda) \| \leq \frac{b_1 \| x_n(\lambda) - x_{n-1}(\lambda) \| + b_2}{\overline{\alpha}_n \overline{\beta}_n} \| x_n(\lambda) - x_{n-1}(\lambda) \|$$

$$(n \geq 1), \quad (4.3.35)$$

$$\| x_{n+1}(\lambda) - x_n(\lambda) \| \leq t_{n+1}(\lambda) - t_n(\lambda) \quad (n \geq 0), \quad (4.3.36)$$

$$\left\| x_n(\lambda) - x^*(\lambda) \right\| \leq r^* - t_n(\lambda) \quad (n \geq 0), \quad (4.3.37)$$

where

$$\bar{\alpha}_n := \bar{\alpha}_n(\lambda) = 1 - b_3 \bar{\gamma}_n \quad (n \geq 0), \tag{4.3.38}$$

$$\bar{\beta}_n := \bar{\beta}_n(\lambda) = 1 - (a_{14} + a_{15}) \|\lambda - \lambda_0\|$$
$$- [a_1(a_4 + a_5) + a_8(a_{11} + a_{12})] \|x_n(\lambda) - x_0(\lambda)\| \quad (n \geq 0), \tag{4.3.39}$$

and

$$\bar{\gamma}_n := \bar{\gamma}_n(\lambda) = \|x_n(\lambda) - x_{n-1}(\lambda)\| \bar{\beta}_n^{-1} \quad (n \geq 1). \tag{4.3.40}$$

Proof. (i) By (4.3.21) and (4.3.30) we deduce $0 \leq t_0(\lambda) \leq t_1(\lambda) \leq r^*$. Let us assume $0 \leq t_{k-1}(\lambda) \leq t_k(\lambda) \leq r^*$ for $k = 1, 2, ..., n$. Then it follows from (4.3.30) and (4.3.31) that $0 \leq t_k(\lambda) \leq t_{k+1}(\lambda)$. Hence, the sequence $\{t_n(\lambda)\}$ $(n \geq 0)$ is monotonically increasing. Moreover by (4.3.31) and the induction hypotheses, we get in turn

$$t_{k+1}(\lambda) \leq t_k(\lambda) + \frac{b_1 r^* + b_2}{b(r^*) - b_3 r^*}(t_k(\lambda) - t_{k-1}(\lambda))$$

$$\leq \cdots$$

$$\leq n + \frac{b_1 r^* + b_2}{b(r^*) - b_3 r^*} r^*$$

$$= T_\lambda(r^*) \leq r^* \quad \text{(by (4.3.21)).}$$

That is the sequence $\{t_n(\lambda)\}$ $(n \geq 0)$ is also bounded above by r^*. Because for each fixed $\lambda \in U(\lambda_0, S)$ r^* is the minimum nonnegative number satisfying (4.3.21), it follows that $\lim_{n \to \infty} t_n(\lambda) = r^*$.

(ii) By hypotheses (4.3.30), (4.3.23), and (4.3.22) it follows that

$$x_1(\lambda) \in U(x_0(\lambda), r^*).$$

Moreover from (4.3.26), we deduce $g^1(x_0(\lambda), \lambda)$, $g^2(x_0(\lambda), \lambda)$, $g^3(x_0(\lambda), \lambda)$, $g^4(x_0(\lambda), \lambda) \in U(x_0(\lambda), r^*)$. Let us assume $x_{k+1}(\lambda)$, $g^1(x_k(\lambda), \lambda)$, $g^2(x_k(\lambda), \lambda)$, $g^3(x_k(\lambda), \lambda)$, $g^4(x_k(\lambda), \lambda) \in U(x_0(\lambda), r^*)$ for $k = 0, 1, 2, ..., n$, and that (4.3.36) is true for $k = 1, 2, ..., n$ (as it is true for $k = 0$ by (4.3.23) and (4.3.30)). Then from (4.3.9) and (4.3.26) we get

$$\left\| g^1(x_k(\lambda), \lambda) - x_0(\lambda) \right\| \leq$$

$$\leq \left\| g^1(x_k(\lambda), \lambda) - g^1(x_0(\lambda), \lambda) \right\| + \left\| g^1(x_0(\lambda), \lambda) - x_0(\lambda) \right\|$$

$$\leq a_4 \|x_k(\lambda) - x_0(\lambda)\| + \left\| g_1^1(x_0(\lambda), \lambda) - x_0(\lambda) \right\| s < r^*.$$

That is, $g^1(x_n(\lambda), \lambda) \in U(x_0(\lambda), r^*)$. Similarly, we obtain $g^2(x_n(\lambda), \lambda)$, $g^3(x_n(\lambda), \lambda)$, $g^4(x_n(\lambda), \lambda) \in U(x_0(\lambda), \lambda)$. Using (4.3.6), (4.3.9), (4.3.10), (4.3.13), (4.3.16), (4.3.17), (4.3.19), and (4.3.20), we obtain

$$\left\| BP\left(\left[g^1\left(x_k\left(\lambda\right),\lambda\right),g^2\left(x_k\left(\lambda\right),\lambda\right);F\right]\right.\right.$$

$$+\left[g^3\left(x_k\left(\lambda\right),\lambda\right),g^4\left(x_k\left(\lambda\right),\lambda\right);G\right]$$

$$-\left[g^1\left(x_0\left(\lambda\right),\lambda_0\right),g^2\left(x_0\left(\lambda\right),\lambda_0\right);F\right]$$

$$\left.\left.-\left[g^3\left(x_0\left(\lambda\right),\lambda_0\right),g^4\left(x_0\left(\lambda\right),\lambda_0\right);G\right]\right)\right\|$$

$$\leq\left\| BP\left(\left[g^1\left(x_k\left(\lambda\right),\lambda\right),g^2\left(x_k\left(\lambda\right),\lambda\right);F\right]\right.\right.$$

$$\left.\left.-\left[g^1\left(x_0\left(\lambda\right),\lambda_0\right),g^2\left(x_0\left(\lambda\right),\lambda_0\right);F\right]\right)\right\|$$

$$+\left\| BP\left(\left[g^3\left(x_k\left(\lambda\right),\lambda\right),g^4\left(x_k\left(\lambda\right),\lambda\right);G\right]\right.\right.$$

$$\left.\left.-\left[g^3\left(x_0\left(\lambda\right),\lambda_0\right),g^4\left(x_0\left(\lambda\right),\lambda_0\right);G\right]\right)\right\|$$

$$\leq\left\| BP\left(\left[g^1\left(x_k\left(\lambda\right),\lambda\right),g^2\left(x_k\left(\lambda\right),\lambda\right);F\right]\right.\right.$$

$$\left.\left.-\left[g^1\left(x_0\left(\lambda\right),\lambda\right),g^2\left(x_0\left(\lambda\right),\lambda\right)\right]\right)\right\|$$

$$+\left\| BP\left(\left[g^1\left(x_0\left(\lambda\right),\lambda\right),g^2\left(x_0\left(\lambda\right),\lambda\right);F\right]\right.\right.$$

$$\left.\left.-\left[g^1\left(x_0\left(\lambda\right),\lambda_0\right),g^2\left(x_0\left(\lambda\right),\lambda_0\right);F\right]\right)\right\|$$

$$+\left\| BP\left(\left[g^3\left(x_k\left(\lambda\right),\lambda\right),g^4\left(x_k\left(\lambda\right),\lambda\right);G\right]\right.\right.$$

$$\left.\left.-\left[g^3\left(x_k\left(\lambda\right),\lambda\right),g^4\left(x_0\left(\lambda\right),\lambda\right)\right]\right)\right\|$$

$$+\left\| BP\left(\left[g^3\left(x_0\left(\lambda\right),\lambda\right),g^4\left(x_0\left(\lambda\right),\lambda\right);G\right]\right.\right.$$

$$\left.\left.-\left[g^3\left(x_0\left(\lambda\right),\lambda_0\right),g^4\left(x_0\left(\lambda\right),\lambda_0\right);G\right]\right)\right\|$$

$$\leq a_1\left(a_4+a_5\right)\left\|x_k\left(\lambda\right)-x_0\left(\lambda\right)\right\|+a_{14}\left\|\lambda-\lambda_0\right\|$$

$$+a_8\left(a_{11}+a_{12}\right)\left\|x_k\left(\lambda\right)-x_0\left(\lambda\right)\right\|+a_{15}\left\|\lambda-\lambda_0\right\|$$

$$\leq\left[a_1\left(a_4+a_5\right)+a_8\left(a_{11}+a_{12}\right)\right]r^*+\left(a_{14}+a_{15}\right)S<1\ \text{by}\ (4.3.25)$$

It follows from the Banach lemma on invertible operators that $A\left(x_k\left(\lambda\right),\lambda\right)$ is invertible and

$$\left\|A\left(x_k\left(\lambda\right),\lambda\right)^{-1}B^{-1}\right\|\leq\bar{\delta}_k s<\delta_k, \tag{4.3.41}$$

where

$$\overline{\gamma}_k = \left(\overline{\alpha}_k \overline{\beta}_k\right)^{-1} \quad \text{and} \quad \delta_k = (\alpha_k \beta_k)^{-1} \quad (k \geq 0).$$

Using (4.3.2) we obtain the approximation

$$
\begin{aligned}
x_{k+1}(\lambda) - x_k(\lambda) &= \\
&= \left(A\left(x_k(\lambda), \lambda\right)^{-1} B^{-1}\right) B \left\{\left[PF\left(x_k(\lambda), \lambda\right) - PF\left(x_{k-1}(\lambda), \lambda\right)\right.\right. \\
&\quad - P\left[g^1\left(x_k(\lambda), \lambda\right), g^2\left(x_k(\lambda), \lambda\right); F\right]\left(x_k(\lambda) - x_{k-1}(\lambda)\right)\right] \\
&\quad + \left[QF\left(x_k(\lambda), \lambda\right) - QF\left(x_{k-1}(\lambda), \lambda\right)\right] + \left[A\left(x_k(\lambda), \lambda\right)\left(z\left(x_k(\lambda), \lambda\right)\right)\right. \\
&\quad \left. - A\left(x_{k-1}(\lambda), \lambda\right)\left(z\left(x_{k-1}(\lambda), \lambda\right)\right)\right] + \left[PG\left(x_k(\lambda), \lambda\right) - PG\left(x_k(\lambda), \lambda\right)\right. \\
&\quad - P\left[g^3\left(x_k(\lambda), \lambda\right), g^4\left(x_k(\lambda), \lambda\right); G\right]\left(x_k(\lambda) - x_{k-1}(\lambda)\right)\right] \\
&\quad \left. + \left[QG\left(x_k(\lambda), \lambda\right) - QG\left(x_{k-1}(\lambda), \lambda\right)\right]\right\} \quad (k \geq 1). \quad (4.3.42)
\end{aligned}
$$

By (4.3.6) we obtain

$$
\begin{aligned}
&\left\|\left(BPF\left(x_k(\lambda), \lambda\right) - PF\left(x_{k-1}(\lambda), \lambda\right)\right.\right. \\
&\qquad \left.\left. - P\left[g^1\left(x_k(\lambda), \lambda\right), g^2\left(x_k(\lambda); F\right)\right]\left(x_k(\lambda) - x_{k-1}\right)\right)\right\| \\
&\leq \left\|BP\left(\left[x_{k-1}(\lambda), x_k(\lambda); F\right]\right)\right. \\
&\qquad \left. - \left[g^1\left(x_{k-1}(\lambda), \lambda\right), g^2\left(x_{k-1}(\lambda), \lambda\right); F\right]\right\|\left\|x_k(\lambda) - x_{k-1}(\lambda)\right\| \\
&\leq a_1\left(\left\|x_{k-1}(\lambda) - g^1\left(x_{k-1}(\lambda)\lambda\right)\right\|\right. \\
&\qquad \left. + \left\|x_k(\lambda) - g^2\left(x_{k-1}(\lambda), \lambda\right)\right\|\right)\left\|x_{k-1}(\lambda)\right\|. \quad (4.3.43)
\end{aligned}
$$

Moreover from (4.3.7), (4.3.8), (4.3.9), and (4.3.10), we obtain the estimates

$$
\begin{aligned}
\left\|x_{k-1}(\lambda) - g^1\left(x_{k-1}(\lambda), \lambda\right)\right\| &\leq \left\|x_{k-1}(\lambda) - x_k(\lambda)\right\| + \left\|x_k(\lambda) - g^1\left(x_k(\lambda), \lambda\right)\right\| \\
&\leq \left\|g^1\left(x_k(\lambda), \lambda\right) - g^1\left(x_{k-1}(\lambda), \lambda\right)\right\| \\
&\leq \left\|x_k(\lambda) - x_{k-1}(\lambda)\right\| + a_2\left\|x_{k+1}(\lambda) - x_k(\lambda)\right\| \\
&\qquad + a_4\left\|x_k(\lambda) - x_{k-1}(\lambda)\right\|, \quad (4.3.44)
\end{aligned}
$$

$$
\begin{aligned}
\left\|x_k(\lambda) - g^2\left(x_{k-1}(\lambda), \lambda\right)\right\| &\\
\leq \left\|x_k(\lambda) - g^2\left(x_k(\lambda), \lambda\right)\right\| &+ \left\|g^2\left(x_k(\lambda), \lambda\right) - g^2\left(x_{k-1}(\lambda), \lambda\right)\right\| \\
\leq a_3\left\|x_{k+1}(\lambda) - x_k(\lambda)\right\| &+ a_5\left\|x_k(\lambda) - x_{k-1}(\lambda)\right\|. \quad (4.3.45)
\end{aligned}
$$

Hence from (4.3.43), (4.3.44), and (4.3.45), we get

$$\Big\| BP(F(x_k(\lambda),\lambda) - F(x_{k-1}(\lambda),\lambda)$$
$$- \Big[g^1(x_k(\lambda)\lambda), g^2 x_k(\lambda), \lambda; F\Big](x_k(\lambda) - x_{k-1}(\lambda))) \Big\|$$
$$\leq a_1(1 + a_4 + a_5) \|x_k(\lambda) - x_{k-1}(\lambda)\|^2$$
$$+ a_1(a_2 + a_3) \|x_{k+1}(\lambda) - x_k(\lambda)\| \|x_k(\lambda) - x_{k-1}(\lambda)\|. \quad (4.3.46)$$

As in (4.3.46) but using (4.3.13), (4.3.14), (4.3.15), (4.3.16), and (4.3.17), we obtain

$$\Big\| BP(G(x_k(\lambda),\lambda) - G(x_{k-1}(\lambda),\lambda)$$
$$- \Big[g^3(x_k(\lambda),\lambda), g^4(x_k(\lambda),\lambda); G\Big](x_k(\lambda) - x_{k-1}(\lambda))) \Big\|$$
$$\leq a_8(1 + a_{11} + a_{12}) \|x_k(\lambda) - x_{k-1}(\lambda)\|^2$$
$$+ a_8(a_9 + a_{10}) \|x_{k+1}(\lambda) - x_k(\lambda)\| \|x_k(\lambda) - x_{k-1}(\lambda)\|. \quad (4.3.47)$$

Furthermore from (4.3.11), (4.3.12), and (4.3.18), we get respectively

$$\|B(QF(x_k(\lambda),\lambda) - QF(x_{k-1}(\lambda),\lambda))\| \leq a_6 \|x_k(\lambda) - x_{k-1}(\lambda)\|, \quad (4.3.48)$$
$$\|B(A(x_k)(\lambda),\lambda)(z(x_k(\lambda),\lambda)) - A(x_{k-1}(\lambda),\lambda)(z(x_{k-1}(\lambda),\lambda))\|$$
$$\leq a_7 \|x_k(\lambda) - x_{k-1}(\lambda)\| \quad (4.3.49)$$

and

$$\|B(QG(x_k(\lambda),\lambda) - QG(x_{k-1}(\lambda),\lambda))\| \leq a_{13} \|x_k(\lambda) - x_{k-1}(\lambda)\|. \quad (4.3.50)$$

Finally from (4.3.31), (4.3.41), (4.3.42), (4.3.46)–(4.3.50), we deduce that estimates (4.3.35) and (4.3.36) are true. By (4.3.36) and part (i) it follows that for each fixed $\lambda \in U(\lambda_0, S)$ iteration $\{x_n(\lambda)\}$ ($n \geq 0$) is Cauchy in a Banach space E and as such it converges to some $x^*(\lambda) \in U(x_0(\lambda), r^*)$. Using hypothesis (c) and letting $n \to \infty$ in (4.3.2), we get $F(x^*(\lambda),\lambda) + G(x^*(\lambda),\lambda) = 0$. That is, $x^*(\lambda)$ is a solution of equation (4.3.1). Estimate (4.3.37) follows immediately from (4.3.36) by using standard majorization techniques.

To show uniqueness when $z = 0$, let us assume $y^*(\lambda) \in U(x_0(\lambda), R)$ is a solution of equation (4.3.1). Then from (4.3.2) we get

$$x_{n+1}(\lambda) - y^*(\lambda) = x_n(\lambda) - y^*(\lambda) - A(x_n(\lambda),\lambda)^{-1}\Big[\big(F(x_n(\lambda),\lambda)$$
$$- F(y^*(\lambda),\lambda)\big) + \big(G(x_n(\lambda),\lambda) - G(y^*(\lambda),\lambda)\big)\Big]. \quad (4.3.51)$$

Analyzing the right-hand side of (4.3.51) as in (4.3.42) with $y^*(\lambda)$ "replacing" $x_k(\lambda)$ and $x_n(\lambda)$ "replacing" $x_{k-1}(\lambda)$, we get

$$\|x_{n+1} - y^*(\lambda)\| \leq c\|x_n(\lambda) - y^*(\lambda)\| \leq \cdots \leq c^{n+1}\|x_0(\lambda) - y^*\| \leq c^{n+1}R. \quad (4.3.52)$$

By letting $n \to \infty$ in (4.3.52) and using (4.3.27) we get $\lim_{n\to\infty} x_{n+1}(\lambda) = y^*(\lambda)$ for each fixed $\lambda \in U(\lambda_0, S)$. By the uniqueness of the limit of the sequence $\{x_n(\lambda)\}$ ($n \geq 0$) we deduce $x^*(\lambda) = y^*(\lambda)$.

Remark 4.3.2. (1) Condition (4.3.6) implies that $F(x(\lambda), \lambda)$ is differentiable on D, whereas condition (4.3.13) does not necessarily imply the differentiability of $G(x(\lambda), \lambda)$ on D.

(2) Inequalities (4.3.21), (4.3.23), (4.3.25), (4.3.26), and (4.3.27) will determine r^*, R, and S.

(3) If $a_2 + a_4 \leq 1$, $a_3 + a_5 \leq 1$, $a_9 + a_{11} \leq 1$ and $a_{10} + a_{12} \leq 1$ for $r^* \neq 0$, condition (4.3.26) is satisfied. Indeed from (4.3.7) we have

$$\left\| g^1(x_0(\lambda), \lambda) - x_0(\lambda) \right\| \leq a_2 \|x_1(\lambda) - x_0(\lambda)\| \leq a_3 r^*,$$

and from (4.3.26) we must have

$$\left\| g^1(x_0(\lambda), \lambda) - x_0(\lambda) \right\| \leq (1 - a_4) r^*.$$

It suffices to show $a_2 r^* \leq (1 - a_4)$ or $a_2 + a_4 \leq 1$ ($r^* \neq 0$), which is true by hypothesis. Similarly, we can argue for the rest.

4.4 Computing zeros of operator satisfying autonomous differential equations

In this section, we are concerned with the problem of approximating a locally unique solution x^* of equation

$$F(x) = 0, \tag{4.4.1}$$

where F is a Fréchet-differentiable operator defined on an open convex subset D of a Banach space X with values in a Banach space Y.

We use the Newton-like method:

$$x_{n+1} = x_n - F'(y_n)^{-1} F(x_n) \quad (n \geq 0) \tag{4.4.2}$$

to generate a sequence approximating x^*.

Here $F'(x) \in L(X, Y)$ denotes the Fréchet derivative.

We are interested in the case when:

$$y_n = \lambda_n x_n + (1 - \lambda_n) z_n \quad (n \geq 0) \tag{4.4.3}$$

where,

$$\lambda_n \in [0, 1], \quad (n \geq 0) \tag{4.4.4}$$

$$z_n = x^* \tag{4.4.5}$$

or

$$z_n = x_n \quad (n \geq 0), \tag{4.4.6}$$

or other suitable choice [170].

We provide a local and a semilocal convergence analysis for method (4.4.2) which compare favorably with earlier results [170], and under the same computational cost.

Convergence for method (4.4.2) for z_n given by (4.4.5) and $\lambda_n = 0$ $(n \geq 0)$

We can show the following local result:

Theorem 4.4.1. *Let $F: D \subseteq X \to Y$ be a Fréchet-differentiable operator. Assume: there exists a solution x^* of equation*

$$F(x) = 0 \text{ such that } F'(x^*)^{-1} \in L(Y, X)$$

and

$$\left\| F'(x^*)^{-1} \right\| \leq b; \tag{4.4.7}$$

$$\left\| F'(x) - F'(x^*) \right\| \leq L_0 \left\| x - x^* \right\| \quad \text{for all } x \in D, \tag{4.4.8}$$

and

$$\overline{U}(x^*, r_0) \subseteq D, \text{ with } r_0 = \frac{2}{bL_0}. \tag{4.4.9}$$

Then sequence $\{x_n\}$ $(n \geq 0)$ generated by Newton-like method (4.4.2) is well defined remains in $U(x^*, r_0)$ for all $n \geq 0$, and converges to x^* provided that $x_0 \in U(x^*, r_0)$.

Moreover the following estimates hold for all $n \geq 0$:

$$\left\| x_n - x^* \right\| \leq \theta_0^{2^n - 1} \left\| x_0 - x^* \right\| \quad (n \geq 1), \tag{4.4.10}$$

where

$$\theta_0 = \tfrac{1}{2} b L_0 \left\| x_0 - x^* \right\|. \tag{4.4.11}$$

Proof. By (4.4.2) and $F(x^*) = 0$ we get for all $n \geq 0$:

$$x_{n+1} - x^* = -F'(x^*)^{-1} \left[\int_0^1 \left(F'(x^* + t(x_n - x^*)) \right. \right.$$

$$\left. \left. - F'(x^*) \right) (x_n - x^*) \right] dt \tag{4.4.12}$$

from which it follows

$$\left\| x_{n+1} - x^* \right\| \leq \tfrac{1}{2} b L_0 \left\| x_n - x^* \right\|^2 \tag{4.4.13}$$

from which (4.4.10) follows.

By (4.4.9) and (4.4.11), $\theta_0 \in [0, 1)$. Hence it follows from (4.4.10) that $x_n \in U(x^*, r_0)$ $(n \geq 0)$ and

$$\lim_{n \to \infty} x_n = x^* \tag{4.4.14}$$

Method (4.4.2) has the advantages of the quadratic convergence of NK method and the simplicity of the modified Newton's method, as the operator $F'(x^*)^{-1}$ is computed only once.

Moreover in order for us to compare Theorem 4.4.1 with earlier results, consider the condition

$$\|F'(x) - F'(y)\| \le L\|x - y\| \quad \text{for all } x \in D \tag{4.4.15}$$

used in [170] instead of (4.4.8). The corresponding radius of convergence is given by

$$r_R = \frac{2}{bL}. \tag{4.4.16}$$

Because

$$L_0 \le L \tag{4.4.17}$$

holds in general, we obtain

$$r_R \le r_0. \tag{4.4.18}$$

Furthermore in case strict inequality holds in (4.4.17), so it does in (4.4.18) (see Chapter 2). Below we give an example of a case where strict inequality holds in (4.4.17) and (4.4.18).

Example 4.4.2. Let $X = Y = R$, $D = U(0, 1)$ and define F on D by

$$F(x) = e^x - 1. \tag{4.4.19}$$

Note that (4.4.19) satisfies (4.4.14) for $T(x) = x + 1$. Using (4.4.7), (4.4.8), (4.4.9), (4.4.15), and (4.4.16), we obtain

$$b = 1, \quad L_0 = e - 1, \quad L = e, \tag{4.4.20}$$

$$r_0 = 1.163953414, \tag{4.4.21}$$

and

$$r_R = .735758882. \tag{4.4.22}$$

In order to keep the iterates inside D we can restrict r_0 and choose

$$r_0 = 1. \tag{4.4.23}$$

In any case (4.4.17) and (4.4.18) holds as a strict inequalities.

We can show the following global result:

Theorem 4.4.3. *Let $F: X \to Y$ be Fréchet-differentiable operator, and G a continuous operator from Y into Y. Assume:*

> condition (4.4.14) holds;
>
> $G(0)^{-1} \in L(Y, X)$ so that (4.4.7) holds;
>
> $F(x) \le c$ for all $x \in X$; $\tag{4.4.24}$

$$\|G(0) - G(z)\| \le a_0 \|z\|, \quad \text{for all} \quad z \in Y, \tag{4.4.25}$$

and

$$h_0 = \alpha_0 bc < 1. \tag{4.4.26}$$

Then, sequence $\{x_n\}$ $(n \ge 0)$ generated by method (4.4.2) is well defined and converges to a unique solution x^* of equation $F(x) = 0$.

Moreover the following estimates hold for all $n \ge 0$:

$$\|x_n - x^*\| \le \frac{h_0^n}{1 - h_0} \|x_1 - x_0\| \quad (n \ge 0). \tag{4.4.27}$$

Proof. It follows from the contraction mapping principle by using (4.4.25), (4.4.26) instead of

$$\|G(v) - G(z)\| \le a \|v - z\| \quad \text{for all} \quad v, z \in Y \tag{4.4.28}$$

and

$$h = abc < 1. \tag{4.4.29}$$

Remark 4.4.4. If F' is L_0 Lipschitz continuous in a ball centered at x^*, then the convergence of method (4.4.2) will be quadratic as soon as

$$bL_0 \|x_0 - x^*\| < 2 \tag{4.4.30}$$

holds with x_0 replaced by an iterate x_n sufficiently close to x^*.

Remark 4.4.5. If (4.4.25) is replaced by the stronger (4.4.28), Theorem 4.4.3 reduces to Theorem 2 in [170]. Otherwise our Theorem is weaker than Theorem 2 in [170] as

$$a_0 < a \tag{4.4.31}$$

holds in general.

We note that if (4.4.25) holds and

$$\|F(x) - F(x_0)\| \le \gamma_0 \|x - x_0\| \tag{4.4.32}$$

then

$$\|F(x)\| \le \|F(x) - F(x_0)\| + \|F(x_0)\|$$
$$\le \gamma_0 \|x - x_0\| + \|F(x_0)\|. \tag{4.4.33}$$

Let $r = \|x - x_0\|$, and define

$$P(r) = a_0 b (\|F(x_0)\| + \gamma_0 r). \tag{4.4.34}$$

If $P(0) = a_0 b \|F(x_0)\| < 1$, then as in Theorem 3 in [170, p. 114] inequality (4.4.26) and the contraction mapping principle we obtain the following semilocal result:

Theorem 4.4.6. *If*

$$q = (1 - a_0 b \, \| F(x_0) \|)^2 - 4b a_0 \gamma_0 \left\| G(0)^{-1} F(x_0) \right\| \geq 0, \qquad (4.4.35)$$

then a solution x^* of equation

$$F(x) = 0$$

exists in $U(x_0, r_1)$, and is unique in $U(x_0, r_2)$, where

$$r_1 = \frac{1 - a_0 b \| F(x_0) \| - \sqrt{q}}{2b a_0 \gamma_0} \qquad (4.4.36)$$

and

$$r_2 = \frac{1 - a_0 b \| F(x_0) \|}{b a_0 \gamma_0}. \qquad (4.4.37)$$

Remark 4.4.7. Theorem 4.4.6 reduces to Theorem 3 in [170, p. 114] if (4.4.25) and (4.4.32) are replaced by the stronger (4.4.28) and

$$\| F(x) - F(y) \| \leq \gamma \, \| x - y \| \qquad (4.4.38)$$

respectively. Otherwise our Theorem is weaker than Theorem 3 in [170].

4.5 The method of tangent hyperbolas

In this section, we are concerned with the problem of approximating a locally unique solution x^* of equation

$$F(x) = 0, \qquad (4.5.1)$$

where F is a twice-Fréchet-differentiable operator on an open convex subset D of a Banach space X with values in a Banach space Y.

The method of tangent hyperbolas (Halley)

$$x_{n+1} = x_n - \left\{ I - \tfrac{1}{2} \Gamma_n F''(x_n) \Gamma_n F(x_n) \right\}^{-1} \Gamma_n F(x_n),$$
$$\Gamma_n = F'(x_n)^{-1} \qquad (n \geq 0) \qquad (4.5.2)$$

is one of the best known cubically convergent iterative procedures for solving non-linear equations like (4.5.1).

Here we provide a semilocal convergence analysis based on Lipschitz and center-Lipschitz conditions on the first and second Fréchet derivatives of F. This way, existing convergence conditions are finer and the information on the location of the solution more precise than before.

We need the following results on majorizing sequences.

Theorem 4.5.1. *Let η, ℓ_i, $i = 0, 1, \ldots, 3$ be nonnegative parameters. Define scalar sequence $\{t_n\}$ $(n \geq 0)$ by*

$$t_0 = 0, \quad t_1 = \frac{2\eta}{2 - \ell_0 \eta} = \eta_0, \quad t_{n+2} - t_{n+1} = b_{n+1} c_{n+1} \eta_{n+1}, \qquad (4.5.3)$$

where,

$$c_n = (1 - \ell_1 t_n)^{-1}, \quad d_n = \ell_0 + \ell_2 t_n, \quad a_n = \frac{1}{2} \frac{\ell_0 + \ell_2 t_n}{(1 - \ell_1 t_n)^2} \eta_n, \tag{4.5.4}$$

$$b_n = (1 - a_n)^{-1}, \quad \eta_{n+1} = \frac{1}{4} d_n c_n \eta_n (t_{n+1} - t_n)^2 + \frac{1}{6} \ell_3 (t_{n+1} - t_n)^3, \tag{4.5.5}$$

and parameter α by

$$\alpha = \left[\frac{1}{2} \frac{\ell_0 + 2\ell_2 \eta_0}{1 - 2\ell_1 \eta_0} + \frac{1}{3} \ell_3 \right] \eta_0^2. \tag{4.5.6}$$

Assume:

$$\ell_0 \eta < 1, \tag{4.5.7}$$

$$2\ell_1 \eta_0 < 1, \tag{4.5.8}$$

$$\frac{1}{4} \frac{\ell_0 + 2\ell_2 \eta_0}{(1 - 2\ell_1 \eta_0)^2} \eta_0 < 1, \tag{4.5.9}$$

and

$$\alpha \leq \min \left\{ 1, (1 - 2\ell_1 \eta_0) \left[1 - \frac{1}{2} \frac{\ell_0 + 2\ell_2 \eta_0}{(1 - 2\ell_1 \eta_0)^2} \eta_0 \right] \right\}. \tag{4.5.10}$$

Then, sequence $\{t_n\}$ $(n \geq 0)$ is nondecreasing, bounded above by

$$t^{**} = 2\eta_0, \tag{4.5.11}$$

and converges to t^* such that

$$0 \leq t^* \leq t^{**}. \tag{4.5.12}$$

Moreover, the following estimates hold for all $n \geq 0$:

$$0 \leq t_{n+2} - t_{n+1} \leq \frac{1}{2} (t_{n+1} - t_n) \leq \left(\frac{1}{2} \right)^{n+1} \eta_0. \tag{4.5.13}$$

Proof. Using induction on k we show:

$$\left[\frac{1}{2} \frac{\ell_2 t_k + \ell_0}{1 - \ell_1 t_k} \eta_k (t_{k+1} - t_k) + \frac{1}{3} \ell_3 (t_{k+1} - t_k)^2 \right] c_{k+1} b_{k+1} \leq 1, \tag{4.5.14}$$

$$t_{k+1} - t_k \geq 0, \tag{4.5.15}$$

$$\frac{1}{2} \frac{\ell_0 + \ell_2 t_{k+1}}{(1 - \ell_1 t_{k+1})^2} \eta_{k+1} < 1, \tag{4.5.16}$$

and

$$1 - \ell_1 t_{k+1} > 0. \tag{4.5.17}$$

For $k = 0$ (4.5.14)–(4.5.17) hold by (4.5.3) and (4.5.7)–(4.5.9). By (4.5.3) we then get

$$t_2 - t_1 \leq \frac{\alpha}{2} (t_1 - t_0) \leq \frac{1}{2} (t_1 - t_0). \tag{4.5.18}$$

Let us assume (4.5.14)–(4.5.17)) hold for all $k \leq n + 1$. We can easily obtain from (4.5.3) that

$$t_{k+1} \leq \frac{1 - \left(\frac{1}{2} \right)^{k+1}}{1 - \frac{1}{2}} \eta_0 = 2 \left[1 - \left(\frac{1}{2} \right)^{k+1} \right] \eta_0 \leq t^{**} \tag{4.5.19}$$

and the left-hand side of (4.5.14) is bounded above by one (see (4.5.9) and (4.5.10)). That completes the induction for (4.5.14). Using (4.5.3) we get

$$t_{k+2} \leq 2\left[1 - \left(\tfrac{1}{2}\right)^{k+2}\right]\eta_0 \leq t^{**}, \tag{4.5.20}$$

which shows (4.5.17) and (4.5.16) (by (4.5.14) and as $\eta_i \leq \tfrac{1}{2}\eta_0$, $i \geq 1$). Finally

$$t_{k+2} - t_{k+1} \geq 0 \tag{4.5.21}$$

follows from (4.5.3), (4.5.14), (4.5.16), and (4.5.17).

The induction for (4.5.14)–(4.5.17) is now complete. Hence, sequence $\{t_n\}$ ($n \geq 0$) is bounded above by t^{**}, nondecreasing and as such it converges to some t^* satisfying (4.5.12).

Similarly, we show the following result on majorizing sequences.

Theorem 4.5.2. *Let η, ℓ_0, ℓ_2, ℓ_3 be nonnegative parameters. Define scalar sequence $\{v_n\}$ ($n \geq 0$) by*

$$v_0 = 0, \quad v_1 = \eta_0^1 = \frac{2\eta}{2 - \ell_0\eta}, \quad v_{n+2} - v_{n+1} = b_{n+1}^1 c_{n+1}^1 \eta_{n+1}^1 \tag{4.5.22}$$

where,

$$c_n^1 = \left[1 - v_n\left(\ell_0 + \frac{\ell_2}{2}v_n\right)\right]^{-1}, \quad a_n^1 = \frac{1}{2}d_n^1 c_n^1 \eta_n^1\left(c_n^1\right)^2, \quad b_n^1 = (1 - a_n^1)^{-1}, \tag{4.5.23}$$

$$\eta_{n+1}^1 = \frac{1}{4}d_n^1 c_n^1 \eta_n^1(v_{n+1} - v_n)^2 + \frac{1}{6}\ell_3(v_{n+1} - v_n)^2, \tag{4.5.24}$$

and parameter α^1 by

$$\alpha^1 = \left[\frac{1}{2}\frac{\ell_0 + 2\ell_2 v_1}{1 - 2\ell_0 v_1 - 2\ell_2 v_1^2} + \frac{1}{3}\ell_3\right]v_1^2. \tag{4.5.25}$$

Assume:

$$2(\ell_0 + \ell_2 v_1)v_1 < 1, \tag{4.5.26}$$

$$\frac{1}{4}\frac{\ell_0 + 2\ell_2 v_1}{1 - 2\ell_0 v_1 - 2\ell_1 v_1^2}v_1 < 1, \tag{4.5.27}$$

and

$$\alpha^1 \leq \min\left\{1, (1 - 2\ell_0 v_1 - 2\ell_2 v_1^2)\left[1 - \frac{1}{2}\frac{\ell_0 + 2\ell_2 v_1}{(1 - 2\ell_0 v_1 - 2\ell_2 v_1^2)^2}v_1\right]\right\}. \tag{4.5.28}$$

Then, sequence $\{v_n\}$ $(n \geq 0)$ is nondecreasing, bounded above by

$$v^{**} = 2v_1, \tag{4.5.29}$$

and converges to v^* such that

$$0 \leq v^* \leq v^{**}. \tag{4.5.30}$$

Moreover the following estimates hold for all $n \geq 0$

$$0 \leq v_{n+2} - v_{n+1} \leq \tfrac{1}{2}(v_{n+1} - v_n) \leq \left(\tfrac{1}{2}\right)^{n+1} v_1. \tag{4.5.31}$$

We can show the main semilocal convergence theorem for method (4.5.2).

Theorem 4.5.3. *Let $F: D \subseteq X \rightarrow Y$ be a twice Fréchet-differentiable operator. Assume: there exist a point $x_0 \in D$ and nonnegative parameters η, ℓ_0, ℓ_1, ℓ_2, ℓ_3 such that*

$$F'(x_0)^{-1} \in L(Y, X), \tag{4.5.32}$$

$$\|F'(x_0)^{-1} F(x_0)\| \leq \eta, \tag{4.5.33}$$

$$\|F'(x_0)^{-1} F''(x_0)\| \leq \ell_0, \tag{4.5.34}$$

$$\|F'(x_0)^{-1} \left[F'(x) - F'(x_0)\right]\| \leq \ell_1 \|x - x_0\|, \tag{4.5.35}$$

$$\|F'(x_0)^{-1} \left[F''(x) - F''(x_0)\right]\| \leq \ell_2 \|x - x_0\| \tag{4.5.36}$$

and

$$\|F'(x_0)^{-1} \left[F''(x) - F''(y)\right]\| \leq \ell_3 \|x - y\| \quad \textit{for all} \ \ x, y \in D. \tag{4.5.37}$$

Moreover hypotheses of Theorem 4.5.1 hold, and

$$\overline{U}(x_0, t^*) \subseteq D. \tag{4.5.38}$$

Then sequence $\{x_n\}$ $(n \geq 0)$ generated by method of tangent hyperbolas (4.5.2) is well defined, remains in $\overline{U}(x_0, t^*)$ for all $n \geq 0$ and converges to a solution $x^* \in \overline{U}(x_0, t^*)$ of equation $F(x) = 0$. Moreover the following estimates hold for all $n \geq 0$:

$$\|x_{n+1} - x_n\| \leq t_{n+1} - t_n \tag{4.5.39}$$

and

$$\|x_n - x^*\| \leq t^* - t_n. \tag{4.5.40}$$

Furthermore, if there exists $R \geq t^*$ such that

$$U(x_0, R) \subseteq D, \tag{4.5.41}$$

and

$$\ell_1(t^* + R) \leq 2, \tag{4.5.42}$$

the solution x^* is unique in $U(x_0, R)$.

Proof. Let us prove that

$$\|x_{k+1} - x_k\| \le t_{k+1} - t_k,$$
(4.5.43)

and

$$\overline{U}(x_{k+1}, t^* - t_{k+1}) \subseteq \overline{U}(x_k, t^* - t_k) \text{ for all } k \ge 0.$$
(4.5.44)

For every $z \in \overline{U}(x_1, t^* - t_1)$,

$$\|z - x_0\| \le \|z - x_1\| + \|x_1 - x_0\| \le t^* - t_1 + t_1 = t^* - t_0,$$
(4.5.45)

implies $z \in \overline{U}(x_0, t^* - t_0)$. Note also that

$$\|x_1 - x_0\| = \left\| \left[I - \tfrac{1}{2} F'(x_0)^{-1} F''(x_0) F'(x_0)^{-1} F(x_0) \right]^{-1} F'(x_0)^{-1} F(x_0) \right\|$$

$$\le \frac{\left\| F'(x_0)^{-1} F(x_0) \right\|}{1 - \tfrac{1}{2} \left\| F'(x_0)^{-1} F''(x_0) \right\| \left\| F'(x_0)^{-1} F(x_0) \right\|} = \overline{\eta}_0 \le \frac{\eta}{1 - \tfrac{1}{2}\ell_0\eta} = \eta_0.$$

Because also

$$\|x_1 - x_0\| = \left\| F'(x_0)^{-1} F(x_0) \right\| \le \eta \le t_1$$
(4.5.46)

(4.5.43) and (4.5.44) hold for $k = 0$. Given they hold for $n = 0, 1, \ldots, k$, then

$$\|x_{k+1} - x_0\| \le \sum_{i=1}^{k+1} \|x_i - x_{i-1}\| \le \sum_{i=1}^{k+1} (t_i - t_{i-1}) = t_{k+1} - t_0 = t_{k+1},$$
(4.5.47)

and

$$\|x_k + \theta(x_{k+1} - x_k) - x_0\| \le t_k + \theta(t_{k+1} - t_k) < t^* \quad \theta \in [0, 1].$$
(4.5.48)

It follows from (4.5.35),

$$\left\| F'(x_0)^{-1} \left[F'(x_{k+1}) - F'(x_0) \right] \right\| \le \ell_1 \|x_{k+1} - x_0\|$$

$$\le \ell_1 t_{k+1} \le 2\ell_1 \eta_0 < 1 \quad \text{(by (4.5.8))},$$
(4.5.49)

and the Banach Lemma on invertible operators that the inverse $F'(x_{k+1})^{-1}$ exists, and

$$\left\| F'(x_{k+1})^{-1} F'(x_0) \right\| \le [1 - \ell_1 \|x_{k+1} - x_0\|]^{-1} \le (1 - \ell_1 t_{k+1})^{-1}.$$
(4.5.50)

Moreover, we have the estimates

$$\left\| F'(x_0)^{-1} F''(x_k) \right\| \le \left\| F'(x_0)^{-1} \left[F''(x_k) - F''(x_0) \right] \right\| + \left\| F'(x_0)^{-1} F''(x_0) \right\|$$

$$\le \ell_2 \|x_k - x_0\| + \ell_0 = \overline{d}_k \le d_k,$$
(4.5.51)

$$\left\| F'(x_0)^{-1} F(x_k) \right\| \le \overline{\eta}_k,$$
(4.5.52)

$$\left\| \left[F'(x_k)^{-1} F'(x_0) \right] \left[F'(x_0)^{-1} F''(x_k) \right] \left[F'(x_k)^{-1} F'(x_0) \right] \left[F'(x_0)^{-1} F(x_k) \right] \right\|$$

$$\leq \frac{(\ell_2 \|x_k - x_0\| + \ell_0)\overline{\eta}_k}{(1 - \ell_1 \|x_k - x_0\|)^2} = 2\overline{a}_k \leq 2a_k < 2, \quad (4.5.53)$$

the inverse of $\left[I - \frac{1}{2} F'(x_k)^{-1} F''(x_k) F'(x_k)^{-1} F(x_k) \right]$ exists, and

$$\left\| \left[I - \frac{1}{2} F'(x_k)^{-1} F''(x_k) F'(x_k)^{-1} F(x_k) \right]^{-1} \right\| \leq \overline{b}_k = (1 - \overline{a}_k)^{-1} \leq (1 - a_k)^{-1}.$$
$$(4.5.54)$$

Furthermore using the expression from (4.5.2) for $x_{k+1} - x_k$, we have

$$F(x_k) + F'(x_k)(x_{k+1} - x_k) + \frac{1}{2} F''(x_k)(x_{k+1} - x_k)^2$$
$$= \frac{1}{4} F''(x_k) F'(x_k)^{-1} F''(x_k) F'(x_k)^{-1} F(x_k)(x_{k+1} - x_k)^2, \quad (4.5.55)$$

so that

$$\left\| F'(x_0)^{-1} F(x_{k+1}) \right\|$$

$$= \left\| F'(x_0)^{-1} \left\{ \frac{1}{4} F''(x_k) F'(x_k)^{-1} F''(x_k) F'(x_k)^{-1} F(x_k)(x_{k+1} - x_k)^2 \right. \right.$$

$$+ \int_0^1 (1 - \theta) F''(x_k + \theta(x_{k+1} - x_k)) d\theta (x_{k+1} - x_k)^2$$

$$\left. \left. - \frac{1}{2} F''(x_k)(x_{k+1} - x_k)^2 \right\} \right\|$$

$$\leq \frac{1}{4} \left\| F'(x_0)^{-1} F''(x_k) \right\| \left\| F'(x_k)^{-1} F'(x_0) \right\| \left\| F'(x_0)^{-1} F(x_k) \right\| \cdot \|x_{k+1} - x_k\|^2$$

$$+ \int_0^1 (1 - \theta) \left\| F'(x_0)^{-1} \left[F''(x_k + \theta(x_{k+1} - x_k)) \right. \right.$$

$$\left. \left. - F''(x_k) \right] \right\| \|x_{k+1} - x_k\|^2 d\theta$$

$$\leq \frac{1}{4} \overline{d}_k \overline{c}_k \overline{\eta}_k \|x_{k+1} - x_k\|^2 + \frac{\ell_3}{6} \|x_{k+1} - x_k\|^3 = \overline{\eta}_{k+1} \leq \eta_{k+1}. \quad (4.5.56)$$

Using (4.5.2), (4.5.32)–(4.5.37), we get

$$\|x_{k+2} - x_{k+1}\| \leq \left\| \left[I - \frac{1}{2} F'(x_{k+1})^{-1} F'(x_0) F'(x_0)^{-1} \right. \right.$$

$$\left. \left. \cdot F''(x_{k+1}) F'(x_{k+1})^{-1} F'(x_0) F'(x_0)^{-1} F(x_{k+1}) \right]^{-1} \right\|$$

$$\cdot \left\| F'(x_{k+1})^{-1} F'(x_0) \right\| \left\| F'(x_0)^{-1} F(x_{k+1}) \right\|$$

$$\leq \overline{b}_{k+1} \overline{c}_{k+1} \overline{\eta}_{k+1} \leq b_{k+1} c_{k+1} \eta_{k+1} = t_{k+2} - t_{k+1}, \quad (4.5.57)$$

which together with (4.5.43) show (4.5.39) for all $n \geq 0$. Thus for every $z \in \overline{U}(x_{k+2}, t^* - t_{k+2})$, we have

$$\|z - x_{k+1}\| \leq \|z - x_{k+2}\| + \|x_{k+2} - x_{k+1}\|$$

$$\leq t^* - t_{k+2} + t_{k+2} - t_{k+1} = t^* - t_{k+1}. \qquad (4.5.58)$$

That is,

$$z \in \overline{U}(x_{k+1}, t^* - t_{k+1}). \qquad (4.5.59)$$

Estimates (4.5.57) and (4.5.59) imply that (4.5.43) and (4.5.44) hold for $n = k + 1$. By induction the proof of (4.5.43) and (4.5.44) is completed.

Theorem 4.5.1 implies $\{t_n\}$ $(n \geq 0)$ is a Cauchy sequence. From (4.5.43) and (4.5.44) $\{x_n\}$ $(n \geq 0)$ becomes a Cauchy sequence, too, and as such it converges to some $x^* \in \overline{U}(x_0, t^*)$ (as $\overline{U}(x_0, t^*)$ is a closed set) such that

$$\|x_k - x^*\| \leq t^* - t_k. \qquad (4.5.60)$$

The combination of (4.5.43) and (4.5.60) yields $F(x^*) = 0$. Finally to show uniqueness let y^* be a solution of equation $F(x) = 0$ in $U(x_0, R)$. It follows from (4.5.35), the estimate

$$\left\| F'(x_0)^{-1} \int_0^1 \left[F'(y^* + \theta(x^* - y^*)) - F'(x_0) \right] d\theta \right\|$$

$$\leq \ell_1 \int_0^1 \|[y^* + \theta(x^* - y^*) - x_0]\| \, d\theta$$

$$\leq \ell_1 \int_0^1 \left[\theta \|x^* - x_0\| + (1 - \theta)\|y^* - x_0\| \right] d\theta$$

$$< \frac{\ell_1}{2}(t^* + R) \leq 1, \quad \text{(by (4.5.42))}, \qquad (4.5.61)$$

and the Banach Lemma on invertible operators that linear operator

$$L = \int_0^1 F'(y^* + \theta(x^* - y^*)) d\theta \qquad (4.5.62)$$

is invertible.

Using the identity

$$0 = F(x^*) - F(y^*) = L(x^* - y^*), \qquad (4.5.63)$$

we deduce

$$x^* = y^*.$$

Similarly, we show the result:

Theorem 4.5.4. *Let $F: D \subseteq X \to Y$ be a twice Fréchet-differentiable operator. Assume there exist a point $x_0 \in D$ and nonnegative parameters η, ℓ_0, ℓ_2, ℓ_3 such that*

$$F'(x_0)^{-1} \in L(Y, X),$$

$$\left\| F'(x_0)^{-1} F'(x_0) \right\| \leq \eta,$$

$$\left\| F'(x_0)^{-1} F''(x_0) \right\| \leq \ell_0,$$

$$\left\| F'(x_0)^{-1} \left[F''(x) - F''(x_0) \right] \right\| \leq \ell_2 \|x - x_0\|,$$

and

$$\left\| F'(x_0)^{-1} \left[F''(x) - F''(y) \right] \right\| \leq \ell_3 \|x - y\| \quad \text{for all} \quad x, y \in D.$$

Moreover, hypotheses of Theorem 4.5.2 hold, and

$$\overline{U}(x_0, v^*) \subseteq D. \tag{4.5.64}$$

Then the sequence $\{x_n\}$ $(n \geq 0)$ generated by method of tangent hyperbolas (4.5.2) is well defined, remains in $\overline{U}(x_0, v^*)$ for all $n \geq 0$ and converges to a solution $x^* \in \overline{U}(x_0, v^*)$ of equation $F(x) = 0$. Moreover the following estimates hold for all $n \geq 0$:

$$\|x_{n+1} - x_n\| \leq v_{n+1} - v_n, \tag{4.5.65}$$

and

$$\|x_n - x^*\| \leq v^* - v_n. \tag{4.5.66}$$

Furthermore, if there exists $R_1 \geq v^*$ such that

$$U(x_0, R_1) \subseteq D, \tag{4.5.67}$$

and

$$\gamma = \frac{1}{2} \left[\ell_0 + \frac{\ell_2}{4} (R_1 + v^*) \right] (R_1 + v^*) \in [0, 1], \tag{4.5.68}$$

the solution x^* is unique in $U(x_0, R_1)$.

Proof. The computation of the inverses $F'(x_k)^{-1}$, $x_k \in \overline{U}(x_0, v^*)$ is carried out using the identity

$$
\begin{aligned}
I - &F'(x_0)^{-1} F'(x_k) \\
&= -F'(x_0)^{-1} \left[F'(x_k) - F'(x) - F''(x_0)(x_k - x_0) + F''(x_0)(x_k - x_0) \right] \\
&= \int_0^1 -F'(x_0)^{-1} \left\{ F'' \left[x_0 + \theta(x_k - x_0) \right] - F''(x_0) \right\} d\theta (x_k - x_0) \\
&\quad - F'(x_0)^{-1} F''(x_0)(x_k - x_0).
\end{aligned}
\tag{4.5.69}
$$

By (4.5.34), (4.5.36), and (4.5.9) we get

$$\|I - F'(x_0)^{-1}F'(x_k)\| \leq \int_0^1 \ell_2\theta d\theta \|x_k - x_0\|^2 + \ell_0\|x_k - x_0\|$$

$$\leq \tfrac{1}{2}\ell_2 v_k^2 + \ell_0 v_k \leq \tfrac{1}{2}\ell_2(2\eta_0)^2 + \ell_0(2\eta_0) < 1 \quad (4.5.70)$$

(by (4.5.26)). Hence

$$\|F'(x_k)^{-1}F'(x_0)\| \leq \bar{c}_n^1 = \left[1 - \left(\tfrac{1}{2}\ell_2\|x_k - x_0\|^2 + \ell_0\|x_k - x_0\|\right)\right]^{-1} \leq c_n^1.$$

$$(4.5.71)$$

The rest of the proof until the uniqueness part is identical to Theorem 4.5.3.

Let y^* be a solution of equation $F(x) = 0$ in $U(x_0, R_1)$. For $z \in U(x_0, R_1)$ we have

$$\left\|F'(x_0)^{-1}\left[F'(z) - F'(x_0)\right]\right\| = \left\|F'(x_0)^{-1}\int_0^1 F''(x_0 + \theta_1(z - x_0))(z - x_0)d\theta_1\right\|$$

$$\leq \left\|F'(x_0)^{-1}\int_0^1 \left[F''(x_0 + \theta_1(z - x_0)) - F''(x_0)\right]d\theta_1\right\| \|z - x_0\|$$

$$+ \left\|F'(x_0)^{-1}F''(x_0)\right\| \cdot \|z - x_0\|$$

$$\leq \ell_2 \int_0^1 \|z - x_0\|^2 + \ell_0\|z - x_0\| \leq \frac{\ell_2}{2}\|z - x_0\|^2 + \ell_0\|z - x_0\|. \quad (4.5.72)$$

Set

$$L = \int_0^1 F'(y^* + \theta(x^* - y^*))d\theta. \quad (4.5.73)$$

Then we have for $z = y^* + \theta(x^* - y^*)$, $\theta \in [0, 1]$:

$$\|z - x_0\| \leq (1-\theta)\|y^* - x_0\| + \theta\|x^* - x_0\| < (1-\theta)R_1 + \theta v^* \leq (1-\theta)R_1 + \theta R_1 = R_1.$$

Hence, we get

$$\left\|F'(x_0)^{-1}\left[L - F'(x_0)\right]\right\| \leq \int_0^1 \left[\frac{\ell_2}{2}\|z - x_0\|^2 + \ell_0\|z - x_0\|\right]d\theta$$

$$< \frac{\ell_2}{8}(R_1 + v^*)^2 + \frac{\ell_0}{2}(R_1 + v^*)$$

$$= \gamma \in [0, 1]. \quad (4.5.74)$$

By the Banach Lemma on invertible operators and (4.5.74) L is invertible.
Using the identity

$$F(x^*) - F(y^*) = L(x^* - y^*), \quad (4.5.75)$$

we get

$$x^* = y^*.$$

Remark 4.5.5. It can easily be seen by using induction on n that if

$$\ell_0 + \ell_2\eta_0 < \ell_1, \qquad (4.5.76)$$

then hypotheses of Theorem 4.5.1 imply those of Theorem 4.5.2 and under the rest of the hypotheses

$$v_{n+1} - v_n < t_{n+1} - t_n \quad (n \geq 1), \qquad (4.5.77)$$
$$v_n < t_n \quad (n \geq 1), \qquad (4.5.78)$$

and

$$v^* \leq t^*. \qquad (4.5.79)$$

Moreover, if equality holds in (4.5.76), then (4.5.77)–(4.5.79) also hold as equalities. Furthermore if

$$\ell_0 + \ell_2\eta_0 > \ell_1, \qquad (4.5.80)$$

then

$$t_{n+1} - t_n < v_{n+1} - v_n \quad (n \geq 1), \qquad (4.5.81)$$
$$t_n < v_n \quad (n \geq 1), \qquad (4.5.82)$$

and

$$t^* \leq v^*. \qquad (4.5.83)$$

Remark 4.5.6. In order for us to compare our results with earlier ones in [123], [211] define sequences $\{\delta_n\}$, $\{M_n\}$, $\{N_n\}$, $\{\beta_n\}$ by

$$\delta_0 = \eta, \quad M_0 = \ell_0, \quad N_0 = \ell_3, \qquad (4.5.84)$$

$$\sigma_n = \left(1 - \frac{1}{2}M_n\delta_n\right)^{-1}, \quad \beta_n = \sigma_n\delta_n, \quad h_n = M_n\beta_n, \qquad (4.5.85)$$

$$\varepsilon_n = \frac{N_n}{M_n^2}, \quad \phi_n(h) = h + \frac{1}{2}\varepsilon_n h^2, \qquad (4.5.86)$$

$$s_n^2 = \left(\frac{1}{6}\varepsilon_n\sigma_n + \frac{1}{4}\right) \Big/ \left(1 - \varphi_n(h_n)\right), \qquad (4.5.87)$$

$$\ell_n^2 = \frac{\phi_n^1(h_n)s_n^2 h_n^2}{1 - \phi_n(h_n)}, \qquad (4.5.88)$$

$$\delta_{n+1} = s_n^2 h_n^2 \delta_n, \quad M_{n+1} = \frac{M_n\phi_n^1(h_n)}{1 - \phi_n(h_n)}, \qquad (4.5.89)$$

$$N_{n+1} = \frac{N_n}{1 - \phi_n(h_n)}, \qquad (4.5.90)$$

function f by

$$f(t) = \frac{1}{6}\ell_3 t^3 + \frac{1}{2}\ell_0 t^2 - t + \eta \qquad (4.5.91)$$

and sequence $\{w_n\}$ $(n \geq 0)$ by

$$w_0 = 0, \quad w_{n+1} = w_n - \left[1 - f''(w_n)f(w_n)/2f'(w_n)^2\right]^{-1} f(w_n)/f'(w_n). \quad (4.5.92)$$

Assume:

$$\delta_0 M_0 < 2, \qquad (4.5.93)$$
$$\phi_0(h_0) < 1, \qquad (4.5.94)$$
$$\ell_0^2 \leq 1, \qquad (4.5.95)$$
$$\overline{U}(x_0, r) \subseteq D, \quad r = \frac{\beta}{1 - s_0^2 h_0^2}, \qquad (4.5.96)$$

or equation

$$f(t) = 0 \qquad (4.5.97)$$

has one negative and two positive roots w^*, w^{**} such that $w^* \leq w^{**}$ and $\overline{U}(x_0, w^*)$ $\subseteq D$ or equivalently

$$\eta \leq \frac{\ell_0^2 + 4\ell_3 - \ell_0\sqrt{\ell_0^2 + 2\ell_3}}{3\ell_3(\ell_0 + \sqrt{\ell_0^2 + 2\ell_3})}, \quad \ell_3 \neq 0, \quad \left(\ell_0\eta \leq \tfrac{1}{2} \text{ if } \ell_3 = 0\right), \qquad (4.5.98)$$

and

$$\overline{U}(x_0, w^*) \subseteq D. \qquad (4.5.99)$$

Then, the method of tangent parabolas $\{x_n\}$ $(n \geq 0)$ generated by (4.5.2) is well defined, remains in $\overline{U}(x_0, w^*)$ for all $n \geq 0$ and converges to a solution $x^* \in \overline{U}(x_0, w^*)$ of equation $F(x) = 0$. Moreover, the following estimates hold for all $n \geq 0$:

$$\|x_{n+1} - x_n\| \leq w_{n+1} - w_n, \qquad (4.5.100)$$

and

$$\|x_n - x^*\| \leq w^* - w_n. \qquad (4.5.101)$$

Furthermore, if: $w^* < w^{**}$ the solution is unique in $U(x_0, w^{**})$ otherwise the solution is unique in $\overline{U}(x_0, w^*)$.

In general we have:

$$\ell_2 \leq \ell_3. \qquad (4.5.102)$$

If strict inequality holds in (4.5.102) using induction on n we can easily show under the hypotheses of Theorem 4.5.4 and (4.5.84)–(4.5.92), (4.5.98), (4.5.99) we get

$$v_{n+1} - v_n < w_{n+1} - w_n \quad (n \geq 1), \qquad (4.5.103)$$
$$v_n < w_n, \quad (n \geq 1) \qquad (4.5.104)$$

and

$$v^* \leq w^*. \qquad (4.5.105)$$

That is our Theorem 4.5.4 provides more precise error bunds and a better information on the location of the solution x^*. In the case of $\ell_2 = \ell_3$, Theorem 4.5.4 reduces to earlier ones mentioned in this remark.

We complete this study with one simple example:

Example 4.5.7. Let $\ell_0 = \ell_2 = 0$, $\eta = 1$ and $\ell_3 = 1$. Then, (4.5.99) is violated since

$$\eta = 1 > \frac{\sqrt{2}}{2}. \tag{4.5.106}$$

Hence the results in [208] cannot be used. However all hypotheses of Theorems 4.5.2, 4.5.4 are satisfied, as (4.5.26)–(4.5.28) hold.

Our idea of using a combination of Lipschitz and center-Lipschitz in the study of iterative processes instead of only Lipschitz conditions, although very recent, has already picked up by several authors [4], [47], [172]. Here in the remaining sections of this chapter we report the results in [47], [172], [4] without proofs.

4.6 A modified secant method and function optimization

In this section, we are concerned with the problem of approximating a locally unique solution x^* of nonlinear equation

$$F'(x) = 0, \tag{4.6.1}$$

where F is a twice differentiable function defined on a convex subset (open or closed) D of the real **R** or complex space **C**.

This study is important especially in the optimization of functions.

Indeed the well-known K-T condition states that if F is differentiable, then the optimal solution of

$$\min F(x)$$

is a solution of equation (4.6.1) [154].

Recently [47] the modified secant method

$$x_{n+1} = \frac{x_n + x_{n-1}}{2} - \frac{[x_n, x_{n-1}]}{2[x_n, x_{n-1}, x_{n-2}]}, \quad (n \geq 0), \quad (x_{-2}, x_{-1}, x_0 \in D) \tag{4.6.2}$$

was proposed to approximate x^* as an alternative to the NK method

$$x_{n+1} = x_n - \frac{F'(x_n)}{F''(x_n)}, \quad (n \geq 0), \quad (x_0 \in D), \tag{4.6.3}$$

or the secant method

$$x_{n+1} = x_n - \frac{x_n - x_{n-1}}{F'(x_n) - F'(x_{n-1})}, \quad (n \geq 0), \quad (x_{-1}, x_0 \in D), \tag{4.6.4}$$

where $[x, y]$, $[x, y, z]$ denote divide differences of order one and two respectively for function F (see Section 1.2).

Methods (4.6.3) and (4.6.4) are being avoided in general because of the evaluations required on the first and second derivatives of F.

We can state the following local convergence result for the method (4.6.2):

Theorem 4.6.1. *Let x^* be a solution of equation (4.6.1) such that $F''(x^*) \neq 0$. Moreover, assume:*

(a) *there exist nondecreasing functions v, w such that*

$$\left| F''(x^*)^{-1} \left(F''(x) - F''(y) \right) \right| \leq v\left(|x - y|\right), \quad \forall x, y \in D, \tag{4.6.5}$$

and

$$\left| F''(x^*)^{-1} \left(F''(x) - F''(x^*) \right) \right| \leq v\left(|x - x^*|\right), \quad \forall x \in D, \tag{4.6.6}$$

(b) *equation*

$$\int_0^1 (1-t)\, v\left((1-t)r\right) dt + \int_0^1 \int_0^1 (1-t)\, v\left((1-s)(1-t)r\right) ds\, dt + w(r) = 1$$

has a minimum positive r_0 such that

$$w(r_0) < 1.$$

(c) $U(x^*, r_1) \subseteq D$,
where r_1 is the solution of equation

$$\int_0^1 w(t)\, dt = r.$$

Then sequence $\{x_n\}$ generated by method (4.6.2) starting from any initial points $x_{-2}, x_{-1}, x_0 \in U(x^*, r_0)$ is well defined, remains in $U(x^*, r_0)$ for all $n \geq 0$ and converges to the unique solution x^* of equation (4.6.1) in $U(x^*, r_1)$ and $r_0 \leq r_1$. Set

$$e_n = |x_n - x^*|, \quad n \geq -2.$$

Moreover, the following error estimates hold for all $n \geq 0$:

$$e_{n+1} \leq \frac{a_n}{b_n},$$

$$e_n \leq c\lambda^n,$$

where

$$a_n = \int_0^1 \int_0^1 (1-t)\, v\left(s(1-t)e_{n-1} + (1-s)(1-t)e_{n-2}\right) ds\, dt$$

$$+ e_{n-1} \int_0^1 \int_0^1 (1-t)\, v\left((1-s)(1-t)e_{n-2}\right) ds\, dt,$$

$$b_n = 1 - 2\int_0^1 \int_0^1 (1-t)\, w\left(te_n + s(1-t)e_{n-1} + (1-s)(1-t)e_{n-2}\right) ds\, dt$$

$$c = \max\left\{e_0, \frac{e_1}{\lambda}\right\}$$

$$\lambda = \frac{a + \sqrt{a^2 + 4b}}{2},$$

$$\theta = \max\{e_0, e_{-1}, e_{-2}\},$$

$$a = \frac{\int_0^1 (1-t)\, v\,((1-t)\,\theta)\, dt}{1 - w\,(r_0)}$$

and

$$b = \frac{\int_0^1 \int_0^1 (1-t)\, v\,((1-s)\,(1-t)\,\theta)\, ds\, dt}{1 - w\,(r_0)}.$$

We complete this section with an example dealing with the Hölder continuous case.

Example 4.6.2. Let us consider the scalar function

$$F(x) = \frac{4}{15} x^{\frac{5}{2}} - \frac{1}{2} x^2, \quad \text{for all } x \in D = [.81, 6.25].$$

Then we get $x^* = \frac{9}{4}$, $F''(x^*) = \frac{1}{2} \neq 0$,

$$\left| F''(x^*)^{-1} \left(F''(x) - F''(y) \right) \right| = 2 \left| x^{\frac{1}{2}} - y^{\frac{1}{2}} \right|$$

$$= 2 \left(\left| x^{\frac{1}{2}} - y^{\frac{1}{2}} \right|^2 \right)^{\frac{1}{2}}$$

$$\leq 2 \left(\left| x^{\frac{1}{2}} - y^{\frac{1}{2}} \right| \left| x^{\frac{1}{2}} + y^{\frac{1}{2}} \right| \right)^{\frac{1}{2}}$$

$$= 2\, |x - y|^{\frac{1}{2}},$$

for all $x, y \in D$, and

$$\left| F''(x^*)^{-1} \left(F''(x) - F''(x^*) \right) \right| = 2 \left| x^{\frac{1}{2}} - \frac{3}{2} \right|^{\frac{1}{2}} \left| x^{\frac{1}{2}} - \frac{3}{2} \right|^{\frac{1}{2}}. \tag{4.6.7}$$

But as

$$4 \left| x^{\frac{1}{2}} - \frac{3}{2} \right| \leq x^{\frac{1}{2}} + \frac{3}{2}$$

holds for all $x \in D$, we get

$$2 \left| x^{\frac{1}{2}} - \frac{3}{2} \right| \leq \left(x^{\frac{1}{2}} - \frac{3}{2} \right)^{\frac{1}{2}}, \quad \text{for all } x \in D.$$

Therefore, (4.6.7) can be written as

$$\left| F''(x^*)^{-1} \left(F''(x) - F''(y) \right) \right| \leq |x - x^*|^{\frac{1}{2}}, \quad \text{for all } x \in D.$$

That is, according to (4.6.5) and (4.6.6), we can set $v(t) = 2t^{\frac{1}{2}}$ and $w(t) = t^{\frac{1}{2}}$.
Then we obtain

$$r_0 = \frac{9}{4} = .183673...$$

and

$$r_1 = \frac{9}{4}.$$

However, using only condition (4.6.5) we obtain

$$\bar{r}_0 = .09 < r_0.$$

That is, new condition (4.6.6) helps us enlarge the radius of convergence of method (4.6.4).

4.7 Local convergence of a King-Werner-type method

In this section, we are concerned with the problem of approximating a locally unique solution x^* of equation (2.1.1), using the King-Werner method [127], [207], [43]

$$x_{n+1} = x_n - F'\left(\frac{x_n + z_n}{2}\right)^{-1} F(x_n) \tag{4.7.1}$$

$$z_{n+1} = x_{n+1} - F'\left(\frac{x_n + z_n}{2}\right)^{-1} F(x_{n+1}), \quad (n \geq 0), \quad (x_0, z_0 \in D).$$

Although the number of function evaluations increases by one when compared with the NK method (2.1.3), the convergence order is raised from 2 to $1 + \sqrt{2}$ [127]. The convergence of this method has been examined under several conditions in [127], [207], [43].

Here a local convergence analysis is provided that compares favorably with the ones mentioned above.

We state the following local convergence theorem for method (4.7.1):

Theorem 4.7.1. *[172] Assume:*
  (a) there exists a solution x^* of equation (2.1.1) such that $F'(x^*) \in L(Y, X)$.
  (b) There exist nondecreasing functions v and w such that

$$\left\| F''(x^*)^{-1} (F''(x) - F''(y)) \right\| \leq v(\|x - y\|), \quad \forall x, y \in D, \tag{4.7.2}$$

and

$$\left\| F''(x^*)^{-1} (F''(x) - F''(x^*)) \right\| \leq v(\|x - x^*\|), \quad \forall x \in D, \tag{4.7.3}$$

  (c) equation

$$q(r) \int_0^1 v[(1 + tq(r))r] dt + w(r) = 1$$

has a minimum zero r_0 and

$$w(r_0) < 1,$$

where

$$q(r) = \frac{\int_0^1 v\left[\left(\left|\frac{1}{2} - t\right| + \frac{1}{2}\right)r\right]dt}{1 - w(r)};$$

(d) $U(x^*, r_0) \subseteq D$.

Then sequence $\{x_n\}$ generated by King-Werner-type method (4.7.1) is well defined, remains in $U(x^*, r_0)$ for all $n \geq 0$, and converges to x^* provided that $x_0, z_0 \in U(x^*, r_0)$.

Moreover, the following error estimates hold for all $n \geq 0$:

$$\|x_{n+1} - x^*\| \leq \frac{a_n}{b_n} \|x_n - x^*\|,$$

$$\|z_{n+1} - x^*\| \leq \frac{c_n}{b_n} \|x_{n+1} - x^*\|,$$

$$a_n = \int_0^1 v\left(\left|\frac{1}{2} - t\right| \|x_n - x^*\| + \frac{1}{2} \|z_n - x^*\|\right)dt,$$

$$b_n = 1 - w\left(\frac{\|x_n - x^*\| + \|z_n - x^*\|}{2}\right),$$

$$c_n = \int_0^1 v\left(\frac{1}{2}\|x_n - x^*\| + \frac{1}{2}\|z_n - x^*\| + t\|x_{n+1} - x^*\|\right)dt.$$

We complete this section with an example:

Example 4.7.2. Let $X = Y = \mathbf{R}$, $D = (-1, 1)$ and define function F on D by

$$F(x) = \sin x.$$

It is easy to see, $x^* = 0$, $F'(x^*) = 1$,

$$\left\|F''(x^*)^{-1}\left(F''(x) - F''(y)\right)\right\| \leq \sin 1\, \|x - y\|,$$

and

$$\left\|F''(x^*)^{-1}\left(F''(x) - F''(x^*)\right)\right\| = 1 - \cos x$$

$$= \frac{x^2}{2!} - \frac{x^4}{4!} + \cdots + (-1)^n \frac{x^{2n}}{(2n)!} + \cdots$$

$$= \left\|\frac{x}{2}\right\|\left[1 - \frac{2x^2}{4!} + \frac{4x^4}{6!} - \cdots + (-1)^{n-1}\frac{2x^{2n-1}}{(2n)!} + \cdots\right]\|x - x^*\|$$

$$\leq \frac{1}{2}\|x - x^*\|,$$

for all $x, y \in D$.

Therefore we can set $v(r) = (\sin 1)r$ and $w(r) = \frac{1}{2}r$.

If we only use (4.7.2) we obtain $\bar{r}_0 = .590173...$, whereas if we use both (4.7.2) and (4.7.3) we get

$$r_0 = .739126... > \bar{r}_0.$$

Note that $\bar{r}_0$ is the value obtained if we were to use the results in the references mentioned in the introduction of this section. That is, we managed to enlarge the radius of convergence for the King-Werner-type method (4.7.1) (if $w(r) < v(r)$).

4.8 Secant-type methods

In this section (see also relevant Section 5.4) we study the local convergence of the secant-type method [4]:

$$x_{n+1} = x_n - \left[x_n, x_n + \alpha_n (x_{n-1} - x_n) \right]^{-1} F(x_n), \quad (n \geq 0), \tag{4.8.1}$$

where $\alpha_n \in [0, 1]$. The advantages of this method have been explained in [4]. In practice, the α_n are computed such that

$$\text{tol}_c \ll \| \alpha_n (x_{n-1} - x_n) \| \leq \text{tol}_u,$$

where tol_c is related with the computer precision and tol_u is a free parameter for the user. The new iterative method is, in general, a good alternative to the NK method, as $\left[x_n, x_n + \alpha_n (x_{n-1} - x_n) \right]$ is always a good approximation to $F'(x_n)$. Moreover, even for semismooth operators, it is superlinearly convergent with Q-factor at least near to 2 and its efficiency index is at least near to $\sqrt{2}$. So, it is more efficient than the classic secant method with efficiency index $\frac{3}{\sqrt{2}}$ in this cases.

Assume there exists a simple zero x^* of equation (2.1.1) and nondecreasing functions f, g such that:

$$\left\| F'(x^*)^{-1} ([x, y] - [y, z]) \right\| \leq f(\|x - z\|) \tag{4.8.2}$$

and

$$\left\| F'(x^*)^{-1} ([x, y] - [y, x^*]) \right\| \leq f(\|x - x^*\|), \quad \text{for all } x, y, z \in D. \tag{4.8.3}$$

Theorem 4.8.1. *Let $F: D \subseteq X \to Y$ be a nonlinear Fréchet-differentiable operator satisfying conditions (4.8.2) and (4.8.3) and let x^* be a simple zero of F. In addition, let us assume that*
    (a) equation

$$f(r) + 2g(r) = 1, \tag{4.8.4}$$

has a minimum positive zero R.
    (b) $\bar{U}(x^*, R) \subset D$.
    Then, the generalized secant's method $\{x_n\}$ (4.8.1) is well defined, remains in $\bar{U}(x^*, R)$ for all $n > 0$ and converges to x^* provided that $x_0 \in \bar{U}(x^*, R)$. Moreover, the following error estimates hold for all $n > 0$:

$$\|x_{n+1} - x^*\| \le \frac{f\left(\|\hat{x}_n - x^*\|\right)}{1 - g\left(\|x_n - x^*\|, \|\hat{x}_n - x^*\|\right)} \|x_n - x^*\|,$$

where

$$\hat{x}_n = x_n + \alpha_n (x_{n-1} - x_n).$$

We now analyze two particular cases.

Case 1 (Lipschitz case). Let us assume that

$$\left\|F'\left(x^*\right)^{-1}\left([x, y; F] - [y, z; F]\right)\right\| \le l_0 \|x - z\|$$

$$\left\|F'\left(x^*\right)^{-1}\left([x, y; F] - [y, x^*; F]\right)\right\| \le l_1 \|x - x^*\|,$$

where $l_1 \le l_0$.

In this case, equation (4.8.4) becomes

$$l_0 r + 2 l_1 r = 1,$$

that has a unique solution $R = 1/(l_0 + 2 l_1)$.

Notice that this radius of convergence is greater than the one R_0 obtained by taking $l_1 = l_0$, which was the case considered previously (for instance, [9]). In fact,

$$R_0 = \frac{1}{3 l_0} \le \frac{1}{l_0 + 2 l_1}.$$

Case 2 (Hölder case). Let us assume that

$$\left\|F'\left(x^*\right)^{-1}\left([x, y; F] - [y, z; F]\right)\right\| \le l_2 \|x - z\|^p$$

$$\left\|F'\left(x^*\right)^{-1}\left([x, y; F] - [y, x^*; F]\right)\right\| \le l_3 \|x - x^*\|^p,$$

with $0 < p < 1$.

In this case, equation (4.8.4) becomes

$$l_2 r^p + 2 l_3 r^p = 1,$$

and the radius of convergence is $R = (l_2 + 2 l_3)^{-1/p}$.

Remark 4.8.2. A modified theorem can be asserted assuming that

(a) There exist functions $f, g : [0, \infty) \to [0, \infty)$ nondecreasing such that

$$\left\|F'\left(x^*\right)^{-1}\left([x, x^*; F] - [x, y; F]\right)\right\| \le f\left(\|x^* - y\|\right)$$

$$\left\|F'\left(x^*\right)^{-1}\left([x^*, x^*; F] - [x, y; F]\right)\right\| \le g\left(\max\left(\|x^* - x\|, \|x^* - y\|\right)\right),$$

for all $x, y \in D$.

(b) Equation

$$g(r) + f(r) = 1$$

has a minimum positive zero R.

In the following example, we compare our results with the previous ones.

Example 4.8.3. Let $D = (-1, 1)$ and $F(x) = e^x - 1$. In this case, we have $x^* = 0$ and $F'(x^*) = 1$. On the other hand, because F is differentiable, we can compute the divided differences by $[u, v; F] = \int_0^1 F'(u + t(v - u)) \, dt$.

Thus,

$$\left\| [u, v; F] - [x, y; F] \right\| = \left\| \int_0^1 F'(u + t(v - u)) - F'(x + t(y - x)) \, dt \right\|$$

$$\leq \int_0^1 e \, \| (u + t(v - u)) - (x + t(y - x)) \| \, dt$$

$$\leq 2e \max \left(\| u - x \|, \| v - y \| \right),$$

that is, the classic Lipschitz constant is $k_2 = 2e$.

Similarly, it easy to check that our Lipschitz constants are $l_0 = e/2$ and $l_1 = e$.

Finally, the convergence conditions (4.8.2) and (4.8.3) considered in our theorem can be modified in order to include non-Fréchet operators.

Let us consider $x^{**} \in U(x^*, \varepsilon)$. Assume the weaker conditions:

(a)

$$\left\| [x^*, x^{**}; F]^{-1} ([x, y; F] - [y, z; F]) \right\| \leq f \left(\| x - z \| \right),$$

$$\left\| [x^*, x^{**}; F]^{-1} ([x, y; F] - [y, x^*; F]) \right\| \leq g \left(\| x - x^* \| \right), \quad \text{for all } x, y, z \in D.$$

(b) Equation

$$f(r) + g(r + \varepsilon) + g(r) = 1$$

has a minimum positive R.

Example 4.8.4. We consider the nondifferentiable system

$$\begin{cases} x^{3/2} - y = 0 \\ y^{3/2} + x = 0. \end{cases}$$

The associated nonlinear operator $F : \mathbf{R}^2 \to \mathbf{R}^2$ is given by

$$F(x_1, x_2) = \begin{pmatrix} F_1(x_1, x_2) \\ F_2(x_1, x_2) \end{pmatrix},$$

where $F_1(x_1, x_2) = x_1^{3/2} - x_2$ and $F_2(x_1, x_2) = x_2^{3/2} + x_1$.

We use the infinity norm $\| x \| = \| x \|_\infty = \max(|x_1|, |x_2|)$, and the associated matrix norms.

We consider the following divided differences of F, $[u, v; F]$ for $u, v \in \mathbf{R}^2$, and $i = 1, 2$.

$$[u, v; F]_{i1} = \frac{F_i(u_1, v_2) - F_i(v_1, v_2)}{u_1 - v_1}.$$

For example, we have

$$[u, v; H] = \begin{pmatrix} \dfrac{u_1^{3/2} - v_1^{3/2}}{u_1 - v_1} & -1 \\ 1 & \dfrac{u_2^{3/2} - v_2^{3/2}}{u_2 - v2} \end{pmatrix}.$$

Considering $X^{**} = (x^{**}, y^{**}) = (1, 1)$ we obtain

$$[X^*, X^{**}; F]^{-1} = \begin{pmatrix} 1/2 & 1/2 \\ -1/2 & 1/2 \end{pmatrix}.$$

Moreover, it is easy to check that

$$\|[X, Y; F] - [Y, Z; F]\| \leq \|Y - Z\|^{1/2}$$

$$\|[X, Y; F] - [Y, X^*; F]\| \leq \|Y - X^*\|^{1/2},$$

and the hypothesis is fulfilled.

Example 4.8.5. We start with a semismooth example. Let us consider

$$f(x) = \begin{cases} x(x+1), & x < 0 \\ -2x(x-1), & x \geq 0. \end{cases}$$

In Table 4.8.1, we can see the advantage of using our modifications. We consider the modified secant method (4.8.1) with $\alpha_n = \text{tol}_u / \|x_{n-1} - x_n\|$ and $\text{tol}_u = 10^{-4}$. In order to obtain a good approximation, secant method needs more than 21 function evaluations and 20 inversions. On the other hand, modified secant method (4.8.1) needs only 16 function evaluations and 8 inversions to arrive at the exact solution.

Iteration	Secant	Modified Secant
2	1.33×10^{-1}	1.04×10^{-1}
4	5.39×10^{-3}	1.96×10^{-4}
6	1.54×10^{-4}	4.08×10^{-15}
8	4.24×10^{-6}	0
10	1.18×10^{-7}	
12	3.27×10^{-9}	
14	9.07×10^{-11}	
16	2.52×10^{-12}	
18	7.01×10^{-14}	
20	1.95×10^{-15}	

Table 4.8.1. Comparison table, case 1.

We study the sytems of nonlinear equations considered in Example 3.3.11. We consider the same choice for the parameters α_n than in the previous example and we obtain similar results than in the 1-D case, as we show in Table 4.8.2.

Iteration	Secant	Modified Secant
1	3.15×10^{-1}	4.19×10^{-2}
2	2.71×10^{-2}	2.51×10^{-3}
3	5.41×10^{-3}	1.11×10^{-5}
4	2.84×10^{-4}	2.53×10^{-9}
5	3.05×10^{-6}	1.11×10^{-16}

Table 4.8.2. Comparison table, case 2.

4.9 Exercises

4.9.1.

(a) Suppose that F', F'' are uniformly bounded by nonnegative constants $\alpha < 1$, K respectively, on a convex subset D of X and the ball

$$\bar{U}\left(x_0, r_0 \equiv \frac{2\|x_0 - F(x_0)\|}{1-\alpha}\right) \subseteq D.$$

Moreover, if

$$h_S = \frac{K}{2} \frac{1+2\alpha}{1-\alpha} \frac{\|x_0 - F(x_0)\|}{1-\alpha} < 1,$$

holds, then Stirling's method

$$x_{n+1} = x_n - \left(I - F'(F(x_n))\right)(x_n - F(x_n)) \quad (n \geq 0) \qquad (4.9.1)$$

converges to the unique fixed point x^* of F in $\bar{U}(x_0, r_0)$. Moreover, the following estimates hold for all $n \geq 0$:

$$\|x_n - x^*\| \leq h_S^{2^n - 1} \frac{\|x_0 - F(x_0)\|}{1-\alpha}. \qquad (4.9.2)$$

(b) Let $F: D \subseteq X \to Y$ be analytic. Assume:

$$\|F'(x)\| \leq \alpha < 1, \quad \text{for all } x \in D,$$

$$x_0 \neq F(x_0), \quad x_0 \in D,$$

$$\frac{\gamma(1+2\alpha)\|x_0 - F(x_0)\|}{(1-\alpha)^2} < 1,$$

$$r_0 < r_1,$$

$$U(x_0, r_1) \subseteq D,$$

and

$$0 \neq \gamma \equiv \sup_{\substack{k>1 \\ x \in D}} \left\| \frac{1}{k!} F^{(k)}(x_0) \right\|^{\frac{1}{k-1}} < \infty,$$

where,

$$r_1 = \frac{1}{\gamma}\left[1 - \sqrt[3]{\frac{\gamma(1+2\alpha)\|x_0 - F(x_0)\|}{(1-\alpha)^2}}\right].$$

Show: sequence $\{x_n\}$ $(n \geq 0)$ generated by Stirling's method (4.9.1) is well defined, remains in $U(x_0, r_0)$ for all $n \geq 0$ and converges to a unique fixed point x^* of operator F at the rate given by (4.9.2) with

$$K = \frac{2\gamma}{(1-\gamma r_0)^3} \cdot$$

(c) Let $X = D = \mathbb{R}$ and define function F on D by

$$F(x) = \begin{cases} -\frac{1}{3}x, & x \le 3 \\ \frac{1}{45}(x^2 - 7x - 33), & 3 \le x \le 4 \\ \frac{1}{3}(x - 7), & x > 4. \end{cases}$$

Using Stirling's method for $x_0 = 3$, we obtain the fixed point $x^* = 0$ of F in one iteration, as $x_1 = 3 - (1 + \frac{1}{3})^{-1}(3 + 1) = 0$.

Show: Newton's method fails to converge.

4.9.2. It is convenient for us to define certain parameters, sequences, and functions. Let $\{t_n\}$ $(n \ge 0)$ be a Fibonacci sequence given by

$$t_0 = t_1 = 1, \quad t_{n+1} = t_n + t_{n-1} \quad (n \ge 1).$$

Let also c, ℓ, η be nonnegative parameters and define:

- the real function f by

$$f(x) = \frac{1}{1-x}, \quad x \in [0, 1),$$

- sequences $\{s_n\}$ $(n \ge -1)$, $\{a_n\}$ $(n \ge -1)$, $\{A_n\}$ $(n \ge -1)$ by

$$s_{-1} = \frac{\eta}{c+\eta}, \quad s_0 = \ell(c + \eta), \quad s_n = f^2(s_{n-1})a_{n-1}a_{n-2} \quad (n \ge 1),$$

$$a_{-1} = a_{-2} = 0, \quad a_{n-2} = \sum_{j=0}^{n-1} c_j, \quad c_j = t_0 + t_1 + \cdots + t_{j+1},$$

$$A_n = [x_n, x_{n-1}; F],$$

for $x_n \in X$, and

- parameters b, d, r_0 by

$$b = \max \left\{ \frac{\ell n}{(1-s_0)^2 s_0}, \frac{s_0}{(1-s_0)^2} \right\}, \quad d = \frac{s_0}{1-s_0}, \quad r_0 = \frac{\eta}{1-d}.$$

Let $F : D \subseteq X \to Y$ be a nonlinear operator. Assume there exist $x_{-1}, x_0 \in D$ and nonnegative parameters c, ℓ, η such that:

$$A_0^{-1} \text{ exists},$$

$$\|x_0 - x_{-1}\| \le c,$$

$$\|A_0^{-1}F(x_0)\| \le \eta,$$

$$\|A_0^{-1}([x, y; F] - [z, w; F])\| \le \ell(\|x - z\| + \|y - w\|),$$

$$\forall x, y, z \in D, x \ne y, w \ne z,$$

$$s_0 < \frac{3-\sqrt{5}}{2},$$

$$\ell n < (1 - s_0)^2 s_0 = \alpha,$$

and
$$\bar{U}(x_0, r_0) \subseteq D.$$

Show: sequence $\{x_n\}$ $(n \geq -1)$ generated by the secant method

$$x_{n+1} = x_n - [x_n, x_{n-1}; F]^{-1} F(x_n) \quad (n \geq 0) \ (x_{-1}, x_0 \in D)$$

is well defined, remains in $U(x_0, r_0)$ for all $n \geq 0$ and converges to a unique solution $x^* \in \bar{U}(x_0, r_0)$ of equation $F(x) = 0$. Moreover the following estimates hold for all $n \geq 1$:

$$\|x_n - x^*\| \leq \tfrac{d^n}{1-d} b^{a_n-2} \|x_1 - x_0\|,$$

and
$$\|L_0^{-1} F(x_{n+1})\| \leq b^{c_n-1} s_0.$$

Furthermore, let
$$r^* = \tfrac{1}{\ell} - r_0 - \eta.$$

Then $r^* > r_0$ and the solution x^* is unique in $U(x_0, r^*)$.

4.9.3. Consider the Stirling method

$$z_{n+1} = z_n - [I - F'(F(z_n))]^{-1} [z_n - F(z_n)]$$

for approximating a fixed point x^* of the equation $x = F(x)$ in a Banach space X.

Show:
(i) If $\|F'(x)\| \leq \alpha < \tfrac{1}{3}$, then the sequence $\{x_n\}$ $(n \geq 0)$ converges to the unique fixed point x^* of equation $x = F(x)$ for any $x_0 \in X$. Moreover, show that:

$$\|x^* - x_n\| \leq \left(\tfrac{2\alpha}{1-\alpha}\right)^n \tfrac{\|x_0 - F(z_0)\|}{1-\alpha} \quad (n \geq 0).$$

(ii) If F' is Lipschitz continuous with constant K and $\|F'(x)\| \leq \alpha < 1$, then NK method converges to x^* for any $x_0 \times X$ such that

$$h_N = \tfrac{1}{2} K \tfrac{\|x_0 - F(x_0)\|}{(1-\alpha)^2} < 1$$

and
$$\|x_n - x^*\| \leq (h_N)^{2^n-1} \tfrac{\|x_0 - F(x_0)\|}{1-\alpha} \quad (n \geq 0).$$

(iii) If F' is Lipschitz continuous with constant K and $\|F'(x)\| \leq \alpha < 1$, then $\{z_n\}$ $(n \geq 0)$ converges to x^* for any $z_0 \in X$ such that

$$h_s = \tfrac{K}{2} \tfrac{1+2\alpha}{1-\alpha} \tfrac{\|x_0 - F(z_0)\|}{1-\alpha} \quad (n \geq 0).$$

4.9.4. Let H be a real Hilbert space and consider the nonlinear operator equation $P(x) = 0$ where $P: U(x_0, r) \subseteq H \rightarrow H$. Let P be differentiable in $U(x_0, r)$ and set $F(x) = \|P(x)\|^2$. Then $P(x) = 0$ reduces to $F(x) = 0$. Define the iteration

$$x_{n+1} = x_n - \tfrac{\|P(x_n)\|^2}{2\|Q(x_n)\|^2} Q(x_n) \quad (n \geq 0)$$

where $Q(x) = P'(x) P(x)$, and the linear operator $P'(x)$ is the adjoint of $P'(x)$. Show that if:

(a) there exist two positive constants B and K such that

$$B^2 K < 4;$$

(b) $\left\| P'(x) y \right\| \geq B^{-1} \|y\|$ for all $y \in H$, $x \in u(x_0, r)$

(c) $\left\| Q'(x) \right\| \leq K$ for all $x \in U(x_0, r)$;

(d) $\|x_1 - x_0\| < \eta_0$ and $r = \frac{2\eta_0}{2 - B\sqrt{K}}$.

The equation $P(x) = 0$ has a solution $x^* \in U(x_0, r)$ and the sequence $\{x_n\}$ $(n \geq 0)$ converges to x^* with

$$\left\| x_n - x^* \right\| \leq \eta_0 \frac{\alpha^n}{1 - \alpha}$$

where

$$\alpha = \tfrac{1}{2} B \sqrt{K}.$$

4.9.5. Consider the equation

$$x = T(x)$$

in a Banach space X, where $T : D \subset X \to X$ and D is convex. Let $T_1(x)$ be another nonlinear continuous operator acting from X into X, and let P be a projection operator in X. Then the operator $PT_1(x)$ will be assumed to be Fréchet-differentiable on D. consider the iteration

$$x_{n+1} = T(x_n) + PT_1'(x_n)(x_{n+1} - x_n) \quad (n \geq 0).$$

Assume:

(a) $\left\| [I - PT_1'(x_0)]^{-1} (x_0 - T(x_0)) \right\| \leq \eta,$

(b) $\Gamma(x) = \Gamma = [I - PT_1'(x)]^{-1}$ exists for all $x \in D$ and $\|\Gamma\| \leq b,$

(c) $PT_1'(x)$, $QT_1(x)$ $(Q = I - P)$ and $T(x) - T_1$ satisfy a Lipschitz condition on D with respective constants M, q and f,

(d) $\bar{U}(x_0, H\eta) \subseteq D$, where

$$H = 1 + \sum_{j=1}^{\infty} \prod_{i=1}^{j} J_i, \quad J_1 = b + \frac{h}{2},$$

$$J_i = b + \tfrac{h}{2} J_1 \cdots J_{i-1}, \quad i \geq 2, \quad J_0 = \eta,$$

(e) $h = BM\eta < 2(1 - b)$, $b = B(q + f) < 1$. Then show that the equation $x = T(x)$ has a solution $x^* \in \bar{U}(x_0, H\eta)$ and the sequence $\{x_n\}$ $(n \geq 0)$ converges to x^* with

$$\left\| x_n - x^* \right\| \subseteq H\eta \prod_{i=1}^{n} J_i.$$

4.9.6. Let H be a real separable Hilbert space. An operator F on H is said to be weakly closed if

(a) x_n converges weakly to x, and

(b) $F(x_n)$ converges weakly to y imply that

$$F(x) = y.$$

Let F be a weakly closed operator defined on $\bar{U}(x_0, r)$ with values in H. Suppose that F maps $\overline{U}(x_0, r)$ into a bounded set in H provided the following conditions is satisfied:

$$(F(x), x) \leq (x, x) \text{ for all } x \in S$$

where $S = \{x \in H \mid \|x\| = r\}$.
Then show that there exists $x^* \in U(x_0, r)$ such that

$$F(x^*) = x^*.$$

4.9.7. Let X be a Banach space, $LB(X)$ the Banach space of continuous linear operators on X equipped with the uniform norm, and B_1 the unit ball. Recall that a nonlinear operator K on X is compact if it maps every bounded set into a set with compact closure. We shall say a family H of operators on X is collectively compact if and only if every bounded set $B \subset X$, $\bigcup_{P \in H} H(B)$ has compact closure.
Show:
(i) If
(a) H is a collectively compact family of operators on X,
(b) K is in the pointwise closure of H,
then K is compact
(ii) If
(a) H is a collectively compact family on X,
(b) H is equidifferentiable on $D \subset X$.
Then for every $x \in D$, the family $\{P'(x) \mid P \in H\}$ is collectively compact.
4.9.8. Consider the equations

$$x - K(x) = 0$$

and

$$x - K_n(x) = 0,$$

where K is a compact operator from a domain D of a Banach space X into X, $\{K_n\}$ $(n \geq 1)$ are collectively compact operators.
Moreover assume:
(a) The family $\{K_n\}$ $(n \geq 1)$ is pointwise convergent to K on D, i.e.,

$$K_n(x) \to K(x) \text{ as } n \to (x) \text{ as } n \to \infty, x \in D.$$

(b) The family $\{K_n\}$ $(n \geq 1)$, has continuous first and bounded second derivatives on $\bar{U}(x_0, r)$.
(c) The linear operator $I - K'(x^*)$ is regular.
Then show there exists a constant $r^*, 0 < r^* \leq r$ such that for all sufficiently large n, equation $x - K_n(x) = 0$ has a unique solution $x_n \in \bar{U}(x^*, r^*)$ and $\lim x_n = x^*$ as $n \to \infty$.

4.9.9. Let X be a regular partially ordered Banach space. Denote the order by $\leq$ and consider the iteration

$$x_{n+1} = x_n - \frac{F(x_n)}{c_1} \qquad c_1 > 0$$

for approximating a solution x^* of equation $F(x) = 0$ in X. Assume that there exist positive numbers c_1 and c_2 such that

$$c_2(x - y) \leq F(x) - F(y) \leq c_1(x - y) \quad \text{for all } x \leq y$$

and

$$\|F(x_0)\| \leq \frac{c_2 r}{2} \quad \text{for some fixed } r > 0.$$

Then show that:
   (i) The sequence $\{x_n\}$ $(n \geq 0)$ converges to a solution x^* of the equation $F(x) = 0$.
   (ii) The following estimates are true:

$$\|x_n - x^*\| \leq \frac{c}{c_2} \|F(x_0)\| c_3^n,$$

and

$$\|x_n - x^*\| \leq cc_3 \|x_n - x_{n-1}\|,$$

where $c_3 = \frac{c_1 - c_2}{c_1}$ and c is such that $\|x\| \leq c \|y\|$ whenever $0 \leq x \leq y$.
   (iii) The sequence $\{x_n\}$ $(n \geq 0)$ belongs to

| the set | $\{x \in X \mid x < x_0, \|x - x_0\| \leq r\}$ | if $0 < F(x_0)$, |
| or the set | $\{x \in X \mid x_0 < x, \|x - x_0\| \leq r\}$ | if $F(x_0) < 0$. |

4.9.10. Let $F: D \subseteq X \to Y$ and let D be an open set. Assume:
   (a) the divided difference $[x, y]$ of F satisfies

$$[x, y](y - x) = F(y) - F(x) \quad \text{for all } x, y \in D$$
$$\|[x, y] - [y, u]\| \leq I_1 \|x - y\|^p + I_2 \|x - y\|^p + I_2 \|y - u\|^p$$

for all $x, y, u \in D$ where $I_1 \geq 0$, $I_2 \geq 0$ are constants which do not depend on x, y and u, while $p \in (0, 1]$;
   (b) $x^* \in D$ is a simple solution of equation $F(x) = 0$;
   (c) there exists $\varepsilon > 0$, $b > 0$ such that $\|[x, y]^{-1}\| \leq b$ for every $x, y \in U(x^*, \varepsilon)$;
   (d) there exists a convex set $D_0 \subset D$ such that $x^* \in D_0$, and there exists $\varepsilon_1 > 0$, with $0 < \varepsilon_1 < \varepsilon$ such that $F'(\cdot) \in H_{D_0}(c, p)$ for every $x, y \in D_0$ and $U(x^*, \varepsilon_1) \subset D_0$.

Let $r > 0$ be such that:

$$0 < r < \min \left\{ \varepsilon_1 (q(p))^{-\frac{1}{p}} \right\}$$

where

$$q(p) = \frac{b}{p+1} \left[2^p (I_1 + I_2)(1 + p) + c \right].$$

Then
(i) if $x_0, x_1 \in \bar{U}(x^*, r)$, the secant iterates are well defined, remain in $\bar{U}(x^*, r)$ for all $n \geq 0$, and converge to the unique solution x^* of equation $F(x) = 0$ in $\bar{U}(x^*, r)$. Moreover, the following estimation:

$$\|x_{n+1} - x^*\| \leq \gamma_1 \|x_{n-1} - x^*\|^p \|x_n - x^*\| + \gamma_2 \|x_n - x^*\|^{1+p}$$

holds for sufficiently large n, where

$$\gamma_1 = b(I_1 + I_2) 2^p \text{ and } \gamma_2 = \frac{bc}{1+p}.$$

(ii) If the above condition hold with the difference that x_0 and x_1 are chosen such that

$$\|x^* - x_0\| \leq a d_0; \ \|x^* - x_1\| \leq \min \left\{ a d_0^{t_1}, \ \|x^* - x_0\| \right\},$$

where $0 < d_0 < 1, a = (q(b))^{-\frac{1}{p}}$, and t_1 is the positive root of the equation:

$$t^2 - t - p = 0,$$

then show that for every $n \in N$, $x_n \in U(x^*, a)$ and

$$\|x_{n+1} - x^*\| \leq a d_0^{t_1^{n+1}} \quad (n \geq 0).$$

4.9.11. Let $F: D \subseteq X \to Y$ and let D be an open set. Assume:
(a) $x_0 \in X$ is fixed, and consider the nonnegative real numbers: $B, v, w, p \in (0, 1], \alpha, \beta, q \geq 1, I_1, I_2$ and I_3, where

$$w = B\alpha \left(I_1 B^p + I_2 \beta^p + I_3 B^p \alpha^p \|F(x_0)\|^{p(q-1)} \right)$$

and

$$v = w^{\frac{1}{p+q-1}} \|F(x_0)\|.$$

Denote $r = \max \{B, \beta\}$ and suppose $\bar{U}(x_0, r^*) \subseteq D$, where

$$r^* = \frac{rv}{w^{\frac{1}{p+q-1}} \left(1 - v^{p+q-1}\right)};$$

(b) Condition (a) of the previous exercise holds with the last I_2 replaced by I_3;
(c) for every $x, y \in \bar{U}(x_0, r^*)$, $[x, y]^{-1}$ exists, and $\|[x, y]^{-1}\| \leq B$;
(d) for every $x \in \bar{U}(x_0, r^*)$, $\|F(g(x))\| \leq \alpha \|F(x)\|^q$ where $g: X \to Y$ is an operator having at least one fixed point that coincides with the solution x^* of equation $F(x) = 0$;
(e) for every $x \in \bar{U}(x_0, r^*)$, $\|x - g(x)\| \leq \beta \|F(x)\|$;
(f) the number v is such that: $0 < v < 1$.

Then show that the Steffensen-type method

$$x_{n+1} = x_n - [x_n, g(x_n)]^{-1} F(x_n) \quad n \geq 0$$

is well defined, remains in $\bar{U}(x_0, r^*)$ for all $n \geq 0$, and converges to a solution x^* of equation $F(x) = 0$ with

$$\|x^* - x_n\| \leq \frac{r w^{(p+q)^n}}{w^{\frac{1}{p+q-1}} (1 - v^{p+q-1})} \quad (n \geq 0).$$

4.9.12. (a) Let F, G be real or complex iteration functions of order $p \geq 1$ and $q > 1$ at x^*, respectively. Suppose that F is continuously differentiable and G is continuous at x^*. Then show: iteration function

$$Q(x) = F(x) - \frac{1}{p} F'(x) [x - G(x)]$$

is order at least $p + 1$.

(b) Suppose F is continuously differentiable at x^*. If $F'(x^*) \neq p$, then show: iteration function

$$H(x) = x - \frac{x - F(x)}{1 - \frac{1}{p} F'(x)}$$

is of order at least $p + 1$.

4.9.13. (a) Consider conditions of the form

$$\|A_0^{-1}([x, y; F] - [z, w; F])\| \leq w(\|x - z\|, \|y - w\|), \tag{4.9.3}$$

$$\|A_0^{-1}([x, y; F] - A_0)\| \leq w_0(\|x - x_{-1}\|, \|y - x_0\|) \tag{4.9.4}$$

for all $x, y, z, w \in D$ provided that $A_0^{-1} \in L(Y, X)$, where $w, w_0 : [0, +\infty) \times [0, +\infty) \to [0, +\infty)$ are continuous nondecreasing functions in two variables. Let $F : D \subseteq X \to Y$ be an operator. Assume: there exists a divided difference of order one such that $[x, y; F] \subseteq L(X, Y)$ for all $x, y \in D$ satisfying (4.9.3), (4.9.4); there exist points $x_{-1}, x_0 \in D$ such that $A_0 = [x_{-1}, x_0; F]^{-1} \in L(Y, X)$ and set

$$\|A_0^{-1} F(x_0)\| \leq \eta;$$

equation

$$t = \left[\frac{c_0(t) c_1(t)}{1 - c(t)} + c_0(t) + 1 \right] \eta$$

has at least one positive zero. Denote by t^* the smallest such zero;

$$w_0(t^* + \eta_0, t^*) < 1;$$
$$c(t^*) < 1;$$

and

$$\bar{U}(x_0, t^*) \subseteq D.$$

Show: sequence $\{x_n\}$ $(n \geq 0)$ generated by the secant method is well defined, remains in $\bar{U}(x_0, t^*)$ for all $n \geq 0$, and converges to a unique solution x^* of equation $F(x) = 0$ in $\bar{U}(x_0, t^*)$.

Moreover, the following estimates hold

$$\|x_2 - x_1\| \leq c_0 \|x_1 - x_0\|$$
$$\|x_{n+1} - x_n\| \leq c \|x_n - x_{n-1}\| \quad (n \geq 3)$$

and

$$\|x_n - x^*\| \leq \frac{c^{n-2}}{1-c} \|x_3 - x_2\| \quad (n \geq 2),$$

where,

$$c_0 = c_0(t^*), \quad c_1 = c_1(t^*), \quad c = c(t^*).$$

(b) Assume: x^* is a simple zero of operator F such that:

$$A_*^{-1} = F'(x^*)^{-1} \in L(Y, X);$$
$$\|A_*^{-1}([x, y; F] - [x, x^*; F])\| \leq v(\|y - x^*\|),$$
$$\|A_*^{-1}([x, y; F] - F'(x^*))\| \leq v_0(\|x - x^*\|, \|y - x^*\|)$$

for all $x, y \in D$ for some continuous nondecreasing functions $v: R_+ \to R_+$ and $v_0: R_+ \times R_+ \to R_+$; equation

$$v_0(q, q_0) + v(q_0) = 1$$

where,

$$q_0 = \|x_{-1} - x^*\|, \quad x_{-1} \in D$$

has at least one positive solution. Denote by q^* the minimum positive one; and

$$\bar{U}(x^*, q^*) \subseteq D.$$

Under the above stated hypotheses, show sequence $\{x_n\}$ $(n \geq 0)$ generated by the secant method is well defined, remains in $\bar{U}(x^*, q^*)$ for all $n \geq 0$, and converges to x^* provided that $x_0 \in U(x^*, q)$ for some $x_{-1} \in D$.

Moreover, the following estimates hold for all $n \geq 0$:

$$\|x_{n+1} - x^*\| \leq \gamma_n \|x_n - x^*\|$$

where,

$$\gamma_n = \frac{v(\|x_{n-1} - x^*\|)}{1 - v_0(\|x_n - x^*\|, \|x_{n-1} - x^*\|)}.$$

4.9.14. (a) Let $x_0, x_{-1} \in D$ with $x_0 \neq x_{-1}$. It is convenient to define the parameters α, n by

$$\alpha = \|x_0 - x_{-1}\|,$$
$$\|L_0^{-1} F(x_0)\| \leq n,$$

and functions a, b, L_n by

$$a(r) = \frac{w(\alpha,r)}{1-w(\alpha,r)},$$

$$b(r) = \frac{2w(\alpha+r,r)}{1-w(\alpha+r,r)},$$

$$L_n = \left[x_{n-1}, x_n; F\right] \quad (n \geq 0).$$

We can state the following semilocal convergence results for the secant method. Let F be a nonlinear operator defined on an open convex subset D of a Banach space X with values in a Banach space Y. Assume:
(1) there exist distinct points x_0, x_{-1} such that $L_0^{-1} \in L(Y, X)$;
(2) condition

$$\left\| \left[x_{-1}, x_0; F\right]^{-1} \left(\left[x, y; F\right] - \left[x_{-1}, x_0; F\right]\right)\right\| \leq w(\|x - x_{-1}\|, \|y - x_0\|),$$

holds for all $x, y \in D$;
(3) there exists a mininum positive zero denoted by r^* such that:

$$r \geq \left[\frac{a(r)b(r)}{1-b(r)} + a(r) + 1\right]\eta \quad \text{for all } r \in (0, r^*];$$

(4)

$$w(\alpha + r^*, r^*) < 1,$$
$$b(r^*) < 1$$

and

$$U(x_0, r^*) \subseteq D.$$

Show: sequence $\{x_n\}$ $n \geq -1$ generated by secant method is well defined, remains in $\bar{U}(x_0, r^*)$ for all $n \geq -1$, and converges to a solution x^* of equation $F(x) = 0$, which is unique in $U(x_0, r^*)$.
(b) Let us consider the two boundary value problem:

$$\begin{aligned}y'' + y^{1+p} &= 0, \\ y(0) = y(1) &= 0,\end{aligned} \quad p \in [0, 1]. \tag{4.9.5}$$

also considered in [43]. As in [43] we divide the interval $[0, 1]$ into m subintervals and let $h = \frac{1}{m}$. We denote the points of subdivision by $t_i = ih$, and $y(t_i) = y_i$. We replace y'' by the standard approximations

$$y''(t) \cong [y(t + h) - 2y(t) + y(t - h)]/h^2$$
$$y''(t_i) \cong (y_{i+1} - 2y_i + y_{i-1})/h^2, \quad i = 1, 2, \ldots, m - 1.$$

System (4.9.5) becomes

$$2y_1 - h^2 y_1^{1+p} - y_2 = 0,$$
$$-y_{i-1} + 2y_i - h^2 y_i^{1+p} - y_{i+1} = 0$$
$$-y_{m-2} + 2y_{m-1} - h^2 y_{m-1}^{1+p} = 0, \quad i = 2, 3, \ldots, m - 2.$$

Define operator $F: \mathbb{R}^{m-1} \to \mathbb{R}^{m-1}$ by

$$F(y) = H(y) - h^2 g(y),$$

where

$$y = (y_1, y_2, \ldots, y_{m-1})^t, \quad g(y) = (y_1^{1+p}, y_2^{1+p}, \ldots, y_{m-1}^{1+p})^t,$$

and

$$H = \begin{bmatrix} -2 & -1 & 0 & \cdots & 0 \\ -1 & 2 & -1 & \cdots & 0 \\ 0 & -1 & 2 & \cdots & 0 \\ \vdots & \vdots & \vdots & & \vdots \\ 0 & 0 & 0 & \cdots & 2 \end{bmatrix}.$$

We apply our Theorem to approximate a solution y^* of equation

$$F(y) = 0. \tag{4.9.6}$$

Let $x \in \mathbb{R}^{m-1}$, and choose the norm $\|x\| = \max\limits_{1 \le i \le m-1} |x_i|$. The corresponding matrix $M \in \mathbb{R}^{m-1} \times \mathbb{R}^{m-1}$ is

$$\|M\| = \max\limits_{1 \le i \le m-1} \sum_{j=1}^{m-1} |m_{ij}|.$$

A standard divided difference at the points $x, y \in \mathbb{R}^{m-1}$ is defined by the matrix whose entries are

$$[x, y; F]_{ij}$$
$$= \frac{1}{x_i - y_i} \left[F_i(x_1, \ldots, x; y_{j+1}, \ldots, y_k) - F_i(x_1, \ldots, x_{j-1}, y; , \ldots, y_k) \right],$$

$$k = m - 1.$$

We can set

$$[x, y; F] = \int_0^1 F'[x + t(y - x)] \, dt.$$

Let $x, v \in \mathbb{R}^{m-1}$ with $|x_i| > 0$, $|v_i| > 0$, $i = 1, 2, \ldots, m - 1$. Using the max-norm we obtain

$$\|F'(x) - F'(v)\|$$
$$= \left\| \operatorname{diag} \left\{ h^2 (1 + p)(v_i^p - x_i^p) \right\} \right\|$$
$$= \max\limits_{1 \le i \le m-1} \left| h^2 (1 + p) (v_i^p - x_i^p) \right| \le (1 + p) h^2 \max\limits_{1 \le i \le m-1} |v_i^p - x_i^p|$$
$$\le (1 + p) h^2 |v_i - x_i|^p = (1 + p) h^2 \|v - x\|^p.$$

Hence, we get

$$\|[x, y; F] - [v, w; F]\|$$

$$\leq \int_0^1 \|F'(x + t(y - x)) - F'(v + t(w - v))\| \, dt$$

$$\leq h^2 \int_0^1 (1 + p) \|(1 - t)(x - v) + t(y - w)\|^p \, dt$$

$$\leq h^2(1 + p) \int_0^1 \left[(1 - t)^p \|x - v\|^p + t^p \|y - w\|^p\right] dt$$

$$= h^2(\|x - v\|^p + \|y - w\|^p).$$

Define the function w by

$$w(r_1, r_2) = \| [y_{-1}, y_0; F]^{-1} \| h^2 (r_1^p + r_2^p),$$

where y_{-1}, y_0 will be the starting points for the secant method

$$y_{n+1} = y_n - [y_{n-1}, y_n; F]^{-1} F(y_n) \quad (n \geq 0)$$

applied to equation $F(y) = 0$ to approximate a solution y^*. Choose $p = \frac{1}{2}$ and $m = 10$, then we obtain 9 equations. Because a solution of (4.9.6) vanishes at the end points and is positive in the interior, a reasonable initial approximation seems to be $135 \sin nt$. This choice gives the following vector

$$z_{-1} = \begin{bmatrix} 41.7172942406179 \\ 79.35100905948387 \\ 109.2172942406179 \\ 128.3926296998458 \\ 135.0000000000000 \\ 128.3926296998458 \\ 109.2172942406179 \\ 79.35100905948387 \\ 41.7172942406179 \end{bmatrix}.$$

Choose y_0 by setting $z_0(t_i) = z_{-1}(t_i) - 10^{-5}$, $i = 1, 2, \ldots, 9$. Using secant method, we obtain after 3 iterations

$$z_2 = \begin{bmatrix} 33.64838334335734 \\ 65.34766285832966 \\ 91.77113354118937 \\ 109.4133887062593 \\ 115.6232519796117 \\ 109.4133887062593 \\ 91.77113354118937 \\ 65.34766285832964 \\ 33.64838334335733 \end{bmatrix}$$

and

$$z_3 = \begin{bmatrix} 33.57498274928053 \\ 65.204528678501265 \\ 91.56893412724006 \\ 109.1710943553677 \\ 115.3666988182897 \\ 109.1710943553677 \\ 91.56893412724006 \\ 65.20452867501265 \\ 33.57498274928053 \end{bmatrix}.$$

Set $y_{-1} = z_2$ and $y_0 = z_3$. Show:
We obtain

$$\alpha = .256553, \quad \eta = .00365901.$$

Moreover, show:

$$\| [y_{-1}, y_0; F]^{-1} \| \le 26.5446,$$

$r^* = .0047$, $w(\alpha + r^*, r^*) = .153875247 < 1$ and $b(r^*) = .363717635$. All hypotheses are satisfied. Hence, equation has a unique solution $y^* \in U(y_0, r^*)$. Note that in [43] they found $r^* = .0043494$.

4.9.15. Let $F: S \subseteq X \to Y$ be a three times Fréchet-differentiable operator defined on an open convex domain S of Banach space X with values in a Banach space Y. Assume $F'(x_0)^{-1}$ exists for some $x_0 \in S$, $\|F'(x_0)^{-1}\| \le \beta$, $\|F'(x_0)^{-1} F(x_0)\| \le \eta$, $\|F''(x)\| \le M$, $\|F'''(x)\| \le N$, $\|F'''(x) - F'''(y)\| \le L\|x - y\|$ for all $x, y \in S$, and $\bar{U}(x_0, r\eta) \subseteq S$, where

$$A = M\beta\eta, \quad B = N\beta\eta^2, \quad C = L\beta\eta^3,$$

$$a_0 = 1 = c_0, \quad b_0 = \frac{2A}{3}, \quad d_0 = \frac{A}{2}(1 + A),$$

$$a_{n+1} = \frac{a_n}{1 - Aa_n(c_n + d_n)}, \quad b_{n+1} = \frac{2A}{3} a_{n+1} c_{n+1},$$

$$c_{n+1} = \frac{32}{2187} \frac{27\left[4 + \left(1 + \frac{3}{2}b_n\right)^2\right] A^3 a_n^2 + 18ABa_n + 17c}{b_n^4 \left(1 + \frac{3}{2}b_n\right)^4} a_{n+1} d_n^4,$$

$$d_{n+1} = \frac{3}{4} b_{n+1}\left(1 + \frac{3}{2}b_{n+1}\right) c_{n+1} \quad (n \ge 0),$$

and

$$r = \lim_{n \to \infty} \sum_{i=0}^{n} (c_i + d_i).$$

If $A \in \left[0, \frac{1}{2}\right]$, $B = \left[0, \frac{1}{18A}(P(A) - 17c)\right]$ and $c \in \left[0, \frac{P(A)}{17}\right]$, where $P(A) = 27(A - 1)(2A - 1)(A^2 + A + 2)(A^2 + 2A + 4)$. Then show [87]: Chebysheff-Halley method given by

$$y_n = x_n - F'(x_n)^{-1} F(x_n)$$

$$H_n = F'(x_n)^{-1} \left[F'\left(x_n + \frac{2}{3}(y_n - x_n)\right) - F'(x_n) \right]$$

$$x_{n+1} = y_n - \frac{3}{4} H_n \left[I - \frac{3}{2} H_n \right] (y_n - x_n)$$

is well defined, remains in $U(x_0, r\eta)$ and converges to a solution $x^* \in \bar{U}(x_0, r\eta)$ of equation $F(x) = 0$. Moreover, the solution x^* is unique in $U\left(x_0, \frac{2}{M\beta} - r\eta\right)$. Furthermore, the following error estimates hold for all $n \geq 0$

$$\|x_n - x^*\| \leq \sum_{i=n}^{\infty} (c_i + d_i) \eta \leq \frac{3}{2A} \left[1 + \frac{A}{2}(1 + A) \right] \frac{b_1}{\gamma^{1/3}} \sum_{i=1}^{\infty} \gamma^{4^{i-1}/3},$$

where $\gamma = \frac{b_2}{b_1}$.

4.9.16. Consider the scalar equation [89]

$$f(x) = 0.$$

Using the degree of logarithmic convexity of f

$$Lf(x) = \frac{f(x) f''(x)}{f'(x)^2},$$

the convex acceleration of Newton's method is given by

$$x_{n+1} = F(x_n) = x_n - \frac{f(x_n)}{f'(x_n)} \left[1 + \frac{Lf(x_n)}{2(1 - Lf(x_n))} \right] \quad (n \geq 0)$$

for some $x_0 \in \mathbb{R}$. Let $k \geq 1754877$, the interval $[a, b]$ satisfying $a + \frac{2k-1}{2(k-1)} \frac{f(b)}{f'(b)} \leq b$ and $x_0 \in [a, b]$ with $f(x_0) > 0$, and $x_0 \geq a + \frac{2k-1}{2(k-1)} \frac{f(b)}{f'(b)}$. If $|Lf(x)| \leq \frac{1}{k}$ and $Lf'(x) \in \left[\frac{1}{k}, 2(k-1)^2 - \frac{1}{k} \right]$ in $[a, b]$, then show: Newton's method converges to a solution x^* of equation $f(x) = 0$ and $x_{2n} \geq x^*$, $x_{2n+1} \leq x^*$ for all $n \geq 0$.

4.9.17. Consider the midpoint method [91]:

$$y_n = x_n = \Gamma_n F(x_n), \quad \Gamma_n = F'(x_n)^{-1},$$

$$z_n = x_n + \frac{1}{2}(y_n - x_n),$$

$$x_{n+1} = x_n - \bar{\Gamma}_n F(x_n), \quad \bar{\Gamma}_n = F'(z_n)^{-1} \quad (n \geq 0),$$

for approximating a solution x^* of equation $F(x) = 0$. Let $F: \Omega \subseteq X \to Y$ be a twice Fréchet-differentiable operator defined on an open convex subset of a Banach space X with values in a Banach space Y. Assume:

(1) $\Gamma_0 \in L(Y, X)$ for some $x_0 \in \Omega$ and $\|\Gamma_0\| \leq \beta$;
(2) $\|\Gamma_0 F(x_0)\| \leq \eta$;
(3) $\|F''(x)\| \leq M \quad (x \in \Omega)$;
(4) $\|F''(x) - F''(y)\| \leq K \|x - y\| \quad (x, y \in \Omega)$.

Denote $a_0 = M\beta\eta$, $b_0 = K\beta\eta^2$. Define sequence $a_{n+1} = a_n f(a_n)^2 g(a_n, b_n)$, $b_{n+1} = b_n f(a_n)^3 g(a_n, b_n)^2$, $f(x) = \frac{2-x}{2-3x}$, and $g(x, y) = \frac{x^2}{(2-x)^2} + \frac{7y}{24}$. If $0 < a_0 < \frac{1}{2}$, $b_0 < h(a_0)$, where

$$h(x) = \frac{96(1-x)(1-2x)}{7(2-x)^2}, \quad \bar{U}(x_0, R\eta) \subseteq \Omega,$$

$$R = \frac{2}{2-a_0} \frac{1}{1-\Delta}, \quad \Delta = f(a_0)^{-1},$$

then show: midpoint method $\{x_n\}$: $(n \geq 0)$ is well defined, remains in $\bar{U}(x_0, R\eta)$, and converges at least R-cubically to a solution x^* of equation $F(x) = 0$. The solution x^* is unique in $U(x_0, \frac{2}{M\beta} - R\eta) \cap \Omega$ and

$$\|x_{n+1} - x^*\| \leq \frac{2}{2-a_0} \gamma^{\frac{3^n-1}{2}} \frac{\Delta^n}{1-\Delta} \eta.$$

4.9.18. Consider the multipoint method [115]:

$$y_n = x_n - \Gamma_n F(x_n), \quad \Gamma_n = F'(x_n)^{-1},$$
$$z_n = x_n + \theta(y_n - x_n),$$
$$H_n = \frac{1}{\theta} \Gamma_n \left[F'(x_n) - F'(z_n) \right], \quad \theta \in (0, 1],$$
$$x_{n+1} = y_n + \frac{1}{2} H_n (y_n - x_n) \quad (n \geq 0),$$

for approximating a solution x^* of equation $F(x) = 0$. Let F be a twice-Fréchet-differentiable operator defined on some open convex subset Ω of a Banach space X with values in a Banach space Y. Assume:

(1) $\Gamma_0 \in L(Y, X)$, for some $x_0 \in X$ and $\|\Gamma_0\| \leq \beta$;
(2) $\|\Gamma_0 F(x_0)\| \leq \eta$;
(3) $\|F''(x)\| \leq M$, $(x \in \Omega)$;
(4) $\|F''(x) - F''(y)\| \leq K \|x - y\|^p$, $(x, y) \in \Omega$, $K \geq 0$, $p \in [0, 1]$.

Denote $a_0 = M\beta\eta$, $b_0 = K\beta\eta^{1+p}$ and define sequence

$$a_{n+1} = a_n f(a_n)^2 g_\theta(a_n, b_n),$$
$$b_{n+1} = b_n f(a_n)^{2+p} g_\theta(a_n, b_n)^{1+p},$$
$$f(x) = \frac{2}{2 - 2x - x^2} \quad \text{and}$$
$$g_\theta(x, y) = \frac{x^3 + 4x^2}{8} + \frac{\left[2 + (p+2)\theta^p\right] y}{2(p+1)(p+2)}.$$

Suppose $a_0 \in \left(0, \frac{1}{2}\right)$ and $b_0 < h_p(a_0, \theta)$, where

$$h_p(x, \theta) = \frac{(p+1)(p+2)}{4[2 + (p+2)\theta^p]}(1 - 2x)\left(8 - 4x^2 - x^3\right).$$

Then, if $\bar{U}(x_0, R\eta) \subseteq \Omega$, where $R = \left(1 + \frac{a_0}{2}\right)\frac{1}{1-\gamma\Delta}$, $\Delta = f(a_0)^{-1}$ show: iteration $\{x_n\}$ $(n \geq 0)$ is well defined, remains in $\bar{U}(x_0, R\eta)$ for all $n \geq 0$, and converges with R-order at least $2 + p$ to a solution x^* of equation $F(x) = 0$. The solution x^* is unique in $U(x_0, \frac{2}{M\beta} - R\eta) \cap \Omega$. Moreover, the following estimates hold for all $n \geq 0$

$$\|x_n - x^*\| \leq \left[1 + \frac{a_0}{2}\gamma^{\left(\frac{(2+p)^n - 1}{1+p}\right)}\right]\gamma^{\left(\frac{(2+p)^n - 1}{1+p}\right)}\frac{\Delta^n}{1 - \gamma^{(2+p)^n}\Delta}\eta,$$

where $\gamma = \frac{a_1}{a_0}$.

4.9.19. Consider the multipoint iteration [118]:

$$y_n = x_n - \Gamma_n F(x_n), \quad \Gamma_n = F'(x_n)^{-1},$$

$$z_n = x_n - \tfrac{2}{3}\Gamma_n F(x_n),$$

$$H_n = \Gamma_n \left[F'(z_n) - F'(x_n)\right],$$

$$x_{n+1} = y_n - \tfrac{3}{4}\left[I + \tfrac{3}{2}H_n\right]^{-1} H_n (y_n - x_n) \quad (n \geq 0),$$

for approximation equation $F(x) = 0$. Let $F: \Omega \subseteq X \to Y$ be a three times Fréchet-differentiable operator defined on some convex subset Ω of a Banach space X with values in a Banach space Y. Assume $F'(x_0)^{-1} \in L(Y, X)$ $(x_0 \in \Omega)$, $\|\Gamma_0\| \leq \alpha$, $\|\Gamma_0 F(x_0)\| \leq \beta$, $\|F''(x)\| \leq M$, $\|F''(x)\| \leq N$, and $\|F'''(x) - F'''(y)\| \leq k\|x - y\|$ for all $x, y \in \Omega$. Denote $\theta = M\alpha\beta$, $w = N\alpha\beta^2$ and $\delta = K\alpha\beta^3$. Define sequences

$$a_0 = c_0 = 1, \quad b_0 = \frac{2}{3}\theta, \quad d_0 = \frac{2 - \theta}{2(1 - \theta)},$$

$$a_{n+1} = \frac{a_n}{1 - \theta a_n d_n}, \quad b_{n+1} = \frac{2}{3}\theta a_{n+1}c_{n+1},$$

$$c_{n+1} = \frac{8(2 - 3b_n)^4}{(4 - 3b_n)^4}\left[\frac{a_n^2}{(2 - 3b_n)^2}\theta^3 + \frac{17}{108}\delta + \frac{wa_n}{3(2 - 3b_n)}\theta\right]a_{n+1}d_n^4$$

and

$$d_{n+1} = \frac{4 - 3b_{n+1}}{4 - 6b_{n+1}}c_{n+1} \quad (n \geq 0).$$

Moreover, assume: $\bar{U}(x_0, R\beta) \subseteq \Omega$, where

$$R = \lim_{n \to \infty}\sum_{i=0}^{n} d_i, \quad \theta \in \left(0, \tfrac{1}{2}\right),$$

$$0 \leq \delta < \frac{27(2\theta - 1)(\theta^3 - 8\theta^2 + 16\theta - 8)}{17(1 - \theta)^2},$$

$$0 \leq w < \frac{3(2\theta - 1)(\theta^3 - 8\theta^2 + 16\theta - 8)}{4\theta(1 - \theta)^2} - \frac{17\delta}{36\theta}.$$

Then show: iteration $\{x_n\}$ $(n \geq 0)$ is well defined, remains in $U(x_0, R\beta)$ for all $n \geq 0$, and converges to a solution x^* of equation $F(x) = 0$. Furthermore, the solution x^* is unique in $U\left(x_0, \frac{2}{\alpha M} - R\beta\right)$ and for all $n \geq 0$

$$\|x_n - x^*\| \leq \sum_{i \geq n} d_i \beta \leq \beta \frac{(2 - \theta)}{2(1 - \theta)} \frac{1}{\gamma^{1/3}} \sum_{j \geq n} \gamma^{4^{i+1}/3},$$

where $\gamma = \frac{b_1}{b_0}$.

4.9.20. Consider the multipoint iteration [90]:

$$y_n = x_n - F'(x_n)^{-1} F(x_n)$$
$$G_n = \left[F'(x_n + p(y_n - x_n)) - F'(x_n)\right](y_n - x_n), \quad p \in (0, 1],$$
$$x_{n+1} = y_n = \frac{1}{2p} F'(y_n)^{-1} G_n \quad (n \geq 0)$$

for approximating a solution x^* of equation $F(x) = 0$. Let $F: \Omega \subseteq X \to Y$ be a continuously Fréchet-differentiable operator in an open convex domain Ω that is a subset of a Banach space X with values in a Banach space Y. Let $x_0 \in \Omega$ such that $\Gamma_0 = F'(x_0)^{-1} \in L(Y, X)$; $\|\Gamma_0\| \leq \beta$, $\|y_0 - x_0\| \leq \eta$, $p = \frac{2}{3}$, and $\|F'(x) - F'(y)\| \leq K \|x - y\|$ for all $x, y \in \Omega$. For $b_0 = K\beta\eta$, define $b_n = b_{n-1} f(b_{n-1})^2 g(b_{n-1})$, where

$$f(x) = \frac{2(1 - x)}{x^2 - 4x + 2} \quad \text{and} \quad g(x) = \frac{x(x^2 - 8x + 8)}{8(1 - x)^2}.$$

If $b_0 < r = .2922...$, where r is the smallest positive root of the polynomial $q(x) = 2x^4 - 17x^3 + 48x^2 - 40x + 8$, and $\bar{U}(x_0, \frac{1}{K\beta}) \subseteq \Omega$, then show: iteration $\{x_n\}$ $(n \geq 0)$ is well defined, remains in $\bar{U}(x_0, \frac{1}{K\beta})$, and converges to a solution x^* of equation $F(x) = 0$, which is unique in $U(x_0, \frac{1}{K\beta})$.

4.9.21. Consider the biparametric family of multipoint iterations [92]:

$$y_n = x_n - \Gamma_n F(x_n), \quad z_n = x_n + p(y_n - x_n), \quad p \in [0, 1],$$
$$H_n = \frac{1}{p} \Gamma_n \left[F'(z_n) - F'(x_n)\right],$$
$$x_{n+1} = y_n - \frac{1}{2} H_n (I + \alpha H_n)(y_n - x_n), \quad (n \geq 0)$$

where $\Gamma_n = F'(x_n)^{-1}$ $(n \geq 0)$ and $\alpha = -2\beta \in \mathbb{R}$. Assume $\Gamma_0 = F'(x_0)^{-1} \in L(Y, X)$ exists at some $x_0 \in \Omega_0 \subseteq X$, $F: \Omega_0 \subseteq X \to Y$ twice Fréchet-differentiable, X, Y Banach spaces, $\|\Gamma_0\| \leq \beta$, $\|\Gamma_0 F(x_0)\| \leq \eta$, $\|F''(x)\| \leq M$, $x \in \Omega_0$ and $\|F''(x) - F''(y)\| \leq K \|x - y\|$ for all $x, y \in \Omega_0$. Denote $a_0 = M\beta\eta$, $b_0 = k\beta\eta^2$. Define sequences

$$a_{n+1} = a_n f(a_n)^2 g(a_n, b_n), \quad b_{n+1} = b_n f(a_n)^3 g(a_n, b_n)^2,$$

where

$$f(x) = 2\left[2 - 2x - x^2 - |\alpha|\,x^3\right]^{-1},$$

and

$$g(x, y) = \frac{|\alpha|^2}{8}x^5 + \frac{|\alpha|}{4}x^4 + \frac{1 + 4\,|\alpha|}{8}x^3 + \frac{|1 + \alpha|}{2}x^2 + \frac{1 - p}{4}xy + \frac{2 + 3p}{12}y$$

for some real parameters α and p. Assume:

$$a_0 \in \left(0, \tfrac{1}{2}\right), \quad b_0 < P = \frac{3\left(8 - 16a_0 - 4a_0^2 + 7a_0^3 + 2a_0^4\right)}{3a_0 + 2},$$

$|\alpha| < \min\{6, r\}$, $p \in (0, 1]$ and $p < h\,(|\alpha|)$, where r is a positive root of

$$h(x) = \frac{1}{6b_0(1 - a_0)}\left[\left(24 - 48a_0 - 12a_0^2 + 21a_0^3 + 6a_0^4 - 2b_0(3a_0 + 2)\right)\right.$$
$$\left. + 6a_0^2\left(2a_0^3 + 3a_0^2 - 6a_0 - 2\right)x + 3a_0^5(2a_0 - 1)x^2\right],$$

$$\bar{U}(x_0, R\eta) \subseteq \Omega_0, \quad R = \frac{\left[1 + \frac{a_0}{2}(1 + |\alpha|\,a_0)\right]}{1 - \gamma\Delta}, \quad \gamma = \frac{a_1}{a_0}, \quad \Delta = f(a_0)^{-1}.$$

Then show: iteration $\{x_n\}$ $(n \geq 0)$ is well defined, remains in $\bar{U}(x_0, R\eta)$ for all $n \geq 0$, and converges to a unique solution x^* of equation $F(x) = 0$ in $U(x_0, \frac{2}{M\beta} - R\eta) \cap \Omega_0$. The following estimates hold for all $n \geq 0$:

$$\|x_n - x^*\| \leq \left[1 + \tfrac{1}{2}\gamma^{\frac{3^n - 1}{2}}a_0\left(1 + |\alpha|\,\gamma^{\frac{3^n - 1}{2}}a_0\right)\right]\gamma^{\frac{3^n - 1}{2}}\frac{\Delta^n}{1 - \gamma^{3^n}\Delta}\eta.$$

4.9.22. Let f be a real function, x^* a simple root of f and G a function satisfying $G(0) = 1$, $G'(0) = \frac{1}{2}$ and $\left|G''(0)\right| < +\infty$. Then show [100]: iteration

$$x_{n+1} = x_n - G(Lf(x_n))\frac{f(x_n)}{f'(x_n)}, \quad (n \geq 0)$$

where

$$Lf(x) = \frac{f(x)\,f''(x)}{f'(x)^2}$$

is of third order for an appropriate choice of x_0. This result is due to Gander. Note that function G can be chosen

$$G(x) = 1 + \frac{x}{2} \qquad \text{(Chebyshev method)};$$

$$G(x) = 1 + \frac{x}{2 - x} \qquad \text{(Halley method)};$$

$$G(x) = 1 + \frac{x}{2(1 - x)} \qquad \text{(Super-Halley method)}.$$

4.9.23. Consider the super-Halley method for all $n \geq 0$ in the form:

$$F(x_n) + F'(x_n)(y_n - x_n) = 0,$$

$$3F(x_n) + 3F'(x_n)\left[x_n + \tfrac{2}{3}(y_n - x_n)\right](y_n - x_n) + 4F'(y_n)(x_{n+1} - y_n) = 0,$$

for approximating a solution x^* of equation $F(x) = 0$. Let $F : \Omega \subseteq X \rightarrow Y$ be a three times Fréchet-differentiable operator defined on an open convex subset Ω of a Banach space X with values in a Banach space Y. Assume:

(1) $\Gamma_0 = F'(x_0)^{-1} \in L(Y, X)$ for some $x_0 \in \Omega$ with $\|\Gamma_0\| \leq \beta$;
(2) $\|\Gamma_0 F(x_0)\| \leq \eta$;
(3) $\|F''(x)\| \leq M \ (x \in \Omega)$;
(4) $\|F'''(x) - F'''(y)\| \leq L\|x - y\| \ (x, y \in \Omega), \ (L \geq 0)$.

Denote by $a_0 = M\beta\eta$, $c_0 = L\beta\eta^3$, and define sequences

$$a_{n+1} = a_n f(a_n)^2 g(a_n, c_n),$$

$$c_{n+1} = c_n f(a_n)^4 g(a_n, c_n)^3,$$

where

$$f(x) = \frac{(1-x)}{x^2 - 4x + 2} \quad \text{and} \quad g(x, y) = \frac{1}{8}\left[\frac{x^3}{(1-x)^2} + \tfrac{17}{27}y\right].$$

Suppose: $a_0 \in \left(0, \tfrac{1}{2}\right)$, $c_0 < h(a_0)$, where

$$h(x) = \frac{27(2x-1)(x-1)\left(x - 3 + \sqrt{5}\right)\left(x - 3 - \sqrt{5}\right)}{17(1-x)^2},$$

$$\bar{U}(x_0, R\eta) \subseteq \Omega, \quad R = \left[1 + \frac{a_0}{2(1-a_0)}\right]\frac{1}{1-\Delta},$$

and $\Delta = f(a_0)^{-1}$. Then show: iteration $\{x_n\}$ $(n \geq 0)$ is well defined, remains in $\bar{U}(x_0, R\eta)$ for all $n \geq 0$, and converges to a solution x^* of equation $F(x) = 0$. The solution x^* is unique in $U(x_0, \frac{2}{M\beta} - R\eta) \cap \Omega$ and

$$\|x_n - x^*\| \leq \left[1 + \frac{a_0\gamma^{\frac{4^n-1}{3}}}{2(1-a_0)}\right]\gamma^{\frac{4^n-1}{3}}\frac{\Delta^n}{1-\gamma^{4^n}\Delta}\eta \quad (n \geq 0),$$

where $\gamma = \frac{a_1}{a_0}$.

4.9.24. Consider the multipoint iteration method [93]:

$$y_n = x_n - F'(x_n)^{-1}F(x_n)$$

$$G_n = F'(x_n)^{-1}\left[F'\left(x_n + \tfrac{2}{3}(y_n - x_n)\right) - F'(x_n)\right],$$

$$x_{n+1} = y_n - \tfrac{3}{4}G_n\left[I - \tfrac{3}{2}G_n\right](y_n - x_n) \quad (n \geq 0),$$

for approximating a solution x^* of equation $F(x) = 0$. Let $F: \Omega \subseteq X \to Y$ be a three times Fréchet-differentiable operator defined on an open convex subset Ω of a Banach space X with values in a Banach space Y. Assume:

(1) $\Gamma_0 = F'(x_0)^{-1} \in L(Y, X)$ exists for some $x_0 \in \Omega$ and $\|\Gamma_0\| \leq \beta$;

(2) $\|\Gamma_0 F(x_0)\| \leq \eta$;

(3) $\|F''(x)\| \leq M \ (x \in \Omega)$;

(4) $\|F'''(x)\| \leq N \ (x \in \Omega)$;

(5) $\|F'''(x) - F'''(y)\| \leq L \|x - y\| \ (x, y \in \Omega), \ (L \geq 0)$.

Denote by $a_0 = M\beta\eta$, $b_0 = N\beta\eta^2$ and $c_0 = L\beta\eta^3$. Define the sequence

$$a_{n+1} = a_n f(a_n)^2 g(a_n, b_n, c_n),$$
$$b_{n+1} = b_n f(a_n)^3 g(a_n, b_n, c_n)^2,$$
$$c_{n+1} = c_n f(a_n)^4 g(a_n, b_n, c_n)^3,$$

where

$$f(x) = \frac{2}{2 - 2x - x^2 - x^3},$$

and

$$g(x, y, z) = \frac{1}{216}\left[27x^3\left(x^2 + 2x + 5\right) + 18xy + 17z\right].$$

If $a_0 \in \left(0, \frac{1}{2}\right)$, $17c_0 + 18a_0 b_0 < p(a_0)$, where

$$p(x) = 27(1 - x)(1 - 2x)\left(x^2 + x + 2\right)\left(x^2 + 2x + 4\right),$$

$$\bar{U}(x_0, R\eta) \subseteq \Omega, \ R = \left[1 + \frac{a_0}{2}(1 + a_0)\right]\frac{1}{1 - \gamma\Delta}, \ \gamma = \frac{a_1}{a_0}, \ \Delta = f(a_0)^{-1},$$

then show: iteration $\{x_n\}$ $(n \geq 0)$ is well defined, remains in $\bar{U}(x_0, R\eta)$ for all $n \geq 0$, and converges to a solution x^* of equation $F(x) = 0$, which is unique in $U(x_0, \frac{2}{M\beta} - R\eta) \cap \Omega$. Moreover, the following error bounds hold for all $n \geq 0$:

$$\|x_n - x^*\| \leq \left[1 + \frac{a_0}{2}\gamma^{\frac{4^n - 1}{3}}\left(1 + a_0\gamma^{\frac{4^n - 1}{3}}\right)\right]\gamma^{\frac{4^n - 1}{3}}\frac{\Delta^n}{1 - \gamma^{4^n}\Delta}\eta.$$

4.9.25. Consider the Halley method [68]

$$x_{n+1} = x_n - [I - L_F(x_n)]^{-1} F'(x_n)^{-1} F(x_n) \ (n \geq 0)$$

where

$$L_F(x) = F'(x)^{-1} F''(x) F'(x)^{-1} F(x),$$

for approximating a solution x^* of equation $F(x) = 0$. Let $F: \Omega \subseteq X \to$ be a twice Fréchet-differentiable operator defined on an open convex subset of a Banach space X with values in a Banach space Y. Assume:

(1) $F'(x_0)^{-1} \in L(Y, X)$ exists for some $x_0 \in \Omega$;

(2) $\|F'(x_0)^{-1} F(x_0)\| \leq \beta$;

(3) $\|F'(x_0)^{-1} F''(x_0)\| \le \gamma$;
(4) $\|F'(x_0)^{-1} [F''(x) - F''(y)]\| \le M \|x - y\|$ $(x, y \in \Omega)$.

If

$$\beta \le \frac{2\left[2\sqrt{\gamma^2 + 2M} + \gamma\right]}{3\left[\sqrt{\gamma^2 + 2M} + \gamma\right]^2}, \quad \bar{U}(x_0, r_1) \subseteq \Omega,$$

$(r_1 \le r_2)$ where r_1, r_2 are the positive roots of $h(t) = \beta - t + \frac{\gamma}{2}t^2 + \frac{M}{6}t^3$, then show: iteration $\{x_n\}$ $(n \ge 0)$ is well defined, remains in $\bar{U}(x_0, r_1)$ for all $n \ge 0$, and converges to a unique solution x^* of equation $F(x) = 0$ in $\bar{U}(x_0, r_1)$. Moreover, the following error bounds hold for all $n \ge 0$:

$$\|x^* - x_{n+1}\| \le (r_1 - t_{n+1}) \left(\frac{\|x^* - x_n\|}{r_1 - t_n}\right)^3,$$

$$\frac{(\lambda_2 \theta)^{3^n}}{\lambda_2 - (\lambda_2 \theta)^{3^n}} (r_2 - r_1) \le r_1 - t_n \le \frac{\lambda_1 \theta}{\lambda_1 - (\lambda_1 \theta)^{3^n}} (r_2 - r_1),$$

$$\theta = \frac{r_1}{r_2}, \quad \lambda_1 = \sqrt{\frac{(r_0 - r_2)^2 + r_0 r_2}{(r_0 - r_1)^2 + r_0 r_1}} \le 1, \quad \lambda_2 = \frac{\sqrt{3}}{2},$$

$-r_0$ is the negative root of h, and $t_{n+1} = H(t_n)$, where

$$H(t) = t - \frac{h(t)/h'(t)}{1 - \frac{1}{2}L_h(t)}, \quad L_h(t) = \frac{h(t)/h''(t)}{h'(t)^2}.$$

4.9.26. Consider the iteration [91]

$$x_{n+1} = x_n - [I + T(x_n)] \Gamma_n F(x_n) \quad (n \ge 0),$$

where $\Gamma_n = F'(x_n)^{-1}$ and $T(x_n) = \frac{1}{2}\Gamma_n A \Gamma_n F(x_n)$ $(n \ge 0)$, for approximating a solution x^* of equation $F(x) = 0$. Here $A: X \times X \to Y$ is a bilinear operator with $\|A\| = \alpha$, and $F: \Omega \subseteq X \to Y$ is a Fréchet-differentiable operator defined on an open convex subset Ω of a Banach space X with values in a Banach space Y. Assume:

(1) $F'(x_0)^{-1} = \Gamma_0 \in L(Y, X)$ exists for some $x_0 \in \Omega$ with $\|\Gamma_0\| \le \beta$;
(2) $\|\Gamma_0 F(x_0)\| \le \eta$;
(3) $\|F'(x) - F'(y)\| \le k \|x - y\|$ $(x, y \in \Omega)$.

Let a, b be real numbers satisfying $a \in \left[0, \frac{1}{2}\right), b \in (0, \sigma)$, where

$$\sigma = \frac{2\left[2a^2 - 3a - 1 + \sqrt{1 + 8a - 4a^2}\right]}{a(1 - 2a)}.$$

Set $a_0 = 1, c_0 = 1, b_0 = \frac{b}{2}$ and $d_0 = 1 + \frac{b}{2}$. Define sequence

$$a_{n+1} = \frac{a_n}{1 - aa_n d_n}, \qquad c_{n+1} = \frac{a_{n+1}}{2}\left[a + \frac{b}{(1 + b_n)^2}\right]d_n^2,$$

$$b_{n+1} = \frac{b}{2}a_{n+1}c_{n+1}, \qquad d_{n+1} = (1 + b_{n+1})c_{n+1}$$

and $r_{n+1} = \sum_{k=0}^{n+1} d_k$ $(n \geq 0)$. If $a = \kappa\beta\eta \in \left[0, \frac{1}{2}\right)$, $\bar{U}(x_0, r\eta) \subseteq \Omega$, $r = \lim_{n\to\infty} r_n$, $\alpha \in \left[0, \frac{\sigma}{\beta\eta}\right)$, then show: iteration $\{x_n\}$ $(n \geq 0)$ is well defined, remains in $\bar{U}(x_0, r\eta)$, and converges to a solution x^* of equation $F(x) = 0$, which is unique in $U(x_0, \frac{2}{\kappa\beta} - r\eta)$. Moreover, the following error bounds hold for all $n \geq 0$

$$\|x_{n+1} - x_n\| \leq d_n \eta$$

and

$$\|x^* - x_{n+1}\| \leq (r - r_n)\eta = \sum_{k=n+1}^{\infty} d_k \eta.$$

4.9.27. Consider the Halley method [106] in the form:

$$y_n = x_n - \Gamma_n F(x_n), \qquad \Gamma_n = F'(x_n)^{-1},$$

$$x_{n+1} = y_n + \tfrac{1}{2}L_F(x_n) H_n (y_n - x_n) \qquad (n \geq 0),$$

$$L_F(x_n) = \Gamma_n F''(x_n) \Gamma_n F(x_n), \qquad H_n = [I - L_F(x_n)]^{-1} \qquad (n \geq 0),$$

for approximating a solution x^* of equation $F(x) = 0$. Let $F: \Omega \subseteq X \to Y$ be a twice Fréchet-differentiable operator defined on an open convex subset Ω of a Banach space X with values in a Banach space Y. Assume:

(1) $\Gamma_0 \in L(Y, X)$ exists for some $x_0 \in \Omega$ with $\|\Gamma_0\| \leq \beta$;
(2) $\|F''(x)\| \leq M$ $(x \in \Omega)$;
(3) $\|F''(x) - F''(y)\| \leq N\|x - y\|$ $(x, y \in \Omega)$;
(4) $\|\Gamma_0 F(x_0)\| \leq \eta$;
(5) the polynomial $p(t) = \frac{k}{3}t^2 - \frac{1}{\beta}t + \frac{\eta}{\beta}$, where $M^2 + \frac{N}{2\beta} \leq k^2$ has two positive roots r_1 and r_2 with $(r_1 \leq r_2)$.

Let sequences $\{s_n\}$, $\{t_n\}$, $(n \geq 0)$ be defined by

$$s_n = t_n - \frac{p(t_n)}{p'(t_n)}, \qquad t_{n+1} = s_n + \frac{1}{2}\frac{L_p(t_n)}{1 - L_p(t_n)}(s_n - t_n) \quad (n \geq 0).$$

If, $\bar{U}(x_0, r_1) \subseteq \Omega$, then show: iteration $\{x_n\}$ $(n \geq 0)$ is well defined, remains in $U(x_0, r_1)$ for all $n \geq 0$, and converges to a solution x^* of equation $F(x) = 0$. Moreover, if $r_1 < r_2$, the solution x^* is unique in $\bar{U}(x_0, r_2)$. Furthermore, the following error bounds hold for all $(n \geq 0)$

$$\|x^* - x_n\| \leq r_1 - t_n = \frac{(r_2 - r_1)\theta^{4^n}}{1 - \theta^{4^n}}, \qquad \theta = \frac{r_1}{r_2}.$$

5

Newton-like Methods

General classes of iterative methods are examined under weaker conditions than before.

5.1 Newton-like methods of "bounded deterioration"

We use Newton-like (NL) method

$$x_{n+1} = x_n - A(x_n)^{-1} F(x_n) \quad (n \geq 0) \tag{5.1.1}$$

to generate a sequence approximating x^*.

A survey of results concerning the convergence of method (5.1.1) can be found in [34], [35], [74] and the references there. Here $A(x) \in L(X, Y)$. We consider $A(x_n)$ (see (5.1.15)) $(n \geq 0)$ as expressing a sequence of Jacobian approximations to $F'(x_n)$ $(n \geq 0)$, which is of "bounded deterioration" [75]. That is, although $A(x_n)$ is not necessarily converging to $F'(x^*)$ as $x_n \to x^*$, the divergence is proportional to the distance between method (5.1.1) and its starting point.

We provide a convergence analysis based on this concept (i.e., (5.1.15)). In particular using the majorant method, and more precise majorizing sequences than before [75], we show under the same hypotheses and computational cost: in the semilocal case, finer error bounds on the distances $\|x_{n+1} - x_n\|$, $\|x_n - x^*\|$ $(n \geq 0)$ and more precise information on the location of the solution x^*; whereas in the local case, again finer error bounds and a larger convergence radius are obtained.

Other favorable comparisons with special Newton-like methods are also given as well as some numerical results.

To first examine the semilocal case, we need the following lemma on majorizing sequences:

Lemma 5.1.1. *Assume there exist nonnegative parameters* K, K_1, η, Δ, $a_n > 0$ $(n \geq 0)$, $b_n \geq 0$, $c_n \geq a_n$ *and* $\delta \in [0, 2)$ *such that for all* $n \geq 0$:

$$0 \leq h_\delta^n = \frac{1}{a_n} \left\{ \left[K \left(\frac{\delta}{2} \right)^n + \frac{2K_1}{2-\delta} \left(1 - \left(\frac{\delta}{2} \right)^{n+1} \right) \right] n + 2(b_n - \Delta) \right\} \leq \delta. \tag{5.1.2}$$

I.K. Argyros, *Convergence and Applications of Newton-type Iterations*,
DOI: 10.1007/978-0-387-72743-1_5, © Springer Science+Business Media, LLC 2008

Then, iteration $\{t_n\}$ $(n \geq 0)$ *given by*

$$t_0 = 0, t_1 = \eta, t_{n+2} = t_{n+1} + \frac{1}{2c_n}\left[K(t_{n+1} - t_n) + 2(b_n - \Delta + K_1 t_n)\right](t_{n+1} - t_n)$$
$$(5.1.3)$$

is nondecreasing, bounded above by $t^{**} = \frac{2\eta}{2-\delta}$, *and converges to some* t^* *such that*

$$0 \leq t^* \leq t^{**}. \qquad (5.1.4)$$

Moreover, the following estimates hold for all $n \geq 0$:

$$0 \leq t_{n+2} - t_{n+1} \leq \frac{\delta}{2}(t_{n+1} - t_n) \leq \left(\frac{\delta}{2}\right)^{n+1}\eta. \qquad (5.1.5)$$

Proof. We shall show using induction on $k \geq 0$:

$$\frac{1}{a_n}\left[K(t_{k+1} - t_k) + (2b_k - \Delta + K_1 t_k)\right] \leq \delta \qquad (5.1.6)$$

and

$$t_{k+1} - t_k \geq 0. \qquad (5.1.7)$$

Estimate (5.1.5) can then follow immediately from (5.1.3), (5.1.6), and (5.1.7). For $k = 0$, (5.1.6) and (5.1.7) hold by (5.1.2) and (5.1.3), respectively. We also get

$$0 \leq t_2 - t_1 \leq \frac{\delta}{2}(t_1 - t_0). \qquad (5.1.8)$$

Let us assume (5.1.5)–(5.1.7) hold for all $k \leq n + 1$.

We can have in turn:

$$\frac{1}{a_{k+1}}\left[K(t_{k+2} - t_{k+1}) + 2(b_{k+1} - \Delta + K_1 t_{k+1})\right] \qquad (5.1.9)$$

$$\leq \frac{1}{a_{k+1}}\left\{K\left(\frac{\delta}{2}\right)^{k+1}\eta + 2(b_{k+1} - \Delta) + 2K_1\left[t_1 + \frac{\delta}{2}(t_1 - t_0)\right.\right.$$

$$\left.\left. + \left(\frac{\delta}{2}\right)^2(t_1 - t_0) + \cdots + \left(\frac{\delta}{2}\right)^k(t_1 - t_0)\right]\right\} = h_s^{k+1} \leq \delta \quad \text{(by (5.1.2))}.$$

We shall show:

$$t_k \leq t^{**} \quad (k \geq 0). \qquad (5.1.10)$$

Clearly (5.1.10) holds for $k = 0, 1, 2$ by the initial conditions. Assume (5.1.10) holds for all $k \leq n$.

It follows from (5.1.3) and (5.1.5)

$$t_{k+2} \leq t_{k+1} + \frac{\delta}{2}(t_{k+1} - t_k) \leq \cdots \leq \eta + \frac{\delta}{2}\eta + \cdots + \left(\frac{\delta}{2}\right)^{k+1}\eta \qquad (5.1.11)$$

$$= \frac{1 - \left(\frac{\delta}{2}\right)^{k+2}}{1 - \frac{\delta}{2}}\eta \leq \frac{2\eta}{2-\delta} = t^{**}.$$

Hence, sequence $\{t_n\}$ $(n \geq 0)$ is bounded above by t^{**}, nondecreasing, and as such it converges to some t^* satisfying (5.1.4).

We can show the following semilocal result for method NL.

Theorem 5.1.2. *Let* $F: D \subseteq X \to Y$ *be a Fréchet-differentiable operator. Assume: there exists an approximation* $A(x) \in L(X, Y)$ *of operator* $F'(x)$ *an open convex subset* D_0 *of* D, $x_0 \in D_0$ *so that* $A^{-1} \in L(Y, X)$, *nonnegative parameters* η, d, K_0, K, K_1, b_n, Δ *and* $\delta \in [0, 2)$ *such that:*

$$\left\| A_0^{-1} F(x_0) \right\| \leq \eta, \quad A_0 = A(x_0), \tag{5.1.12}$$

$$\left\| A_0^{-1} \left[F'(x) - F'(x_0) \right] \right\| \leq K_0 \|x - x_0\|, \tag{5.1.13}$$

$$\left\| A_0^{-1} \left[F'(x) - F'(y) \right] \right\| \leq K \|x - y\|, \tag{5.1.14}$$

for all $x, y \in D_0$,

$$\left\| A_0^{-1} \left[A(x_n) - F'(x_n) \right] \right\| \leq b_n - \Delta + K_1 \sum_{j=1}^{n} \|x_j - x_{j-1}\|, \tag{5.1.15}$$

$$\left\| A_0^{-1} \left(A(x_0) - F'(x_0) \right) \right\| \leq d; \tag{5.1.16}$$

condition (5.1.2) holds for

$$a_n = 1 - \left[d + b_n - \Delta + \frac{K_2 \eta}{2-\delta} \left(1 - \left(\frac{\delta}{2} \right)^{n+1} \right) \right], \quad K_2 = 2(K_0 + K_1); \tag{5.1.17}$$

$$q_n = d + b_n - \Delta + \frac{K_2 \eta}{2-\delta} \left(1 - \left(\frac{\delta}{2} \right)^{n+1} \right) < 1 \quad (n \geq 0); \tag{5.1.18}$$

and

$$\overline{U}(x_0, t^*) \subseteq D_0, \tag{5.1.19}$$

where t^* *is defined in Lemma 5.1.1.*

  Then sequence $\{x_n\}$ $(n \geq 0)$ *generated by NL method is well defined, remains in* $U(x_0, t^*)$ *for all* $n \geq 0$, *and converges to a solution* $x^* \in \overline{U}(x_0, t^*)$ *of equation* $F(x) = 0$.

  Moreover, the following estimates hold for all $n \geq 0$:

$$\|x_{n+1} - x_n\| \leq t_{n+1} - t_n \tag{5.1.20}$$

and

$$\|x_n - x^*\| \leq t^* - t_n, \tag{5.1.21}$$

where sequence $\{t_n\}$ *is given by (5.1.3) for*

$$c_n = 1 - \left(d + b_n - \Delta + \frac{K_2}{2} t_n \right) \quad (n \geq 0). \tag{5.1.22}$$

Furthermore the solution x^* *is unique in* $\overline{U}(x_0, t^*)$ *if*

$$0 < b_n + d - \Delta. \tag{5.1.23}$$

Finally if there exists $t_1^* > t^*$ *such that*

$$U(x_0, t_1^*) \subseteq D_0, \tag{5.1.24}$$

and

$$\frac{K_0}{2} \left(t^* + t_1^* \right) \leq 1, \tag{5.1.25}$$

then, the solution x^* *is unique in* $U(x_0, t_1^*)$.

Proof. We shall show:

$$\|x_{k+1} - x_k\| \leq t_{k+1} - t_k \qquad (5.1.26)$$

$$x_{k+1} \in \bar{U}(x_0, t^*) \qquad (5.1.27)$$

and

$$\sum_{i=0}^{k} \|x_{i+1} - x_i\| \leq t^* \qquad (5.1.28)$$

for all $k \geq 0$. Estimates (5.1.26)–(5.1.28) hold for $k = 0$ by the initial conditions. Assume they hold for all $k \leq n$. Then, we have:

$$\|x_{k+1} - x_0\| \leq \|x_{k+1} - x_k\| + \|x_k - x_{k-1}\| + \cdots + \|x_1 - x_0\|$$
$$\leq (t_{k+1} - t_k) + (t_k - t_{k-1}) + \cdots + (t_1 - t_0) = t_{k+1} \leq t^*, \quad (5.1.29)$$

and

$$\sum_{i=0}^{k} \|x_{i+1} - x_i\| \leq t^*. \qquad (5.1.30)$$

By (5.1.13), (5.1.15)–(5.1.18), and the induction hypotheses, we obtain in turn:

$$\left\| A_0^{-1} \left(A(x_{k+1}) - A(x_0) \right) \right\| \leq$$

$$\leq \left\| A_0^{-1} \left(A(x_{k+1}) - F'(x_{k+1}) \right) \right\|$$

$$+ \left\| A_0^{-1} \left(F'(x_{k+1}) - F'(x_0) \right) \right\| + \left\| A_0^{-1} \left(A(x_0) - F'(x_0) \right) \right\|$$

$$\leq b_{k+1} - \Delta + K_1 \sum_{j=1}^{n+1} \|x_j - x_{j-1}\| + K_0 \|x_{k+1} - x_0\| + d$$

$$\leq b_{k+1} - \Delta + (K_0 + K_1) t_{k+1} + d \leq q_{k+1} < 1. \qquad (5.1.31)$$

It follows from (5.1.31), and the Banach Lemma on invertible operators that $A(x_{k+1})^{-1}$ exists so that:

$$\left\| A(x_{k+1})^{-1} A_0 \right\| \leq c_{k+1}^{-1} \leq a_{k+1}^{-1}. \qquad (5.1.32)$$

Using NL, (5.1.3), (5.1.14), (5.1.15), and the approximation

$$A_0^{-1} F(x_{k+1}) =$$

$$= A_0^{-1} \left[F(x_{k+1}) - F(x_k) - A(x_k)(x_{k+1} - x_k) \right]$$

$$= A_0^{-1} \left\{ \int_0^1 \left[F'(x_{k+1} + \theta(x_k - x_{k+1})) - F'(x_k) \right] (x_{k+1} - x_k) \, d\theta \right.$$

$$\left. \times \left[F'(x_k) - A(x_k) \right] (x_{k+1} - x_k) \right\}, \qquad (5.1.33)$$

we get

$$\left\| A_0^{-1} F\left(x_{k+1}\right) \right\| \leq$$

$$\leq \tfrac{1}{2} K \left\| x_{k+1} - x_k \right\|^2 + \left(b_k - \Delta + K_1 \sum_{j=1}^{k} \left\| x_j - x_{j-1} \right\| \right) \left\| x_{k+1} - x_k \right\|$$

$$\leq \tfrac{1}{2} K \left(t_{k+1} - t_k\right)^2 + \left(b_k - \Delta + K_1 t_k\right) \left(t_{k+1} - t_k\right). \tag{5.1.34}$$

Moreover by NL, (5.1.32), and (5.1.34) we obtain

$$\left\| x_{k+2} - x_{k+1} \right\| = \left\| \left[A\left(x_{k+1}\right)^{-1} A_0 \right] \left[A_0^{-1} F\left(x_{k+1}\right) \right] \right\|$$

$$\leq \left\| A\left(x_{k+1}\right)^{-1} A_0 \right\| \cdot \left\| A_0^{-1} F\left(x_{k+1}\right) \right\|$$

$$\leq \frac{\tfrac{1}{2} K \left(t_{k+1} - t_k\right)^2 + \left(b_k - \Delta + K_1 t_k\right)\left(t_{k+1} - t_k\right)}{c_{k+1}} = t_{k+2} - t_{k+1}, \tag{5.1.35}$$

which completes the induction for (5.1.26).

Furthermore we have:

$$\sum_{i=0}^{k+1} \left\| x_{i+1} - x_i \right\| \leq t_{k+1} \leq t^*, \tag{5.1.36}$$

and

$$\left\| x_{k+2} - x_0 \right\| \leq t_{k+2} \leq t^*, \tag{5.1.37}$$

which complete the induction for (5.1.27) and (5.1.28).

It follows from (5.1.26) and (5.1.27) that sequence $\{x_n\}$ $(n \geq 0)$ is Cauchy in a Banach space X and as such it converges to some $x^* \in \overline{U}\left(x_0, t^*\right)$ (as $\overline{U}\left(x_0, t^*\right)$ is a closed set). By letting $K \to \infty$ in (5.1.39) we obtain $F\left(x^*\right) = 0$. Estimate (5.1.21) follows from (5.1.20) by using standard majorization techniques.

To show uniqueness in $\overline{U}\left(x_0, t^*\right)$, let y^* be a solution of equation $F\left(x\right) = 0$. By (5.1.13), (5.1.18), and (5.1.23) we have

$$\left\| A_0^{-1} \int_0^1 \left[F'\left(y^* + \theta\left(x^* - y^*\right)\right) - F'\left(x_0\right) \right] d\theta \right\|$$

$$\leq K_0 \int_0^1 \left\| y^* + \theta\left(x^* - y^*\right) - x_0 \right\| d\theta$$

$$\leq K_0 \int_0^1 \left[\theta \left\| x^* - x_0 \right\| + \left(1 - \theta\right) \left\| y^* - x_0 \right\| \right] d\theta \tag{5.1.38}$$

$$\leq L t^* < 1. \tag{5.1.39}$$

It follows again from (5.1.35) and the Banach Lemma on invertible operators that linear operator

$$L = \int_0^1 F'\left(y^* + \theta\left(x^* - y^*\right)\right) d\theta \tag{5.1.40}$$

is invertible. Using the identity

$$0 = F\left(x^*\right) - F\left(y^*\right) = L\left(x^* - y^*\right),\qquad(5.1.41)$$

we deduce

$$x^* = y^*.\qquad(5.1.42)$$

Similarly if $y^* \in U\left(x_0, t_1^*\right)$, we obtain again that operator L is invertible as by (5.1.25) and (5.1.39)

$$\left\|A_0^{-1}\left[L - F'\left(x_0\right)\right]\right\| < \frac{K_0}{2}\left(t^* + t_1^*\right) \le 1.\qquad(5.1.43)$$

Hence, again we get (5.1.43).

Remark 5.1.3. (a) If

$$K_0 = K,\qquad(5.1.44)$$
$$K_1 = \sigma K,\qquad(5.1.45)$$

and

$$\delta = 1\qquad(5.1.46)$$

where $\sigma \ge 1$, (and Δ) are given in [74, p. 441], then Theorem 5.1.2 reduces to essentially Theorems 2.4–2.5 and 3.2 in [74].

However in general

$$K_0 \le K\qquad(5.1.47)$$

holds. Hence if strict inequality holds in (5.1.47), we obtain immediately under the hypotheses of Theorem 5.1.2 and the ones in [74]

$$\|x_{n+1} - x_n\| \le t_{n+1} - t_n < s_{n+1} - s_n\quad(n \ge 0),\qquad(5.1.48)$$
$$\|x_n - x^*\| \le t^* - t_n \le s^* - s_n,\qquad(5.1.49)$$
$$t_n < s_n\quad(n \ge 1),\qquad(5.1.50)$$

and

$$t^* \le s^*\qquad(5.1.51)$$

where s_n, s^* used in [74], and are given by

$$s_{n+i} = s_n + \frac{f\left(t_n\right)}{c_n},\quad s_0 = 0,\qquad(5.1.52)$$

$$f\left(t\right) = \tfrac{1}{2}\sigma K t^2 - \Delta t + b_0 \eta,\qquad(5.1.53)$$

and s^* is the smallest zero of equation

$$f\left(t\right) = 0.\qquad(5.1.54)$$

That is, we obtain finer estimates on the distances involved and a more precise information on the location of the solution under the same hypotheses and computational cost. Note that in practice the computation of Lipschitz constant K also involves the computation of K_0. Moreover, note that our hypotheses hold whenever the corresponding ones in [74] hold but not necessarily vice versa (unless if (5.1.40)–(5.1.42) hold).

(b) In the special case of Newton's method, i.e., when

$$A(x) = F'(x) \quad (x \in 1) \tag{5.1.55}$$

we have

$$K_1 = 0, \quad \sigma = 1, \quad b_n = \Delta \quad (n \geq 0) \text{ and } d = 0.$$

In order for us to compare Theorem 5.1.2 with the corresponding ones in [74], assume δ is given by (5.1.46). Then it can easily be seen that the conditions of Lemma 5.1.1 and Theorem 5.1.2 are satisfied if

$$h_1 = (K + K_0)\eta \leq 1, \tag{5.1.56}$$

whereas the conditions in [74] reduce to the famous Newton-Kantorovich hypothesis.

The advantages of this approach have been explained in Section 2.2.

The rest of the results mentioned there can also be improved along the same lines. To avoid repetitions, we leave these details to the motivated reader and we study the local convergence of NL method instead (not considered in [74]).

In what follows, we study the local convergence for NL method.

Theorem 5.1.4. *Let $F: D \subseteq X \to Y$ be a Fréchet-differentiable operator. Assume: there exist an approximation $A(x) \in L(X, Y)$ of operator $F'(x)$, an open convex subset D_0 of D, a solution x^* of equation $F(x) = 0$ such that $A(x^*)^{-1} \in L(Y, X)$, and nonnegative parameters, $\bar{b}_n, \bar{\Delta}, b, L_i, i = 0, 1, ..., 6$ such that the following conditions hold for all $x_n, x, y \in D_0$ $(n \geq 0)$:*

$$\left\| A(x^*)^{-1}[F'(x) - F'x^*] \right\| \leq L_0 \|x - x^*\|, \tag{5.1.57}$$

$$\left\| A(x^*)^{-1}[F'(x) - F'(y)] \right\| \leq L_1 \|x - y\|, \tag{5.1.58}$$

$$\left\| A(x^*)^{-1}[A(x) - A(x^*)] \right\| \leq L_2 \|x - x^*\| + L_3, \tag{5.1.59}$$

$$\left\| A(x^*)^{-1}[A(x_n) - F'(x_n)] \right\| \leq \bar{b}_n - \bar{\Delta} + L_4 \sum_{j=1}^{n} \|x_j - x_{j-1}\|, \tag{5.1.60}$$

and

$$\left\| A(x^*)^{-1}[F'(x^*) - A(x)] \right\| \leq L_5 \|x - x^*\| + L_6; \tag{5.1.61}$$

$$\bar{b}_n - \bar{\Delta} \leq b; \tag{5.1.62}$$

equation

$$\left(\tfrac{1}{2}L_1 + L_2\right)r + \frac{L_4}{1 - \alpha(r)}\eta + b + L_3 - 1 = 0 \tag{5.1.63}$$

has a minimal nonnegative zero r^* satisfying

$$L_2 r + L_3 < 1 \qquad (5.1.64)$$

and

$$\alpha(r) < 1; \qquad (5.1.65)$$

$$\overline{U}(x^*, r^*) \subseteq D_0. \qquad (5.1.66)$$

Then, sequence $\{x_n\}$ ($n \geq 0$) generated by NL method is well defined, remains in $U(x^*, r^*)$ for all $n \geq 0$, and converges to x^* provided that $x_0 \in U(x^*, r^*)$.

Moreover, the following estimates hold for all $n \geq 0$:

$$\|x_{n+2} - x_{n+1}\|$$
$$\leq \frac{\left[\int_0^1 L_0 \|(1-\theta)(x_{n+1}-x^*) + \theta(x_n - x^*)\| d\theta + L_6 + L_5 \|x_n - x^*\|\right] \|x_{n+1} - x_n\|}{1 - (L_3 + L_2 \|x_{n+1} - x^*\|)} \qquad (5.1.67)$$

and

$$\|x_{n+1} - x^*\| \leq \frac{\left[\frac{1}{2} L_1 \|x_n - x^*\| + \left(\overline{b}_n - \overline{\Delta} + L_4 \sum_{j=1}^{n} \|x_j - x_{j-1}\|\right)\right] \|x_n - x^*\|}{1 - (L_3 + L_2 \|x_n - x^*\|)}. \qquad (5.1.68)$$

Proof. By hypothesis $x_0 \in U(x^*, r^*)$. Let $x \in U(x^*, r^*)$. Using (5.1.59) and (5.1.65), we get

$$\left\| A(x^*)^{-1} \left[A(x) - A(x^*) \right] \right\| \leq L_3 + L_2 \|x - x^*\| \leq L_3 + L_2 r^* < 1. \qquad (5.1.69)$$

It follows from (5.1.69), and the Banach Lemma on invertible operators that $A(x)^{-1}$ exists with

$$\left\| A(x)^{-1} A(x^*) \right\| \leq \left[1 - (L_3 + L_2 \|x - x^*\|) \right]^{-1}. \qquad (5.1.70)$$

Moreover in (5.1.46) using (5.1.57)–(5.1.62) induction on n (5.1.69) for $x = x_{n+1}$, and the approximations

$$x_{n+2} - x_{n+1}$$

$$= \left[A(x_{n+1})^{-1} A(x^*) \right] A(x^*)^{-1}$$

$$\times \int_0^1 \left[F'(x_{n+1} + \theta(x_n - x_{n+1})) - F'(x^*) \right] (x_n - x_{n+1}) d\theta$$

$$+ \left[A(x_{n+1})^{-1} A(x^*) \right] A(x^*)^{-1} \left[F'(x^*) - A(x_n) \right] (x_n - x_{n+1}), \qquad (5.1.71)$$

and

$$x_{n+1} - x^*$$

$$= \left[A\,(x_n)^{-1}\,A\,(x^*) \right] A\,(x^*)^{-1}$$

$$\times \int_0^1 \left[F'\,(x_n + \theta\,(x^* - x_n)) - F'\,(x_n) \right] (x^* - x_n)\,d\theta$$

$$+ \left[A\,(x_n)^{-1}\,A\,(x^*) \right] A\,(x^*)^{-1}\,(F'\,(x_n) - A\,(x_n))\,(x^* - x_n), \qquad (5.1.72)$$

we obtain (5.1.67) and (5.1.68), respectively.

Furthermore by (5.1.63)–(5.1.66) we obtain

$$\| x^* - x_{n+1} \| < \| x^* - x_n \| < x^* - x_0, \qquad (5.1.73)$$

$$\sum_{j=1}^{n} \| x_j - x_{j-1} \| \le \eta + \alpha\,(r^*)\,\eta + \alpha^2\,(r^*)\,\eta + \cdots + \alpha^{n-1}\,(r^*)\,\eta$$

$$\le \frac{1 - \alpha^n\,(r^*)}{1 - \alpha(r)}\eta \le \frac{\eta}{1 - \alpha(r)} \qquad (5.1.74)$$

Hence, we deduce $x_n \in U\,(x^*, r^*)\ (n \ge 0)$, and $\lim_{n \to \infty} = x^*$.

Remark 5.1.5. In order for us to compare Theorem 5.1.4 with results already in the literature we consider again Newton's method. We can choose:

$$L_0 = L_2 = L_5, \quad \overline{b}_n = \overline{\Delta}, \quad L_4 = L_6 = 0 \qquad (5.1.75)$$

Then hypotheses of Theorem 5.1.2 are satisfied if

$$r^* = \frac{2}{2L_0 + L_1}. \qquad (5.1.76)$$

Rheinboldt in [175] in this case using only (5.1.58) obtained

$$r_R = \frac{2}{3L_1} \quad \text{(see also Section 2.4)}. \qquad (5.1.77)$$

5.2 Weak conditions for the convergence of a certain class of iterative methods

In this section, we are concerned with the problem of approximating a locally unique solution x^* of the nonlinear equation

$$F\,(x) + G\,(x) = 0, \qquad (5.2.1)$$

where F, G are operator defined on an open subset Q a Banach space X with values in a Banach space Y. Operator F is Fréchet-differentiable on $\overline{U}\,(z, R)$, and the differentiability of G is not assumed.

Recently in [35], we used the Newton-like method

$$x_0 \in U\,(z, R), \quad x_{n+1} = x_n - A\,(x_n)^{-1}\,[F\,(x_n) + G\,(x_n)]\ (n \ge 0) \qquad (5.2.2)$$

to generate a sequence approximating x^*. Here, $A(v) \in L(X, Y)$ $(v \in X)$.

Throughout this study, we assume there exists $z \in X$, $R > 0$, $a \geq 0$, $b \geq 0$, $\eta \geq 0$ with $A(z)^{-1} \in L(Y, X)$, and for any $x, y \in \overline{U}(z, r) \subseteq \overline{U}(z, R) \subseteq Q$

$$\left\| A(z)^{-1}[A(x) - A(x_0)] \right\| \leq w_0(\|x - x_0\|) + a, \tag{5.2.3}$$

$$\left\| A(z)^{-1}[F'(x + t(y - x)) - A(x)] \right\|$$
$$\leq w(\|x - z\| + t\|x - y\|) - w_1(\|x - z\| + b, t \in [0, 1]) \tag{5.2.4}$$

$$\left\| A(z)^{-1}[G(x) - G(y)] \right\| \leq w_2(r)\|x - y\|, \tag{5.2.5}$$

$$\left\| A(z)^{-1}[F(z) + G(z)] \right\| \leq \eta, \tag{5.2.6}$$

where, $w_0(r)$, $w_1(z)$, $w_2(r)$, $w(r)$, $w(r + t) - w_1(r)$ $(t \geq 0)$ are nondecreasing, nonnegative functions on $[0, R]$ with $w(0) = w_0(0) = w_1(0) = w_2(0) = 0$, and parameters a, b satisfy

$$a + b < 1. \tag{5.2.7}$$

Using (5.2.3)–(5.2.7) instead of the less flexible conditions considered in [58], we showed in [35] that the following can be obtained under the same computational cost

(a) weaker sufficient convergence conditions for method (5.2.2);
(b) finer estimates on the distances

$$\|x_{n+1} - x_n\|, \quad \|x_n - x^*\| \quad (n \geq 0);$$

(c) more precise information on the location of the solution.

Here we continue the work in [35] to show how to improve even further on (a)–(c).

We study the semilocal convergence analysis for method (5.2.2) (see also Chapter 2).

It is convenient to define scalar iteration $\{t_n\}$ $n \geq 0$ for some $r_0 \in [0, r]$, $c \geq 0$

$$t_0 = r_0, \quad t_1 = r_0 + c, \tag{5.2.8}$$

$$t_{n+2} = t_{n+1} + \frac{\left\{ \int_0^1 w[t_n + \theta(t_{n+1} - t_n)]d\theta - w_1(t_n) + b \right\}(t_{n+1} - t_n) + \int_{t_n}^{t_{n+1}} w_2(\theta)d\theta}{1 - a - w_0(t_{n+1})}$$

$$(n \geq 0).$$

Iteration $\{t_n\}$ plays a crucial role in the study of the convergence of method (5.2.2). It turns out that under certain conditions, $\{t_n\}$ is a majorizing sequence for $\{x_n\}$, [35], [58]. Here we try to weaken the earlier conditions and further improve estimates on the error bounds and location of the solution x^*.

Clearly if

$$t_n < w_0^{-1}(1 - a) \quad (n \geq 0) \tag{5.2.9}$$

then it follows from (5.2.8) that sequence $\{t_n\}$ is nondecreasing and bounded above by $w_0^{-1}(1-a)$, and as such it converges to some $t^* \in \left[0, w_0^{-1}(1-a)\right]$.

We can provide stronger but more manageable conditions that imply (5.2.9).

We need the following general result on majorizing sequences for method (5.2.2).

Lemma 5.2.1. *Assume there exist constant* $d \geq 0$, *sequences* $a_n \in [0, 1)$, $b_n \geq 0$, $c_n \geq 0$, *and* $\bar{d}_n \geq 0$ *such that for*

$$a_n = a + w_0\left(\bar{d}_n\right), \quad b_n = (1 - a_n)^{-1},$$

$$c_n = \left\{\int_0^1 w\left[t_n + \theta\left(t_{n+1} - t_n\right)\right]d\theta - w_1(t_n) + b + w_2(t_{n+1})\right\}b_n, \quad (5.2.10)$$

$$\bar{d}_0 = d_0 = r_0, \quad \bar{d}_1 = d_1 = r_0 + c,$$

$$\bar{d}_n = r_0 + c + c_1(t_1 - t_0) + c_2(t_2 - t_1) + \cdots + c_{n-1}(t_{n-1} - t_{n-2}) \quad (n \geq 2),$$
$$(5.2.11)$$

the following conditions hold for all $n \geq 0$:

$$w_0\left(\bar{d}_n\right) \leq w_0(d_n) \leq w_0(d) < 1 - a. \quad (5.2.12)$$

Then sequence $\{t_n\}$ *generated by iteration (5.2.8) is well defined, nondecreasing bounded above by* $w_0^{-1}(1-a)$, *and converges to some* t^*.

Moreover, the following estimates hold:

$$t_n \leq \bar{d}_n \quad (n \geq 0) \quad (5.2.13)$$

and

$$t_{n+1} - t_n = c_n(t_n - t_{n-1}) \quad (n \geq 1). \quad (5.2.14)$$

Proof. It suffices to show hypotheses of the Lemma imply condition (5.2.9). Indeed using (5.2.8), (5.2.10)–(5.2.12) we can have in turn for all $n \geq 2$ (as (5.2.9) holds for $n = 0, 1$ by the initial conditions):

$$t_{n+2} \leq t_{n+1} + c_{n+1}(t_{n+1} - t_n) = t_n + c_n(t_n - t_{n-1}) + c_{n+1}(t_{n+1} - t_n)$$
$$\leq \cdots + r_0 + c + c_1(t_1 - t_0) + \cdots + c_{n+1}(t_{n+1} - t_n) = \bar{d}_{n+2} \leq d_{n+2},$$
$$(5.2.15)$$

which shows (5.2.13) for all $n \geq 0$. Moreover by (5.2.12), we obtain

$$w_0(t_n) \leq w_0(d_n) < 1 - a \text{ for all } n \geq 0, \quad (5.2.16)$$

which shows (5.2.9). Moreover, (5.2.14) follows from (5.2.8) and (5.2.11).

For simplicity next, we provide some choices of functions and parameters defined above in the special case of NK method. That is we choose

$$A(x) = F'(x), \quad G(x) = 0 \quad (x \in U(z, R)), \quad z = x_0 \text{ and } r_0 = 0. \quad (5.2.17)$$

Remark 5.2.2. Assume the Lipschitz choices:

$$w_0(r) = \ell_0 r, \quad w(r) = w_1(r) = \ell r \ (r \in [0, R]), \ \text{and} \ a = b = 0, \qquad (5.2.18)$$

where

$$0 \le \ell_0 \le \ell \qquad (5.2.19)$$

holds in general. Special choices of sequences appearing in Lemma 5.2.1 are given below.

(a) The Newton-Kantorovich case. Assume $\ell_0 = \ell$, and $h = 2\ell c \le 1$.
    Define $d_n, d \ (n \ge 0)$ by

$$d_n = c + \frac{1}{2^1} h^{2^1-1} c + \cdots + \frac{1}{2^{n-1}} h^{2^n-1} c,$$

and

$$d = \frac{1-\sqrt{1-h}}{\ell} \quad (\ell \ne 0).$$

Then it follows from the proof of the Newton-Kantorovich's theorem (see Chapter 2 Section 2.2) that

$$a_n < 1,$$

and condition (5.2.9) hold.
(b) Assume that any of conditions conditions (2.2.48)–(2.2.50) hold. Then by Theorem 3 in [35] conditions (5.2.9) hold for

$$d_n = \left[1 + \frac{\delta}{2} + \cdots + \left(\frac{\delta}{2}\right)^{n-1}\right] c \quad (n \ge 1)$$

and

$$d = \frac{2c}{2-\delta}$$

Moreover other alternatives which imply condition (5.2.9) are given in Remarks 5.2.3, 5.2.5, and Lemma 5.2.4 that follow:

Remark 5.2.3. Assume there exist parameters $\alpha_1 \in [0, 1-a)$, $b \in [0, 1]$, α_2 (depending on b and α_1) such that

$$w_0(r_0 + c) \le \alpha_1 < 1 - a, \qquad (5.2.20)$$

$$\alpha_1 \le \alpha_2, \qquad (5.2.21)$$

$$q(\alpha_2) \le b \quad \text{for} \ \ b \in [0, 1), \qquad (5.2.22)$$

or

$$q(\alpha_2) < 1 \quad \text{for} \ \ b = 1, \qquad (5.2.23)$$

where

$$q(\alpha) = \frac{\int_0^1 w\left[w_0^{-1}(\alpha+\theta c)\right]d\theta - w_1\left(w_0^{-1}(\alpha)\right) + b + w_2\left(w_0^{-1}(\alpha)\right)}{1-a-\alpha}. \qquad (5.2.24)$$

Then, function

$$d(b) = r_0 + \left(1 + b + b^2 + \cdots b^n + \cdots\right)c, \qquad (5.2.25)$$

is well defined on interval $I_b = [\alpha_1, \alpha_2]$ $(b \neq 1)$.

Moreover, assume there exists $\alpha^* \in I_b$ such that

$$w_0(d(\alpha^*)) \leq \alpha^*. \qquad (5.2.26)$$

Then using induction on $n \geq 0$, we can show condition (5.2.5). Indeed (5.2.9) holds for $n = 0, 1$ by the initial conditions. By (5.2.8), we have

$$t_2 - t_1 \leq q(\alpha^*)(t_1 - t_0),$$

then,

$$w_0(t_2) \leq w_0\left[t_1 + q(\alpha^*)(t_1 - t_0)\right] \leq w_0(d(\alpha^*)) \leq \alpha^* < 1.$$

If

$$w_0(t_n) \leq \alpha^* < 1 - a, \text{ then } t_{n+1} - t_n \leq q(\alpha^*)(t_n - t_{n-1})$$

then

$$
\begin{aligned}
w_0(t_{n+1}) &\leq w_0\left[t_n + q(\alpha^*)(t_n - t_{n-1})\right] \\
&\leq w_0\left[r_0 + \left(1 + \alpha^* + (\alpha^*)^2 + \cdots + (\alpha^*)^{n-1}\right), c\right] \\
&\leq w_0(d(\alpha^*)) \leq \alpha^* < 1 - a,
\end{aligned}
$$

which completes the induction.

Hence, we showed:

Lemma 5.2.4. *Under the stated hypotheses:*

(a) condition (5.2.9) holds;

(b) sequence $\{t_n\}$ is nondecreasing and converges to some t^* such that

$$w_0(t_n) \leq w_0(t^*) \leq 1 - a; \qquad (5.2.27)$$

(c) the following estimates hold for all $n \geq 0$:

$$0 \leq t_{n+2} - t_{n+1} \leq q(\alpha^*)(t_{n+1} - t_n) \leq b(t_{n+1} - t_n) \leq b^{n+1}c, \qquad (5.2.28)$$

and

$$0 \leq t^* - t_n \leq \frac{b^n c}{1-b}. \qquad (5.2.29)$$

Remark 5.2.5. (a) For $b = 1$, condition (5.2.23) together with (5.2.8) implies

$$0 \leq t_{n+1} - t_n < t_n - t_{n-1} \quad (n \geq 1) \qquad (5.2.30)$$

Hence, we deduce again $t^* = \lim_{n \to \infty} t_n$ exists.

Moreover if we replace (5.2.9) by

$$w_0(t^*) < 1 - a \qquad (5.2.31)$$

conclusions (a) and (b) of Lemma 5.2.4 hold, whereas for estimates of the form (5.2.28) we use (5.2.30).

(b) It can easily be seen from (5.2.22)–(5.2.25) that conditions (5.2.22) and (5.2.23)
can be replaced by

$$q_1(\alpha_2) \le b \quad \text{for} \quad b \in [0, 1), \tag{5.2.32}$$

or

$$q_1(\alpha_2) < 1 \quad \text{for} \quad b = 1 \tag{5.2.33}$$

respectively, where

$$q_1(\alpha) = q\left(\frac{c}{1-b}\right). \tag{5.2.34}$$

We provide the main semilocal convergence theorem for method (5.2.2), which
improves our earlier result (see Theorem 3, [35]).

Theorem 5.2.6. *Assume:*
hypotheses (5.2.3)–(5.2.8) and (5.2.9) hold for

$$r_0 \in [0, r], \quad c = r_1 - r_0, \quad r \in [0, R], \tag{5.2.35}$$

$$w_0^{-1}(1 - a) + r_0 \le r, \quad \overline{U}(z, r) \subseteq Q, \tag{5.2.36}$$

and

$$x_0 \in D(t^*) \tag{5.2.37}$$

where

$$t^* = \lim_{n \to \infty} t_n, \tag{5.2.38}$$

$\{t_n\}$ is given by (5.2.8) above, and r_1, $D(t^*)$ are defined by (12), (14) in [35], respectively (see also (5.3.74)–(5.3.75)).

Then, iteration $\{x_n\}$ ($n \ge 0$) generated by method (5.2.2) is well defined, remains in $\overline{U}(z, t^*)$ for all $n \ge 0$, and converges to a solution x^* of equation $F(x) + G(x) = 0$.

Moreover, the following error bounds hold for all $n \ge 0$:

$$\|x_{n+1} - x_n\| \le t_{n+1} - t_n \tag{5.2.39}$$

and

$$\|x_n - x^*\| \le t^* - t_n. \tag{5.2.40}$$

Furthermore the solution x^* is unique in $\overline{U}(z, t^*)$ if

$$\int_0^1 \left[w\left((1 + 2t)t^*\right) - w_1(t^*) \right] dt + w_2(3t^*) + w_0(t^*) + a + b < 1, \tag{5.2.41}$$

and in $\overline{U}(z, R_0)$ for $R_0 \in (t^*, r]$ if

$$\int_0^1 \left[w\left(t^* + t(t^* + R_0)\right) - w_1(t^*) \right] dt + w_2(2t^* + R_0) + w_0(t^*) + a + b < 1. \tag{5.2.42}$$

Proof. Simply repeat the corresponding proof of Theorem 3 in [35] but use (5.2.9) above instead of conditions (54)–(57) in [35] (see also the similar proof of Theorem 5.3.3 that follows).

Remark 5.2.7. Our condition (5.2.9) is weaker than all earlier ones [58], [74], [125], [147] in general. Moreover, our error bounds are finer than the corresponding ones in Theorem 3 [35, p. 664], which in turn were shown in the same paper to be finer than the ones given by Chen and Yamamoto in [58]. Furthermore the information on the location of the solution x^* is more precise than the corresponding ones in [35] or [58].

Remark 5.2.8. Assume the Newton-Mysovskii-type conditions [43]:

$$\left\| A\,(w)^{-1} \left[F'\,(x + t\,(y - x)) - A\,(x) \right] \right\|$$

$$\leq \overline{w}\,(\|x - z\| + t\,\|y - x\|) - \overline{w}_1\,(\|x - z\|) + \overline{b} \quad (5.2.43)$$

and

$$\left\| A\,(w)^{-1} \left[G\,(x) - G\,(y) \right] \right\| \leq \overline{w}_2\,(r)\,\|x - y\|,$$

for all

$$x, y, w \in \overline{U}\,(z, r) \subseteq \overline{U}\,(z, R) \subseteq D, \quad t \in [0, 1], \quad (5.2.44)$$

where parameter $\overline{b}$, functions $\overline{w}$, $\overline{w}_1$, and $\overline{w}_2$ are as b, w, w_1, and w_2, respectively. Replace conditions (5.2.3)–(5.2.5), by (5.2.43), (5.2.44), condition (5.2.7) by $b < 1$, and set $b_n = 1$ for all $n \geq 0$, $(a = 0)$. Then clearly all results obtained here hold in this setting. All the above justify the claims (a)–(c) made at the introduction.

Example 5.2.9. Let $X = Y = \mathbf{R}$, $x_0 = -.6$, $D = [-1, 2]$ and define F on D by $F\,(x) = \frac{1}{3}x^3 + .897462$. Set $w_0^{(r)} = \ell_0 r$, $w\,(r) = \ell r$, $a = b = 0$. Then we obtain $c = .049295$, $\ell_0 = 3.\bar{8}$ and $\ell = 11.\bar{1}$. The NK hypothesis is violated as $h = 2\ell c = 1.0 - 5\bar{4} > 1$. However it can be easily seen that conditions of Remark 5.2.3 hold in this case.

5.3 Unifying convergence analysis for two-point Newton methods

In this section, we are concerned with the problem of approximating a locally unique solution x^* of the nonlinear equation

$$F\,(x) + G\,(x) = 0, \quad (5.3.1)$$

where F, G are operators define on a closed ball $\overline{U}\,(w, R)$ centered at point w and of radius $R > 0$, which is a subset of a Banach space X with values in a Banach space Y. F is Fréchet-differentiable on $\overline{U}\,(w, R)$, and the differentiability of operator G is not assumed.

We use the two-point Newton method

$$y_{-1}, y_0 \in \overline{U}(w, R), \quad y_{n+1} = y_n - A(y_{n-1}, y_n)^{-1}[F(y_n) + G(y_n)] \quad (n \geq 0)$$
$$(5.3.2)$$

to generate a sequence converging to x^*. Here $A(x, y) \in L(X, Y)$. We provide a local as well as a semilocal convergence analysis for method (5.3.2) under very general Lipschitz-type hypotheses (see (5.3.3), (5.3.4)).

Our new idea is to use center-Lipschitz conditions instead of Lipschitz conditions for the upper bounds on the inverses of the linear operators involved. It turns out that this way we obtain more precise majorizing sequences. Moreover, despite the fact that our conditions are more general than related ones already in the literature, we can provide weaker sufficient convergence conditions and finer error bounds on the distances involved.

We first study the semilocal case. In order for us to show that these observations hold in a more general setting, we first need to introduce the following:

Let $R > 0$ be given. Assume there exist $v, w \in X$ such that $A(v, w)^{-1} \in L(Y, X)$, and for any $x, y, z \in \overline{U}(w, r) \subseteq \overline{U}(w, R)$, $t \in [0, 1]$, the following hold:

$$\left\| A(v, w)^{-1}[A(x, y)] \right\| \leq h_0(\|x - v\|, \|y - w\|) + a, \qquad (5.3.3)$$

and

$$\left\| A(v, w)^{-1}\left\{[F'(y + t(z - y)) - A(x, y)](z - y) + G(z) - G(y)\right\} \right\| \quad (5.3.4)$$
$$\leq [h_1(\|y - w\| + t\|z - y\|) - h_2(\|y - w\|) + h_3(\|z - x\|) + b]\|z - y\|,$$

where, $h_0(r, s), h_1(r + \bar{r}) - h_2(r) \ (\bar{r} \geq 0), h_2(r), h_3(r)$ are monotonically increasing functions for all $r.s$ on $[0, R]$ with $h_0(0, 0) = h_1(0) = h_2(0) = h_3(0) = 0$, and the constants a, b satisfy $a \geq 0, b \geq 0$. Define parameters c_{-1}, c, c_1 by

$$\|y_{-1} - v\| \leq c_{-1}, \quad \|y_{-1} - y_0\| \leq c, \quad \|v - w\| \leq c_1. \qquad (5.3.5)$$

Remark 5.3.1. Conditions similar to (5.3.3)–(5.3.4) but less flexible were considered by Chen and Yamamoto in [58] in the special case when $A(x, y) = A(x)$ for all $x, y \in \overline{U}(w, R) \ (A(x) \in L(X, Y))$ (see also Theorem 5.3.9). However, we also want the choice of operator A to be more flexible and be related to the difference $G(z) - G(y)$ for all $y, z \in \overline{U}(w, R)$. It has already been shown in special cases [43], [54] for method (5.3.2) is improved (see also Application 5.3.17). Note also that if we choose:

$$A(x, y) = F'(x), G(x) = 0, w = d_0, h_0(r, r) = \gamma_0 r, \qquad (5.3.6)$$
$$h_1(r) = h_2(r) = \gamma_1 r, \ h_3(r) = 0,$$

for all $x, y \in \overline{U}(w, R), r \in [0, R]$, and $a = b = 0$ then conditions (5.3.3), (5.3.4) reduce to the ones for NM (see Chapter 2). Other choices of operators, functions, and constants appearing in (5.3.3) and (5.3.4) can be found in the applications that follow.

With the above choices, we show the following result on majorizing sequences for method (5.3.2).

Lemma 5.3.2. *Assume:*

there exist parameters $\eta \geq 0$, $a \geq 0$, $b \geq 0$, $c_{-1} \geq 0$, $c \geq 0$, $\delta \in [0, 2)$, $r_0 \in [0, r]$, $r \in [0, R]$ *such that:*

$$2\left[\int_0^1 h_1\left(r_0 + \theta\eta\right) d\theta - h_2\left(r_0\right) + b + h_3\left(c + \eta\right)\right] + \left[a + h_0\left(c + c_{-1}, \eta + r_0\right)\right]$$

$$\leq \delta, \tag{5.3.7}$$

$$\frac{2\eta}{2 - \delta} + r_0 + c \leq r, \tag{5.3.8}$$

$$h_0\left[\frac{1 - \left(\frac{\delta}{2}\right)^{n+1}}{1 - \frac{\delta}{2}}\eta + c + c_{-1}, \frac{1 - \left(\frac{\delta}{2}\right)^{n+2}}{1 - \left(\frac{\delta}{2}\right)}\eta + r_0\right] + a < 1, \tag{5.3.9}$$

and

$$2\int_0^1 h_1\left[\frac{1 - \left(\frac{\delta}{2}\right)^{n+1}}{1 - \frac{\delta}{2}}\eta + \theta\left(\frac{\delta}{2}\right)^{n+1}\eta + r_0\right] d\theta - 2h_2\left[\frac{1 - \left(\frac{\delta}{2}\right)^{n+1}}{1 - \frac{\delta}{2}}\eta + r_0\right]$$

$$+ 2h_3\left[\left(\frac{\delta}{2}\right)^n \left(1 + \frac{\delta}{2}\right)\eta\right] + \delta h_0\left[\frac{1 - \left(\frac{\delta}{2}\right)^{n+1}}{1 - \frac{\delta}{2}}\eta + c + c_{-1}, \frac{1 - \left(\frac{\delta}{2}\right)^{n+2}}{1 - \frac{\delta}{2}}\eta + r_0\right]$$

$$\leq \delta \tag{5.3.10}$$

for all $n \geq 0$.

   Then, iteration $\{t_n\}$ $(n \geq -1)$ *given by*

$$t_{-1} = r_0, \quad t_0 = c + r_0, \quad t_1 = c + r_0 + \eta, \tag{5.3.11}$$

$$t_{n+2} = t_{n+1} + \frac{1}{1 - a - h_0\left(t_n - t_{-1} + c_{-1}, t_{n+1} - t_0 + r_0\right)}$$

$$\times \left\{\int_0^1 \left[h_1\left(t_n - t_0 + r_0 + \theta\left(t_{n+1} - t_n\right)\right) - h_2\left(t_n - t_0 + r_0\right) + b\right] d\theta\right.$$

$$\left. + h_3\left(t_{n+1} - t_{n-1}\right)\right\}\left(t_{n+1} - t_n\right)$$

is monotonically increasing, bounded above by

$$t^{**} = \frac{2\eta}{2 - \delta} + r_0 + c, \tag{5.3.12}$$

and converges to some t^* *such that*

$$0 \leq t^* \leq t^{**} \leq r. \tag{5.3.13}$$

Moreover, the following estimates hold for all $n \geq 0$:

$$0 \leq t_{n+2} - t_{n+1} \leq \frac{\delta}{2}\left(t_{n+1} - t_n\right) \leq \left(\frac{\delta}{2}\right)^{n+1}\eta. \tag{5.3.14}$$

Proof. We must show:

$$2\left\{\int_0^1 \left[h_1\left(t_k - t_0 + r_0 + \theta\left(t_{k+1} - t_k\right)\right) - h_2\left(t_k - t_0 + r_0\right) + b\right]d\theta\right.$$

$$\left. + h_3\left(t_{k+1} - t_{k-1}\right)\right\} + \delta\left[a + h_0\left(t_k - t_{-1} + c_{-1}, t_{k+1} - t_0 + r_0\right)\right]$$

$$\leq \delta, \tag{5.3.15}$$

$$0 \leq t_{k+1} - t_k, \tag{5.3.16}$$

and

$$h_0\left(t_k - t_{-1} + c_{-1}, t_{k+1} - t_0 + r_0\right) + a < 1 \tag{5.3.17}$$

for all $k \geq 0$.

Estimate (5.3.14) can then follow from (5.3.15)–(5.3.17) and (5.3.11).

Using induction on the integer $k \geq 0$, we get for $k = 0$

$$2\left[\int_0^1 h_1\left(r_0 + \theta\eta\right) - h_2\left(r_0\right) + b + h_3\left(c + \eta\right)\right]$$

$$+ \delta\left[a + h_0\left(c + c_{-1}, \eta + r_0\right)\right] \leq \delta,$$

$$0 \leq t_1 - t_0,$$

$$h_0\left(c + c_{-1}, \eta + r_0\right) + a < 1,$$

which hold by (5.3.7) and the definition of t_1.

By (5.3.11) we get

$$0 \leq t_2 - t_1 \leq \tfrac{\delta}{2}\left(t_1 - t_0\right).$$

Assume (5.3.15)–(5.3.17) hold for all $k \leq n + 1$. Using (5.3.15)–(5.3.17) we obtain in turn

$$2\left\{\int_0^1 \left[h_1\left(t_{k+1} - t_0 + r_0 + \theta\left(t_{k+2} - t_{k+1}\right)\right) - h_2\left(t_{k+1} - t_0 + r_0\right) + b\right]d\theta\right.$$

$$\left. + h_3\left(t_{k+2} - t_k\right)\right\} + \delta\left[a + h_0\left(t_{k+1} - t_{-1} + c_{-1}, t_{k+2} - t_0 + r_0\right)\right]$$

$$\leq 2\left\{\int_0^1 h_1\left[\left(\frac{1-\left(\frac{\delta}{2}\right)^{k+1}}{1-\frac{\delta}{2}} + \theta\left(\frac{\delta}{2}\right)^{k+1}\right)\eta + r_0\right] - h_2\left[\frac{1-\left(\frac{\delta}{2}\right)^{k+1}}{1-\frac{\delta}{2}}\eta + r_0\right]\right.$$

$$\left. + b + h_3\left[\left(\frac{\delta}{2}\right)^{k+1}\eta + \left(\frac{\delta}{2}\right)^k\eta\right]\right\}$$

$$+ \delta\left[a + h_0\left(\frac{1-\left(\frac{\delta}{2}\right)^{k+1}}{1-\frac{\delta}{2}}\eta + c + c_{-1}, \frac{1-\left(\frac{\delta}{2}\right)^{k+2}}{1-\frac{\delta}{2}}\eta + r_0\right)\right]$$

$$\leq \delta$$

by (5.3.7) and (5.3.10). Hence we showed (5.3.15) holds for $k = n + 2$. Moreover, we must show:

$$t_k \leq t^{**} \tag{5.3.18}$$

$$t_{-1} = r_0 \leq t^{**}, \quad t_0 = r_0 + c \leq t^{**}, \quad t_1 = c + r_0 + \eta \leq t^{**},$$

$$t_2 \leq c + r_0 + \eta + \tfrac{\delta}{2}\eta = \tfrac{2+\delta}{2}\eta + r_0 + c \leq t^{**}.$$

Assume (5.3.18) holds for all $k \leq n + 1$. It follows from (5.3.11), (5.3.15)–(5.3.17):

$$t_{k+2} \leq t_{k+1} + \tfrac{\delta}{2}\,(t_{k+1} - t_k) \leq t_k + \tfrac{\delta}{2}\,(t_k - t_{k-1}) + \tfrac{\delta}{2}\,(t_{k+1} - t_k)$$

$$\leq \cdots \leq c + r_0 + \eta + \tfrac{\delta}{2}\eta + \left(\tfrac{\delta}{2}\right)^2 \eta + \cdots + \left(\tfrac{\delta}{2}\right)^{k+1}\eta$$

$$= \frac{1 - \left(\tfrac{\delta}{2}\right)^{k+2}}{1 - \tfrac{\delta}{2}}\eta + r_0 + c \leq \tfrac{2\eta}{2-\delta} + r_0 + c = t^{**}. \tag{5.3.19}$$

Hence sequence $\{t_n\}$ $(n \geq -1)$ is bounded above by t^{**}. Inequality (5.3.17) holds for $k = n + 2$ by (5.3.8) and (5.3.9). Moreover (5.3.16) holds for $k = n + 2$ by (5.3.19) and as (5.3.15) and (5.3.17) also hold for $k = n + 2$. Furthermore, sequence $\{t_n\}$ $(n \geq 0)$ is monotonically increasing by (5.3.16) and as such it converges to some t^* satisfying (5.3.13).

We provide the main result on the semilocal convergence of method (5.3.2) using majorizing sequence (5.3.11).

Theorem 5.3.3. *Assume:*
hypotheses of Lemma 5.3.2 hold, and there exist

$$y_{-1} \in \overline{U}\,(w, r), \; y_0 \in \overline{U}\,(w, r_0)\,, \; r \in [0, R]\,, \tag{5.3.20}$$

such that

$$\left\| A\,(y_{-1}, y_0)^{-1}\,[F\,(y_0) + G\,(y_0)] \right\| \leq \eta. \tag{5.3.21}$$

Then, sequence $\{y_n\}$ $(n \geq -1)$ generated by method (5.3.2) is well defined, remains in $\overline{U}\,(w, t^*)$ for all $n \geq -1$, and converges to a solution x^* of equation $F\,(x) + G\,(x) = 0$. Moreover, the following estimates hold for all $n \geq -1$:

$$\|y_{n+1} - y_n\| \leq t_{n+1} - t_n \tag{5.3.22}$$

and

$$\|y_n - x^*\| \leq t^* - t_n. \tag{5.3.23}$$

Furthermore the solution x^* is unique in $\overline{U}\,(w, t^*)$ if

$$\int_0^1 h_1\,((1 + 2t)\,t^*)\,dt - h_2\,(t^*) + h_3\,(2t^*) + h_0\,(t^* + c_1, t^*) + a + b < 1, \tag{5.3.24}$$

and in $\overline{U}\,(w, R_0)$ for $R_0 \in (t^*, r]$ if

$$\int_0^1 h_1\,(t^* + (t^* + R_0)t)\,dt - h_2\,(t^*) + h_3\,(R_0 + t^*) + h_0\,(t^* + c_1, t^*) + a + b < 1. \tag{5.3.25}$$

Proof. We first show estimate (5.3.22), and $y_n \in \overline{U}(w, t^*)$ for all $n \geq -1$. For $n = -1, 0$, (5.3.22) follows from (5.3.5), (5.3.11), and (5.3.21). Suppose (5.3.22) holds for all $n = 0, 1, ..., k + 1$; this implies in particular (using (5.3.5), (5.3.20))

$$
\begin{aligned}
\|y_{k+1} - w\| &\leq \|y_{k+1} - y_k\| + \|y_k - y_{k-1}\| + \cdots + \|y_1 - y_0\| + \|y_0 - w\| \\
&\leq (t_{k+1} - t_k) + (t_k - t_{k-1}) + \cdots + (t_1 - t_0) + r_0 \\
&= t_{k+1} - t_0 + r_0 \leq t_{k+1} \leq t^*.
\end{aligned}
$$

That is, $y_{k+1} \in \overline{U}(w, t^*)$.

We show (5.3.22) holds for $n = k + 2$. By (5.3.3) and (5.3.11), we obtain for all $x, y \in \overline{U}(w, t^*)$

$$
\left\| A(v, w)^{-1} \left[A(x, y) - A(v, w) \right] \right\| \leq h_0 (\|x - v\|, \|y - w\|) + a. \tag{5.3.26}
$$

In particular for $x = y_k$ and $y = y_{k+1}$, we get using (5.3.3), (5.3.5),

$$
\left\| A(v, w)^{-1} \left[A(y_k, y_{k+1}) - A(v, w) \right] \right\|
$$

$$
\leq h_0 (\|y_k - v\|, \|y_{k+1} - w\|) + a
$$

$$
\leq h_0 (\|y_k - y_{-1}\| + \|y_{-1} - v\|, \|y_{k+1} - x_0\| + \|y_0 - w\|) + a
$$

$$
\leq h_0 (t_k - t_{-1} + c_{-1}, t_{k+1} - t_0 + r_0) + a
$$

$$
\leq h_0 \left[\frac{1 - \left(\frac{\delta}{2}\right)^k}{1 - \frac{\delta}{2}} \eta + c + c_{-1}, \frac{1 - \left(\frac{\delta}{2}\right)^{k+1}}{1 - \frac{\delta}{2}} \eta + r_0 \right] + a < 1, \text{ (by (5.3.9)). } \tag{5.3.27}
$$

It follows from (5.3.27) and the Banach Lemma on invertible operator that $A(y_k, y_{k+1})^{-1}$ exists, and

$$
\left\| A(y_k, y_{k+1})^{-1} A(v, w) \right\|
$$

$$
\leq \left[1 - a - h_0 (t_k - t_{-1} + c_{-1}, t_{k+1} - t_0 + r_0) \right]^{-1} \tag{5.3.28}
$$

$$
\leq \bar{b}_0 = \left[1 - a - h_0 (R - t_{-1} + c_{-1}, t_{k+1} - t_0 + r_0) \right]^{-1}.
$$

Using (5.3.2), (5.3.4), (5.3.11), (5.3.28) we obtain in turn

$$
\|y_{k+2} - y_{k+1}\| = \left\| A(y_k, y_{k+1})^{-1} \left[F(y_{k+1}) + G(y_{k+1}) \right] \right\|
$$

$$
= \left\| A(y_k, y_{k+1})^{-1} \left[F(y_{k+1}) + G(y_{k+1}) \right. \right.
$$

$$
\left. \left. - A(y_{k-1}, y_k)(y_{k+1} - y_k) - F(y_k) - G(y_k) \right] \right\|
$$

$$
\leq \left\| A(y_k, y_{k+1})^{-1} A(v, w) \right\| \left\| \left[A(v, w)^{-1} F(y_{k+1}) - F(y_k) \right. \right.
$$

$$
\left. \left. - A(y_{k-1}, y_k)(y_{k+1} - y_k) + G(y_{k+1}) - G(y_k) \right] \right\|
$$

$$\leq \frac{\left\{\int_0^1 [h_1(\|y_k-w\|+t\|y_{k+1}-y_k\|)-h_2(\|y_k-w\|)+b]dt+h_3(\|y_{k+1}-y_{k-1}\|)\right\}\|y_{k+1}-y_k\|}{1-a-h_0(t_k-t_{-1}+c_{-1},t_{k+1}-t_0+r_0)}$$

$$\leq \frac{\left\{\int_0^1 [h_1(t_k-t_0+r_0+t(t_{k+1}-t_k))-h_2(t_k-t_0+r_0)+b]dt+h_3(t_{k+1}-t_{k-1})\right\}(t_{k+1}-t_k)}{1-a-h_0(t_k-t_{-1}+c_{-1},t_{k+1}-t_0+r_0)} \quad (5.3.29)$$

$$= t_{k+2} - t_{k+1},$$

which shows (5.3.22) for all $n \geq 0$.

Note also that

$$\|y_{k+2} - w\| \leq \|y_{k+2} - y_{k+1}\| + \|y_{k+1} - w\|$$
$$\leq t_{k+2} - t_{k+1} + t_{k+1} - t_0 + r_0$$
$$= t_{k+2} - t_0 + r_0 \leq t_{k+2} \leq t^*. \quad (5.3.30)$$

That is, $y_{k+2} \in \overline{U}(z, t^*)$.

It follows from (5.3.22) that $\{y_n\}$ $(n \geq -1)$ is a Cauchy sequence in a Banach space X, and as such it convergence to some $x^* \in \overline{U}(w, t^*)$. We can have as above

$$\|y_{k+2} - y_{k+1}\| \leq \bar{b}_0 \left\|A(v,w)^{-1}\left[F(y_{k+1}) + G(y_{k+1})\right]\right\| \leq \bar{b}\|y_{k+1} - y_k\| \quad (5.3.31)$$

where,

$$\bar{b} = \bar{b}_0 \bar{b}_1$$

and

$$\bar{b}_1 = \int_0^1 h_1(1 + 2t)R \, dt - h_2(R) + h_3(2R) + b. \quad (5.3.32)$$

By letting $k \to \infty$ in (5.3.29), using (5.3.28), and the continuity of the operators F, G we obtain

$$\bar{b}_0 \left\|A(v,w)^{-1}\left[F(x^*) + G(x^*)\right]\right\| = 0, \quad (5.3.33)$$

from which we obtain $F(x^*) + G(x^*) = 0$ (as $\bar{b}_0 > 0$). Estimate (5.3.23) follows from (5.3.22) by using standard majorization techniques.

To show uniqueness in $\overline{U}(w, t^*)$, let y^* be a solution of equation (5.3.1) in $\overline{U}(w, t^*)$. Then as in (5.3.29) we obtain the identity:

$$y^* - y_{k+1}$$
$$= y^* - y_k +^{-1}(F(y_k) + G(y_k)) - A(y_{k-1}, y_k)^{-1}\left(F(y^*) + G(y^*)\right)$$
$$= -\left[A(y_{k-1}, y_k)^{-1}A(v,w)\right]A(v,w)^{-1}$$
$$\times \left[F(y^*) - F(y_k) - A(y_{k-1}, y_k)(y^* - y_k) + G(y^*) - G(y_k)\right] \quad (5.3.34)$$

Using (5.3.34) we obtain in turn

$$\|y^* - y_{k+1}\| \leq \frac{\left[\int_0^1 h_1(\|y_k-w\|+t\|y^*-y_k\|)dt-h_2(\|y_k-w\|)+h_3(\|y^*-y_{k-1}\|)+b\right]\|y^*-y_k\|}{1-a-h_0(\|y_{k-1}-v\|,\|y_n-w\|)}$$

$$\leq \frac{\int_0^1 h_1[(1+2t)t^*]dt-h_2(t^*)+h_3(2t^*)+b}{1-a-h_0(t^*+c_1,t^*)} \|y^* - y_k\|$$

$$< \|y^* - y_k\|. \quad (5.3.35)$$

That is, $x^* = y^*$, as $\lim_{n\to\infty} y_n = y^*$.

If $y^* \in \overline{U}(w, R_0)$ then as in (5.3.35) we get

$$\|y^* - y_{k+1}\| \leq \frac{\int_0^1 h_1[t^* + (t^* + R_0)t]dt - h_2(t^*) + h_3(R_0 + t^*) + b}{1 - a - h_0(t^* + c_1, t^*)} \|y^* - y_k\|$$

$$< \|y^* - y_k\|. \tag{5.3.36}$$

Hence again we get $x^* = y^*$.

Remark 5.3.4. Conditions (5.3.9), (5.3.10) can be replaced by the stronger but easier to check

$$h_0\left[\frac{2\eta}{2-\delta} + c + c_{-1}, \frac{2\eta}{2-\delta} + r_0\right] + a < 1, \tag{5.3.37}$$

and

$$2\int_0^1 h_1\left[\frac{2\eta}{2-\delta} + \theta\frac{\delta}{2} + r_0\right]d\theta - 2h_2\left[\frac{2\eta}{2-\delta} + r_0\right]$$
$$+ 2h_3\left[\left(1 + \frac{\delta}{2}\right)\eta\right] + \delta h_0\left[\frac{2\eta}{2-\delta} + c + c_{-1}, \frac{2\eta}{2-\delta} + r_0\right]$$
$$\leq \delta \tag{5.3.38}$$

respectively. Conditions (5.3.7)–(5.3.10) can be weakened even further along the lines of Section 4.2.

Application 5.3.5. *Let us consider some special cases of operator A, functions $h_i i = 0, 1, 2, 3$, parameters a, b and points v, w.*

Define

$$A(x, y) = F'(y) + [x, y; G], \tag{5.3.39}$$

$$v = y_{-1}, \quad w = y_0, \tag{5.3.40}$$

and set

$$r_0 = 0, \tag{5.3.41}$$

where $F', [\cdot, \cdot; G]$ denote the Fréchet derivatiove of F and the divided difference of order one for operator G. Hence, we consider method (5.3.2) in the form

$$y_{n+1} = y_n - \left(F'(y_n) + [y_{n-1}, y_n; G]\right)^{-1}\left(F(y_n) + G(y_n)\right) \quad (n \geq 0) \tag{5.3.42}$$

The method was studied in [43], [54]. It is shown to be of order $\frac{1+\sqrt{5}}{2} \approx 1.618...$ (same as the order of chord), but higher than the order of

$$z_{n+1} = z_n - F'(z_n)^{-1}\left(F(z_n) + G(z_n)\right) \quad (n \geq 0), \tag{5.3.43}$$

and

$$w_{n+1} = w_n - A(w_n)^{-1}\left(F(w_n) + G(w_n)\right) \quad (n \geq 0). \tag{5.3.44}$$

Assume:

$$\left\| A(y_{-1}, y_0)^{-1} \left[F'(y) - F'(y_0) \right] \right\| \leq \gamma_2 \|y - y_0\|, \qquad (5.3.45)$$

$$\left\| A(y_{-1}, y_0)^{-1} \left[F'(x) - F'(y) \right] \right\| \leq \gamma_3 \|x - y\|, \qquad (5.3.46)$$

$$\left\| A(y_{-1}, y_0)^{-1} \left([x, y; G] - [y_{-1}, y_0; G] \right) \right\| \leq \gamma_4 (\|x - y_{-1}\| + \|y - y_0\|), \qquad (5.3.47)$$

and

$$\left\| A(y_{-1}, y_0)^{-1} \left([x, y; G] - [z, x; G] \right) \right\| \leq \gamma_5 \|z - y\| \qquad (5.3.48)$$

for some nonnegative parameters γ_i, $i = 2, 3, 4, 5$ *and all* $x, y \in \overline{U}(y_0, r) \subseteq \overline{U}(y_0, R)$.

Then we can define

$$a = b = 0, \quad h_1 = h_2, h_1(q) = \gamma_3(q), h_3(q) = \gamma_5 q, \quad \text{and}$$
$$h_0(q_1, q_2) = \gamma_4 q_1 + (\gamma_2 + \gamma_4) q_2. \qquad (5.3.49)$$

If the hypotheses of Theorem 5.3.3 hold for the above choices, the conclusions follow.

Note that conditions (5.3.45)–(5.3.48) are weaker than the corresponding ones in [54, pp. 48–49]. Indeed, conditions

$$\left\| F'(x) - F'(y) \right\| \leq \gamma_6 \|x - y\|, \quad \left\| A(x, y)^{-1} \right\| \leq \gamma_7, \quad \|[x, y, z; G]\| \leq \gamma_8,$$

and

$$\|[x, y; G] - [z, w; G]\| \leq \gamma_9 (\|x - z\| + \|y - w\|)$$

for all $x, y, z, w \in \overline{U}(y_0, r)$ are used there instead of (5.3.45)–(5.3.48), where $[x, y, z; G]$ denotes a second-order divided difference of G at (x, y, z), and γ_i, $i = 6, 7, 8, 9$ are nonnegative parameters.

Application 5.3.6. *Returning back to Remark 5.3.1 and (5.3.6), iteration (5.3.2) reduces to the famous NK method (see Chapter 2).*

In order to compare with earlier results, we consider the case when $x = y$ and $v = w$ (single-step methods). We can then prove along the same lines to Lemma 5.3.2 and Theorem 5.3.3, respectively, the following results by assuming: there exists $w \in X$ such that $A(w)^{-1} \in L(Y, X)$, for any $x, y \in \overline{U}(w, r) \subseteq \overline{U}(w, R)$, $t \in [0, 1]$:

$$\left\| A(w)^{-1} [A(x) - A(w)] \right\| \leq g_0(\|x - w\|) + \alpha \qquad (5.3.50)$$

and

$$\left\| A(w)^{-1} \{ [F(x + t(y - x)) - A(x)](y - x) + G(y) - G(x) \} \right\|$$
$$\leq [g_1(\|x - w\| + t \|y - x\|) - g_2(\|x - w\|) + g_3(r) + \beta] \|y - x\|, \qquad (5.3.51)$$

where $g_0, g_1, g_2, g_3, \alpha, \beta$ are as h_0, (one variable) h_1, h_2, h_3, a, and b, respectively.

Then we can show the following result on majorizing sequences.

Lemma 5.3.7. *Assume:*
there exist $\eta \geq 0$, $\alpha \geq 0$, $\beta \geq 0$, $\delta \in [0, 2)$, $r_0 \in [0, r]$, $r \in [0, R]$ *such that:*

$$\overline{h}_\delta = 2\left[\int_0^1 g_1(r_0 + \theta\eta)\, d\theta - g_2(r_0) + g_3(r_0 + \eta) + \beta\right]$$

$$+ \delta[\alpha + g_0(r_0 + \eta)]$$

$$\leq \delta, \tag{5.3.52}$$

$$\tfrac{2\eta}{2-\delta} + r_0 \leq r, \tag{5.3.53}$$

$$g_0\left[\tfrac{2\eta}{2-\delta}\left(1 - \left(\tfrac{\delta}{2}\right)^{n+1}\right) + r_0\right] + \alpha < 1, \tag{5.3.54}$$

$$2\int_0^1 g_1\left[\tfrac{2\eta}{2-\delta}\left(1 - \left(\tfrac{\delta}{2}\right)^{n+1}\right) + r_0 + \theta\left(\tfrac{\delta}{2}\right)^{n+1}\eta\right] d\theta$$

$$- 2g_2\left[\tfrac{2\eta}{2-\delta}\left(1 - \left(\tfrac{\delta}{2}\right)^{n+1}\right) + r_0\right] + 2g_3\left[\tfrac{2\eta}{2-\delta}\left(1 - \left(\tfrac{\delta}{2}\right)^{n+1}\right) + r_0\right]$$

$$+ \delta g_0\left[\tfrac{2\eta}{2-\delta}\left(1 - \left(\tfrac{\delta}{2}\right)^{n+1}\right) + r_0\right]$$

$$\leq \delta \tag{5.3.55}$$

for all $n \geq 0$.
Then, iteration $\{s_n\}$ $(n \geq 0)$ *given by*

$$s_0 = r_0,$$

$$s_1 = r_0 + \eta,$$

$$s_{n+2} = s_{n+1} + \frac{\int_0^1 \{g_1(s_n + \theta(s_{n+1} - s_n)) - g_2(s_n) + \beta\}\, d\theta(s_{n+1} - s_n) + \int_{s_n}^{s_{n+1}} g_3(\theta)\, d\theta}{1 - \alpha - g_0(s_{n+1})} \tag{5.3.56}$$

is monotonically increasing, bounded above by

$$s^{**} = \tfrac{2\eta}{2-\delta} + r_0, \tag{5.3.57}$$

and converges to some s^* *such that*

$$0 \leq s^* \leq s^{**}. \tag{5.3.58}$$

Moreover, the following estimates hold for all $n \geq 0$

$$0 \leq s_{n+2} - s_{n+1} \leq \tfrac{\delta}{2}(s_{n+1} - s_n) \leq \left(\tfrac{\delta}{2}\right)^{n+1}\eta. \tag{5.3.59}$$

Theorem 5.3.8. *Assume:*
hypotheses of Lemma 5.3.7 hold there exists $y_0 \in \overline{U}(w, r_0)$ *such that*

$$\left\|A(y_0)^{-1}[F(y_0) + G(y_0)]\right\| \leq \eta. \tag{5.3.60}$$

Then, sequence $\{w_n\}$ $(n \geq 0)$ *generated by method (5.3.44) is well-defined, remains in* $\overline{U}(w, s^*)$ *for all* $n \geq 0$, *and converges to a solution* x^* *of equation* $F(x) + G(x) = 0$. *Moreover, the following error bounds hold for all* $n \geq 0$

$$\|w_{n+1} - w_n\| \leq s_{n+1} - s_n \tag{5.3.61}$$

and

$$\|w_n - x^*\| \leq s^* - s_n. \tag{5.3.62}$$

Furthermore the solution x^* is unique in $\overline{U}(w, s^*)$ if

$$\int_0^1 \left[g_1\left(s^* + 2\theta s^*\right) - g_2\left(s^*\right)\right] d\theta + g_3\left(s^*\right) + g_0\left(s^*\right) + \alpha + \beta < 1, \tag{5.3.63}$$

or in $\overline{U}(w, R_0)$ if $s^* < R_0 \leq r$, and

$$\int_0^1 \left[g_1\left(s^* + \theta(s^* + R_0)\right) - g_2\left(s^*\right)\right] d\theta + g_3\left(s^*\right) + g_0\left(s^*\right) + \alpha + \beta < 1. \tag{5.3.64}$$

We state the relevant results due to Chen and Yamamoto [58, p. 40]. We assume: $A(w)^{-1}$ exists, and for any $x, y, \in \overline{U}(w, r) \subseteq \overline{U}(w, R)$:

$$0 < \left\|A(w)^{-1}(F(w) + G(w))\right\| \leq \overline{\eta}, \tag{5.3.65}$$

$$\left\|A(w)^{-1}(A(x) - A(w))\right\| \leq \overline{g}_0(\|x - w\|) + \overline{\alpha}, \tag{5.3.66}$$

$$\left\|A(w)^{-1}\left[F'(x + t(y - x)) - A(x)\right]\right\|$$

$$\leq \overline{g}_1(\|x - w\|) + t\|y - x\| - \overline{g}_0(\|x - w\|) + \overline{\beta}, t \in [0, 1], \tag{5.3.67}$$

$$\left\|A(w)^{-1}[G(x) - G(y)]\right\| \leq g_3(r)\|x - y\|, \tag{5.3.68}$$

where $\overline{g}_0, \overline{g}_1, \overline{\alpha}, \overline{\beta}$ are as g_0, g_1, α, β, respectively, but $\overline{g}_0$ is also differentiable with $\overline{g}_0'(r) > 0$, is also differentiable with $\overline{g}_0(r) > 0, r \in [0, R]$ and $\overline{\alpha} + \overline{\beta} < 1$.
As in [58] set:

$$\varphi(r) = \overline{\eta} - r + \int_0^r \overline{g}_1(t) \, dt, \quad \psi(r) = \int_0^r g_3(t) \, dt, \tag{5.3.69}$$

$$\chi(r) = \phi(r) + \psi(r) + \left(\overline{\alpha} + \overline{\beta}\right) r. \tag{5.3.70}$$

denote the minimal value of $\chi(r)$ on $[0, R]$ by χ^*, and the minimal point by r^*. If $\chi(R) \leq 0$, denote the unique zero of χ by $r_0^* \in (0, r^*]$. Define scalar sequence $\{r_n\}$ $(n \geq 0)$ by

$$r_0 \in [0, R], r_{n+1} = r_n + \frac{u(r_n)}{g(r_n)} \quad (n \geq 0), \tag{5.3.71}$$

where

$$u(r) = \chi(r) - x^*, \tag{5.3.72}$$

and

$$g(r) = 1 - \overline{g}_0(r) - \overline{\alpha}. \tag{5.3.73}$$

With the above notation they showed:

Theorem 5.3.9. *Suppose* $\chi(R) \leq 0$. *Then equation (5.3.1) has a solution* $x^* \in \overline{U}(w, r_0^*)$, *which is unique in*

$$\overline{U} = \begin{cases} \overline{U}(w, R) & \text{if } \chi(R) = 0 \text{ or } \psi(R) = 0, \text{ and } r_0^* < R. \\ U(w, R) & \text{if } \chi(R) = 0 \text{ and } r_0^* < R. \end{cases} \tag{5.3.74}$$

Let

$$D^* = \overline{U}_{r \in [0, r^*)} \left\{ y \in \overline{U}(w, r) \mid \left\| A(y)^{-1} [F(y) + G(y)] \right\| \leq \frac{u(r)}{g(r)} \right\}. \tag{5.3.75}$$

Then, for any $y_0 \in D$, *sequence* $\{y_n\}$ $(n \geq 0)$ *generated by method (5.3.44) is well defined, remains in* $\overline{U}(w, r^*)$, *and satisfies*

$$\|y_{n+1} - y_n\| \leq r_{n+1} - r_n, \tag{5.3.76}$$

and

$$\|y_n - x^*\| < r^* - r_n \tag{5.3.77}$$

provided that r_0 *is chosen as in (5.3.71) so that* $r_0 \in Ry_0$, *where for* $y \in D^*$

$$R_y = \left\{ r \in [0, r^*) \mid \left\| A(y)^{-1} (F(y) + G(y)) \right\| \leq \frac{u(r)}{y(r)}, \|y - z\| \leq r \right\}. \tag{5.3.78}$$

Remark 5.3.10.
(a) Hypothesis on $\overline{g}_0$ is stronger than the corresponding one on g_0.
(b) Iteration (5.3.71) converges to r^* (even if $r_0 = 0$) not r_0^*.
(c) Choices of y_{-1}, y_0 other than the ones in Theorems 5.3.3, 5.3.8 can be given by (5.3.75) and (5.3.76)

Remark 5.3.11. The conclusions of Theorem 5.3.9 hold if the more general conditions replace (5.3.66)–(5.3.68), and

$$\overline{g}_0(r) \leq g_2(r), \quad r \in [0, R], \tag{5.3.79}$$

is satisfied. Moreover if strict inequality holds in (5.3.79) we obtain more precise error bounds. Indeed, define the sequence $\{\overline{r}_n\}$ $(n \geq 0)$, using (5.3.51), g_2 instead of (5.3.67), $\overline{g}_0$, respectively (with $\overline{g}_1 = g_1, \alpha = \overline{\alpha}, \beta = \overline{\beta}$), by

$$\overline{r}_0 = r_0, \ \overline{r}_1 = r_1, \tag{5.3.80}$$

$$\overline{r}_{n+1} - \overline{r}_n = \frac{u(\overline{r}_n) - u(\overline{r}_{n-1}) + (1 - g_2(\overline{r}_{n-1}) - \overline{\alpha})(\overline{r}_n - \overline{r}_{n-1})}{g(\overline{r}_n)} \ (n \geq 1)$$

It can easily be seen using induction on n (see also the proof of Proposition 5.3.13 that follows) that

$$\overline{r}_{n+1} - \overline{r}_n < r_{n+1} - r_n \tag{5.3.81}$$

$$\overline{r}_n < r_n \tag{5.3.82}$$

$$\overline{r}^* - \overline{r}_n \leq r^* - r_n, \quad \overline{r}^* = \lim_{n \to \infty} \overline{r}_n, \tag{5.3.83}$$

and $\overline{r}^* \leq r^*$.

Furthermore condition (5.3.51) allows us more flexibility in choosing functions and constants.

Remark 5.3.12. Our error bounds (5.3.61), (5.3.62) are finer than the corresponding ones (5.3.76) and (5.3.77), respectively, in many interesting cases. Let us choose:

$$\alpha = \bar{\alpha}, \quad \beta = \bar{\beta}, \quad g_0(r) = \bar{g}_0(r), \quad g_1(r) = g_2(r) = \bar{g}_1(r), \text{ and} \quad (5.3.84)$$
$$\bar{g}_3(r) = g_3(r) \quad \text{for all } r \in [0, R].$$

Then we can show:

Proposition 5.3.13. *Under the hypotheses of Theorem 5.3.8 and 5.3.9, further assume:*

$$s_1 < r_1 \qquad (5.3.85)$$

Then, the following hold:

$$s_n < r_n \quad (n \geq 1), \qquad (5.3.86)$$
$$s_{n+1} - s_n < r_{n+1} - r_n \quad (n \geq 0), \qquad (5.3.87)$$
$$s^* - s_n \leq r^* - r_n \quad (n \geq 0), \qquad (5.3.88)$$

and

$$s^* \leq r^*. \qquad (5.3.89)$$

Proof. It suffices to show (5.3.86) and (5.3.87), as then (5.3.88) and (5.3.89) respectively can easily follow. Inequality (5.3.86) holds for $n = 1$ by (5.3.85). By (5.3.56) and (5.3.71) we get in turn

$$s_2 - s_1 = \frac{\int_0^1 \{g_1(s_0+\theta(s_1-s_0))d\theta - g_2(s_0)+\alpha\}(s_1-s_0)+\int_{s_0}^{s_1} g_3(\theta)d\theta}{1-\beta-g_0(s_1)}$$
$$< \frac{\int_0^1 \{\bar{g}_1(r_0+\theta(r_1-r_0))d\theta - \bar{g}_2(r_0)+\bar{\alpha}\}(r_1-r_0)+\int_{r_0}^{r_1} \bar{g}_3(\theta)d\theta}{1-\bar{\beta}-\bar{g}_0(r_1)}$$
$$= \frac{u(r_1)-u(r_0)+g(r_0)(r_1-r_0)}{1-\bar{\beta}-\bar{g}_0(r_1)} = \frac{u(r_1)}{g(r_1)} = r_2 - r_1. \qquad (5.3.90)$$

Assume:

$$s_{k+1} < r_{k+1}, \qquad (5.3.91)$$

and

$$s_{k+1} - s_k < r_{k+1} - r_k \qquad (5.3.92)$$

hold for all $k \leq n$.

Using (5.3.56), (5.3.62), and (5.3.92), we obtain

$$s_{k+2} - s_{k+1}$$
$$= \frac{\int_0^1 \{g_1(s_k+\theta(s_{k+1}-s_k))d\theta - g_2(s_k)+\alpha\}(s_{k+1}-s_k)+\int_{s_k}^{s_{k+1}} g_3(\theta)d\theta}{1-\beta-g_0(s_{k+1})}$$
$$< \frac{\int_0^1 \{\bar{g}_1(r_k+\theta(r_{k+1}-r_k))d\theta - \bar{g}_2(r_k)+\bar{\alpha}\}(r_{k+1}-r_k)+\int_{r_k}^{r_{k+1}} \bar{g}_3(\theta)d\theta}{1-\bar{\beta}-\bar{g}_0(r_{k+1})}$$
$$= \frac{u(r_{k+1})-u(r_k)+g(r_k)(r_{k+1}-r_k)}{g(r_{k+1})} = \frac{u(r_{k+1})}{g(r_{k+1})} = r_{k+2} - r_{k+1}.$$

In order for us to include a case where operator G is nontrivial, we consider the following example:

Example 5.3.14. Let $X = Y = C[0, 1]$ the space of continuous on $[0, 1]$ equipped with the sup-norm, and $R > 0$. Consider the integral equation on $\overline{U}\left(x_0, \frac{r}{2}\right)$ given by

$$x(t) = \int_0^1 k(t, s, x(s)) \, ds, \tag{5.3.93}$$

where the kernel $k(t, s, x(s))$ with $(t, s) \in [0, 1] \times [0, 1]$ is a nondifferentiable operator on $\overline{U}\left(x_0, \frac{r}{2}\right)$. Define operators F, G on $\overline{U}\left(x_0, \frac{r}{2}\right)$ by

$$F(x)(t) = I x(t) \quad (I \text{ the identity operator}) \tag{5.3.94}$$

$$G(x)(t) = -\int_0^1 k(t, s, x(s)) \, ds. \tag{5.3.95}$$

Choose $x_0 = 0$, and assume there exists a constant $\theta_0 \in [0, 1)$, a real function $\theta_1(t, s)$ such that

$$\|k(t, s, x) - k(t, s, y)\| \le \theta_1(t, s) \|x - y\| \tag{5.3.96}$$

and

$$\sup_{t \in [0,1]} \int_0^1 \theta_1(t, s) \, ds \le \theta_0 \tag{5.3.97}$$

for all $t, s \in [0, 1]$, $x, y \in \overline{U}\left(x_0, \frac{r}{2}\right)$.

Moreover choose in Theorem 5.3.8: $r_0 = 0$, $y_0 = y_{-1}$, $A(x, y) = I(x)$, $g_0(r) = r$, $\alpha = \beta = 0$, $g_1(r) = g_2(r) = 0$, and $g_3(r) = \theta_0$ for all $x, y \in \overline{U}\left(x_0, \frac{r}{2}\right)$, $r, s \in [0, 1]$. It can easily be seen that the conditions of Theorem 5.3.8 hold if

$$t^* = \frac{\eta}{1-\theta_0} \le \frac{r}{2}. \tag{5.3.98}$$

We now study the local convergence of method (5.3.2).

In order to cover the local case, let us assume x^* is a zero equation (5.3.1), $A(x^*, x^*)^{-1}$ exists and for any $x, y \in \overline{U}(x^*, r) \subseteq \overline{U}(x^*, R)$, $t \in [0, 1]$:

$$\left\| A\left(x^*, x^*\right)^{-1} \left[A(x, y) - A\left(x^*, x^*\right)\right] \right\| \le \overline{h}_0 \left(\|x - x^*\|, \|y - x^*\| \right) + \overline{a}, \tag{5.3.99}$$

and

$$\left\| A\left(x^*, x^*\right)^{-1} \left[\left(F'\left(x^* + t\left(y - x^*\right)\right) - A(x, y)\right)\left(y - x^*\right) + G(y) - G\left(x^*\right)\right] \right\|$$
$$\le \left[\overline{h}_1 \left(\|y - x^*\| (1 + t)\right) - \overline{h}_2 \left(\|y - x^*\| + \overline{h}_3 \left(\|x - x^*\|\right) + \overline{b}\right)\right] \|y - x^*\|, \tag{5.3.100}$$

where, $\overline{h}_0, \overline{h}_1, \overline{h}_2, \overline{h}_3, \overline{a}, \overline{b}$ are as h_0, h_1, h_2, h_3, a, b, respectively. In order for us to compare our results with earlier ones, we only consider the case $r_0 = 0$, $x_{-1} = v$, $x_0 = w$ in (5.3.2) and call the corresponding sequence $\{x_n\}$ instead of $\{y_n\}$. Then exactly as in (5.3.34) but using (5.3.99), (5.3.100), instead of (5.3.3), (5.3.4), we can show the following local result for method (5.3.2).

Theorem 5.3.15. *Assume:*
there exists a solution of equation

$$f(\lambda) = 0, \tag{5.3.101}$$

in $[0, r]$ where

$$f(\lambda) = \int_0^1 \left[\overline{h}_1((1+t)\lambda) - \overline{h}_2(\lambda)\right] dt + \overline{h}_0(\lambda, \lambda) + \overline{a} + \overline{b} - 1. \tag{5.3.102}$$

Denote by λ_0 the smallest of the solutions in $[0, r]$. Then, sequence $\{x_n\}$ $(n \geq -1)$ generated by method (5.3.2) is well defined, remains in $\overline{U}(x^*, \lambda_0)$ for all $n \geq 0$, and converges to x^* provided that $x_{-1}, x_0 \in \overline{U}(x^*, \lambda)$.
Moreover the following estimates hold for all $n \geq 0$:

$$\|x^* - x_{n+1}\| \leq p_n, \tag{5.3.103}$$

where,

$$p_n = \frac{\left\{\int_0^1 [\overline{h}_1((1+t)\|x_n - x^*\|) - \overline{h}_2(\|x_n - x^*\|)] dt + \overline{a} + \overline{h}_3(\|x_{n-1} - x^*\|)\right\}}{1 - \overline{b} - \overline{h}_0(\|x_n - x^*\|)} \|x_n - x^*\|. \tag{5.3.104}$$

Application 5.3.16. *Let us again consider Newton's method, i.e., $F'(x) = A(x, y)$, $G(x) = 0$, and assume:*

$$\left\|F'(x^*)^{-1}[F'(x) - F'(x^*)]\right\| \leq \lambda_1 \|x - x^*\|, \tag{5.3.105}$$

and

$$\left\|F'(x^*)^{-1}[F'(x) - F'(y)]\right\| \leq \lambda_2 \|x - y\| \tag{5.3.106}$$

for all $x, y \in \overline{U}(x^*, r) \subseteq \overline{U}(x^*, R)$. Then we can set:

$$\overline{a} = \overline{b} = 0,$$

$$\overline{h}_1(r) = \overline{h}_2(r) = \lambda_2 r,$$

and

$$\overline{h}_0(r, r) = \lambda_1 r \text{ for all } r \in [0, R]. \tag{5.3.107}$$

Using (5.3.105), (5.3.106) we get:

$$\lambda_0 = \frac{2}{2\lambda_1 + \lambda_2}. \tag{5.3.108}$$

Then, see Section 2.4.

Application 5.3.17. *Notice that in Example 3.3.11 we provided a numerical result where our approach here compare favorably to the one given by Zabrejko [146], Chen and Yamamoto [58].*

5.4 On a two-point method of convergent order two

In this section, we are concerned with the problem of approximating a solution x^* of the nonlinear equation

$$F(x) = 0, \tag{5.4.1}$$

where F is a Fréchet-differentiable operator defined on an open subset D of a Banach space X with values in a Banach space Y.

We introduce the two-point method

$$x_{n+1} = x_n - A_n^{-1} F(x_n), \tag{5.4.2}$$

$$A_n = \left[(1 + \lambda_n) x_n - \lambda_n x_{n-1}, x_{n-1} \right], \quad (x_{-1}, x_0 \in D) \, (n \geq 0)$$

to generate a sequence approximating x^*. Here the numbers λ_n are chosen (if possible) so that iterates $(1 + \lambda_n) x_n - \lambda_n x_{n-1}$ $(n \geq 0)$ stay in D, whereas $[x, y; F]$ or simply $[x, y]$ belongs in $L(X, Y)$ so that:

$$[x, y] \, (x - y) = F(x) - F(y) \quad \text{for all } x, y \in D. \tag{5.4.3}$$

Linear operator $[x, y]$ is called a divided difference of order one on D. Clearly, iteration (5.4.2) has a geometrical interpretation similar to the secant method (see also (5.4.78)).

We provide a local as well as a semilocal convergence analysis for method (5.4.2) based on majorizing sequences and the corresponding majorant principle. It turns out that method (5.4.2) is essentially of quadratic order, and uses two previous iterates at every step as the secant method, which is only of order $1.618\ldots$ Moreover it is faster than the corresponding three-point method given by Potra in the elegant paper [163], which is only of order $1.839\ldots$ (see also (5.4.79)). Some numerical examples are also provided to show:

(a) how to choose linear operator A;

(b) that our iteration compares favorably with other methods using divided differences of order one and two previous iterates at every step.

Finally, the monotone convergence of method (5.4.2) is examined on partially ordered topological spaces or POTL-spaces (see Chapter 1).

We can show the following local convergence result for method (5.4.2):

Theorem 5.4.1. *Let F be a nonlinear operator defined on a convex subset D of a Banach space X with values in a Banach space Y. Assume:*
equation $F(x) = 0$ has a solution $x^* \in D$ at which the Fréchet derivative exists and $F'(x^*)^{-1} \in L(Y, X)$;
operator F is Fréchet-differentiable on $D_0 \subseteq D$ with divided differences of order one on D denoted by $[x, y]$ and satisfying (5.4.3) for $x, y \in D_0$;
there exist nondecreasing functions $a, b, c: [0, +\infty) \to [0, +\infty)$ and function $\lambda: X^2 \to R$ such that for all $x, y \in D_0$:

$$[1 + \lambda(x, y)] y - \lambda(x, y)x \in D_0, \tag{5.4.4}$$

$$\left\| F'(x^*)^{-1} \left[F'(x) - F'(x^*) \right] \right\| \le a(\|x - x^*\|), \tag{5.4.5}$$

$$\left\| F'(x^*)^{-1} ([x, y] - [x, x^*]) \right\| \le b(\|y - x^*\|), \tag{5.4.6}$$

$$\left\| F'(x^*)^{-1} ([y, y] - [(1 + \lambda(x, y))y - \lambda(x, y)x, x]) \right\| \le c(\|y - x\|); \tag{5.4.7}$$

equation

$$a(r) + b(r) + 2c(2r) - 1 = 0 \tag{5.4.8}$$

has a minimum positive zero r^*,
and

$$U(x^*, r^*)\} \subseteq D_0. \tag{5.4.9}$$

Then, sequence $\{x_n\}$ $(n \ge -1)$ *generated by method (5.4.2) is well defined, remains in* $U(x^*, r^*)$ *for all* $n \ge 0$, *and converges to* x^* *provided that:*

$$x_{-1}, x_0 \in U(x^*, r^*). \tag{5.4.10}$$

Moreover, the following estimates hold for all $n \ge 0$:

$$\|x_{n+1} - x^*\| \le \frac{b(\|x_n - x^*\|) + c(\|x_n - x_{n-1}\|)}{1 - [a(\|x_n - x^*\|) + c(\|x_n - x_{n-1}\|)]} \|x_n - x^*\|. \tag{5.4.11}$$

Proof. We shall first show:

$$F'(x) = [x, x] \quad \text{for all } x \in U(x^*, r^*). \tag{5.4.12}$$

By the Fréchet-differentiability of F, there exists $d > 0$ such that

$$\left\| F'(x^*) \right\| \le d. \tag{5.4.13}$$

Using (5.4.6) for $x \in U(x^*, r^*)$, we obtain in turn

$$\|F(x + \Delta x) - F(x) - [x, x](\Delta x)\|$$
$$= \left\| F'(x^*)F'(x^*)^{-1} ([x + \Delta x, x] - [x, x]) \Delta x \right\|$$
$$\le d \left[\left\| F'(x^*)^{-1} ([x + \Delta x, x] - [x^*, x]) \Delta x \right\| \right.$$
$$\left. + \left\| F'(x^*)^{-1} ([x^*, x] - [x, x]) \Delta x \right\| \right]$$
$$\le d \left[b(r^* + \|\Delta x\|) + b(r^*) \right] \|\Delta x\|. \tag{5.4.14}$$

If $b(r^*) \ne 0$, by letting $\Delta x \to 0$ we obtain (5.4.12). However if $b(r^*) = 0$, then by (5.4.6) there is an operator L in $L(X, Y)$ such that $[x, y] = L$ for all $x, y \in D_0$. Hence, from (5.4.3) we can set $F'(x) = L$ for all $x \in D_0$.

Let us denote by L the linear operator

$$L = [(1 + \lambda(x, y))y - \lambda(x, y)x, x].$$ (5.4.15)

Assume $x_{n-1}, x_n \in U(x^*, r^*)$. We shall show L is invertible on $U(x^*, r^*)$, and for

$$\lambda_n = \lambda(x_{n-1}, x_n), \quad L_n = [(1 + \lambda_n)x_n - \lambda_n x_{n-1}, x_{n-1}]$$ (5.4.16)

$$\|L_n^{-1} F'(x^*)\| \le [1 - a(\|x_n - x^*\|) - c(\|x_n - x_{n-1}\|)]^{-1}$$
$$\le [1 - a(r^*) - c(2r^*)]^{-1}.$$ (5.4.17)

Using (5.4.2), (5.4.4), (5.4.5), (5.4.7)–(5.4.10) and (5.4.12), we get in turn:

$$\left\| F'(x^*)^{-1} [F'(x^*) - L_n] \right\|$$
$$= \left\| F'(x^*)^{-1} [([x^*, x^*] - [x_n, x_n])\right.$$
$$\left. + ([x_n, x_n] - [(1 + \lambda_n)x_n - \lambda_n x_{n-1}, x_{n-1}])] \right\|$$
$$\le a(\|x_n - x^*\|) + c(\|x_n - x_{n-1}\|) \le a(r^*) + c(\|x_n - x^*\| + \|x^* - x_{n-1}\|)$$
$$\le a(r^*) + c(2r^*) < 1,$$ (5.4.18)

by the choice of r^*.

It follows from the Banach Lemma on invertible operators and (5.4.18) that L_n^{-1} exists, so that estimate (5.4.17) holds. Moreover by (5.4.6) and (5.4.7), we get:

$$\left\| F'(x^*)^{-1} ([x_n, x^*] - L_n) \right\|$$
$$= \left\| F'(x^*)^{-1} [([x_n, x^*] - [x_n, x_n]) + ([x_n, x_n] - L_n)] \right\|$$
$$\le \left\| F'(x^*)^{-1} [([x_n, x^*] - [x_n, x_n]) + ([x_n, x_n] - L_n)] \right\|$$
$$\le b(\|x_n - x^*\|) + c(\|x_n - x_{n-1}\|)$$
$$\le b(r^*) + c(\|x_n - x^*\| + \|x^* - x_{n-1}\|) \le b(r^*) + c(2r^*).$$ (5.4.19)

Furthermore, estimate (5.4.11) follows from (5.4.2), (5.4.17), (5.4.19), and the approximation

$$\|x_{n+1} - x^*\| = \left\| -L_n^{-1} ([x_n, x^*] - L_n) (x_n - x^*) \right\|$$
$$\le \left\| L_n^{-1} F'(x^*) \right\| \left\| F'(x^*)^{-1} ([x_n, x^*] - L_n) \right\| \|x_n - x^*\|.$$ (5.4.20)

Estimate (5.4.20) and the choice of r^* imply

$$\|x_{n+1} - x^*\| < \|x_n - x^*\| < r^* \quad (n \ge 0).$$ (5.4.21)

Hence, we deduce: $\lim x_n = x^*$ and $x_n \in U(x^*, r^*)$ $(n \ge 0)$.

We can show the following result on majorizing sequences for method (5.4.2).

Lemma 5.4.2. *Assume there exist nondecreasing, nonnegative functions* α_i, $i = 1, \ldots, 5$, *nonnegative parameters* β, γ, η *and* $\delta \in [0, 1)$ *such that:*

$$h_\delta^0 = \alpha_4(\eta) + \alpha_5(\gamma) + \delta\left[\alpha_1(\eta) + \alpha_2(0) + \alpha_3(\eta) + \beta\right] \leq \delta, \quad \text{and for all } k \geq 0$$

$$(5.4.22)$$

$$h_\delta^k = \alpha_4(\delta^k\eta) + \alpha_5(\delta^{k-1}\eta) + \delta\left[\alpha_1\left(\frac{1-\delta^{k+1}}{1-\delta}\eta\right) + \alpha_2\left(\frac{1-\delta^k}{1-\delta}\eta\right) + \alpha_3(\delta^k\eta) + \beta\right]$$

$$(5.4.23)$$

$$\leq \delta,$$

$$p_\delta^k = \alpha_1\left(\frac{1-\delta^{k+1}}{1-\delta}\eta\right) + \alpha_2\left(\frac{1-\delta^k}{1-\delta}\eta\right) + \alpha_3(\delta^k\eta) + \beta < 1. \qquad (5.4.24)$$

Then sequence $\{t_n\}$ $(n \geq -1)$ *given by*

$$t_{-1} = 0, \quad t_0 = \gamma, \quad t_1 = \gamma + \eta,$$

$$t_{n+2} = t_{n+1} + \frac{\alpha_4(t_{n+1}-t_n)+\alpha_5(t_n-t_{n-1})}{1-[\beta+\alpha_1(t_{n+1}-t_0)+\alpha_2(t_n-t_0)+\alpha_3(t_{n+1}-t_n)]}(t_{n+1} - t_n) \qquad (5.4.25)$$

is nondecreasing, bounded above by

$$t^{**} = \gamma + \frac{\eta}{1-\delta} \qquad (5.4.26)$$

and converges to some t^* *such that*

$$0 \leq t^* \leq t^{**}. \qquad (5.4.27)$$

Moreover, the following estimates hold for all $n \geq 0$:

$$0 \leq t_{n+2} - t_{n+1} \leq \delta(t_{n+1} - t_n) \leq \delta^{n+1}\eta. \qquad (5.4.28)$$

Proof. We shall show for all $k \geq 0$

$$\alpha_4(t_{k+1} - t_k) + \alpha_5(t_k - t_{k-1}) + \delta\left[\alpha_1(t_{k+1} - t_0) + \alpha_2(t_k - t_0) \right.$$
$$\left. + \alpha_3(t_{k+1} - t_k) + \beta\right] \leq \delta, \qquad (5.4.29)$$
$$\alpha_1(t_{k+1} - t_0) + \alpha_2(t_k - t_0) + \alpha_3(t_{k+1} - t_k) + \beta < 1, \qquad (5.4.30)$$

and

$$0 \leq t_{k+1} - t_k. \qquad (5.4.31)$$

Estimate (5.4.28) can then follow from (5.4.25) and (5.4.29)–(5.4.31). Inequalities (5.4.29)–(5.4.31)) hold for $k = 0$ by (5.4.22), (5.4.23), and (5.4.25). Let us assume (5.4.29)–(5.4.31) hold for all $k \leq n + 1$. We can have in turn:

$$\alpha_4(t_{k+2} - t_{k+1}) + \alpha_5(t_{k+1} - t_k) + \delta\left[\alpha_1(t_{k+2} - t_0) + \right.$$
$$\left. + \alpha_2(t_{k+1} - t_0) + \alpha_3(t_{k+2} - t_{k+1}) + \beta\right] \leq$$
$$\leq \alpha_4(\delta^{k+1}\eta) + \alpha_5(\delta^k\eta)$$
$$+ \delta\left[\alpha_1\left(\frac{1-\delta^{k+2}}{1-\delta}\eta\right) + \alpha_2\left(\frac{1-\delta^{k+1}}{1-\delta}\eta\right) + \alpha_3(\delta^{k+1}\eta) + \beta\right]$$
$$= h_\delta^{k+1} \leq \delta, \quad \text{(by (5.4.23))} \qquad (5.4.32)$$

and

$$\alpha_1(t_{k+2} - t_0) + \alpha_2(t_{k+1} - t_0) + \alpha_3(t_{k+2} - t_{k+1}) + \beta \leq$$

$$\leq \alpha_1 \left(\frac{1-\delta^{k+2}}{1-\delta}\eta\right) + \alpha_2 \left(\frac{1-\delta^{k+1}}{1-\delta}\eta\right) + \alpha_3(\delta^{k+1}\eta) + \beta = p_\delta^{k+1} < 1, \quad (5.4.33)$$

which together with (5.4.25) imply estimates (5.4.29)–(5.4.31) hold for all $k \geq 0$.
    Moreover we shall show

$$t_k \leq t^{**} \quad (k \geq -1). \tag{5.4.34}$$

For $k = -1, 0, 1, 2$ we have:

$$t_{-1} = 0 \leq t^{**}, \; t_0 = \gamma \leq t^{**}, \; t_1 = \gamma + \eta \leq t^{**}, \; t_2 = \gamma + \eta + \delta\eta \leq \gamma + \frac{\eta}{1-\delta} \leq t^{**}. \tag{5.4.35}$$

It follows from (5.4.25), (5.4.29)–(5.4.31) that for all $k \geq 0$

$$t_{k+2} \leq t_{k+1} + \delta(t_{k+1} - t_k) \leq \cdots \leq t_1 + \delta(t_1 - t_0) + \cdots + \delta(t_{k+1} - t_k)$$

$$\leq \gamma + \eta + \delta\eta + \cdots + \delta^{k+1}\eta = \gamma + \frac{1-\delta^{k+2}}{1-\delta}\eta \leq \gamma + \frac{\eta}{1-\delta} = t^{**}. \tag{5.4.36}$$

Hence, sequence $\{t_n\}$ $(n \geq -1)$ is nondecreasing, bounded above by t^{**} and as such it converges to some t^* satisfying (5.4.27).

Remark 5.4.3. Conditions (5.4.22), (5.4.23), and (5.4.24) can be replaced by (5.4.37), and (5.4.38), respectively, so that they can be independent of k, say, e.g.,

$$h_\delta = \alpha_4(\eta) + \alpha_5(\eta) + \alpha_1\left(\frac{\eta}{1-\delta}\right) + \alpha_2\left(\frac{\eta}{1-\delta}\right) + \alpha_3(\eta) + \beta \leq \delta, \tag{5.4.37}$$

$$p_\delta = \alpha_1\left(\frac{\eta}{1-\delta}\right) + \alpha_2\left(\frac{\eta}{1-\delta}\right) + \alpha_3(\eta) + \beta < 1. \tag{5.4.38}$$

Conditions of the form (5.4.22)–(5.4.24) or (5.4.37) and (5.4.38) are standard in the study of Newton-type methods. In the special case:

$$\alpha_i(r) = \theta_i r, \; i = 1, 2, 3, 4, \; \alpha_5(r) = \theta_5 r^2 \quad \text{for some } \theta_i \geq 0, \; i = 1, \ldots, 5 \tag{5.4.39}$$

it can easily be seen from (5.4.25) that there exist $n_0 \geq 0, \theta \geq 0$ such that

$$0 \leq t_{n+2} - t_{n+1} \leq \theta(t_{n+1} - t_n)^2 \quad (n \geq n_0). \tag{5.4.40}$$

Hence the order of convergence for sequence $\{t_n\}$ $(n \geq -1)$ is essentially quadratic (under the hypotheses of Lemma 5.4.2). Let $x, y, z \in D_0$, and define the divided difference of order two of operator F at the points $x, y,$ and z denoted by $[x, y, z]$ by:

$$[x, y, z](y - z) = [x, y] - [x, z]. \tag{5.4.41}$$

    We can show the following result for the semilocal convergence of method (5.4.2).

Theorem 5.4.4. *Let F be a nonlinear operator defined on a subset D of a Banach space X with values in a Banach space Y. Assume:*
Operator F has divided differences of order one and two on $D_0 \subseteq D$;
there exist points x_{-1}, x_0 in D_0 such that $A_0 = \left[(1 + \lambda_0)x_0 - \lambda_0 x_{-1}, x_{-1}\right]$ is invertible;
conditions (5.4.4), (5.4.22)–(5.4.24) hold;
Define constants β, γ, and η by

$$\|x_0 - x_{-1}\| \le \gamma \tag{5.4.42}$$

$$\|A_0^{-1}([(1 + \lambda_0)x_0 - \lambda_0 x_{-1}, x_{-1}, x_0]\lambda_0 - [x_0, x_{-1}, x_0])(x_0 - x_{-1})\| \le \beta \tag{5.4.43}$$

and

$$\|A_0^{-1} F(x_0)\| \le \eta; \tag{5.4.44}$$

there exist nondecreasing, nonnegative functions α_i, $i = 1, 2, \ldots, 6$ such that for all $x, y \in D_0$:

$$\|A_0^{-1}([x_0, x_0] - [y, x_0])\| \le \alpha_1(\|y - x_0\|), \tag{5.4.45}$$

$$\|A_0^{-1}([y, x_0] - [y, x])\| \le \alpha_2(\|x - x_0\|), \tag{5.4.46}$$

$$\|A_0^{-1}([y, x] - [(1 + \lambda(x, y))y - \lambda(x, y)x, x])\| \le \alpha_3(\|y - x\|), \tag{5.4.47}$$

$$\|A_0^{-1}([y, x] - [x, x])\| \le \alpha_4(\|y - x\|), \tag{5.4.48}$$

$$\|A_0^{-1}([y, x, y] - [(1 + \lambda(x, y))y - \lambda(x, y)x, x, y]\lambda(x, y))(y - x)\|$$
$$\le \alpha_5(\|y - x\|), \tag{5.4.49}$$

$$\|A_0^{-1}(A_0 - [x, x])\| \le \alpha_6(\|(1 + \lambda_0)x_0 - \lambda_0 x_{-1} - x\| + \|x_{-1} - x\|), \tag{5.4.50}$$

$$\int_0^1 \alpha_6 \left[2\gamma + 2(1 + 2t)t^*\right] dt < 1, \tag{5.4.51}$$

and

$$\overline{U}(x_0, t^*) \subseteq D_0, \tag{5.4.52}$$

where t^* was defined in Lemma 5.4.2.
  Then sequence $\{x_n\}$ $(n \ge -1)$ generated by method (5.4.2) is well defined, remains in $\overline{U}(x_0, t^*)$ for all $n \ge 0$, and converges to a unique solution x^* of equation $F(x) = 0$ in $\overline{U}(x_0, t^*)$.
  Moreover, the following estimates hold for all $n \ge 0$:

$$\|x_{n+1} - x_n\| \le t_{n+1} - t_n \tag{5.4.53}$$

and

$$\|x_n - x^*\| \le t^* - t_n, \tag{5.4.54}$$

where sequence $\{t_n\}$ is given by (5.4.25).
Furthermore if there exists $R > t^*$ such that:

$$U(x_0, R) \subseteq D_0 \tag{5.4.55}$$

and

$$\int_0^1 \alpha_6 (2\gamma + 2R + 2t(t^* + R)) \, dt \le 1, \tag{5.4.56}$$

then the solution x^* is unique in $U(x_0, R)$.

Proof. We shall show using induction on $k \ge 0$:

$$x_k \in \overline{U}(x_0, t^*), \tag{5.4.57}$$

and

$$\|x_{k+1} - x_k\| \le t_{k+1} - t_k. \tag{5.4.58}$$

Estimates (5.4.57), (5.4.58) hold for $k = -1, 0$ by the initial conditions and because $t_{-1} \le t^*$, $t_0 \le t^*$. Assume (5.4.57) and (5.4.58) hold for all $n \le k + 1$. Using (5.4.42), (5.4.45)–(5.4.47) we get

$$\|A_0^{-1}(A_0 - A_{k+1})\|$$

$$= \|A_0^{-1}([(1 + \lambda_0)x_0 - \lambda_0 x_{-1}, x_{-1}] - [x_0, x_{-1}] + [x_0, x_{-1}]$$
$$- [x_0, x_0] + [x_0, x_0] - [x_{k+1}, x_0] + [x_{k+1}, x_0] - [x_{k+1}, x_k]$$
$$+ [x_{k+1}, x_k] - [(1 + \lambda_{k+1})x_{k+1} - \lambda_{k+1}x_{k+1}, x_k])\|$$

$$= \|A_0^{-1}(([(1 + \lambda_0)x_0 - \lambda_0 x_{-1}, x_{-1}, x_0] \lambda_0 - [x_0, x_{-1}, x_0])(x_0 - x_{-1})$$
$$+ ([x_0, x_0] - [x_{k+1}, x_0]) + ([x_{k+1}, x_0] - [x_{k+1}, x_k])$$
$$+ ([x_{k+1}, x_k] - [(1 + \lambda_{k+1})x_{k+1} - \lambda_{k+1}x_{k+1}, x_k]))\|$$

$$\le \beta + \alpha_1(\|x_{k+1} - x_0\|) + \alpha_2(\|x_k - x_0\|) + \alpha_3(\|x_{k+1} - x_k\|)$$

$$\le \beta + \alpha_1(t_{k+1} - t_0) + \alpha_2(t_k - t_0) + \alpha_3(t_{k+1} - t_k) < 1. \tag{5.4.59}$$

It follows by the Banach lemma on invertible operators and (5.4.59) that A_{k+1}^{-1} exists, and

$$\|A_{k+1}^{-1}A_0\| \le [1 - (\beta + \alpha_1(t_{k+1} - t_0) + \alpha_2(t_k - t_0) + \alpha_3(t_{k+1} - t_k))]^{-1}. \tag{5.4.60}$$

By (5.4.48) and (5.4.49), we can also have:

$$\|A_0^{-1}([x_{k+1}, x_k] - A_k)\|$$

$$= \|A_0^{-1}([x_{k+1}, x_k] - [x_k, x_k] + [x_k, x_k]$$
$$- [x_k, x_{k-1}] + [x_k, x_{k-1}] - [(1 + \lambda_k)x_k - \lambda_k x_{k-1}, x_{k-1}])\|$$

$$= \|A_0^{-1}(([x_{k+1}, x_k] - [x_k, x_k]) + ([x_k, x_{k-1}, x_k]$$
$$- [(1 + \lambda_k)x_k - \lambda_k x_{k-1}, x_{k-1}, x_k] \lambda_k)(x_k - x_{k-1}))\|$$

$$\le \alpha_4(\|x_{k+1} - x_k\|) + \alpha_5(\|x_k - x_{k-1}\|)$$

$$\le \alpha_4(t_{k+1} - t_k) + \alpha_5(t_k - t_{k-1}). \tag{5.4.61}$$

By (5.4.2), (5.4.60), and (5.4.61), we obtain in turn

$$\|x_{k+2} - x_{k+1}\|$$

$$= \|A_{k+1}^{-1} F(x_{k+1})\| = \|A_{k+1}^{-1}(F(x_{k+1}) - F(x_k) - A_k(x_{k+1} - x_k))\|$$

$$\leq \|A_{k+1}^{-1} A_0\| \, \|A_0^{-1}([x_{k+1}, x_k] - A_k)\| \, \|x_{k+1} - x_k\|$$

$$\leq \frac{\alpha_4(t_{k+1} - t_k) + \alpha_5(t_k - t_{k-1})}{1 - [\beta + \alpha_1(t_{k+1} - t_0) + \alpha_2(t_k - t_0) + \alpha_3(t_{k+1} - t_k)]}(t_{k+1} - t_k)$$

$$= t_{k+2} - t_{k+1}, \qquad (5.4.62)$$

which shows (5.4.58) for all $n \geq 0$. We can also get

$$\|x_{k+2} - x_0\| \leq \sum_{j=0}^{k+1} \|x_{j+1} - x_j\| \leq \sum_{j=0}^{k+1} (t_{j+1} - t_j) = t_{k+2} - t_0 \leq t^*. \quad (5.4.63)$$

That is $x_n \in \overline{U}(x_0, r_0)$ for all $n \geq 0$. It follows from (5.4.58) that sequence $\{x_n\}$ $(n \geq -1)$ is Cauchy in a Banach space X, and as such it converges to some $x^* \in \overline{U}(x_0, t^*)$. By letting $k \to \infty$ in (5.4.62), we obtain $F(x^*) = 0$.

To show uniqueness in $\overline{U}(x_0, t^*)$ let y^* be a solution of equation (5.4.1) in $\overline{U}(x_0, t^*)$. By (5.4.42), (5.4.50), and (5.4.51) we have for

$$M = \int_0^1 \left[y^* + t(x^* - y^*), y^* + t(x^* - y^*) \right] dt, \qquad (5.4.64)$$

$$\|A_0^{-1}(A_0 - M)\|$$

$$\leq \int_0^1 \alpha_6([\|(1 + \lambda_0)x_0 - \lambda_0 x_{-1} - y^* - t(x^* - y^*)\|$$
$$+ \|x_{-1} - y^* - t(x^* - y^*)\|]) \, dt$$

$$\leq \int_0^1 \alpha_6 [\|x_0 - x_{-1}\| + \|x_0 - y^*\| + t(\|x_0 - y^*\| + \|x_0 - x^*\|)$$
$$+ \|x_0 - x_{-1}\| + \|x_0 - y^*\| + t(\|x_0 - x^*\| + \|x_0 - y^*\|)] \, dt \quad (5.4.65)$$

$$\leq \int_0^1 \alpha_6 \left[2\gamma + 2(1 + 2t)t^* \right] dt < 1. \qquad (5.4.66)$$

It follows from the Banach Lemma on invertible operators and (5.4.65) that M^{-1} exists.

We deduce from (5.4.64) and the identity

$$F(x^*) - F(y^*) = M(x^* - y^*) \qquad (5.4.67)$$

that

$$x^* = y^*. \qquad (5.4.68)$$

Finally to show uniqueness in $U(x_0, R)$, let us assume $y^* \in U(x_0, R)$ is a solution of equation (5.4.1). As in (5.4.65) we obtain again

$$\|A_0^{-1}(A_0 - M)\| < \int_0^1 \alpha_6 \left[2\gamma + 2R + 2t(t^* + R)\right] dt \leq 1, \qquad (5.4.69)$$

which shows (5.4.68).

Remark 5.4.5. (a) In the special case

$$a(r) = \bar{a}r, \quad b(r) = \bar{b}r, \quad c(r) = \bar{c}r^2 \text{ for } \bar{a} \geq 0, \bar{b} \geq 0, \bar{c} \geq 0 \qquad (5.4.70)$$

then, equation (5.4.8) in Theorem 5.4.1 gives

$$r^* = \frac{4}{\bar{a} + \bar{b} + \sqrt{(\bar{a} + \bar{b})^2 + 32\bar{c}}} \qquad (5.4.71)$$

and as in Remark 5.4.3 we see that the convergence of sequence $\{x_n\}$ $(n \geq -1)$ is essentially quadratic. Note that in Section 1.2 we showed how to choose constants $\bar{a}$, $\bar{b}, \bar{c}$.

(b) Conditions (5.4.5) and (5.4.6) can be combined in the stronger

$$\|F'(x^*)^{-1}([x, y] - [z, w])\| \leq a_1(\|x - z\| + \|y - w\|) \qquad (5.4.72)$$

for all $x, y, z, w \in D_0$ and some nondecreasing, nonnegative function a_1. However note that

$$a(r) \leq a_1(2r), \quad r = \|x - x^*\|, \qquad (5.4.73)$$

and

$$b(r_0) \leq a_1(r_0), \quad r_0 = \|y - x^*\|. \qquad (5.4.74)$$

(c) In order for us to compare method (5.4.2) with others [163], [196] using divided differences consider the conditions

$$\|F'(x^*)^{-1}([y, x, y] - [(1 + \lambda(x, y))y - \lambda(x, y)x, x, y]\lambda(x, y))(y - x)\|$$
$$\leq c_1(\|y - x\|) \qquad (5.4.75)$$

or even

$$\|F'(x^*)^{-1}([u, x, y] - [v, x, y]\lambda(x, y))(y - x)\| \leq c_2(\|y - x\|), \qquad (5.4.76)$$

where c_1, c_2 are nondecreasing, nonnegative functions (or simply nonnegative constants). We can write, e.g., in (5.4.18)

$$[x_n, x_n] - [(1 + \lambda_n)x_n - \lambda_n x_{n-1}, x_{n-1}]$$
$$= ([x_n, x_n] - [x_n, x_{n-1}]) + ([x_n, x_{n-1}] - [(1 + \lambda_n)x_n - \lambda_n x_{n-1}, x_{n-1}])$$
$$= ([x_n, x_{n-1}, x_n] - [(1 + \lambda_n)x_n - \lambda_n x_{n-1}, x_{n-1}, x_n]\lambda_n)(x_n - x_{n-1}) \qquad (5.4.77)$$

and consequently replace $c(\|y - x\|)$ in (5.4.7) etc. by $c_1(\|y - x\|)$ or $c_2(\|y - x\|)$.
(d) The secant method

$$x_{n+1} - x_n - [x_{n-1}, x_n]^{-1} F(x_n) \quad (x_{-1}, x_0 \in D) \qquad (5.4.78)$$

also uses two previous iterates. However, it is only of order $1.618\ldots$.
(e) Potra's three-point method [163]

$$x_{n+1} = x_n - ([x_n, x_{n-1}] + [x_{n-2}, x_n] - [x_{n-2}, x_{n-1}])^{-1} F(x_n) \quad (x_{-2}, x_{-1}, x_0 \in D)$$
$$(5.4.79)$$

uses (5.4.72) and (5.4.76) (for c_2 being a constant) to obtain a convergence radius for method (5.4.79), which however is smaller than ours (see (5.4.71) above and (5.4.22) in [163, p. 87]). Moreover method (5.4.79) is only of order $1.839\ldots$.
(f) The radius of convergence for NK method given by Rheinboldt [175] using (5.4.72) for a_1 being a constant is given by $r_R = \frac{1}{3a_1}$. However, we showed in Section 2.4 that $\frac{a_1}{a}$, $\frac{a_1}{b}$, $\frac{a_1}{c}$ can be arbitrarily large. Hence r_R can be smaller than r^*.
(g) Condition (5.4.4) automatically holds if $D = X$, or it can be dropped if divided differences are defined on the entire space X instead of just D. In practice, we choose numbers $\lambda(x, y)$ so that (5.4.4) is satisfied. Note also that (5.4.4) is required to hold only for the iterates x_n and not all points in D (see Example 5.4.7).

The choice $\lambda(x, y) = 1$ for all $x, y \in D$ seems to be very realistic and promising. However, other cases may also be convenient (see also Example 5.4.7). For example if $\lambda(x, y) = -.5$ for all $x, y \in D$ then it can easily be shown using induction on the integer n that all iterates remain in the balls $U(x^*, r^*)$ (in the local case) or $U(x_0, t^*)$ (in the semilocal case) provided that the initial guesses x_{-1}, x_0 are inside those balls. That is, in this case delicate condition (5.4.4) is automatically satisfied.

There is another stronger but more practical way to satisfy (5.4.4).
First: In the local case: Assume

$$\lambda^* = \max_{x, y \in D_0^2} (|1 + \lambda(x, y)| + |\lambda(x, y)|)$$

exists, and is finite, and

$$U_1 = U(x^*, R^*) \subseteq D_0 \text{ with } R^* = (|1 + \lambda^*| + |\lambda^*|) r^*. \qquad (5.4.80)$$

Then it follows from the proof of Theorem 5.4.1 that the condition (5.4.80) can replace (5.4.4) and (5.4.9) in Theorem 5.4.1. Indeed, for $x_{n-1}, x_n \in U(x^*, r^*)$ we get

$$\|(1 + \lambda_n) x_n - \lambda_n x_{n-1} - x^*\|$$
$$\leq |1 + \lambda_n| \|x_n - x^*\| + |\lambda_n| \|x_{n-1} - x^*\|$$
$$\leq |1 + \lambda_n| r^* + |\lambda_n| r^* \leq \lambda^* r^* = R^*.$$

That is, $(1 + \lambda_n) x_n - \lambda_n x_{n-1} \in U_1$ $(n \geq 0)$. In case $\lambda(x, y) = 1$, then $R^* = 3r^*$.
Second: In the semilocal case: Replace (5.4.4) and (5.4.52) in Theorem 5.4.4 by

$$U_2 = U(x_0, R_0) \subseteq D_0 \text{ with } R_0 = \lambda^* t^*.$$

Indeed, for $x_{n-1}, x_n \in U(x_0, t^*)$ $(n \geq 0)$ we get

$$\|(1 + \lambda_n) x_n - \lambda_n x_{n-1} - x_0\|$$
$$\leq |1 + \lambda_n| \|x_n - x_0\| + |\lambda_n| \|x_{n-1} - x_0\|$$
$$\leq |1 + \lambda_n| t^* + |\lambda_n| t^* \leq R_0.$$

Note again that if $\lambda(x, y) = 1$, then $R_0 = 3t^*$.

Remark 5.4.6. According to (5.4.40) the order of convergence of iteration $\{x_n\}$ $(n \geq -1)$ is essentially quadratic.

Comments similar to Remark 5.4.5 for the semilocal case can now follow. However, we leave the details to the motivated reader and conclude this section with some numerical examples.

A simple numerical example follows to show:

(a) how to choose divided difference in method (5.4.2);
(b) method (5.4.2) is faster than the secant method (5.4.78).
(c) method (5.4.2) can be at least as fast as NK method (5.4.80).

Note that the analytical representation of $F'(x_n)$ may be complicated, which makes the use of method (5.4.2) very attractive.

Example 5.4.7. Let $X = Y = \mathbf{R}$, and define function F on $D_0 = D = (.4, 1.5)$ by

$$F(x) = x^2 - 6x + 5. \tag{5.4.81}$$

Moreover define divided difference of order one appearing in method (5.4.2) for $\lambda(x, y) = 1$ for all $x, y \in D$ by

$$[2y - x, x] = \frac{F(2y - x) - F(x)}{2(y - x)}. \tag{5.4.82}$$

In this case method (5.4.2) becomes

$$x_{n+1} = \frac{x_n^2 - 5}{2(x_n - 3)}, \tag{5.4.83}$$

and coincides with NK method (5.4.80) applied to F. Furthermore secant method (5.4.78) becomes:

$$x_{n+1} = \frac{x_{n-1} x_n - 5}{x_{n-1} + x_n - 6}. \tag{5.4.84}$$

Choose $x_{-1} = .6$ and $x_0 = .7$. Then we obtain:

We conclude this section with an example involving a nonlinear integral equation:

n	Method (5.4.2)	Secant method (5.4.84)
1	.980434783	.96875
2	.999905228	.997835498
3	.999999998	.99998323
4	$1 = x^*$	.999999991
5	—	1

Example 5.4.8. Let $H(x, t, x(t))$ be a continuous function of its arguments that is sufficiently many times differentiable with respect to x. It can easily be seen that if operator F in (5.4.1) is given by

$$F(x(s)) = x(s) - \int_0^1 H(s, t, x(t))dt, \tag{5.4.85}$$

then divided difference of order one appearing in (5.4.2) can be defined as

$$h_n(s, t) = \frac{H(s,t,(1+\lambda_n)x_n(t)-\lambda_n x_{n-1}(t))-H(s,t,x_{n-1}(t))}{(1+\lambda_n)(x_n(t)-x_{n-1}(t))},$$
$$\lambda_n = \lambda_n(t) = \lambda(x_{n-1}(t), x_n(t)) \tag{5.4.86}$$

provided that if for $t = t_m$ we get $x_n(t) = x_{n-1}(t)$, then the above function equals $H'_x(s, t_m, x_n(t_m))$. Note that this way $h_n(s, t)$ is continuous for all $t \in [0, 1]$ provided that $1 + \lambda_n \neq 0$ ($n \geq 0$) and, e.g., sequence $|1 + \lambda_n|$ is bounded below by a positive number.

The monotone convergence of method (5.4.2) is examined in the next result.

Theorem 5.4.9. *Let F be a nonlinear operator defined on an open subset of a regular POTL-space X with values in a POTL-space Y. Let x_0, y_0, y_{-1} be points of $D \subseteq X$ such that:*

$$x_0 \leq y_0 \leq y_{-1}, \quad D_0 = \langle x_0, y_{-1} \rangle \subseteq D, \quad F(x_0) \leq 0 \leq F(y_0). \tag{5.4.87}$$

Moreover assume: there exist, a function $\lambda: D^2 \to R$, a divided difference $[\cdot, \cdot]: D \to L(X, Y)$ such that for all $(x, y) \in D_0^2$ with $x \leq y$:

$$(1 + \lambda(x, y))y - \lambda(x, y)x \in D_0, \tag{5.4.88}$$

and

$$F(y) - F(x) \leq [x, (1 + \lambda(x, y))y - \lambda(x, y)x] (y - x). \tag{5.4.89}$$

Furthermore, assume that for any $(x, y) \in D_0^2$ with $x \leq y$, and $(x, (1 + \lambda(x, y))y - \lambda(x, y)x) \in D_0^2$ the linear operator $[x, (1 + \lambda(x, y))y - \lambda(x, y)x]$ has a continuous nonsingular, nonnegative left subinverse.
Then there exist two sequences $\{x_n\}$ ($n \geq 1$), $\{y_n\}$ ($n \geq 1$), and two points x^*, y^* of X such that for all $n \geq 0$:

$$F(y_n) + \left[y_{n-1}, (1+\lambda_n)y_n - \lambda_n y_{n-1}\right](y_{n+1} - y_n) = 0, \quad \lambda_n = \lambda(y_{n-1}, y_n),$$

$$(5.4.90)$$

$$F(x_n) + \left[y_{n-1}, (1+\lambda_n)y_n - \lambda_n y_{n-1}\right](x_{n+1} - x_n) = 0, \quad (5.4.91)$$

$$F(x_n) \leq 0 \leq F(y_n), \quad (5.4.92)$$

$$x_0 \leq x_1 \leq \cdots \leq x_n \leq x_{n+1} \leq y_{n+1} \leq y_n \leq \cdots \leq y_1 \leq y_0, \quad (5.4.93)$$

$$\lim_{n \to \infty} x_n = x^*, \quad \lim_{n \to \infty} y_n = y^*. \quad (5.4.94)$$

Finally, if linear operators $A_n = \left[y_{n-1}, (1+\lambda_n)y_n - \lambda_n y_{n-1}\right]$ are inverse nonnegative, then any solution of the equation $F(x) = 0$ from the interval D_0 belongs to the interval $\langle x^*, y^* \rangle$ (i.e., $x_0 \leq v \leq y_0$ and $F(v) = 0$ imply $x^* \leq v \leq y^*$).

Proof. Let $\overline{A}_0$ be a continuous nonsingular, nonnegative left subinverse of A_0. Define the operator $Q: \langle 0, y_0 - x_0 \rangle \to X$ by

$$Q(x) = x - \overline{A}_0\left[F(x_0) + A_0(x)\right].$$

It is easy to see that Q is isotone and continuous. We also have:

$$Q(0) = -\overline{A}_0 F(x_0) \geq 0,$$

$$Q(y_0 - x_0) = y_0 - x_0 - \overline{A}_0(F(y_0)) + \overline{A}_0(F(y_0) - F(x_0) - A_0(y_0 - x_0))$$

$$\leq y_0 - x_0 - \overline{A}_0(F(y_0)) \leq y_0 - x_0.$$

According to Kantorovich's theorem concerning fixed points on POTL-spaces (see Section 1.2), operator Q has a fixed point $w \in \langle 0, y_0 - x_0 \rangle$. Set $x_1 = x_0 + w$. Then we get

$$F(x_0) + A_0(x_1 - x_0) = 0, \quad x_0 \leq x_1 \leq y_0. \quad (5.4.95)$$

By (5.4.89) and (5.4.95) we deduce:

$$F(x_1) = F(x_1) - F(x_0) + A_0(x_0 - x_1) \leq 0.$$

Consider the operator $H: \langle 0, y_0 - x_1 \rangle \to X$ given by

$$H(x) = x + \overline{A}_0(F(y_0) - A_0(x)).$$

Operator H is clearly continuous, isotone, and we have:

$$H(0) = \overline{A}_0 F(y_0) \geq 0,$$

$$H(y_0 - x_1) = y_0 - x_1 + \overline{A}_0 F(x_1) + \overline{A}_0\left[F(y_0) - F(x_1) - A_0(y_0 - x_1)\right]$$

$$\leq y_0 - x_1 + \overline{A}_0 F(x_1) \leq y_0 - x_1.$$

By Kantorovich's theorem on fixed points, there exists a point $z \in \langle 0, y_0 - x_1 \rangle$ such that $H(z) = z$. Set $y_1 = y_0 - z$ to obtain

$$F(y_0) + A_0(y_1 - y_0) = 0, \quad x_1 \leq y_1 \leq y_0. \quad (5.4.96)$$

Using (5.4.89), (5.4.96), we get:

$$F(y_1) = F(y_1) - F(y_0) - A_0(y_1 - y_0) \geq 0.$$

Proceeding by induction, we can show that there exist two sequences $\{x_n\}$ ($n \geq 1$), $\{y_n\}$ ($n \geq 1$) satisfying (5.4.90)–(5.4.93) in a regular space X, and as such they converge to points $x^*, y^* \in X$, respectively. We obviously have $x^* \leq y^*$. If $x_0 \leq u \leq y_0$ and $F(u) = 0$, then we can write

$$A_0(y_1 - u) = A_0(y_0) - F(y_0) - A_0(u) = A_0(y_0 - u) - (F(y_0) - F(u)) \geq 0,$$

and

$$A_0(x_1 - u) = A_0(x_0) - F(x_0) - A_0(u) = A_0(x_0 - u) - (F(x_0) - F(u)) \leq 0.$$

If the operator A_0 is inverse nonnegative, then it follows that $x_1 \leq u \leq y_1$. Proceeding by induction, we deduce that $x_n \leq u \leq y_n$ holds for all $n \geq 0$. Hence we conclude

$$x^* \leq u \leq y^*.$$

In what follows, we give some natural conditions under which the points x^* and y^* are solutions of equation $F(x) = 0$.

Proposition 5.4.10. *Under the hypotheses of Theorem 5.4.9, assume that F is continuous at x^* and y^* if one of the following conditions is satisfied:*
(a) $x^* = y^*$;
(b) X is normal, and there exists an operator $T: X \to Y$ ($T(0) = 0$) that has an isotone inverse continuous at the origin and such that $A_n \leq T$ for sufficiently large n;
(c) Y is normal and there exists an operator $Q: X \to Y$ ($Q(0) = 0$) continuous at the origin and such that $A_n \leq Q$ for sufficiently large n;
(d) operators A_n ($n \geq 0$) are equicontinuous.
Then we deduce

$$F(x^*) = F(y^*) = 0. \tag{5.4.97}$$

Proof. (a) Using the continuity of F and (5.4.92), we get

$$F(x^*) \leq 0 \leq F(y^*).$$

Hence, we conclude

$$F(x^*) = 0.$$

(b) Using (5.4.90)–(5.4.93), we get

$$0 \geq F(x_n) = A_n(x_n - x_{n+1}) \geq T(x_n - x_{n+1}),$$
$$0 \leq F(y_n) = A_n(y_n - y_{n+1}) \leq T(y_n - y_{n+1}).$$

Therefore, it follows:

$$0 \geq T^{-1}F(x_n) \geq x_n - x_{n+1}, \quad 0 \leq T^{-1}F(y_n) \leq y_n - y_{n+1}.$$

By the normality of X, and

$$\lim_{n \to \infty}(x_n - x_{n+1}) = \lim_{n \to \infty}(y_n - y_{n+1}) = 0,$$

we get $\lim_{n \to \infty}T^{-1}F(x_n)) = \lim_{n \to \infty}T^{-1}(F(y_n)) = 0$. Using the continuity of F we obtain (5.4.97).

(c) As before for sufficiently large n

$$0 \geq F(x_n) \geq Q(x_n - x_{n+1}), \quad 0 \leq F(y_n) \leq Q(y_n - y_{n+1}).$$

By the normality of Y and the continuity of F and Q, we obtain (5.4.97).

(d) It follows from the equicontinuity of operator A_n that $\lim_{n \to \infty}A_n v_n = 0$ whenever $\lim_{n \to \infty}v_n = 0$. Therefore, we get $\lim_{n \to \infty}A_n(x_n - x_{n+1}) = \lim_{n \to \infty}A_n(y_n - y_{n+1}) = 0$. By (5.4.90), (5.4.91), and the continuity of F at x^* and y^*, we obtain (5.4.97).

Remark 5.4.11. Hypotheses of Theorem 5.4.9 can be weakened along the lines of Remarks 5.4.3, 5.4.5, 5.4.6 above and the works in [163, pp. 102–105], [43], [199] on the monotone convergence of Newton-like methods. However, we leave the details to the motivated reader.

5.5 Exercises

5.5.1. Introduce the method

$$x_{n+1} = x_n - \left[2x_n - x_{n-1}, x_{n-1}\right]^{-1} F(x_n) \quad (x_{-1}, x_0 \in D) \quad (n \geq 0)$$

for approximating x^*.

Let F be a nonlinear operator defined on an open set D of a Banach space X with values in a Banach space Y. Assume:

operator F has divided differences of order one and two on D;
there exist points x_{-1}, x_0 in D such that $2x_0 - x_{-1} \in D$ and
$A_0 = \left[2x_0 - x_{-1}, x_{-1}\right]$ is invertible on D;
Set $A_n = \left[2x_n - x_{n-1}, x_{n-1}\right]$ $(n \geq 0)$.
There exist constants α, β such that:

$$\|A_0^{-1}([x, y] - [u, v])\| \leq \alpha(\|x - u\| + \|y - v\|),$$

$$\|A_0^{-1}([y, x, y] - [2y - x, x, y])\| \leq \beta\|x - y\|, \text{ for all } x, y, u, v \in D,$$

and for all $x, y \in D \Longrightarrow 2y - x \in D$.

Define constants γ, δ by

$$\|x_0 - x_{-1}\| \leq \gamma,$$
$$\|A_0^{-1} F(x_0)\| \leq \delta,$$
$$2\beta\gamma^2 \leq 1;$$

Moreover define θ, r, h by

$$\theta = \{(\alpha + \beta\gamma)^2 + 3\beta(1 - \beta\gamma^2)\}^{1/2},$$
$$r = \frac{1-\beta\gamma^2}{\alpha+\beta\gamma+\theta},$$

and

$$h(t) = -\beta t^3 - (\alpha + \beta\gamma)t^2 + (1 - \beta\gamma^2)t,$$
$$\delta \leq h(r) = \frac{1}{3} \frac{\alpha+\beta\gamma+2\theta}{1-2\beta\gamma^2} r^2;$$
$$U_0 = U(x_0, r_0) \subseteq D,$$

where $r_0 \in (0, r]$ is the unique solution of equation

$$h(t) = (1 - 2\beta\gamma^2)\delta$$

on interval $(0, r]$.

Then show: sequence $\{x_n\}$ $(n \geq -1)$ is well defined, remains in $U(x_0, r_0)$ for all $n \geq -1$, and converges to a solution x^* of equation $F(x) = 0$. Moreover, the following estimates hold for all $n \geq -1$

$$\|x_{n+1} - x_n\| \leq t_n - t_{n+1},$$

and

$$\|x_n - x^*\| \leq t_n,$$

where,

$$t_{-1} = r_0 + \gamma, \quad t_0 = r_0,$$
$$\gamma_0 = \alpha + 3\beta r_0 + \beta\gamma, \quad \gamma_1 = 3\beta r_0^2 - 2\gamma_0 r_0 - \beta\gamma^2 + 1,$$

and for $n \geq 0$

$$t_{n+1} = \frac{\gamma_0 t_n - (t_n - t_{n-1})^2 \beta - 2\beta t_n^2}{\gamma_1 + 2\gamma_0 t_n - (t_n - t_{n-1})^2 - 3\beta t_n^2} \cdot t_n.$$

Furthermore if D is a convex set and

$$2\alpha(\gamma + 2r_0) < 1,$$

x^* is the unique solution of equation in $\overline{U}(x_0, r_0)$.

5.5.2. Let F be a nonlinear operator defined on an open convex subset D of a Banach space X with values in a Banach space Y and let $A(x) \in L(X, Y)$ $(x \in D)$. Assume:

- there exists $x_0 \in D$ such that $A(x_0)^{-1} \in L(Y, X)$;
- there exist nondecreasing, nonnegative functions a, b such that:

$$\left\| A(x_0)^{-1} [A(x) - A(x_0)] \right\| \leq a(\|x - x_0\|),$$
$$\left\| A(x_0)^{-1} [F(y) - F(x) - A(x)(y - x)] \right\| \leq b(\|x - y\|)\|x - y\|,$$
$$\text{for all } x, y \in D;$$

- there exist $\eta \geq 0$, $r_0 > \eta$ such that

$$\|A(x_0)^{-1} F(x_0)\| \leq \eta,$$
$$a(r) < 1,$$

and
$$d(r) < 1, \quad \text{for all } r \in (0, r_0],$$

where
$$c(r) = (1 - a(r))^{-1},$$

and
$$d(r) = c(r)b(r);$$

- r_0 is the minimum positive root of equation $h(r) = 0$ on $(0, r_0]$, where

$$h(r) = \frac{\eta}{1 - d(r)} - r;$$

- $\bar{U}(x_0, r_0) \subseteq D$.

Show: sequence $\{x_n\}$ $(n \geq 0)$ generated by Newton-like method

$$x_{n+1} = x_n - A(x_n)^{-1} F(x_n) \quad (n \geq 0)$$

is well defined, remains in $U(x_0, r_0)$ for all $n \geq 0$, and converges to a solution $x^* \in \bar{U}(x_0, r_0)$ of equation $F(x) = 0$.

5.5.3. (a) Let F be a Fréchet-differentiable operator defined on some closed convex subset D of a Banach space X with values in a Banach space Y; let $A(x) \in L(X, Y)$ $(x \in D)$. Assume: there exists $x_0 \in D$ such that $A(x_0) \in L(X, Y)$, $A(x_0)^{-1} \in L(Y, X)$, and

$$\left\| A(x_0)^{-1} [F'(y) - A(x)] \right\| < \varepsilon_0, \quad \text{for all } x, y \in U(x_0, \delta_0).$$

Then, show:
(1) for all $\varepsilon_1 > 0$ there exists $\delta_1 > 0$ such that

$$\left\| [A(x)^{-1} - A(x_0)^{-1}] A(x_0) \right\| < \varepsilon_1, \quad \text{for all } x \in U(x_0, \delta_1).$$

Set $\delta = \min\{\delta_0, \delta_1\}$ and $\varepsilon = \max\{\varepsilon_0, \varepsilon_1\}$.

(2) for $\varepsilon > 0$ there exist $\delta > 0$ as defined above such that

$$\left\| A(x_0)^{-1} \left[F'(y) - A(x) \right] \right\| < \varepsilon$$

and

$$\left\| \left[A(x)^{-1} - A(x_0)^{-1} \right] A(x_0) \right\| < \varepsilon,$$

for all $x, y \in U(x_0, \delta)$.

(b) Let operators F, A, point $x_0 \in D$, and parameters ε, δ be as in (1). Assume there exist $\eta \geq 0$, $c \geq 0$ such that

$$\left\| A(x_0)^{-1} F(x_0) \right\| \leq \eta,$$

$$(1 + \varepsilon)\varepsilon \leq c < 1,$$

$$\frac{\eta}{1 - c} \leq \delta,$$

and

$$\bar{U}(x_0, \delta) \subseteq D.$$

Show: sequence $\{x_n\}$ $(n \geq 0)$ generated by Newton-like method is well defined, remains in $U(x_0, \delta)$ for all $n \geq 0$, and converges to a solution $x^* \in \bar{U}(x_0, \delta)$ of equation $F(x) = 0$. Moreover, if linear operator

$$L = \int_0^1 F'(x + t(y - x)) dt$$

is invertible for all $x, y \in D$, then x^* is the unique solution of equation $F(x) = 0$ in $\bar{U}(x_0, \delta)$. Furthermore, the following estimates hold for all $n \geq 0$

$$\| x_{n+1} - x_n \| \leq c^n \| x_1 - x_0 \| \leq c^n \eta$$

and

$$\| x_n - x^* \| \leq \frac{c^n}{1 - c} \| x_1 - x_0 \|.$$

(c) Let $X = Y = \mathbb{R}$, $D \supseteq U(0, .3)$, $x_0 = 0$,

$$F(x) = \frac{x^2}{2} + x - .04.$$

Set $A(x) = F'(x)$ $(x \in D)$, $\delta_3 = \delta_4 = \varepsilon_3 = \varepsilon_4 = .3$. Then we obtain

$$c_3 = \frac{6}{7} < 1,$$

$$\frac{\eta}{1 - c_3} = .28 < \delta = \delta_3.$$

The conclusions of (b) hold and

$$x^* = .039230485 \in U(x_0, \delta).$$

(d) Let $x^* \in D$ be a simple zero of equation $F(x) = 0$. Assume:

$$\left\| A(x^*)^{-1} \left[F'(y) - A(x) \right] \right\| < \varepsilon_{11}, \quad \text{for all } x \in U(x^*, \delta_{11}),$$

$$\left\| \left[A(x)^{-1} - A(x^*)^{-1} \right] A(x^*) \right\| < \varepsilon_{12}, \quad \text{for all } x \in U(x^*, \delta_{12}).$$

Set

$$\delta_{13} = \min\{\delta_{11}, \delta_{12}\}, \quad c_8 = (1 + \varepsilon_{12})\varepsilon_{11}.$$

Further, assume:

$$0 < c_8 < 1$$
$$x_0 \in \bar{U}(x^*, \delta_{13}),$$

and

$$\bar{U}(x^*, \delta_{13}) \subseteq D.$$

Show: sequence, $\{x_n\}$ $(n \geq 0)$ generated by Newton-like method is well defined, remains in $U(x^*, \delta_{13})$ for all $n \geq 0$ and converges to x^* with

$$\|x_{n+1} - x^*\| \leq c_8 \|x_n - x^*\|, \quad \text{for all } n \geq 0.$$

5.5.4. Consider the equation $F(x) + G(x) = 0$ and the iteration

$$x_{n+1} = x_n - A(x_n)^{-1}(F(x_n) + G(x_n)) \quad (n \geq 0).$$

Assume:
(a) $\left\| A(x_n)^{-1}(A(x) - A(x_n)) \right\| \leq v_n(r) + b_n,$

$$\left\| A(x_n)^{-1}\left(F'(x + t(y - x)) - A(x) \right) \right\|$$
$$\leq w_n(r + t\|y - x\|) - v_n(r) + c_n$$

and

$$\left\| A(x_n)^{-1}(G(x) - G(y)) \right\| \leq e_n(r) \|x - y\|$$

for all $x_n, x, y \in \bar{U}(x_0, r) \subseteq \bar{U}(x_0, R)$, $t \in [0, 1]$, where $w_n(r + t) - v_n(r) t \geq 0$ and $e_n(r)$ $(n \geq 0)$ are nondecreasing, nonnegative functions with $w_n(0) = v_n(0) = e_n(0) = 0$ $(n \geq 0)$, $v_n(r)$ are differentiable, $v_n'(r) > 0$ $(n \geq 0)$ for all $r \in [0, R]$, and the constants b_n, c_n satisfy $b_n \geq 0, c_n \geq 0$ and $b_n + c_n < 1$ for all $n \geq 0$. Introduce for all $n, i \geq 0$, $a_n = \|A(x_n)^{-1}(F(x_n) + G(x_n))\|$, $\varphi_{n,i}(r) = a_i - r + c_{n,i} \int_0^r w_n(t)\, dt$, $z_n(r) = 1 - v_n(r) - b_n$, $\psi_n(r) = c_{n,i} \int_0^r e_n(t)\, dt$, $c_{n,i} = z_n(r_i)^{-1}$, $h_{n,i}(r) = \varphi_{n,i}(r) + \psi_{n,i}(r)$, $r_n = \|x_n - x_0\|$, $a_n = \|x_{n+1} - x_n\|$, the equations

$$r = a_n + c_{0,n}\left(\int_0^r (w_0(r_n + t) + e_n(r_n + t))\, dt + (b_n + c_n - 1)r \right)$$

$$(5.5.1)$$

$$r = a_n + c_{n,n} \left(\int_0^r \left(w_n \left(r_n + t \right) + e_n \left(r_n + t \right) \right) dt + \left(b_n - c_n - 1 \right) r \right)$$

$$\text{(5.5.2)}$$

$$a_n = r + c_{0,n} \left(\int_0^r \left(w_0 \left(t_n + t \right) + e_n \left(r_n + t \right) \right) dt + \left(b_n - c_n - 1 \right) r \right)$$

$$\text{(5.5.3)}$$

$$a_n = r + c_{n,n} \left(\int_0^r \left(w_n \left(r + t \right) + e_n \left(r_n + t \right) \right) dt + \left(b_n - c_n - 1 \right) r \right)$$

$$\text{(5.5.4)}$$

and the scalar iterations

$$s_{0,n} = s_{n,n}^0 = 0, \ s_{k+1,n}^0 = s_{k,n}^0 + \frac{h_{0,n} \left(s_{k,n}^0 + r_n \right)}{c_{0,n} z_0 \left(s_{k,n}^0 + r_n \right)} \quad (k \geq 0)$$

$$s_{k+1,n} = s_{k,n} + \frac{h_{n,n} \left(s_{k,n} + r_n \right)}{c_{n,n} z_{n,n} \left(s_{k,n} + r_n \right)} \quad (k \geq n);$$

(b) The function $h_{0,0} \left(r \right)$ has a unique zero s_0^* in the interval $[0, R]$ and $h_{0,0} \left(R \right) \leq 0$;

(c) The following estimates are true:

$$\frac{h_{n,n} \left(r + r_n \right)}{c_{n,n} z_{n,n} \left(r + r_n \right)} \leq \frac{h_{0,n} \left(r + r_n \right)}{c_{0,n} z_n \left(r + r_n \right)}$$

for all $r \in [0, R - r_n]$ and for each fixed $n \geq 0$.

Then show:

(i) The scalar iterations $\left\{ s_{k+1,n}^0 \right\}$ and $\left\{ s_{k+1,n} \right\}$ for $k \geq 0$ are monotonically increasing and converge to s_n^* and s_n^{**} for each fixed $n \geq 0$, which are the unique solutions of equations (5.5.1) and (5.5.2) in $[0, R - s_n]$, respectively, with $s_n^{**} \leq s_n^*$ $(n \geq 0)$.

(ii) The iteration $\{x_n\}$ is well defined, remains in $\bar{U} \left(x_0, s^* \right)$ for all $n \geq 0$, and converges to a solution x^* of the equation $F \left(x \right) + G \left(x \right) = 0$, which is unique in $\bar{U} \left(x_0, R \right)$.

(iii) The following error estimates are true:

$$\| x_{n+1} - x_n \| \leq s_{n+1,n+1} - s_{n,n} \leq s_{n+1,n+1}^0 - s_{n,n}^0,$$

$$\| x^* - x_n \| \leq s_n^{**} - s_{n,n} \leq s_n^* - s_{n,n}^0 \leq s_0^* - s_{n,0}^0$$

$$\| x^* - x_n \| \geq I_n^*, \quad \| x^* - x_n \| \geq I_n^{**}$$

and

$$I_n^{**} \leq I_n^* \quad (n \geq 0)$$

where I_n^* and I_n^{**} are the solutions of the equations (5.5.3) and (5.5.4), respectively, for all $n \geq 0$.

The above approach shows how to improve upon the results given in [58] by Yamamoto and Chen.

(iv) Define the sequence $\{s_n^1\}$ $(n \geq 0)$ by

$$s_0^1 = 0, \quad s_{n+1}^1 + \frac{h_{n,n} \left(s_n^1 + r_n\right)}{c_{n,n}} \quad (n \geq 0).$$

Then under the hypotheses (5.5.1)–(5.5.3) above show that:

$$\|x_{n+1} - x_n\| \leq s_{n+1}^1 - s_n^1 \leq s_{n+1,n+1} - s_{n,n} \leq s_{n+1,n+1}^0 - s_{n,n}^0$$

and

$$\left\|x^* - x_n\right\| \leq t^* - s_n^1 \leq s_n^{**} - s_{n,n} \leq s_n^* - s_{n,n}^0 \leq s_0^* - s_{n,0}^0 \quad (n \geq 0),$$

where

$$t^* = \lim_{n \to \infty} s_n^1.$$

5.5.5. Consider the Newton-like method. Let $A: D \to L(X, Y)$, $x_0 \in D$, $M_{-1} \in L(X, Y)$, $X \subset Y$, $L_{-1} \in L(X, X)$. For $n \geq 0$ choose $N_n \in L(X, X)$ and define $M_n = M_{n-1} N_n + A(x_n) L_{n-1}$, $L_n = L_{n-1} + L_{n-1} N_n$, $x_{n+1} = x_n + L_n(y_n)$, y_n being a solution of $M_n(y_n) = -[F(x_n) + z_n]$ for a suitable $z_n \in y$.
Assume:
(a) F is Fréchet-differentiable on D.
(b) There exist nonnegative numbers α, α_0 and nondecreasing functions w, $w_0: \mathbb{R}^+ \to \mathbb{R}^+$ with $w(0) = w_0(0) = 0$ such that

$$\|F(x_0)\| \leq \alpha_0, \quad \|R_0(y_0)\| \leq \alpha,$$
$$\|A(x) - A(x_0)\| \leq w_0(\|x - x_0\|)$$

and

$$\left\|F'(x + t(y - x)) - A(x)\right\| \leq w(\|x - x_0\| + t\|x - y\|)$$

for all $x, y \in \bar{U}(x_0, R)$ and $t \in [0, 1]$.
(c) Let M_{-1} and L_{-1} be such that M_{-1} is invertible,

$$\left\|M_{-1}^{-1}\right\| \leq \beta, \quad \|L_{-1}\| \leq \gamma \text{ and } \|M_{-1} - A(x_0) L_{-1}\| \leq \delta.$$

(d) There exist nonnegative sequence $\{a_n\}$, $\{\bar{a}_n\}$, $\{b_n\}$ and $\{c_n\}$ such that for all $n \geq 0$

$$\|N_n\| \leq a_n,$$
$$\|I + N_n\| \leq \bar{a}_n,$$
$$\left\|M_{-1}^{-1}\right\| \cdot \|M_{-1} - M_n\| \leq b_n < 1$$

and

$$\|z_n\| \leq c_n \|F(x_n)\|.$$

(e) The scalar sequence $\{t_n\}$ $(n \geq 0)$ given by

$$t_{n+1} = t_{n+1} + e_{n+1}d_{n+1}(1 + c_{n+1})$$

$$\left[I_n + \sum_{i=1}^{n} h_i w(t_i)(t_i - t_{i-1}) + w(t_{n+1})(t_{n+1} - t_n) \right]$$

$(n \geq 0)$, $t_0 = 0$, $t_1 = \alpha$ is bounded above by a t_0^* with $0 < t_0^* \leq R$, where

$$e_0 = \gamma \bar{a}_0, \quad e_n = I_{n-1}\bar{a}_n \ (n \geq 1), \quad d_n = \frac{\beta}{1-d_n} \ (n \geq 0)$$

$$I_n = \varepsilon_n \varepsilon_{n-1} \ldots \varepsilon_0 \alpha_0 \ (n \geq 0) \quad \varepsilon_n = p_n d_n (1 + c_n) + c_n,$$

$$p_n = q_{n-1}a_n \ (n \geq 1), \quad p_0 = \delta a_0,$$

$$q_n = p_n + w_0(t_{n+1})e_n \ (n \geq 1)$$

and

$$h_i = \prod_{m=i}^{n} \varepsilon_m \quad (i \leq n).$$

(f) The following estimate is true $\varepsilon_n \leq \varepsilon < 1$ $(n \geq 0)$.
Then show:
  (i) The scalar sequence $\{t_n\}$ $(n \geq 0)$ is nondecreasing and converges to a t^* with $0 < t^* \leq t_0^*$ as $n \to \infty$.
  (ii) The Newton-like method is well defined, remains in $\bar{U}(x_0, t^*)$, and converges to a solution x^* of equation $F(x) = 0$.
  (iii) The following estimates are true:

$$\|x_{n+1} - x_n\| \leq t_{n+1} - t_n$$

and

$$\|x_n - x^*\| \leq t^* - t_n \quad (n \geq 0).$$

5.5.6. (a) Let $F: D \subseteq X \to Y$ be a Fréchet-differentiable operator and $A(x) \in L(X, Y)$ $(x \in D)$. Assume there exists a point $x_0 \in D$, $\eta \geq 0$ and nonnegative continuous functions a, b, c such that

$$A(x_0)^{-1} \in L(Y, X),$$

$$\|A(x_0)^{-1}F(x_0)\| \leq \eta,$$

$$\|A(x_0)^{-1}[F'(x) - F'(x_0)]\| \leq a(\|x - x_0\|),$$

$$\|A(x_0)^{-1}[F'(x_0) - A(x)]\| \leq b(\|x - x_0\|),$$

$$\|A(x_0)^{-1}[A(x) - A(x_0)]\| \leq c(\|x - x_0\|)$$

for all $x \in D$;
equation

$$\int_0^1 a[(1-t)r]r \, dt + [b(r) + c(r) - 1]r + \eta = 0$$

has nonnegative solutions. Denote by r_0 the smallest.
Point r_0 satisfies

$$a(r_0) + b(r_0) + c(r_0) < 1,$$

and

$$\bar{U}(x_0, r_0) =\subseteq D.$$

Then show sequence $\{x_n\}$ ($n \geq 0$) generated by Newton-like method is well defined, remains in $\bar{U}(x_0, r_0)$ for all $n \geq 0$, and converges to a unique solution $x^* \in \bar{U}(x_0, r_0)$ of equation $F(x) = 0$. Moreover, the following estimates hold for all $n \geq 0$

$$\|x_{n+2} - x_{n+1}\| \leq q\|x_{n+1} - x_n\|$$

and

$$\|x_n - x^*\| \leq \tfrac{\eta}{1-q}q^{n+1}$$

where,

$$q = \tfrac{a(r_0)+b(r_0)}{1-c(r_0)} .$$

Furthermore, x^* is unique in $U(x_0, R)$ for $R > t^*$ and

$$U(x_0, R) \subseteq D$$

if

$$\int_0^1 a\,[(1 - t)r_0 + t R]\,dt + b(0) \leq 1.$$

(b) Let $F: D \subseteq X \subseteq Y$ be a Fréchet-differentiable operator and $A(x) \in L(X, Y)$. Assume: there exist a simple zero x^* of F and nonnegative continuous functions α, β, γ such that

$$A(x^*)^{-1} \in L(Y, X),$$

$$\|A(x^*)^{-1}\left[F'(x) - F'(x^*)\right]\| \leq \alpha(\|x - x^*\|),$$
$$\|A(x^*)^{-1}\left[F'(x^*) - A(x)\right]\| \leq \beta(\|x - x^*\|),$$
$$\|A(x^*)^{-1}\left[A(x) - A(x^*)\right]\| \leq \gamma(\|x - x^*\|)$$

for all $x \in D$;
equation

$$\int_0^1 \alpha\,[(1 - t)r]\,dt + \beta(r) + \gamma(r) = 1$$

has nonnegative solutions. Denote by r^* the smallest; and

$$\bar{U}(x^*, r^*) \subseteq D.$$

Show: Under the above stated hypotheses: sequence $\{x_n\}$ ($n \geq 0$) generated by the Newton-like method is well defined, remains in $\bar{U}(x^*, r^*)$ for all $n \geq 0$, and converges to x^*, provided $x_0 \in U(x^*, r^*)$.

Moreover the following estimates hold for all $n \geq 0$:

$$\|x_{n+1} - x^*\| \leq \delta_n \|x_n - x^*\|,$$

where

$$\delta_n = \frac{\int_0^1 \alpha[(1-t)\|x_n - x^*\|]dt + \beta(\|x_n - x^*\|)}{1 - \gamma(\|x_n - x^*\|)}.$$

5.5.7. (a) Assume:
there exist parameters $K \geq 0$, $M \geq 0$, $L \geq 0$, $\ell \geq 0$, $\mu \geq 0$, $\eta \geq 0$, $\lambda_1, \lambda_2, \lambda_3 \in [0, 1]$, $\delta \in [0, 2)$ such that:

$$h_q = K\eta^{\lambda_1} + (1 + \lambda_1)\left[M\left(\frac{\eta}{1-q}\right)^{\lambda_2} + \mu\right] + \left[\ell + L\left(\frac{\eta}{1-q}\right)^{\lambda_3}\right]\delta \leq \delta,$$

and

$$\ell + L\left(\frac{\eta}{1-q}\right)^{\lambda_3} \leq 1,$$

where,

$$q = \frac{\delta}{1+\lambda_1}.$$

Then, show: iteration $\{t_n\}$ $(n \geq 0)$ given by

$$t_0 = 0, \quad t_1 = \eta, \quad t_{n+2} = t_{n+1} + \frac{K(t_{n+1} - t_n)^{\lambda_1} + (1+\lambda_1)\left[Mt_n^{\lambda_2} + \mu\right]}{(1+\lambda_1)\left[1 - \ell - Lt_{n+1}^{\lambda_3}\right]} \cdot (t_{n+1} - t_n) \quad (n \geq 0)$$

is nondecreasing, bounded above by

$$t^{**} = \frac{\eta}{1-q},$$

and converges to some t^* such that

$$0 \leq t^* \leq t^{**}.$$

Moreover, the following estimates hold for all $n \geq 0$

$$0 \leq t_{n+2} - t_{n+1} \leq q(t_{n+1} - t_n) \leq q^{n+1}\eta.$$

(b) Let $\lambda_1 = \lambda_2 = \lambda_3 = 1$. Assume:
there exist parameters $K \geq 0$, $M \geq 0$, $L \geq 0$, $\ell \geq 0$, $\mu \geq 0$, $\eta \geq 0$, $\delta \in [0, 1]$ such that:

$$h_\delta = \left(K + L\delta + \frac{4M}{2-\delta}\right)\eta + \delta\ell + 2\mu \leq \delta,$$

$$\ell + \frac{2L\eta}{2-\delta} \leq 1,$$

$$L \leq K,$$

and

$$\ell + 2\mu < 1,$$

then, show: iteration $\{t_n\}$ ($n \geq 0$) is nondecreasing, bounded above

$$t^{**} = \frac{2\eta}{2 - \delta}$$

and converges to some t^* such that

$$0 \leq t^* \leq t^{**}.$$

Moreover, the following estimates hold for all $n \geq 0$

$$0 \leq t_{n+2} - t_{n+1} \leq \frac{\delta}{2}(t_{n+1} - t_n) \leq \left(\frac{\delta}{2}\right)^{n+1} \eta.$$

(c) Let $F: D \subseteq X \to Y$ be a Fréchet-differentiable operator. Assume:
(1) there exist an approximation $A(x) \in L(X, Y)$ of $F'(x)$, an open convex
    subset D_0 of D, $x_0 \in D_0$, parameters $\eta \geq 0$, $K \geq 0$, $M \geq 0$, $L \geq 0$, $\mu \geq 0$,
    $\ell \geq 0$, $\lambda_1 \in [0, 1]$, $\lambda_2 \in [0, 1]$, $\lambda_3 \in [0, 1]$ such that:

$$A(x_0)^{-1} \in L(Y, X),$$

$$\|A(x_0)^{-1} F(x_0)\| \leq \eta,$$

$$\|A(x_0)^{-1}[F'(x) - F'(y)]\| \leq K\|x - y\|^{\lambda_1},$$

$$\|A(x_0)^{-1}[F'(x) - A(x)]\| \leq M\|x - x_0\|^{\lambda_2} + \mu,$$

and

$$\|A(x_0)^{-1}[A(x) - A(x_0)]\| \leq L\|x - x_0\|^{\lambda_3} + \ell \quad \text{for all } x, y \in D_0;$$

(2) hypotheses of (a) or (b) hold;
(3)
$$\bar{U}(x_0, t^*) \subseteq D_0.$$

Then, show sequence $\{x_n\}$ ($n \geq 0$) generated by Newton-like method is well
defined, remains in $\bar{U}(x_0, t^*)$ for all $n \geq 0$, and converges to a solution $x^* \in$
$\bar{U}(x_0, t^*)$ of equation $F(x) = 0$.
Moreover, the following estimates hold for all $n \geq 0$:

$$\|x_{n+1} - x_n\| \leq t_{n+1} - t_n$$

and

$$\|x_n - x^*\| \leq t^* - t_n.$$

Furthermore the solution x^* is unique in $\bar{U}(x_0, t^*)$ if

$$\frac{1}{1 - \ell - L(t^*)^{\lambda_3}} \left[\frac{K}{1 + \lambda_1}(t^*)^{1+\lambda_1} + M(t^*)^{\lambda_2} + \mu\right] < 1,$$

or in $U(x_0, R_0)$ if $R_0 > t^*$, $U(x_0, R_0) \subseteq D_0$, and

$$\frac{1}{1 - \ell - L(t^*)^{\lambda_3}} \left[\frac{K}{1 + \lambda_1}(R + t^*)^{1+\lambda_1} + M(t^*)^{\lambda_2} + \mu\right] \leq 1.$$

(d) Let $F: D \subseteq X \to Y$ be a Fréchet-differentiable operator. Assume:

(a) there exist an approximation $A(x) \in L(X, Y)$ of $F'(x)$, a simple solution $x^* \in D$ of equation $F(x) = 0$, a bounded inverse $A(x^*)$ and parameters $\bar{K}$, $\bar{L}, \bar{M}, \bar{\mu}, \bar{\ell} \geq 0$, $\lambda_4, \lambda_5, \lambda_6 \in [0, 1]$ such that:

$$\|A(x^*)^{-1} [F'(x) - F'(y)]\| \leq \bar{K} \|x - y\|^{\lambda_4},$$
$$\|A(x^*)^{-1} [F'(x) - A(x)]\| \leq \bar{M} \|x - x^*\|^{\lambda_5} + \bar{\mu},$$

and

$$\|A(x^*)^{-1} [A(x) - A(x^*)]\| \leq \bar{L} \|x - x^*\|^{\lambda_6} + \bar{\ell}$$

for all $x, y \in D$;

(b) equation

$$\frac{\bar{K}}{1+\lambda_4} r^{\lambda_4} + \bar{L} r^{\lambda_6} + \bar{M} r^{\lambda_5} + \bar{\mu} + \bar{\ell} - 1 = 0$$

has a minimal positive zero r_0, which also satisfies:

$$\bar{L} r_0^{\lambda_6} + \bar{\ell} < 1$$

and

$$U(x^*, r_0) \subseteq D.$$

Then, show: sequence $\{x_n\}$ $(n \geq 0)$ generated by Newton-like method is well defined, remains in $U(x^*, r_0)$ for all $n \geq 0$, and converges to x^* provided that $x_0 \in U(x^*, r_0)$. Moreover, the following estimates hold for all $n \geq 0$:

$$\|x_{n+1} - x^*\| \leq$$
$$\leq \frac{1}{1 - \bar{L}\|x_n - x^*\|^{\lambda_6} - \bar{\ell}} \left[\frac{\bar{K}}{1+\lambda_4} \|x_n - x^*\|^{\lambda_4} + \bar{M}\|x_n - x^*\|^{\lambda_5} + \bar{\mu} \right] \|x_n - x^*\|.$$

5.5.8. Let F be a nonlinear operator defined on an open convex subset D of a Banach space X with values in a Banach space Y and let $A(x) \in L(X, Y)$ $(x \in D)$. Assume:

(a) there exists $x_0 \in D$ such that $A(x_0)^{-1} \in L(Y, X)$;

(b) there exist nondecreasing, nonnegative functions a, b such that:

$$\|A(x_0)^{-1} [A(x) - A(x_0)]\| \leq a(\|x - x_0\|),$$
$$\|A(x_0)^{-1} [F(y) - F(x) - A(x)(y - x)]\| \leq b(\|x - y\|)\|x - y\|$$

for all $x, y \in D$;

(c) there exist $\eta \geq 0$, $r_0 > \eta$ such that

$$\|A(x_0)^{-1} F(x_0)\| \leq \eta,$$
$$a(r) < 1,$$

and

$$d(r) < 1 \quad \text{for all } r \in (0, r_0],$$

where

$$c(r) = (1 - a(r))^{-1},$$

and

$$d(r) = c^2(r)b(r);$$

(d) r_0 is the minimum positive root of equation $h(r) = 0$ on $(0, r_0]$ where,

$$h(r) = \frac{\eta}{1-d(r)} - r.$$

(e) $\bar{U}(x_0, r_0) \subseteq D$.

Then show: sequence $\{x_n\}$ $(n \geq 0)$ generated by Newton-like method is well defined, remains in $U(x_0, r_0)$ for all $n \geq 0$, and converges to a solution $x^* \in \bar{U}(x_0, r_0)$ of equation $F(x) = 0$.

5.5.9. (a) Let $F : U(z, R) \subseteq X \to Y$ be a Fréchet-differentiable operator for some $z \in X$, $R > 0$, and $A(x) \in L(X, Y)$. Assume: $A(z)^{-1} \in L(Y, X)$ and for any $x, y \in \bar{U}(z, r) \subseteq \bar{U}(z, R)$

$$\|A(z)^{-1}[A(x) - A(x_0)]\| \leq w_0(\|x - x_0\|) + a,$$

$$\|A(z)^{-1}[F'(x + t(y - x)) - A(x)]\| \leq w(\|x - z\| + t\|x - y\|)$$
$$- w_1(\|x - z\|) + b, \; t \in [0, 1],$$

$$\|A(z)^{-1}[G(x) - G(y)]\| \leq w_2(r)\|x - y\|,$$

$$0 < \|A(z)^{-1}[F(z) + G(z)]\| \leq \eta,$$

where $w(r+t) - w_1(r)$ $(t \geq 0)$, $w_1(r)$ and $w_2(r)$ are nondecreasing, nonnegative functions with $w(0) = w_0(0) = w_1(0) = w_2(0) = 0$, w_0 is differentiable, $w_0'(r) > 0$, $r \in [0, R]$,

$$w_0(r) \leq w_1(r) \quad r \in [0, R] \tag{5.5.5}$$

and parameters a, b satisfy

$$a \geq 0, \quad b \geq 0, \quad a + b < 1.$$

Define functions $\varphi_1, \varphi_2, \varphi$ by

$$\varphi_1(r) = \eta - r + \int_0^r w(t)dt,$$

$$\varphi_2(r) = \int_0^r w_2(t)dt,$$

$$\varphi(r) = \varphi_1(r) + \varphi_2(r) + (a + b)r,$$

iteration $\{r_n\}$ $(n \geq 0)$ by

$$r_0 \in [0, R], \quad r_{n+1} = r_n + \frac{\varphi(r_n) - \varphi_0}{1 - a - w_0(r_n)} \quad (n \geq 0),$$

where φ_0 is the minimal value of φ on $[0, R]$;

(b)

$$\varphi(R) \leq 0.$$

Then show iteration $\{x_n\}$ $(n \geq 0)$ generated by Newton-like method is well defined, remains in $\bar{U}(z, r^*)$ for any

$$x_0 \in D(r^*) = \bigcup_{r \in [0, r^*)} \left\{ x \in \bar{U}(z, r) \mid \|A(x)^{-1}(F(x) + G(x))\| \leq \frac{\varphi(r) - \varphi_0}{1 - a - w_0(r)} \right\},$$

and converges to a solution $x^* \in \bar{U}(z, r_0^*)$, which is unique in

$$\tilde{U} = \begin{cases} \bar{U}(z, R) & \text{if } \varphi(R) < 0 \text{ or } \varphi(R) = 0 \text{ and } r_0^* = R, \\ U(z, R) & \text{if } \varphi(R) = 0 \text{ and } r_0^* < R. \end{cases}$$

where r^* is the minimal point, r_0^* is the unique zero on $(0, r^*]$, and

$$r^* = \lim_{n \to \infty} r_n.$$

Moreover, sequence $\{r_n\}$ $(n \geq 0)$ is monotonically increasing and converges to r^*. Furthermore, the following estimates hold for all $n \geq 0$

$$\|x_{n+1} - x_n\| \leq r_{n+1} - r_n,$$
$$\|x_n - x^*\| \leq r^* - r_n,$$

provided that $r_0 \in R_{x_0}$, where for $x \in D(r^*)$

$$R_x = \left\{ r \in [0, r^*) : \|A(x)^{-1} [F(x) + G(x)]\| \leq \frac{\varphi(r) - \varphi_0}{1 - a - w_0(r)}, \|x - z\| \leq r \right\},$$

and

$$\bar{U}\left(z, \frac{|\varphi_0|}{2 - a}\right) \subseteq D(r^*).$$

In the next result, we show how to improve on the error bounds.

(c) Under the hypotheses of (a), show the conclusions hold with $\{r_n\}$ and $D(r^*)$ replaced by $\{t_n\}$ $(n \geq 0)$, $D(t^*)$ given by

$$t_0 = r_0, \quad t_1 = r_1,$$

$$t_{n+1} = t_n + \frac{\int_0^{t_{n-1}} [w(t_{n-1} + t(t_n - t_{n-1})) - w_1(t_{n-1})] dt (t_n - t_{n-1}) + b(t_n - t_{n-1}) + \int_{t_{n-1}}^{t_n} w_2(t) dt}{1 - a - w_0(t_n)}$$

$$(n \geq 1),$$

$$t^* = \lim_{n \to \infty} t_n.$$

Moreover iteration $\{t_n\}$ is monotonically increasing and converges to t^*. Furthermore, the following hold for all $n \geq 1$:

$$t_{n+1} - t_n \leq r_{n+1} - r_n, \tag{5.5.6}$$
$$t_n \leq r_n, \tag{5.5.7}$$
$$t^* - t_n \leq r^* - r_n, \tag{5.5.8}$$

and

$$t^* \leq r^*.$$

If strict inequality holds in (5.5.5), so it does in (5.5.6)–(5.5.8).
If

$$w_0(r) = w_1(r) \quad r \in [0, R]$$

our resolving reduces to Theorem 1 in [58]. If (5.5.5) holds, then our error bounds are at least as fine as the ones given by Chen and Yamamoto [58] (under the same hypotheses and a more general condition). Moreover according to our error bounds once finer.

5.5.10. (a) Let $F: D \subseteq X \to Y$ be differentiable. Assume:
There exist functions $f_1: [0, 1] \times [0, \infty)^2 \to [0, \infty)$, $f_2, f_3: [0, \infty) \to [0, \infty)$, nondecreasing on $[0, \infty)^2$, $[0, \infty)$, $[0, \infty)$ such that

$$\left\| A(x_0)^{-1} \left[F'(x + t(y - x)) - F'(x) \right] \right\| \leq$$

$$\leq f_1(t, \|x - y\|, \|x - x_0\|, \|y - x_0\|),$$

$$\left\| A(x_0)^{-1} \left[F'(x) - A(x) \right] \right\| \leq f_2(\|x - x_0\|),$$

$$\left\| A(x_0)^{-1} \left[A(x) - A(x_0) \right] \right\| \leq f_3(\|x - x_0\|),$$

hold for all $t \in [0, 1]$ and $x, y \in D$;
For $\left\| A(x_0)^{-1} F(x_0) \right\| \leq \eta$, equation

$$\eta + b_0 \eta + \frac{b_0 b_1 \eta}{1 - b(r)} = r$$

has nonnegative solutions, and denote by r_0 the smallest one. In addition, r_0 satisfies:

$$\bar{U}(x_0, r_0) \subseteq D,$$

and

$$\int_0^1 f_1(t, b_1 b_0 r_0, r_0, r_0) \, dt + f_2(r_0) + f_3(r_0) < 1,$$

where

$$b_0 = \frac{\int_0^1 f_1(t, \eta, 0, \eta) dt + f_2(0)}{1 - f_3(\eta)},$$

$$b_1 = \frac{\int_0^1 f_1(t, b_0\eta, \eta, \eta + b_0\eta) dt + f_2(\eta)}{1 - f_3(\eta + b_0\eta)},$$

and

$$b = b(r) = \frac{\int_0^1 f_1(t, b_1 b_0\eta, r, r) dt + f_2(r)}{1 - f_3(r)},$$

Then, show iteration $\{x_n\}$ $(n \geq 0)$ generated by Newton-like method is well defined, remains in $\bar{U}(x_0, r_0)$ for all $n \geq 0$, and converges to a solution $x^* \in \bar{U}(x_0, r_0)$ of equation. Moreover, the following estimates hold

$$\|x_2 - x_1\| \leq b_0 \eta$$
$$\|x_3 - x_2\| \leq b_1 \|x_2 - x_1\|.$$
$$\|x_{n+1} - x^*\| \leq b_2 \|x_n - x_{n-1}\|, \quad (n \geq 3)$$

and

$$\|x^* - x_n\| \leq \frac{b_0 b_1 b_2^{n-2} \eta}{1 - b_2}, \quad (n \geq 3),$$

$$\|x^* - x_n\| \leq \frac{b_2^n}{1 - b_2}, \quad (n \geq 0)$$

where $b_2 = b(r_0)$.
Furthermore, if r_0 satisfies

$$\int_0^1 f_1(t, 2r_0, r_0, r_0)\, dt + f_2(r_0) + f_3(r_0) < 1,$$

x^* is the unique solution of equation $F(x) = 0$ in $\overline{U}(x_0, r_0)$.
Finally, if there exists a minimum nonnegative number R satisfying equation

$$\int_0^1 f_1(t, r + r_0, r_0, r)\, dt + f_2(r_0) + f_3(r_0) = 1,$$

such that $U(x_0, R) \subseteq D$, then the solution x^* is unique in $U(x_0, R)$.
(b) There exist a simple zero x^* of F and continuous functions $f_4 : [0, 1] \times [0, \infty) \to [0, \infty)$, $f_5, f_6 : [0, \infty) \to [0, \infty)$, nondecreasing on $[0, \infty)$ such that

$$\left\| A(x^*)^{-1} \left[F'(x + t(x^* - x)) - F'(x) \right] \right\| \leq f_4(t, \|x^* - x\|),$$

$$\left\| A(x^*)^{-1} \left[F'(x) - A(x) \right] \right\| \leq f_5(\|x^* - x\|),$$

and

$$\left\| A(x^*)^{-1} \left[A(x) - A(x^*) \right] \right\| \leq f_6(\|x^* - x\|),$$

hold for all $t \in [0, 1]$ and $x \in D$;
Equation

$$\int_0^1 f_4(t, r)\, dt + f_5(r) + f_6(r) = 1$$

has a minimum positive zero r^*.

$$\overline{U}(x^*, r^*) \subseteq D.$$

Then, show iteration $\{x_n\}$ $(n \geq 0)$ generated by Newton-like sequence is well defined, remains in $U(x^*, r^*)$ for all $n \geq 0$, and converges to x^* provided that $x_0 \in U(x^*, r^*)$. Moreover, the following estimates hold for all $n \geq 0$

$$\|x_{n+1} - x^*\| \leq \frac{\int_0^1 f_4(t, \|x_n - x^*\|)dt + f_5(\|x_n - x^*\|)}{1 - f_6(\|x_n - x^*\|)} \|x_n - x^*\|$$

$$\leq \gamma \|x_n - x^*\|,$$

where

$$\gamma = \frac{\int_0^1 f_4(t, \|x_0 - x^*\|)dt + f_5(\|x_0 - x^*\|)}{1 - f_6(\|x_0 - x^*\|)} < 1.$$

(c) Let $X = Y = \mathbb{R}$, $D = (-1, 1)$, $x^* = 0$ and define function F on D by

$$F(x) = x + \frac{x^{p+1}}{p+1}, \quad p > 1.$$

For the case $A = F'$ show

$$r_R = \frac{2}{3p} < r^* = \frac{2}{2+p},$$

where r_R stands for Rheinboldt's radius (see Section 5.1) where r_R is the convergence radius given by Rheinboldt's [175].

(d) Let $X = Y = C[0, 1]$, the space of continuous functions defined on $[0, 1]$ equipped with the max-norm. Let $D = \{\phi \in C[0, 1]; \|\phi\| \leq 1\}$ and F defined on D by

$$F(\phi)(x) = \phi(x) - 5 \int_0^1 xt\phi(t)^3 \, dt$$

with a solution $\phi^*(x) = 0$ for all $x \in [0, 1]$.
In this case, for each $\phi \in D$, $F'(\phi)$ is a linear operator defined on D by the following expression:

$$F'(\phi)[v](x) = v(x) - 15 \int_0^1 xt\phi(t)^2 v(t) \, dt, \quad v \in D.$$

In this case and by considering again $A = F'$,

$$r_R = \frac{2}{45} < r^* = \frac{1}{15}.$$

5.5.11. Consider inexact Newton methods for solving equation $F(x) = 0$, $F: D \subseteq \mathbb{R}^N \to \mathbb{R}^N$ of the general form

$$x_{n+1} = x_n + s_n \quad (n \geq 0),$$

where $s_n \in \mathbb{R}^N$ satisfies the equation

$$F'(x_n) s_n = -F(x_n) + r_n \quad (n \geq 0)$$

for some sequence $\{r_n\} \subseteq \mathbb{R}^N$. Let $F \in F_\lambda(\sigma) \equiv \{F: U(x^*, \sigma) \to \mathbb{R}^N$ with $U(x^*, \sigma) \subseteq D$, where $F(x^*) = 0$, F is Fréchet-differentiable on $U(x^*, \sigma)$ and $F'(x)^{-1}$ exists for all $x \in U(x^*, \sigma)$, F' is continous on $U(x^*, \sigma)$, and there exists $\mu_\lambda \geq 0$ such that for all $y, z \in U(x^*, \sigma)$

$$\|F'(x^*)^{-1}[F'(y) - F'(z)]\| \leq \mu_\lambda \|y - z\|^\lambda\}.$$

Suppose the inexact Newton method $\{x_n\}$ $(n \geq 0)$ satisfy:

$$\frac{\left\|s_n + F'(x_n)^{-1} F(x_n)\right\|}{\left\|F'(x_n)^{-1} F(x_n)\right\|} \equiv \frac{\left\|F'(x_n)^{-1} r_n\right\|}{\left\|F'(x_n)^{-1} F(x_n)\right\|} \leq v_n \ (n \geq 0)$$

for some $\{v_n\} \subseteq \mathbb{R}$ $(n \geq 0)$. If $x_i \in U(x^*, \sigma)$, then show:

(a) $\|x_{n+1} - x^*\| \leq w_n \|x_n - x^*\|$,

$$w_n \equiv v_n + \frac{(1+v_n)\mu_\lambda \|x_n - x^*\|^\lambda}{(1-\lambda)\left[1 - \mu_\lambda \|x_n - x^*\|^\lambda\right]} \ (n \geq 0);$$

(b) if $v_n \leq v < 1$ $(n \geq 0)$, then there exists $w \in (0, 1)$ given by

$$w \equiv v + \frac{(1+v)\mu_\lambda \|x_0 - x^*\|^\lambda}{(1+\lambda)\left[1 - \mu_\lambda \|x_0 - x^*\|^\lambda\right]}$$

such that $\|x_{n+1} - x^*\| \leq w \|x_n - x^*\|$ and $\lim_{n \to \infty} x_n = x^*$;

(c) if $\{x_n\}$ $(n \geq 0)$ converges to x^* and $\lim_{n \to \infty} v_n = 0$, then $\{x_n\}$ converges Q-superlinearly;

(d) if $\{x_n\}$ $(n \geq 0)$ converges to x^* and $\lim_{n \to \infty} v_n^{(1+\lambda)^{-1}} < 1$, then $\{x_n\}$ converges with R-order at least $1 + \lambda$.

5.5.12 Consider the two-point method [86]:

$$y_n = x_n - F'(x_n)^{-1} F(x_n),$$
$$H_n = \frac{1}{p} F'(x_n)^{-1}\left[F'(x_n + p(y_n - x_n)) - F'(x_n)\right], \quad p \in (0, 1],$$
$$x_{n+1} = y_n - \frac{1}{2} H_n [I + H_n]^{-1} (y_n - x_n) \quad (n \geq 0),$$

for approximating a solution x^* of equation $F(x) = 0$. Let $F: \Omega \subseteq X \to Y$ be a twice-Fréchet-differentiable operator defined on an open convex subset Ω of a Banach space X with values in a Banach space Y. Assume:

(1) $\Gamma_0 = F'(x_0)^{-1} \in L(Y, X)$ exists for some $x_0 \in \Omega$ with $\|\Gamma_0\| \leq \beta$;
(2) $\|\Gamma_0 F(x_0)\| \leq \eta$;
(3) $\left\|F''(x)\right\| \leq M$ $(x \in \Omega)$;
(4) $\left\|F''(x) - F''(y)\right\| \leq K \|x - y\|$, $\quad x, y \in \Omega$.

Denote by $a_0 = M\beta\eta$, $b_0 = K\beta\eta^2$. Define sequences

$$a_{n+1} = a_n f(a_n)^2 g_p(a_n, b_n),$$
$$b_{n+1} = b_n f(a_n)^3 g_p(a_n, b_n)^2,$$

where $f(x) = \frac{2(1-x)}{x^2 - 4x + 2}$ and $g_p(x, y) = \frac{3x^3 + 2y(1-x)[(1-6p)x + (2+3p)]}{24(1-x)^2}$. If $a_0 \in \left(0, \frac{1}{2}\right)$, $b_0 < h_p(h_0)$, where

$$h_p(x) = \frac{3(2x-1)(x-2)\left(x - 3 + \sqrt{5}\right)\left(x - 3 - \sqrt{5}\right)}{2(1-x)[(1-6p)x + 2 + 3p]}, \qquad \bar{U}(x_0, \tfrac{\eta}{a_0}) \subseteq \Omega$$

then show: iteration $\{x_n\}$ $(n \geq 0)$ is well defined, remains in $\bar{U}(x_0, \frac{\eta}{a_0})$ for all $n \geq 0$, and converges to a solution x^* of equation $F(x) = 0$, which is unique in $U(x_0, \frac{\eta}{a_0})$. Moreover, the following estimates hold for all $n \geq 0$:

$$\|x^* - x_n\| \leq \left[1 + \frac{a_0\gamma^{\frac{3^n-1}{2}}}{2(1-a_0)}\right]\gamma^{\frac{3^n-1}{2}}\frac{\Delta^n}{1-\Delta}\eta,$$

where $\gamma = \frac{a_1}{a_0}$ and $\Delta = f(a_0)^{-1}$.

5.5.13. Consider the two-step method:

$$y_n = x_n - F'(x_n)^{-1}F(x_n)$$

$$x_{n+1} = y_n - F'(x_n)^{-1}F(y_n) \quad (n \geq 0);$$

for approximating a solution x^* of equation $F(x) = 0$. Let $F: \Omega \subseteq X \subseteq Y$ be a Fréchet-differentiable operator defined on an open convex subset Ω of a Banach space X with values in a Banach space Y. Assume:

(1) $\Gamma_0 = f'(X_0)^{-1} \in L(Y, X)$ for some $x_0 \in \Omega$, $\|\Gamma_0\| \leq \beta$;

(2) $\|\Gamma_0 F(x_0)\| \leq \eta$;

(3) $\|F'(x) - F'(y)\| \leq K\|x - y\|$ $(x, y \in \Omega)$.

Denote $a_0 = k\beta\eta$ and define the sequence $a_{n+1} = f(a_n)^2 g(a_n) a_n$ $(n \geq 0)$, where $f(x) = \frac{2}{2-2x-x^2}$ and $g(x) = x^2(x+4)/8$. If $a_0 \in \left(0, \frac{1}{2}\right)$, $\bar{U}(x_0, R\eta)$ $\subseteq \Omega$, $R = \frac{1+\frac{a_0}{2}}{1-\gamma\Delta}$, $\gamma = \frac{a_1}{a_0}$ and $\Delta = f(a_0)^{-1}$, then show: iteration $\{x_n\}$ $(n \geq 0)$ is well defined, remains in $\bar{U}(x_0, R\eta)$ for all $n \geq 0$ and converges to a solution x^* of equation $F(x) = 0$, which is unique in $U(x_0, \frac{2}{K\beta} - R\eta) \cap \Omega$. Moreover, the following estimates hold for all $n \geq 0$

$$\|x_n - x^*\| \leq \left[1 + \frac{a_0}{2}\gamma^{\frac{3^n-1}{2}}\right]\gamma^{\frac{3^n-1}{2}}\frac{\Delta^n}{1-\gamma^{3^n}\Delta}\eta \quad (n \geq 0).$$

5.5.14. Let X be a Banach space, and let Y be a closed subspace. Assume F is a completely continuous operator defined on $\bar{D} \subseteq X$, D an open set, and assume the values $F(x) \in Y$ for all $x \in \bar{D}$. Let X_n be a sequence of finite-dimensional subspace of X such that

$$\inf_{x \in X_n} \|y - x\| \to 0 \text{ as } n \to \infty \text{ for all } y \in Y.$$

Let F_n be a sequence of projections associated with X_n:

$$F_n: X \to X_n \quad (n \geq 1).$$

Assume that when restricted to Y, the projections are uniformly bounded:

$$\sup_n \|F_n \mid Y\| \leq a < \infty.$$

Then projection method for solving

$$x = F(x)$$

becomes

$$x_n = F_n F(x_n)$$

Suppose that $x^* \in D$ if a fixed point of nonzero index for F. Then show for all sufficiently large n the equation $x_n = F_n(x_n)$ has at least one solution $x_n \in X_n \cap D$ such that

$$\lim_{n \to \infty} \|x_n - x^*\| = 0.$$

Let $F: U(x_0, r) \subseteq X \to X$ be differentiable and continuous on $\overline{U}(x_0, r)$ such that $I - P'(x)$ is compact. Suppose $L \in L(X)$ is such that

$$\|LF(x_0)\| \leq a$$
$$\|I - LF'(x_0)\| \leq b < 1,$$
$$\|L[F'(x) - F'(y)]\| \leq c \|x - y\| \text{ for all } x \in U(x_0, r)$$
$$h = \frac{ac}{(1-b)^2} \leq \tfrac{1}{2}$$
$$r_0 = \frac{a}{1-b} f(h) \leq r, \text{ where } f(h) = \frac{1-\sqrt{1-2h}}{h}; \ f(0) = 1.$$

Then show:
(a) equation $F(x) = 0$ has a solution x^* in $U(x_0, r_0)$; x^* is unique in $\overline{U}(x_0, r)$ if, for $r_1 = \frac{a}{1-b} f_1(h)$, where $f_1(h) = \frac{1+\sqrt{1-2h}}{h}$

$$r < r_1 \quad \text{for } h < \tfrac{1}{2};$$
$$r \leq r_1 \quad \text{for } h = \tfrac{1}{2}.$$

(b) Furthermore NK method is well defined, remains in $U(x_0, r)$ for all $n \geq 0$, and converges to x^*.

5.5.15. Let Fibonacci sequence $\{a_n\}$ be defined by $aa_n + ba_{n-1} + ca_{n-2} = 0$ where $a_0 = 0$ and $a_1 = 1$. If the characteristic polynomial $p(x) = ax^2 + bx + c$ has zeros r_1 and r_2 with $|r_1| > |r_2|$, then show:
(a) $a_n \neq 0 \ (n > 0)$;
(b) $\lim_{n \to \infty} \frac{a_{n+1}}{a_n} = r_1$;
(c) Newton $\left(\frac{a_{n+1}}{a_n}\right) = \frac{a_{2n+1}}{a_{2n}}$
and
(d) secant $\left(\frac{a_{m+1}}{a_m}, \frac{a_{n+1}}{a_n}\right) = \frac{a_{m+n+1}}{a_{m+n}}$,
where,

$$x_n = \text{Newton}(x_{n-1}) = x_{n-1} - \frac{F(x_{n-1})}{F'(x_{n-1})} \quad (n \geq 1)$$

and

$$x_n = \text{secant}(x_{n-1}, x_{n-2}) = x_{n-1} - \frac{F(x_{n-1})(x_{n-1} - x_{n-2})}{F(x_{n-1}) - F(x_{n-2})}. \quad (n \geq 1)$$

6

Analytic Computational Complexity: We Are Concerned with the Choice of Initial Approximations

6.1 The general problem

Approximate solution of equation involves a complex problem: choice of an initial approximation x_0 sufficiently close to the true solution. The method of random choice is often successful. Another frequently used method is to replace

$$F'(x) = 0 \qquad (6.1.1)$$

by a "similar" equation and to regard the exact solution of the latter as the initial approximation x_0. Of course, there are no general "prescriptions" for admissible initial approximations. Nevertheless, one can describe various devices suitable for extensive classes of equations.

As usual, let F map X into Y. To simplify the exposition, we shall assume that F is defined throughout X. Assume that $G(x.\lambda)$ $(x \in X; 0 \leq \lambda \leq 1)$ is an operator with values in Y such that

$$G(x; 1) \equiv Fx \quad (x \in X), \qquad (6.1.2)$$

and the equation

$$G(x; 0) = 0 \qquad (6.1.3)$$

has an obvious solution x^0. For example, the operator $G(x; \lambda)$ might be defined by

$$G(x; \lambda) = Fx - (1 - \lambda) Fx^0. \qquad (6.1.4)$$

Consider the equation

$$G(x; \lambda) = 0. \qquad (6.1.5)$$

Suppose that equation (6.1.5) has a continuous solution $x = x(\lambda)$, defined for $0 \leq \lambda \leq 1$ and satisfying the condition

$$x(0) = x^0. \qquad (6.1.6)$$

Were the solution $x(\lambda)$ known,

I.K. Argyros, *Convergence and Applications of Newton-type Iterations*,
DOI: 10.1007/978-0-387-72743-1_6, © Springer Science+Business Media, LLC 2008

$$x^* = x\,(1) \tag{6.1.7}$$

would be a solution of equation (6.1.1). One can thus find a point x_0 close to x^* by approximating $x\,(\lambda)$.

Our problem is thus to approximate the implicit function defined by (6.1.5) and the initial conditions (6.1.6). Global propositions are relevant, here theorems on implicit functions defined on the entire interval [0, 1]. The theory of implicit functions of this type is at present insufficiently developed.

The idea of extending solutions with respect to a parameter is due to S.N. Bernstein [192]; it has found extensive application in various theoretical and applied problems.

Assume that the operator $G\,(x;\lambda)$ is differentiable with respect to both x and λ, in the sense that there exist linear operators $G'_x\,(x;\lambda)$ mapping X and Y and elements $G'_x\,(x;\lambda) \in Y$ such that

$$\lim_{\|h\|+|\Delta\lambda|\to 0} \frac{\|G\,(x+h;\lambda+\Delta\lambda) - G\,(x;\lambda) - G'_x\,(x;\lambda)\,h - G'_x\,(x;\lambda)\,\Delta\lambda\|}{\|h\|+|\Delta\lambda|} = 0.$$

The implicit function $x\,(\lambda)$ is then a solution of the differential equation

$$G'_x\,(x;\lambda)\,\tfrac{dx}{d\lambda} + G'_x\,(x;\lambda) = 0, \tag{6.1.8}$$

satisfying the initial condition (6.1.6). Conditions for existence of a solution of this Cauchy problem defined on [0, 1] are precisely conditions for existence of the implicit function. Assuming the existence of a continuous operator

$$\Gamma\,(x;\lambda) = \left[G'_x\,(x;\lambda)\right]^{-1}, \tag{6.1.9}$$

we can rewrite equation (6.1.8) as

$$\tfrac{dx}{d\lambda} = -\Gamma\,(x;\lambda)\,G'_x\,(x;\lambda). \tag{6.1.10}$$

One must bear in mind that Peano's Theorem is false for ordinary differential equations in Banach spaces. Therefore, even in the local existence theorem for equation (6.1.10) with condition (6.1.6), one must assume that the right-hand side of the equation satisfies certain smoothness conditions. However, there are no sufficient smoothness conditions for the existence of a global extension of the solution to the entire interval $0 \leq \lambda \leq 1$. We shall only mention a trivial fact: if the equation

$$\tfrac{dx}{d\lambda} = f\,(x;\lambda) \tag{6.1.11}$$

in a Banach space satisfies the local existence theorem for some initial condition, and

$$\|f\,(x;\lambda)\| \leq a + b\,\|x\| \quad (0 \leq \lambda \leq 1; x \in X), \tag{6.1.12}$$

then every solution of equation (6.1.11) can be extended to the entire interval $0 \leq \lambda \leq 1$. Thus, if

$$\left\| \Gamma\left(x;\lambda\right) G_y'\left(x;\lambda\right) \right\| \leqq a + b\left\| x \right\|, \tag{6.1.13}$$

then equation (6.1.5) defines an implicit function which satisfies (6.1.6) and is defined for $0 \leq \lambda \leq a$. Consequently, condition (6.1.13) guarantees that equation (6.1.1) is solvable and its solution can be constructed by integrating the differential equation (6.1.10).

To approximate a solution $x\left(\lambda\right)$ of equation (6.1.10), one can use, for example, Euler's method. To this end, divide the interval $[0, 1]$ into m subintervals by points

$$\lambda_0 = 0 < \lambda_1 < \cdots < \lambda_m = 1. \tag{6.1.14}$$

The approximate values $x\left(\lambda_i\right)$ of the implicit function $x\left(\lambda\right)$ are then determined by the equalities $x\left(\lambda_0\right) = x^0$ and

$$x\left(\lambda_{i+1}\right) = x\left(\lambda_i\right) - \Gamma\left[x\left(\lambda_i\right);\lambda_i\right] G_\lambda'\left[x\left(\lambda_i\right);\lambda_i\right]\left(\lambda_{i+1} - \lambda_i\right). \tag{6.1.15}$$

The element $x\left(\lambda_m\right)$ is in general close to the solution x^* of equation (6.1.1) and one may therefore expect it to fulfill the demands imposed on initial approximations for iterative solution of equation (6.1.1). We emphasize that (6.1.15) does not describe an iterative process; it only yields a finite sequence of operations, whose result is an element that may be a suitable initial approximation for iterative solution of equation (6.1.1).

Other constructions may be used to approximate the implicit function $x\left(\lambda\right)$. Partition the interval $[0, 1]$ by the points (6.1.14). The point $x\left(\lambda_1\right)$ is a solution of the equation $G\left(x,\lambda_1\right) = 0$. Now $x\left(\lambda_0\right) = x^0$ is a suitable initial approximation to $x\left(\lambda_1\right)$. Approximate $x\left(\lambda_1\right)$ by performing a fixed number of steps of some iterative process. The result is an element x_1 that should be fairly close to $x\left(\lambda_1\right)$. This element x_1 is obtained from x^0 by a certain operator

$$x_1 = W\left[x^0; G\left(x;\lambda_1\right)\right].$$

Now regard x_1 as an initial approximation to the solution $x\left(\lambda_2\right)$ of the equation $G\left(x;\lambda_2\right)$, and proceed as before. The result is an element x_2:

$$x_2 = W\left[x_1; G\left(x,\lambda_2\right)\right].$$

Continuing in this way, we obtain a finite set of points

$$x_{i+1} = W\left[x_i; G\left(x,\lambda_i\right)\right] \quad (i = 0, 1, \ldots, m-1), \tag{6.1.16}$$

the last of which x_m, may be regarded as an initial approximation for iterative solution of equation (6.1.1)

If the operator W represents one iteration of the method, formula (6.1.16) is

$$x_{i+1} = x_i - \left[G_x'\left(x_i;\lambda_{i+1}\right)\right]^{-1} G\left(x_i;\lambda_{i+1}\right) \quad (i = 1, \ldots, m-1). \tag{6.1.17}$$

6.2 Obtaining good starting points for Newton's method

In this section, we are concerned with the problem of approximating a locally unique solution x^* of equation

$$F(x) = 0, \qquad (6.2.1)$$

where F is a Fréchet-differentiable operator defined on an open convex subset D of a Banach space X with values in a Banach space Y.

The most popular method for generating a sequence approximation x^* is undoubtedly Newton's method:

$$x_{n+1} = x_n - F'(x_n)^{-1} F(x_n) \quad (n \geq 0) \quad (x_0 \in D). \qquad (6.2.2)$$

In particular, the famous Newton-Kantorovich theorem guarantees the quadratic convergence of method (6.2.2) if the initial guess x_0 is "close enough" to the solution x^* (see Chapter 2).

However, we recently showed that the Newton-Kantorovich hypothesis (6.2.13) can always be replaced by the weaker (6.2.7) (under the same computational cost) [35] (see also Section 2.2). In particular, using the algorithm proposed by H.T. Kung [131] (see also [192]), we show that the number of steps required to compute a good starting point x_0 (to be precised later) can be significantly reduced.

This observation is very important in computational mathematics.

In Section 2.2 we showed the following semilocal convergence theorem for Newton's method (6.2.2), which essentially states the following:

If

$$F'(x_0)^{-1} \text{ exists}, \ \left\| F'(x_0)^{-1} \right\| \leq \beta_0, \qquad (6.2.3)$$

$$\left\| F'(x_0)^{-1} F(x_0) \right\| \leq \xi_0, \qquad (6.2.4)$$

$$\left\| F'(x) - F(x_0) \right\| \leq K_0 \|x - x_0\|, \qquad (6.2.5)$$

$$\left\| F'(x) - F'(y) \right\| \leq K \|x - y\|, \qquad (6.2.6)$$

for all $x, y \in \overline{U}(x_0, r)$

$$h_0 = \beta_0 L \xi_0 < \tfrac{1}{2} \qquad (6.2.7)$$

where,

$$L = \tfrac{K_0 + K}{2}, \qquad (6.2.8)$$

$$2\xi_0 \leq r, \qquad (6.2.9)$$

and

$$\overline{U}(x_0, r) \subseteq D, \qquad (6.2.10)$$

then sequence $\{x_n\}$ $(n \geq 0)$ generated by NK method (6.2.2) is well defined, remains in $\overline{U}(x_0, r_0)$ for all $n \geq 0$, and converges quadratically to a unique solution $x^* \in U(x_0, r)$ of equation $F(x) = 0$. Moreover we have

$$\|x_0 - x^*\| \leq 2\xi_0. \qquad (6.2.11)$$

Remark 6.2.1. In general

$$K_0 \leq K \tag{6.2.12}$$

holds. If equality holds in (6.2.12), then the result stated above reduces to the famous Newton-Kantorovich theorem and (6.2.7) to the Newton-Kantorovich hypothesis (6.2.13). If strict inequality holds in (6.2.12), then (6.2.7) is weaker than the Newton-Kantorovich hypothesis

$$\overline{h}_0 = \beta_0 K \xi_0 < \tfrac{1}{2}. \tag{6.2.13}$$

Moreover, the error bounds on the distances $\|x_{n+1} - x_n\|$, $\|x_n - x^*\|$ $(n \geq 0)$ are finer and the information on the location of the solution more precise.

Note also that the computational cost of obtaining (K_0, K) is the same as the one for K as in practice evaluating K requires finding K_0.

Hence all results using (6.2.13) instead of (6.2.7) can now be challenged to obtain more information. That is exactly what we are doing here. In particular, motivated by the elegant work of H.T. Kung [131] on good starting points for NK method, we show how to improve on these results if we use our theorem stated above instead of the Newton-Kantorovich theorem.

Definition 6.2.2. *We say x_0 is a good starting point for approximating x^* by NK method or a good starting point for short if conditions (6.2.3)–(6.2.10) hold.*

Note that the existence of a good starting point implies the existence of a solution x^* of equation $F(x) = 0$ in $U(x_0, 2\xi_0)$.

We provide the following theorem / Algorithm that improves the corresponding ones given in [131, Thm. 4.1] to obtain good starting points.

Theorem 6.2.3. *Let $F: D \subseteq X \to Y$ be a Fréchet-differentiable operator. If F' satisfies center-Lipschitz, Lipschitz conditions (6.2.5), (6.2.6), respectively, on $U(x_0, 2r)$*

$$\|F(x_0)\| \leq \eta_0$$

$$\left\| F'(x)^{-1} \right\| \leq \beta \text{ for all } x \in U(x_0, 2r), \tag{6.2.14}$$

$$U(x_0, 2r) \subseteq D, \tag{6.2.15}$$

and

$$\beta \eta_0 < \tfrac{r}{2}, \tag{6.2.16}$$

then there exists a solution x^* of equation $F(x) = 0$ in $U(x_0, 2r)$.

Proof. Simply use L instead of K in the proof of Theorem 4.1 in [131, p. 11] including the algorithm there, which is essentially repeated here with some modifications:

Algorithm A: The goal of this algorithm is to produce starting point for approximating x^*.

1. Set $h_0 \longleftarrow \beta^2 L \eta_0$ and $i \longleftarrow 0$. Choose any number δ in $\left(0, \tfrac{1}{2}\right)$.

2. If $h_1 < \frac{1}{2}$, x_i is a good starting point for approximating x^* and algorithn A terminates.

3. Set $\lambda_i \longleftarrow \left(\frac{1}{2} - \delta\right) / h_i$, and

$$F_i(x) \longleftarrow [F(x) - F(x_i)] + \lambda_i F(x_i). \qquad (6.2.17)$$

4. (It is shown in the proof that x_i is a good starting point for approximating a zero, denoted by x_{i+1}, of F_i, Apply NK method to F_i, starting from x_i, to find x_{i+1}.

5. (Assume that the exact x_{i+1} is found.) Set $\eta_{i+1} \longleftarrow \|F(x_{i+1})\|$ and

$$h_{i+1} \longleftarrow \beta^2 K \eta_{i+1}.$$

6. Set $i \longleftarrow i + 1$, and return back to step 2.

In the following, we prove algorithm works. First we note that $\lambda_i \in (0, 1)$ and by (6.2.17)

$$\eta_{i+1} = (1 - \lambda_i) \eta_i. \qquad (6.2.18)$$

We shall prove by induction that

$$\|x_i - x_{i-1}\| \le 2\beta \lambda_{i-1} \eta_{i-1}, \qquad (6.2.19)$$

and

$$\|x_i - x_0\| \le r. \qquad (6.2.20)$$

They trivially hold for $i = 0$.

Suppose that (6.2.19) and (6.2.20) hold and $h_i \ge \frac{1}{2}$. By (6.2.17)

$$\beta^2 L \|f_i(x_i)\| \le \beta^2 L \lambda_i \eta_i = \lambda_i h_i = \frac{1}{2} - \delta, \qquad (6.2.21)$$

and by (6.2.18)

$$2\beta \|f_i(x_i)\| \le 2\beta \lambda_i \eta_i < 2\beta \eta_i \le 2\beta \eta_0 < r.$$

Further, by (6.2.20), we have $U(x_i, r) \subseteq U(x_0, 2r)$. Hence x_i is a good starting point for approximating the zero x_{i+1} of f_i. From (6.2.15), we know

$$\|x_{i+1} - x_i\| \le 2\beta \lambda_i \eta_i. \qquad (6.2.22)$$

Hence (6.2.19) holds with i replaced by $i + 1$. By (6.2.22), (6.2.18), and (6.2.16), we have

$$\|x_{i+1} - x_0\| \le \|x_{i+1} - x_i\| + \|x_i - x_{i-1}\| + \cdots + \|x_1 - x_0\| \qquad (6.2.23)$$
$$\le 2\beta \left(\lambda_i \eta_i + \lambda_{i-1} \eta_{i-1} + \cdots + \lambda_0 \eta_0\right)$$
$$\le 2\beta \left((1 - \lambda_{i-1}) \eta_{i-1} + \lambda_{i-1} \eta_{i-1} + \cdots + \lambda_0 \eta_0\right)$$
$$= 2\beta \left(\eta_{i-1} + \lambda_{i-2} \eta_{i-2} + \cdots + \lambda_0 \eta_0\right)$$
$$\le \cdots$$
$$\le 2\beta \eta_0 < r,$$

i.e., (6.2.20) holds with i replaced by $i + 1$.

We now assume that (6.2.19) and (6.2.20) hold and $h_1 < \frac{1}{2}$. By (6.2.16) and (6.2.18), $2\beta \| f(x_i) \| = 2\beta \eta_i < 2\beta \eta_0 < r$. Further by (6.2.20), $s_r(x_i) \subseteq s_{2r}(x_0)$. Hence x_i is a good starting point for approximating α.

It remains to show that the loop starting from step 2 is finite. Suppose that $h_0 \geq \frac{1}{2}$. Because $\lambda_i \in (0, 1)$ for all i, we have

$$\lambda_i = \frac{\frac{1}{2} - \delta}{\beta^2 L \eta_i} = \frac{\frac{1}{2} - \delta}{\beta^2 L (1 - \lambda_{i-1}) \eta_{i-1}} > \frac{\frac{1}{2} - \delta}{\beta^2 L \eta_{i-1}} = \lambda_{i-1}, \text{ for all } i.$$

Hence by (6.2.18)

$$\eta_{i+1} = (1 - \lambda_i) \eta_i < (1 - \lambda_0) \eta_i$$
$$< \cdots$$
$$< (1 - \lambda_0)^{i+1} \eta_0.$$

This implies that $h_i < \frac{1}{2}$ when $\beta^2 L (1 - \lambda_0)^i \eta_0 < \frac{1}{2}$, i.e., when

$$(1 - \lambda_0)^i < \frac{1}{2h_0}. \tag{6.2.24}$$

Because $1 - \lambda_0 < 1$, (6.2.24) is satisfied for large i. Therefore when i is large enough, $h_i < \frac{1}{2}$ and hence Algorithm A terminates.

Remark 6.2.4. As already noted in [131] Theorem 6.2.3 is trivial for the scalar case ($f: R \to R$), as the mean value theorem can be used. Some of the assumptions of Theorem 6.2.3 can be weakened. Avila for example in [196, Theorem 4.3] instead of (6.2.16) used a more complicated condition involving β, K, and η_0. However, the idea algorithm is basically different from Algorithm A. Note also that if $K_0 = K$, then our Theorem 6.2.3 reduces to Theorem 4.1 in [131]. We now modify Algorithm A to make it work in Banach spaces without necessarily assuming that the exact zero of x_{i+1} of F_i can be found using NK method (6.2.2).

Theorem 6.2.5. *Under the hypotheses of Theorem 6.2.3, a good starting point for approximating solution x^* of equation $F(x) = 0$ can be obtained in $N(\delta, K_0, K)$ Newton steps, δ is any number in $\left(0, \frac{1}{2}\right)$,*

$$N(\delta, K_0, K) = \begin{cases} 0, & \text{if } h_0 = \beta^2 L \eta_0 \leq \frac{1}{2} - \delta \\ I(\delta, K_0, K) \cdot J(\delta, K_0, K), & \text{otherwise,} \end{cases}$$

where, $I(\delta, K_0, K)$ is the smallest integer i such that:

$$\left[1 - \frac{\frac{1}{2} - \delta}{h_0} \right]^i \leq \left[\frac{1}{2} - \delta \right] / h_0, \tag{6.2.25}$$

and $J(\delta, K_0, K)$ is the smallest integer j such that:

$$2^{\frac{1}{j-1}} (1 - 2\delta)^{2^j - 1} (a + \beta \eta_0) \le r - 2\beta \eta_0, \tag{6.2.26}$$

$$\frac{1}{2^j} (1 - 2\delta)^{2^j - 1} (a + \beta \eta_0) \le a, \tag{6.2.27}$$

where,

$$a = \min \left(\frac{r}{2} - \beta \eta_0, \frac{\delta}{2\beta L} \right). \tag{6.2.28}$$

Proof. Simply use L instead of K in the proof of Theorem 4.2 in [131, p. 16], and the following algorithm:

Algorithm B.

1. Set $h_0 \longleftarrow \beta^2 L \eta_0, \bar{x}_0 \longleftarrow x_0$ and $i \longleftarrow 0$. Choose any number δ in $\left(0, \frac{1}{2} \right)$.

2. If $h_i \le \frac{1}{2} - \delta, \bar{x}_i$ is a good starting point for approximating x^* and Algorithm B terminates.

3. Set $\lambda_i \longleftarrow \left(\frac{1}{2} - \delta \right) / h_i,$

$$F_i (x) \longleftarrow [F (x) - \eta_i F (x_0) / \eta_0] + \lambda_i \eta_i F (x_0) / \eta_0,$$

and

$$\eta_{i+1} \longleftarrow (1 - \lambda_i) \eta_i. \tag{6.2.29}$$

4. Apply NK method to F_i, starting from $\bar{x}_i$, to find an approxiation $\bar{x}_{i+1}$ to a zero x_{i+1} of F_i such that

$$\|\bar{x}_{i+1} - x_{i+1}\| \le r - 2\beta \eta_0, \tag{6.2.30}$$

and

$$\left\| F_i' (\bar{x}_{i+1})^{-1} F_i (\bar{x}_{i+1}) \right\| \le \min \left(\frac{r}{2} - \beta \eta_{i+1}, \frac{\delta}{2\beta L} \right). \tag{6.2.31}$$

5. Set $h_{i+1} \longleftarrow \beta^2 L \eta_{i+1}.$
6. Set $i \longleftarrow i + 1$ and return to step 2.

Note that the $h_i, \lambda_i, \eta_i, f_i, x_i$ in Algorithm A are the same $h_i, \lambda_i, \eta_i, f_i, x_i$ in Algorithm B. Note also that by (6.2.30) and (6.2.23) we have

$$\|\bar{x}_i - x_0\| \le \|\bar{x}_i - x_i\| + \|x_i - x_0\| \tag{6.2.32}$$
$$\le (r - 2\beta \eta_0) + 2\beta \eta_0 = r, \quad \forall i.$$

It is clear that if $h_0 < \frac{1}{2} - \frac{\delta}{2}, \bar{x}_0$ is a good starting point for approximating α. Now suppose $h_0 > \frac{1}{2} - \frac{\delta}{2}$. Because $\bar{x}_0 = x_0$, in the proof of Theorem 6.2.3, we have shown that $\bar{x}_0$ is a good starting point for approximating x_1 a zero of f_0. Let z_j denote the jth NK iterate starting from $\bar{x}_0$ for approximating x_1. Because

$$\beta^2 L \| f_0 (\bar{x}_0)\| = \beta^2 L \lambda_1 \eta_0 = \frac{1}{2} - \delta,$$

it is known (see, e.g., Section 2.2) that

$$\|z_j - x_1\| \le \frac{1}{2^{j-1}} (1 - 2\delta)^{2^j - 1} \left\| [f'(x_0)]^{-1} f(x_0) \right\|$$

and

$$\left\| [f_0'(z_j)]^{-1} f_0(z_j) \right\| \le \frac{1}{2^j} (1 - 2\delta)^{2^j - 1} \left\| [f'(x_0)]^{-1} f(x_0) \right\|.$$

Hence we may let $\bar{x}_1$ be z_j for j large enough, say, $j = j(\delta)$, then

$$\|\bar{x}_1 - x_1\| \le r - 2\beta\eta_0,$$

and

$$\left\| [f_0'(\bar{x}_1)]^{-1} f_0(\bar{x}_1) \right\| \le \min\left(\frac{r}{2} - \beta\eta_1, \frac{\delta}{2\beta L} \right),$$

i.e., (6.2.30) and (6.2.31) hold for $i = 0$.

Suppose that (6.2.30) and (6.2.31) hold. Then

$$\left\| [f'(\bar{x}_{i+1})]^{-1} f(\bar{x}_{i+1}) \right\| \le \tag{6.2.33}$$

$$\le \left\| [f_i'(\bar{x}_{i+1})]^{-1} f_i(\bar{x}_{i+1}) \right\| + \left\| [f_i'(\bar{x}_{i+1})]^{-1} [f_i'(\bar{x}_{i+1}) - f_i(\bar{x}_{i+1})] \right\|.$$

$$\le \min\left(\frac{r}{2} - \beta\eta_{i+1}, \frac{\delta}{2\beta L} \right) + \beta\eta_{i+1},$$

and

$$\left\| [f_{i+1}'(\bar{x}_{i+1})]^{-1} f_{i+1}(\bar{x}_{i+1}) \right\| \le \tag{6.2.34}$$

$$\le \left\| [f'(\bar{x}_{i+1})]^{-1} [f(\bar{x}_{i+1}) - \eta_{i+1} f(x_0) / \eta_0] \right\|$$

$$\le \left\| [f'(\bar{x}_{i+1})]^{-1} \lambda_{i+1} \eta_{i+1} f(x_0) / \eta_0 \right\|$$

$$\le \left\| [f'(\bar{x}_{i+1})]^{-1} f_i(\bar{x}_{i+1}) \right\| + \lambda_{i+1} \beta\eta_{i+1}.$$

Suppose that $h_{i+1} < \frac{1}{2} - \delta$. We want to prove that $\bar{x}_{i+1}$ is a good starting point for approximating α. By (6.2.33)

$$\beta L \left\| [f'(\bar{x}_{i+1})]^{-1} f(\bar{x}_{i+1}) \right\| \le$$

$$\le \beta L \cdot \frac{\delta}{2\beta L} + h_{i+1} < \frac{\delta}{2} + \frac{1}{2} - \delta = \frac{1}{2} - \frac{\delta}{2}.$$

Let $a = \left\| [f'(\bar{x}_{i+1})]^{-1} f(\bar{x}_{i+1}) \right\|$. If $x \in U(\bar{x}_{i+1}, 2a)$, then

$$\|x - x_0\| \le \|x - \bar{x}_{i+1}\| + \|\bar{x}_{i+1} - x_0\| \tag{6.2.35}$$

$$\le 2a + r$$

$$\le 2\left(\frac{r}{2} - \beta\eta_{i+1} + \beta\eta_{i+1} \right) + r = 2r,$$

i.e., $x \in U(x_0, 2r)$. Hence $\overline{x}_{i+1}$ is a good starting point for approximating α. We now assume that $h_{i+1} > \frac{1}{2} - \delta$, and want to prove that $\overline{x}_{i+1}$ is a good starting point for approximating x_{i+2}, a zero of f_{i+1}.

We have by (6.2.34) and (6.2.31),

$$\beta L \left\| \left[f'_{i+1}(\overline{x}_{i+1}) \right]^{-1} f_{i+1}(\overline{x}_{i+1}) \right\|$$

$$\leq \frac{\delta}{2} + \lambda_{i+1}\beta^2 L \eta_{i+1}$$

$$= \frac{\delta}{2} + \frac{1}{2} - \delta = \frac{1}{2} - \frac{\delta}{2}.$$

Let $b = \left\| \left[f'_{i+1}(\overline{x}_{i+1}) \right]^{-1} f_{i+1}(\overline{x}_{i+1}) \right\|$. If $x \in U(\overline{x}_{i+1}, 2b)$, as in (6.2.35) we can prove that $x \in U(x_0, 2r)$. Hence $\overline{x}_{i+1}$ is a good starting point for approximating x_{i+2}. By the same argument as used for obtaining $\overline{x}_0$ and by (6.2.34), one can prove that if $\overline{x}_{i+2}$ is set to be the $J(\delta)$th Newton iterate starting from $\overline{x}_{i+1}$, then

$$\|\overline{x}_{i+2} - x_{i+1}\| \leq r - 2\beta\eta_0,$$

and

$$\left\| \left[f'_{i+1}(\overline{x}_{i+2}) \right]^{-1} f_{i+1}(\overline{x}_{i+2}) \right\| \leq \min\left(\frac{r}{2} - \beta\eta_{i+2}, \frac{\delta}{2\beta L} \right),$$

i.e., (6.2.30) and (6.2.31) hold with i replaced by $i + 1$. This shows that we need to perform at most $J(\delta)$ Newton steps at step 4 of Algorithm B to obtain each $\overline{x}_{i+1}$. Therefore, for any $\delta \in \left(0, \frac{1}{2}\right)$, to obtain a good starting point we need to perform at most $N(\delta) = I(\delta) \cdot J(\delta)$ Newton steps.

Remark 6.2.6. As noted in [131] δ should not be chosen to minimize the complexity of Algorithm B. Instead, δ should be chosen to minimize the complexity of algorithm:

1. Search Phase: Perform Algorithm B.
2. Iteration Phase: Perform NK method starting from the point obtained by Algorithm B.

An upper bound on the complexity of the iteration phase is the time needed to carry out $T(\delta, K_0, K, \varepsilon)$ is the smallest integer K such that

$$\frac{1}{2^{K-1}} (1 - 2\delta)^{2^K - 1} (a + \beta\eta_0) \leq \varepsilon. \tag{6.2.36}$$

Note also that if $K_0 = K$ our Theorem 6.2.5 reduces to Theorem 4.2 in [131, p. 15]. Hence we showed the following result:

Theorem 6.2.7. *Under the hypotheses of Theorem 6.2.5, the time needed to find a solution x^* of equation $F(x) = 0$ inside a ball of radius ε is bounded above by the time needed to carry out $R(\delta, K_0, K, \varepsilon)$ Newton steps, where*

$$R\left(\varepsilon, K_0, K\right) = \min_{0 < \delta < \frac{1}{2}} \left[N\left(\delta, K_0, K, \varepsilon\right) + T\left(\delta, K_0, K, \varepsilon\right)\right], \tag{6.2.37}$$

where $N\left(\delta, K_0, K, \varepsilon\right)$ and $T\left(\delta, K_0, K, \varepsilon\right)$ are given by (6.2.17) and (6.2.36), respectively.

Remark 6.2.8. If $K_0 = K$ Theorem 6.2.7 reduces to Theorem 4.3 in [131, p. 20]. In order for us to compare our results with the corresponding ones in [131], we computed the values of $R\left(\varepsilon, K_0, K\right)$ for F satisfying the conditions of Theorem 4.3 in [131] and Theorem 6.2.7 above with

$$\beta \eta_0 \le .4r, \tag{6.2.38}$$

and

$$1 \le h_0 = \beta^2 L \eta_0 \le 10, \tag{6.2.39}$$

and for ε equal to $10^{-i}r$, $1 \le i \le 10$.

The following table gives the results for $\varepsilon = 10^{-6}r$. Note that by I we mean $I\left(\delta_0, K, K\right)$, $I_{\alpha K}$ we mean $I\left(\delta_0, K_0, K\right)$ with $K_0 = \alpha K$, $\alpha \in [0, 1]$. Similarly for J, N, and T.

Comparison Table 5.2.9

h_0	δ_0	I	J	N	T	R	$I_{.9K}$	$N_{.9K}$	$R_{.9K}$	$I_{.5K}$	$N_{.5K}$	$R_{.5K}$	I_{0K}	N_{0K}	R_{0K}
1	.165	3	2	6	5	11	3	6	11	2	4	9	1	2	8
2	.103	8	3	24	6	30	7	21	27	5	15	21	2	6	12
3	.118	16	3	48	6	54	14	42	48	10	30	36	5	15	21
4	.129	25	3	75	6	81	23	69	75	16	48	54	9	27	33
5	.137	35	3	105	6	111	33	99	105	30	90	96	13	39	45
6	.144	47	3	141	5	146	44	132	137	31	93	98	17	51	56
7	.149	59	3	177	5	182	55	165	170	40	120	125	22	66	71
8	.154	72	3	216	5	221	67	201	206	41	123	128	28	84	89
9	.159	85	3	255	5	260	80	240	245	58	174	179	33	99	104
10	.163	99	3	295	5	302	93	279	284	68	204	209	39	117	122

Remark 6.2.9. It follows from the table that our results significantly improve the corresponding ones in [131] and under the same computational cost. Suppose for example that $h_0 = 9$, $\delta = .159$. Kung found that the search phase can be done in 255 NK steps and the iteration phase in 5 NK steps. That is, a root can be located inside a ball of radius $10^{-6}r$ using 260 NK steps. However for $K_0 = .9K$, $K_0 = .5K$, and $K_0 = 0$, the corresponding NK steps are 245, 179, and 104, respectively, which constitute a significant improvement.

At the end of his paper, Kung asked whether the number of NK steps used by this procedure is close to the minimum. It is now clear from our approach that the answer is no (in general).

Finally, Kung proposed the open question: Suppose that the conditions of the Newton-Kantorovich theorem hold: Is NK method optimal or close to optimal, in terms of the numbers of function and derivative equations required to approximate the solution x^* of equation $F\left(x\right) = 0$ to within a given tolerance ε?

Clearly according to our approach the answer is no.

6.3 Exercises

6.3.1 Let g be an algorithm for finding a solution x^* of equation $F(x) = 0$, and x the approximation to x^* computed by g. Define the error for approximating x by

$$d(g, F) = \|x - x^*\|.$$

Consider the problem of approximating x^* when F satisfies some conditions. Algorithms based on these conditions cannot differentiate between operators in the class C of all operators satisfying these conditions. We use the class C instead of specific operators from C. Define

$$d_i = \inf_{g \in A} \sup_{F \in C} d(g, F)$$

where A is the class of all algorithms using i units of time. The time t needed to approximate x^* to with in error tolerance $\varepsilon > 0$ is the smallest i such that $d_i \le \varepsilon$, and an algorithm is said to be optimal if

$$\sup_{F \in C} d(g, F) = d_t.$$

If for any algorithm using i units of time, there exist functions F_1, F_2 in C such that:
the minimum distance between any solution of F, and any solution of F_2 is greater or equal to 2ε then, show:

$$d_i \ge \varepsilon.$$

6.3.2 With the notation introduced in Exercise 6.3.1, assume:
(1) $F:[a, b] \to \mathbf{R}$ is continuous;
(2) $F(a) < 0$, $F(b) > 0$,
Then, show:

$$d_i = \frac{b - a}{2^{i+1}}.$$

6.3.3 With the notation introduced in Exercise 6.3.1, assume:
(1) $F:[a, b] \to \mathbf{R}$, $F'(x) \ge \alpha > 0$ for all $x \in [a, b]$;
(2) $F(a) < 0$, $F(b) > 0$.
Then, show:

$$d_i = \frac{b - a}{2^{i+1}}.$$

6.3.4 With the notation introduced in Exercise 6.3.1, assume:
(1) $F:[a, b] \to \mathbf{R}$, $F'(x) \le b$ for all $x \in [a, b]$;
(2) $F(a) < 0$, $F(b) > 0$.
Then, show:

$$d_i = \frac{b - a}{2^{i+1}}.$$

6.3.5 With the notation introduced in Exercise 6.3.1, assume:

(1) $F: [a, b] \to \mathbf{R}, b \geq F'(x) \geq \alpha > 0$ for all $x \in [a, b]$;

(2) $F(a) < 0, F(b) > 0$.

Then, show:

$$d_i \geq (b - a) \left[\frac{\left(1 - \frac{\alpha}{b}\right)^2}{2} \right]^{i+1}.$$

6.3.6 Assume:

(1) hypotheses of Exercise 6.3.5 hold;

(2) $\left\| F''(x) \right\| \leq \gamma$ for all $\gamma \in [a, b]$.

Then show that the problem of finding a solution x^* of equation $F(x) = 0$ can be solved superlinearly.

7

Variational Inequalities

7.1 Variational inequalities and partially relaxed monotone mapping

There are numerous iterative methods available in the literature on the approximation-solvability of the general class of nonlinear inequality (NVI) problems, for instance the auxiliary problem principle. Marcotte and Wu [136] applied an iterative procedure similar to that of the auxiliary problem principle to the solvability of a class of variational inequalities involving cocoercive mappings in $\mathbf{R}^n$, and Verma extended and generalized this iterative process of Marcotte and Wu [136] and applied to the solvability of a certain class of variational inequalities involving partially relaxed monotone mappings a weaker class than the cocoercive and strongly monotone mappings and computation-oriented. In this section, we intend to discuss the approximation-solvability of a class of nonlinear variational inequalities involving multivalued partially relaxed monotone mappings. The estimate for the approximate solutions seems to be of interest in the sense that these are not only helpful to the convergence analysis, but it could be equally important to some numerical computations in $\mathbf{R}^n$ as well.

Let H be a real Hilbert space with inner product $\langle \cdot, \cdot \rangle$ and norm $\|\cdot\|$. Let $P(H)$ denote the power set of H. Let $T: K \rightarrow P(H)$ be a multivalued mapping and K a closed convex subset of H. We consider a class of nonlinear variational inequality (NVI) problems: find an element $x^* \in K$ and $u^* \in T(x^*)$ such that

$$\langle u^*, x - x^* \rangle \geq 0 \text{ for all } x \in K. \tag{7.1.1}$$

For an arbitrary element $x^0 \in K$, we consider an iterative algorithm generated as:

$$\langle u^0 + x^1 - x^0, x - x^1 \rangle \geq 0, \text{ for all } x \in K \text{ and } u^0 \in T(x^0).$$

$$\vdots \tag{7.1.2}$$

$$\langle u^k + x^{k+1} - x^k, x - x^{k+1} \rangle \geq 0, \text{ for all } x \in K, \text{ and for } u^k \in T(x^k).$$

The iterative procedure (7.1.2) can be characterized as a projection equation

I.K. Argyros, *Convergence and Applications of Newton-type Iterations*,
DOI: 10.1007/978-0-387-72743-1_7, © Springer Science+Business Media, LLC 2008

$$x^{k+1} = P_K \left[x^k - u^k \right] \text{ for } k \geq 0, \qquad (7.1.3)$$

where P_K is the projection of H onto K.

A mapping $T: H \to H$ is said to be α-cocoercive if for all $x, y \in H$, we have

$$\|x - y\|^2 \geq \alpha^2 \|T(x) - T(y)\|^2 + \|\alpha (T(x) - T(y)) - (x - y)\|^2,$$

where $\alpha > 0$ is a constant.

A mapping $T: H \to H$ is called α-cocoercive if there exists a constant $\alpha > 0$ such that

$$\langle T(x) - T(y), x - y \rangle \geq \alpha \|T(x) - T(y)\|^2 \text{ for all } x, y \in H.$$

We note that if T is α-cocoercive and expanding, then T is α-strongly monotone. Also, if T is α-strongly and β-Lipschitz continuous, then T is $\left(\alpha/\beta^2\right)$-cocoercive for $\beta > 0$. Clearly every α-cocoercive mapping T is $(1/\alpha)$-Lipschitz continuous. Most importantly, both notions of the cocoercivity are equivalent.

A mapping $T: H \to P(H)$ is called r-strongly monotone if for all $x, y \in H$, we have

$$\langle u - v, x - y \rangle \geq r \|x - y\|^2 \text{ for } u \in T(x) \text{ and } v \in T(y),$$

where $r > 0$ is a constant.

This implies that the mapping T is r-∂-expansive, that is,

$$\partial (T(x), T(y)) \geq r \|x - y\| \text{ for all } x, y \in H,$$

where $\delta(A, B) = \sup \{\|a - b\| : a \in A. b \in B\}$ for any $A, B \in P(H)$. When $r = 1$, T is called a ∂-expanding mapping. The class satisfies the following implications:

$$r\text{-strongly monotone}$$
$$\downarrow$$
$$r\text{-}\partial\text{-expansive}$$
$$\downarrow$$
$$\partial\text{-expansive}$$

A mapping $T: H \to P(H)$ is said to be β-∂-Lipschitz continuous if

$$\partial (T(x), T(y)) \leq \beta \|x - y\| \text{ for all } x, y \in H,$$

where $\partial(A, B) = \sup \{\|a - b\| : a \in A, b \in B\}$ for any $A, B \in P(H)$ and $\beta \geq 0$ is a constant.

A multivalued mapping $T: H \to P(H)$ is said to be α-∂-cocoercive if for all $x, y \in H$, we have

$$\langle u - v, x - y \rangle \geq \alpha \left[\partial (T(x), T(y))\right]^2 \text{ for all } x, y \in H,$$

where $\partial(A, B) = \sup \{\|a - b\| : a \in A, b \in B\}$ for any $A, B \in P(H)$ and $\alpha > 0$ is a constant.

A mapping $T: H \rightarrow P(H)$ is called β-∂-Lipschitz continuous if there exists a constant $\beta \geq 0$ such that

$$\partial (T(x), T(y)) \leq \beta \|x - y\| \text{ for all } x, y \in H,$$

where $\partial (A, B) = \sup \{\|a - b\| : a \in A, b \in B\}$ for any $A, B \in P(H)$.

A mapping $T: H \rightarrow P(H)$ is said to be α-partially relaxed monotone if for all $x, y, z \in H$ we have

$$\langle u - v, z - y \rangle \geq -\alpha \|z - x\|^2 \text{ for } u \in T(x) \text{ and } v \in T(y).$$

The partially relaxed monotone mappings are weaker than the cocoercive and strongly monotone mappings and, on the top of that, computation-oriented.

Lemma 7.1.1. *For all $v, w \in H$, we have*

$$\|v\|^2 + \langle v, w \rangle \geq - (1/4) \|w\|^2.$$

Lemma 7.1.2. *Let $v, w \in H$. Then we have*

$$\langle v, w \rangle = (1/2) \left[\|v + w\|^2 - \|v\|^2 - \|w\|^2 \right].$$

Lemma 7.1.3. *Let K be a nonempty subset of a real Hilbert space H, and $T: K \rightarrow P(H)$ a multivalued mapping. Then the NVI problem has a solution (x^*, u^*) if and only if x^* is a fixed point of the mapping $F: K \rightarrow P(K)$ defined by*

$$F(x) = \bigcup_{u \in T(x)} \{P_K [x - \rho u]\} \text{ for all } x \in K,$$

where $\rho > 0$ is a constant.

Theorem 7.1.4. *Let H be a real (finite) Hilbert space and $T: K \rightarrow P(H)$ an α-partially relaxed monotone and β-∂-Lipschitz continuous mapping from a non-empty closed convex subset K of H into the power set $P(H)$ of mH. Suppose that (x^*, u^*) is a solution of the NVI problem (7.1.1). Then the sequences $\{x^k\}$ and $\{u^k\}$ generated by the iterative algorithm (7.1.2) satisfy the estimate*

$$\left\| x^{k+1} - x^* \right\|^2 \leq \left\| x^k - x^* \right\|^2 - [1 - 2\rho\alpha] \left\| x^k - x^{k+1} \right\|^2,$$

and converges to x^* and u^*, respectively, a solution of the NVI problem (7.1.1), for $0 < \rho < 1/2\alpha$.

Proof. To show that the sequences $\{x^k\}$ and $\{u^k\}$ generated by the iterative algorithm (7.1.2) converge, respectively, to x^* and u^*, a solution of the NVI problem (7.1.1), we proceeed as follows: because x^{k+1} satisfies the iterative algorithm (7.1.2), we have for a constant $\rho > 0$ that

$$\left\langle \rho u^k + x^{k+1} - x^k, x - x^{k+1} \right\rangle \geq 0 \text{ for all } x \in K \text{ and for } u^k \in T\left(x^k \right). \quad (7.1.4)$$

On the other hand, for constant $\rho > 0$, we have

$$\langle \rho u^*, x - x^* \rangle \geq 0. \tag{7.1.5}$$

Replacing x by x^* in (7.1.4) and x by x^{k+1} in (7.1.5), and adding, we obtain

$$0 \leq \langle \rho \left(u^k - u^* \right), x^* - x^{k+1} \rangle + \langle x^{k+1} - x^k, x^* - x^{k+1} \rangle$$
$$= -\rho \langle u^k - u^*, x^{k+1} - x^* \rangle + \langle x^{k+1} - x^k, x^* - x^{k+1} \rangle.$$

Because T is α-partially relaxed monotone, it implies that

$$0 \leq \rho \alpha \left\| x^{k+1} - x^k \right\|^2 + \langle x^{k+1} - x^k, x^* - x^{k+1} \rangle. \tag{7.1.6}$$

Taking $v = x^{k+1} - x^k$ and $w = x^* - x^{k+1}$ in Lemma 7.1.2, and applying to (7.1.6), we have

$$0 \leq (\rho\alpha) \left\| x^{k+1} - x^k \right\|^2$$
$$+ \frac{1}{2} \left[\left\| x^* - x^k \right\|^2 - \left\| x^{k+1} - x^k \right\|^2 - \left\| x^* - x^{k+1} \right\|^2 \right].$$

It follows that

$$\left\| x^{k+1} - x^* \right\|^2 \leq \left\| x^k - x^* \right\|^2 - [1 - 2\rho\alpha] \left\| x^{k+1} - x^k \right\|^2. \tag{7.1.7}$$

Therefore, we have

$$\left\| x^k - x^* \right\|^2 - \left\| x^{k+1} - x^* \right\|^2 \geq [1 - 2\rho\alpha] \left\| x^{k+1} - x^k \right\|^2.$$

This implies that $\left\{ \left\| x^k - x^* \right\|^2 \right\}$ is a strictly decreasing sequence for $1 - 2\rho\alpha > 0$ and the difference of two successive terms tends to zero. As a result, we have

$$\lim_{k \to \infty} \left\| x^{k+1} - x^k \right\| = 0.$$

Let x' be a cluster point of the sequence $\{x^k\}$. Then there exists a subsequence $\{x^{k_i}\}$ such that $\{x^{k_i}\}$ converges to x'. Finally, the continuity of the projection mapping (7.1.3) and Lemma 7.1.3 imply that x' is a fixed point of (7.1.3). Because $u^{k_i} \in T\left(x^{k_i}\right)$, $u' \in T\left(x^{k_i}\right)$, and T is β-∂-Lipschitz continuous, it implies that

$$\left\| u^{k_i} - u' \right\| \leq \partial \left(T\left(x^{k_i}\right), T\left(x'\right) \right) \leq \beta \left\| x^{k_i} - x' \right\| \to 0,$$

that means, $u^{k_i} \to u'$. Thus, the entire sequences $\{x^k\}$ and $\{u^k\}$ must converge, respectively, to x' and u'. Hence, (x', u') is a solution of the NVI problem (7.1.1).

An application to Theorem 7.1.4, based on an iterative procedure introduced and studied by Marcotte and Wu, to a variational inequality in $\mathbf{R}^n$ is as follows: find an element $x^* \in X$ and $u^* \in F(x^*)$ such that

$$(u^*)\, T\, (x - x^*) \geq 0 \text{ for all } x \in X, \tag{7.1.8}$$

where $F: \mathbf{R}^n \to P(\mathbf{R}^n)$ is a multivalued α-partially relaxed monotone mapping, X a closed convex subset of $\mathbf{R}^n m$, and u^T denotes the transpose of u. The iterative scheme is characterized as a variational inequality as follows: for an arbitrarily chosen initial element $x^0 \in X$,

$$\left[u^k + D\left(x^{k+1} - x^k\right) \right]^T \left(x - x^{k+1} \right) \geq 0 \text{ for all } x \in X \text{ and for } u^k \in F\left(x^k\right), \tag{7.1.9}$$

where D denotes a fixed positive-definite matrix. When the matrix D is symmetric, the above variational inequality iteration is equivalent to the projection formula

$$x^{k+1} = P_D\left[x^k - D^{-1} u^k \right], \tag{7.1.10}$$

where P_D denotes the projection on the set X with the Euclidean matrix norm $\|\cdot\|_D$ induced by a symmetric, positive-definite matrix D, $\|x\|_D = \left(x^T D x\right)^{1/2}$ and $\|x\|$ denotes the Euclidean norm.

Theorem 7.1.5. *Let $F: \mathbf{R}^n \to P(\mathbf{R}^n)$ be an α-partially relaxed monotone and b-∂-Lipschitz continuous mapping. Suppose that (x^*, u^*) is a solution of the variational inequality (7.1.8), the sequences $\{x^k\}$ and $\{u^k\}$ are generated by (7.1.9) and D is a positive-definite and symmetric matrix. Then the sequences $\{x^k\}$ and $\{u^k\}$ satisfy the estimate*

$$\left\| x^{k+1} - x^* \right\|_D^2 \leq \left\| x^k - x^* \right\|_D^2 - [1 - (2\rho\alpha/\lambda \min(D))] \left\| x^k - x^{k+1} \right\|_D^2,$$

and converge to x^* and u^*, respectively, a solution of the NVI problem (7.1.1), for

$$0 < \rho < \lambda \min(D)/2\alpha,$$

where $\lambda \min(D)$ denotes the smallest eigenvalue of D.

Proof. The proof is similar to that of Theorem 7.1.4.

In this section, we provide examples of α-partially relaxed monotone mappings; b-∂-Lipschitz continuous mappings; and an application of Theorem 7.1.5.

Theorem 7.1.6. *Let $P: \mathbf{R}^n \to \mathbf{R}^n$ be given by*

$$P(x) = cI(x) + v,$$

where $c > 0$, $x, v \in \mathbf{R}^n$ with v fixed, and I is the $n \times n$ identity matrix. Then the following conclusions hold
(a) P is an α-partially relaxed monotone mapping for $c = \alpha$.
(b) P is a b-Lipschitz continuous mapping if and only if $b = c$.
(c) If P is an α-partially relaxed monotone mapping then $c < 4\alpha$.

Proof. (a) For all x, y, $z \in \mathbf{R}^n$, we have

$$\|y - z\|^2 + \|y - x\|^2 + \|x - z\|^2 \geq 0$$

from which it follows in turn that

$$\langle y - z, y - z \rangle + \langle y - x, y - x \rangle + \langle x - z, x - z \rangle \geq 0,$$

or

$$-\langle x, y \rangle - \langle y, z \rangle + \langle y, y \rangle + \langle z, z \rangle - \langle z, x \rangle + \langle x, x \rangle \geq 0,$$

or

$$\alpha \left[\langle x - y, z - y \rangle + \langle z - x, z - x \rangle \right] \geq 0$$

or

$$\langle \alpha x - \alpha y, z - y \rangle + c \|z - x\|^2 \geq 0 \ (\text{as } \alpha = c),$$

or

$$\langle P(x) - P(y), z - y \rangle + \left[\alpha \|z - x\|^2 \right] \geq 0,$$

which shows that P is an α-partially relaxed monotone mapping.

(b) The result follows immediately from

$$P(x) - P(y) = cI(x - y).$$

(c) For $x \neq y$, $x \neq 0$, set $y = px$ and $y = qx$ for some $p, q > 0$. It follows from the hypothesis that

$$c(1 - p)(q - p)\langle x, x \rangle + \alpha(q - 1)^2 \langle x, x \rangle \geq 0$$

or

$$cp^2 - c(1 + q)p + \alpha(q - 1)^2 + cq \geq 0.$$

Because $c > 0$, the above inequality will always hold as long as the discriminant of the corresponding quadratic equation in p is negative. The discriminant becomes

$$c(c - 4\alpha)(q - 1)^2 < 0,$$

which holds for $c < 4\alpha$. It can be easily seen from the above proof that the above result holds in an arbitrary space with a real symmetric inner product.

With the above choice of P and for $n = 1$, $v = 0$, we obtain the following application of Theorem 7.1.5.

Example 7.1.7. It can easily be seen that the inequality (7.1.9) for $D = d > 0$ becomes

$$\left[\alpha x^k + d \left(x^{k+1} - x^k \right) \right]^T \left(x - x^{k+1} \right) \geq 0 \text{ for all } x \in \mathbf{R}^n, \ P\left(x^k \right) = \alpha x^k,$$

which leads to

$$\rho \alpha x^k + d \left(x^{k+1} - x^k \right) = 0,$$

or

$$x^{k+1} = \left[(d - \rho \alpha)/d \right] x^k.$$

The above iteration converges for $|(d - \rho\alpha)/d| < 1$ or $\rho < 2d/\alpha$, which is implied by the hypothesis of Theorem 7.1.5 that gives $\rho < d/2\alpha$ in this case. Hence, sequences $\{x^k\}$ and $\{u^k\}$ converge to $(x^*, u^*) = (0, 0)$, a solution of the NVI problem (7.1.1) in this case.

7.2 Monotonicity and solvability of nonlinear variational inequalities

Just recently, Argyros [43] and Verma [201]–[204] applied inexact Newton-like iterative procedures to the approximation-solvability of a class of nonlinear equations in a Banach space setting.

The generalized partial relaxed monotonicity is more general than the other notions of strong monotonicity and cocoercivity.

This section deals with a discussion of the approximation-solvability of the NVIP, based on a general version of the existing auxiliary problem principle (APP) introduced by Cohen [65] and later generalized by Verma [202]. This general version of auxiliary problem principle (GAPP) is stated as follows:

GAPP: For a given iterate x^k, determine an x^{k+1} such that (for $k \geq 0$)

$$\left\langle \rho T\left(x^k\right) + h'\left(x^{k+1}\right) - h'\left(x^k\right), \eta\left(x, x^{k+1}\right)\right\rangle + \rho\left[f(x) - f\left(x^{k+1}\right)\right] \geq$$
$$\geq \left(-\sigma^k\right), \quad \text{for all } x \in K, \tag{7.2.1}$$

where $K' = K \cap \{x : \|x\| \leq c, \text{a large constant}\}$, $h: \mathbf{R}^n \to \mathbf{R}$ is continuously Fréchet-differentiable, $\rho > 0$, a parameter and the sequence $\{\sigma^k\}$ satisfies

$$\sigma^k \geq 0, \sum_{k=1}^{\infty} \sigma^k < \infty. \tag{7.2.2}$$

If K is bounded, then $K = K'$.

Next, we recall some auxiliary results crucial to the approximation-solvability of the NVIP.

Let, $h: Y \to \mathbf{R}$ be a continuously Fréchet-differentiable mapping. It follows that $h'(x) \in L(Y, \mathbf{R})$ the space of bounded linear operators from Y into $\mathbf{R}$. From now on, we denote the real number $h'(x)(y)$ by $\langle h'(x), y\rangle$ for $x, y \in Y$.

Lemma 7.2.1. *Let X and Y be two Banach spaces and K be a nonempty convex subset of X. Suppose that the following assumptions hold:*

(i) There exist an $x^* \in K$ and numbers $\alpha \geq 0, b > 0, r > 0$ such that for all $x \in K_0, t \in [0, 1]$

$$\langle h'\left(x^* + t\eta\left(x, x^*\right)\right) - h'\left(x^*\right), \eta\left(x, x^*\right)\rangle \geq t\alpha \left\|\eta\left(x, x^*\right)\right\|^2,$$

where $h: K \to \mathbf{R}$ is a continuously Fréchet-differentiable mapping, and $\eta: K \times K \to Y$, satisfies:

$$\|\eta\,(x, x^*)\| \leq r; \text{ and } \|\eta\,(x, x^*)\| \geq b\,\|x - x^*\|\,.$$

(ii) The set S_0 defined by

$$S_0 = \{(h, \eta) : h'\,(x^* + t\,(x - x^*))\,(x - x^*) \geq \langle h'\,(x^* + t\eta\,(x, x^*))\,,\eta\,(x, x^*)\rangle\}$$

is nonempty.

(iii) The set

$$K_0 = \overline{U}\,(x^*, r)\,.$$

Then, for all $k \in K_0$ and $(h, \eta) \in S_0$, the following estimate holds

$$h\,(x) - h\,(x^*) - \langle h'\,(x^*)\,, \eta\,(x, x^*)\rangle \geq \frac{\alpha b^2}{2}\,\|x - x^*\|^2\,.$$

Proof. Let $x \in K_0$ and $(h, \eta) \in S_0$. Then we obtain

$$h\,(x) - h\,(x^*) - \langle h'\,(x^*)\,, \eta\,(x, x^*)\rangle =$$

$$= \int_0^1 \left[h'\,(x^* + t\,(x - x^*))\,(x - x^*)\,dt \right] \langle h'\,(x^*)\,, \eta\,(x, x^*)\rangle$$

$$\geq \int_0^1 \langle [h'\,(x^* + t\eta\,(x, x^*))\,, \eta\,(x, x^*)] \rangle\,dt - \langle h'\,(x^*)\,, \eta\,(x, x^*)\rangle$$

$$\geq \int_0^1 \langle [h'\,(x^* + t\eta\,(x, x^*)) - h'\,(x^*)]\,, \eta\,(x, x^*)\rangle\,dt$$

$$\geq \alpha \int_0^1 t\,\|\eta\,(x, x^*)\|^2\,dt$$

$$\geq \frac{\alpha b^2}{2}\,\|x - x^*\|^2\,.$$

This completes the proof.

Lemma 7.2.2. *Let X and Y be two Banach spaces and K be a nonempty convex subset of X. Suppose that the following assumptions hold:*

(i) *There exist an $x^* \in K$ and numbers $\alpha \geq 0$, $b = 1$, $r > 0$ such that for all $x \in K_0, t \in [0, 1]$*

$$\langle h'\,(x^* + t\eta\,(x, x^*)) - h'\,(x^*)\,, \eta\,(x, x^*)\rangle \geq t\alpha\,\|\eta\,(x, x^*)\|^2\,,$$

where $h\colon K \to \mathbf{R}$ is a continuously Fréchet-differentiable mapping, and $\eta\colon K \times K \to Y$, satisfies:

$$\|\eta\,(x, x^*)\| \leq r; \text{ and } \|\eta\,(x, x^*)\| \geq b\,\|x - x^*\|\,.$$

(ii) *The set S_0 defined by*

$$S_0 = \{(h, \eta) : h'\,(x^* + t\,(x - x^*))\,(x - x^*) \geq \langle h'\,(x^* + t\eta\,(x, x^*))\,, \eta\,(x, x^*)\rangle\}$$

is nonempty.

(iii) The set

$$K_0 = \overline{U}\left(x^*, r\right).$$

Then, for all $x \in K_0$ and $(h, \eta) \in S_0$, the following estimate holds

$$h\left(x\right) - h\left(x^*\right) - \langle h'\left(x^*\right), \eta\left(x, x^*\right)\rangle \geqq (\alpha/2)\left\|x - x^*\right\|^2.$$

For $Y = \mathbf{R}$ in Lemma 7.2.2, we arrive at

Lemma 7.2.3. *Let X be a Banach space and K be a nonempty convex subset of X. Suppose that the following assumptions hold:*

(i) There exist an $x^* \in K$ and numbers $\alpha \geqq 0, b = 1, r > 0$ such that for all $x \in K_0, t \in [0, 1]$

$$\langle h'\left(x^* + t\eta\left(x, x^*\right)\right) - h'\left(x^*\right), \eta\left(x, x^*\right)\rangle \geqq t\alpha\left\|\eta\left(x, x^*\right)\right\|^2,$$

where $h \colon K \to \mathbf{R}$ is a continuously Fréchet-differentiable mapping, and $\eta \colon K \times K \to R$, satisfies:

$$\left\|\eta\left(x, x^*\right)\right\| \leqq r; \ \ and \ \ \left\|\eta\left(x, x^*\right)\right\| \geqq b\left\|x - x^*\right\|.$$

(ii) The set S_0 defined by

$$S_0 = \left\{(h, \eta) \colon h'\left(x^* + t\left(x - x^*\right)\right)\left(x - x^*\right) \geqq \langle h'\left(x^* + t\eta\left(x, x^*\right)\right), \eta\left(x, x^*\right)\rangle\right\}$$

is nonempty.

(iii) The set

$$K_0 = \overline{U}\left(x^*, r\right).$$

Then, for all $x \in K_0$ and $(h, \eta) \in S_0$, the following estimate holds

$$h\left(x\right) - h\left(x^*\right) - \langle h'\left(x^*\right), \eta\left(x, x^*\right)\rangle \geqq (\alpha/2)\left\|x - x^*\right\|^2.$$

The, following is a more specialized version of Lemma 7.2.3, more suitable for problems on hand.

Lemma 7.2.4. *Let K be a nonempty convex subset of $\mathbf{R}^n$. Suppose that the following assumptions hold:*

(i) There exist an $x^* \in K$ and numbers $\alpha \geqq 0, b = 1, r > 0$ such that for all $x \in K_0, t \in [0, 1]$

$$\langle h'\left(x^* + t\eta\left(x, x^*\right)\right) - h'\left(x^*\right), \eta\left(x, x^*\right)\rangle \geqq t\alpha\left\|\eta\left(x, x^*\right)\right\|^2,$$

where $h \colon K \to \mathbf{R}$ is a continuously Fréchet-differentiable mapping, and $\eta \colon K \times K \to R$, satisfies:

$$\left\|\eta\left(x, x^*\right)\right\| \leqq r; \ \ and \ \ \left\|\eta\left(x, x^*\right)\right\| \geqq \left\|x - x^*\right\|.$$

(ii) The set S_0 defined by

$$S_0 = \left\{(h, \eta) \colon h'\left(x^* + t\left(x - x^*\right)\right)\left(x - x^*\right) \geqq \langle h'\left(x^* + t\eta\left(x, x^*\right)\right), \eta\left(x, x^*\right)\rangle\right\}$$

is nonempty.
   (iii) The set

$$K_0 = \overline{U}\left(x^*, r\right).$$

Then, for all $x \in K_0$ and $(h, \eta) \in S_0$, the following estimate holds

$$h\left(x\right) - h\left(x^*\right) - \langle h'\left(x^*\right), \eta\left(x, x^*\right)\rangle \geq (\alpha/2)\left\|x - x^*\right\|^2.$$

Lemma 7.2.5. *Let X and Y be two Banach space and K be a nonempty invex subset of X. Suppose that the following assumptions hold:*
(i) There exist an $x^* \in K$ and numbers $\partial \geq 0$, $p > 0$, $q > 0$ such that for all $x \in K_1, t \in [0, 1]$

$$\langle h'\left(x^* + t\eta\left(x, x^*\right)\right) - h'\left(x^*\right), \eta\left(x, x^*\right)\rangle \geq t\partial\left\|\eta\left(x, x^*\right)\right\|^2,$$

where $h: K \to \mathbf{R}$ is a continuously Fréchet-differentiable mapping, and $\eta: K \times K \to Y$, satisfies:

$$\left\|\eta\left(x, x^*\right)\right\| \leq p\left\|x - x^*\right\| \leq q.$$

(ii) The set S_1 defined by

$$S_1 = \left\{(h, \eta) : h'\left(x^* + t\left(x - x^*\right)\right)\left(x - x^*\right) \leq \langle h'\left(x^* + t\eta\left(x, x^*\right)\right), \eta\left(x, x^*\right)\rangle\right\}$$

is nonempty.
(iii) The set

$$K_1 = \overline{U}\left(x, q\right) \subset K.$$

Then, for all $x \in K_1$ and $(h, \eta) \in S_1$, the following estimate holds

$$h\left(x\right) - h\left(x^*\right) - \langle h'\left(x^*\right), \eta\left(x, x^*\right)\rangle \leq \frac{\partial p^2}{2}\left\|x - x^*\right\|^2.$$

Proof. Let $x \in K$ and $(h, \eta) \in S$. Then we obtain

$$h\left(x\right) - h\left(x^*\right) - \langle h'\left(x^*\right), \eta\left(x, x^*\right)\rangle$$

$$= \int_0^1 \left[h'\left(x^* + t\left(x - x^*\right)\right)\left(x - x^*\right) dt\right] - \langle h'\left(x^*\right), \eta\left(x, x^*\right)\rangle$$

$$\leq \int_0^1 \left\langle\left[h'\left(x^* + t\eta\left(x, x^*\right)\right), \eta\left(x, x^*\right)\right]\right\rangle dt - \langle h'\left(x^*\right), \eta\left(x, x^*\right)\rangle$$

$$\leq \int_0^1 \left\langle\left[h'\left(x^* + t\eta\left(x, x^*\right)\right) - h'\left(x^*\right)\right], \eta\left(x, x^*\right)\right\rangle dt$$

$$\leq \partial \int_0^1 t\left\|\eta\left(x, x^*\right)\right\|^2 dt$$

$$\leq \frac{\partial p^2}{2}\left\|x - x^*\right\|^2,$$

which completes the proof.

We are just about ready to present, based on the GAPP, the approximation-solvability of the NVIP.

Theorem 7.2.6. *Let* $T: K \to \mathbf{R}^n$ *be* η-γ-μ-*partially relaxed monotone from a nonempty closed invex subset* K *of* $\mathbf{R}^n$ *into* $\mathbf{R}^n$. *Let* $f: K \to \mathbf{R}$ *be proper, invex, and lower semicontinuous on* K *and* $h: K \to \mathbf{R}$ *be a continuously Fréchet-differentiable on* K. *Suppose that there exist an* $x' \in K$ *and nonnegative numbers* α, ∂, $\sigma^k \in K$ $(k \geq 1)$ *such that for all* $t \in [0, 1]$, *and* $x \in K_0 \cap K_1$, *we have*

$$\langle h'(x' + t\eta(x, x')) - h'(x'), \eta(x, x') \rangle \geq t\alpha \|x - x'\|^2, \qquad (7.2.3)$$

$$\langle h'(x' + t\eta(x, x')) - h'(x'), \eta(x, x') \rangle \leq t\partial \|x - x'\|^2, \qquad (7.2.4)$$

and

$$\sum_{k=1}^{\infty} \sigma^k < \infty, \qquad (7.2.5)$$

where $\eta: K \times K \to \mathbf{R}^n$ *is* λ-*Lipschitz continuous with the following assumptions:*

(i) $\eta(x, y) + \eta(y, x) = 0$
(ii) For each fixed $y \in K$, *map* $x \to \eta(y, x)$ *is sequentially continuous from the weak topology to the weak topology in the second variable.*
(iii) η *is expanding.*
(iv) The set S *defined by*

$$S = \{(h, \eta) : h'(x' + t(x - x'))(x - x') \geq \langle h'(x' + t\eta(x, x')), \eta(x, x') \rangle\}$$

is nonempty.

If in addition, $x^* \in K$ is any fixed solution of the NVIP and

$$0 < \rho < (\alpha/2\gamma),$$

then the sequence $\{x^k\}$ converges strongly to x^*.

Proof. To show the sequences $\{x^k\}$ converges to x^*, a solution of the NVIP, we need to compute the estimates. Let us define a function Λ^* by

$$\Lambda^*(x) := h(x^*) - h(x) - \langle h'(x), \eta(x^*, x) \rangle.$$

Then, by Lemma 7.2.4, we have

$$\Lambda^*(x) := h(x^*) - h(x) - \langle h'(x), \eta(x^*, x) \rangle \geq (\alpha/2) \|x^* - x\|^2 \text{ for } x \in K,$$
$$(7.2.6)$$

where x^* is any fixed solution of the NVIP. It follows that

$$\Lambda^*(x^{k+1}) = h(x^*) - h(x^{k+1}) - \langle h'(x^{k+1}), \eta(x^*, x^{k+1}) \rangle. \qquad (7.2.7)$$

Now we can write

$$\Lambda^* \left(x^k \right) - \Lambda^* \left(x^{k+1} \right) = \tag{7.2.8}$$

$$= h \left(x^{k+1} \right) - h \left(x^k \right) - \left\langle h' \left(x^k \right), \eta \left(x^{k+1}, x^k \right) \right\rangle$$

$$+ \left\langle h' \left(x^{k+1} \right) - h' \left(x^k \right), \eta \left(x^*, x^{k+1} \right) \right\rangle$$

$$\geqq (\alpha/2) \left\| x^{k+1} - x^k \right\|^2 + \left\langle h' \left(x^{k+1} \right) - h' \left(x^k \right), \eta \left(x^*, x^{k+1} \right) \right\rangle$$

$$\geqq (\alpha/2) \left\| x^{k+1} - x^k \right\|^2 + \rho \left\langle t \left(x^K \right), \eta \left(x^{k+1}, x^* \right) \right\rangle$$

$$+ \rho \left(f \left(x^{k+1} \right) - f \left(x^* \right) \right) - \sigma^k$$

for $x = x^*$ in (7.2.1).

If we replace x by x^{k+1} above and combine with (7.2.8), we obtain

$$\Lambda^* \left(x^k \right) - \Lambda^* \left(x^{k+1} \right) \geqq$$

$$\geqq [\alpha/2] \left\| x^{k+1} - x^k \right\|^2 + \rho \left\langle T \left(x^k \right), \eta \left(x^{k+1}, x^* \right) \right\rangle$$

$$- \rho \left\langle T \left(x^* \right), \eta \left(x^{k+1}, x^* \right) \right\rangle - \sigma^k$$

$$= [\alpha/2] \left\| x^{k+1} - x^k \right\|^2 + \rho \left\langle T \left(x^k \right) - T \left(x^* \right), \eta \left(x^{k+1}, x^* \right) \right\rangle - \sigma^k.$$

Because T is η-γ-μ-partially relaxed monotone, it implies that

$$\Lambda^* \left(x^k \right) - \Lambda^* \left(x^{k+1} \right) \geqq$$

$$\geqq [\alpha/2] \left\| x^{k+1} - x^k \right\|^2 - \rho\gamma \left\| x^{k+1} - x^k \right\|^2 + \rho\mu \left\| x^k - x^* \right\|^2 - \sigma^k$$

$$\geqq [\alpha/2] \left\| x^{k+1} - x^k \right\|^2 - \rho\gamma \left\| x^{k+1} - x^k \right\|^2 + \rho\mu \left\| x^k - x^* \right\|^2 - \sigma^k$$

$$= (1/2) [\alpha - 2\rho\gamma] \left\| x^{k+1} - x^k \right\|^2 + \rho\mu \left\| x^k - x^* \right\|^2 - \sigma^k$$

$$\geqq (1/2) [\alpha - 2\rho\gamma] \left\| x^{k+1} - x^k \right\|^2 - \sigma^k \tag{7.2.9}$$

$$\geqq \left(-\sigma^k \right) \text{ for } \alpha - 2\rho\gamma > 0. \tag{7.2.10}$$

That is,

$$\Lambda^* \left(x^k \right) - \Lambda^* \left(x^{k+1} \right) \geqq \left(-\sigma^k \right). \tag{7.2.11}$$

It follows that

$$\Lambda^* \left(x^{k+1} \right) - \Lambda^* \left(x^k \right) \leqq \sigma^k. \tag{7.2.12}$$

If we sum from $k = 1, 2, ..., N$, we arrive at

$$\sum_{k=1}^{N} \left[\Lambda^* \left(x^{k+1} \right) - \Lambda^* \left(x^k \right) \right] \leqq \sum_{k=1}^{\infty} \sigma^k.$$

As a result of this, we can get

$$\Lambda^* \left(x^{N+1} \right) - \Lambda^* \left(x^1 \right) \leqq \sum_{k=1}^{\infty} \sigma^k. \tag{7.2.13}$$

It follows using (7.2.7) from (7.2.13) that

$$[\alpha/2] \left\| x^{N+1} - x^* \right\|^2 \leqq \Lambda^* \left(x^1 \right) + \sum_{k=1}^{\infty} \sigma^k. \tag{7.2.14}$$

Under the hypotheses of the theorem, it follows from (7.2.14) that the sequence $\{x^k\}$ is bounded and

$$\lim_{k \to \infty} \left\| x^k - x^* \right\| = 0.$$

Thus, sequence $\{x^k\}$ converges to x^*, a solution of the NVIP.

When $\eta(x, y) = x - y$ in Theorem 7.2.6, we arrive at:

Theorem 7.2.7. *Let $T: K \to \mathbf{R}^n$ be γ-μ-partially relaxed monotone from a non-empty closed invex subset K of $\mathbf{R}^n$ into $\mathbf{R}^n$. Let $f: K \to \mathbf{R}$ be proper, invex, and lower semicontinuous on K and $h: K \to \mathbf{R}$ be continously Fréchet-differentiable on K. Suppose that there exist an $x' \in K_0 \cap K_1$ and $t \in [0, 1]$, $(h, \eta) \in S$, we have*

$$\langle h'(x' + t(x - x')) - h'(x'), x - x' \rangle \geqq t\alpha \left\| x - x' \right\|^2,$$

$$\langle h'(x' + t(x - x')) - h'(x'), x - x' \rangle \leqq t\partial \left\| x - x' \right\|^2,$$

and

$$\sum_{k=1}^{\infty} \sigma^k < \infty.$$

If in addition, $x^* \in K$ is any fixed solution of the NVIP and

$$0 < \rho < (\alpha/2\gamma),$$

then the sequence $\{x^k\}$ converges strongly to x^*.

Remark 7.2.8. The set S is nonempty in many interesting cases, for example, take

$$\eta(x, x') = x - x', h: \mathbf{R} \to \mathbf{R} \text{ and } \eta: \mathbf{R} \times \mathbf{R} \to \mathbf{R}.$$

Remark 7.2.9. In order for us to have some insight into the structure of K, let us assume $\langle z, z \rangle^{1/2} = \|z\|$ and consider the Cauchy-Schwarz inequality

$$|\langle x, y \rangle| \leqq \|x\| \|y\|$$

in the first estimate hypotheses in Lemma 7.2.1. Moreover, assume:

$$\left\| h'(x) - h'(y) \right\| \leqq \ell^2 \|x - y\| \text{ for all } mx, y \in K \text{ and some } \ell \geqq 0.$$

We then have in turn that

$$t\alpha \left\| \eta \left(x,x^* \right)^2 \right\| \leqq \langle h' \left(x^* + t\eta \left(x,x^* \right) \right) - h' \left(x,x^* \right), \eta \left(x,x^* \right) \rangle$$
$$\leqq \sqrt{\left\| h' \left(x^* + t\eta \left(x,x^* \right) \right) - h' \left(x,x^* \right) \right\| \, \left\| \eta \left(x,x^* \right) \right\|}$$
$$\leqq \sqrt{\ell^2 t \left\| \eta \left(x,x^* \right) \right\| \, \left\| \eta \left(x,x^* \right) \right\|}$$
$$\leqq \left(\ell\sqrt{t} \right) \left\| \eta \left(x,x^* \right) \right\|$$

or

$$\sqrt{t}\alpha \left\| \eta \left(x,x^* \right) \right\| \leqq \ell \text{ for } \left\| \eta \left(x,x^* \right) \right\| \neq 0.$$

But

$$b\sqrt{t}\alpha \left\| x - x^* \right\| \leqq \sqrt{t}\alpha \left\| \eta \left(x,x^* \right) \right\| \leqq \ell,$$

which shows that definitely K should be a subset of $\overline{U}\left(x^*, r^* \right)$, where

$$r^* = \ell/b\alpha \text{ for } b, \alpha \neq 0.$$

7.3 Generalized variational inequalities

Let M, $\langle \cdot, \cdot \rangle$ denote the dual, inner product and norm of a Hilbert space H, respectively. Let C be a closed convex subset of H. For $G, F: H \to H$ continuous operators, we study the problem of approximating $x \in H$ such that

$$\langle G(x), F(y) - F(x) \rangle \geq 0 \quad \text{for all } F(x), F(y) \in C. \tag{7.3.1}$$

This is the so-called general nonlinear variational inequality problem. Special cases of this problem have already been studied: in [149], when $F(x) = x \in C$; if $C^* = \{x \in H, \langle x, y \rangle \geq 0, y \in C\}$ is a polar cone of the convex cone C in H; in [149], when $C = H$; and in [151] under stronger conditions than ours.

It is well-known that if C is a convex subset of H, then $x \in H$ is a solution of (7.3.1) if and only if x satisfies

$$F(x) = P_C \left[F(x) - \rho G(x) \right], \tag{7.3.2}$$

where $\rho > 0$ is a constant and P_C is a projection of H into H. Hence (7.3.1) can be seen as a fixed point problem of the form

$$x = Q(x), \tag{7.3.3}$$

where

$$Q(x) = x - F(x) + P_C \left[F(x) - \rho G(x) \right]. \tag{7.3.4}$$

That is, (7.3.4) suggests the following iterative procedure: given $x_0 \in H$, find x_{n+1} using the approximation

$$x_{n+1} = x_n - F(x_n) + P_C \left[F(x_n) - \rho G(x_n) \right] \quad (n \geq 0). \tag{7.3.5}$$

We assume:

$$\|F(x^*) - F(y)\| \leq c_1 \|x^* - y\|^{\lambda_1}, \quad c_1 \geq 0, \ \lambda_1 \geq 1, \ y \in H \quad (7.3.6)$$

$$\|G(x^*) - G(y)\| \leq c_2 \|x^* - y\|^{\lambda_2}, \quad c_2 \geq 0, \ \lambda_2 \geq 1, \ y \in H \quad (7.3.7)$$

$$\langle G(x^*) - G(y), x^* - y \rangle \geq c_3 \|x^* - y\|^2, \quad c_3 \geq 0, \ y \in H \quad (7.3.8)$$

$$\langle F(x^*) - F(y), x^* - y \rangle \geq c_4 \|x^* - y\|^2, \quad c_4 \geq 0, \ y \in H, \quad (7.3.9)$$

where, x^* is a solution of (7.3.1).

Define the parameter θ by

$$\theta = 2\sqrt{1 + c_1^2 c_0^{2(\lambda_1 - 1)} - 2c_4} + \sqrt{1 + \rho^2 c_2^2 c_0^{2(\lambda_2 - 1)} - 2\rho c_3}, \quad (7.3.10)$$

where for given $x_0 \in H$,

$$\|x_0 - x^*\| \leq c_0. \quad (7.3.11)$$

We can now show the following convergence result for general iterative procedure (7.3.5).

Theorem 7.3.1. *Assume:*

(i) Operators F, G satisfy (7.3.6)–(7.3.9);
(ii) x^*, x_{n+1} ($n \geq 0$) solve (7.3.1), (7.3.5), respectively; and
(iii) $\theta \in [0, 1)$, where θ is given by (7.3.10) for sufficiently small c_0 and ρ.

Then, general iterative procedure $\{x_n\}$ ($n \geq 0$) generated by (7.3.5) is well defined for all $n \geq 0$ and converges (strongly in H) to x^*.

Proof. It follows from (7.3.4) and (7.3.5) by using (7.3.6)–(7.3.11)

$$\|x_{n+1} - x^*\| =$$
$$= \|x_n - x^* - (F(x_n) - F(x^*))$$
$$+ P_C [F(x_n) - \rho G(x_n)] - P_C [F(x^*) - \rho G(x^*)]\|$$
$$\leq \|x_n - x^* - (F(x_n) - F(x^*))\|$$
$$+ \|P_C [F(x_n) - \rho G(x_n)] - P_C [F(x^*) - \rho F(x^*)]\|$$
$$\leq 2 \|x_n - x^* - (F(x_n) - F(x^*))\| + \|x_n - x^* - \rho (G(x_n) - G(x^*))\|$$
$$\leq 2\sqrt{1 + c_1^2 \|x_n - x^*\|^{2(\lambda_1 - 1)} - 2c_4} \|x_n - x^*\|$$
$$+ \sqrt{1 + \rho^2 c_2^2 \|x_n - x^*\|^{2(\lambda_2 - 1)} - 2\rho c_3} \|x_n - x^*\|$$
$$\leq \theta \|x_n - x^*\| \leq \theta^{n+1} c_0. \quad (7.3.12)$$

Hence, by (iii) and (7.3.12), we get $\lim_{n \to \infty} x_n = x^*$.

Remark 7.3.2. As mentioned in the introduction, special choices of F, G can reduce Theorem 7.3.1 to earlier ones. For example, let $\lambda_1 = \lambda_2 = 1$ and assume the stronger hypotheses $(7.3.6)'–(7.3.9)'$ where x^* is replaced by any $x \in H$. In the special case Theorem 7.3.1 becomes replaced by Theorem 3.1 in [151]. Moreover if $F = I$, the identity operator, then (7.3.4) becomes the classical problem studied in [149], [150]. Furthermore, in these special cases, (iii) from Theorem 7.3.1 can be dropped and be replaced by

$$\left| \rho - \frac{c_3}{c_2^2} \right| < \sqrt{\frac{c_3^2 - c_2^2(d - d^2)}{c_2^2}}, \quad c_3 > c_2\sqrt{d(d-2)},$$

$$d < 1, \quad d = 2\sqrt{1 - 2c_4 + c_1^2},$$

and

$$0 < \rho < \frac{2c_3}{c_2^2},$$

respectively.

Remark 7.3.3. Condition (iii) can be dropped in other cases not covered by Remark 7.3.2. For example:
Assume:

(1) $1 - 2c_4 \geq 0, 1 - 2\rho c_3 \geq 0, \lambda_1 > 1, \lambda_2 > 1$;
(2) choose $d_1 \geq 0, d_2 \geq 0$ such that:

$$2\sqrt{d_1} + \sqrt{d_2} \equiv d_3 < 1,$$
$$d_1 \geq 1 - 2c_4, \quad d_2 \geq 1 - 2\rho c_3$$
$$1 + c_1^2 c_0^{2(\lambda_1 - 1)} - 2c_4 \leq d_1, \quad 1 + \rho^2 c_2^2 c_0^{2(\lambda_2 - 1)} - 2\rho c_3 \leq d_2;$$

where,

$$d_4 = \left(\frac{2c_4 + d_1 - 1}{c_1^2} \right)^{\frac{1}{2(\lambda_1 - 1)}}, \quad \text{and} \quad d_5 = \left(\frac{2\rho c_3 + d_2 - 1}{c_2^2} \right)^{\frac{1}{2(\lambda_2 - 1)}}.$$

Then, $\theta = d_3 \in [0, 1)$.

Remark 7.3.4. Parameters c_1, c_2, c_3, c_4 appearing in (7.3.6)–(7.3.9) are smaller (in general) than the corresponding ones in [150]. Hence, the ratio of convergence is smaller also. That is under our weaker hypotheses, sequence $\{x_n\}$ ($n \geq 0$) converges faster to x^* than in [150].

7.4 Semilocal convergence

In this section, we are concerned with the problem of approximating a locally unique solution x^* of the variational inequality

$$F(x) + \partial\varphi(x) \ni 0, \tag{7.4.1}$$

where F is a Gâteaux-differentiable operator defined on a closed convex subset D of a Hilbert space H with values in H; $\varphi: H \rightarrow (-\infty, \infty]$ is a lower semicontinuous convex function. Problems of the form (7.4.1) have important applications in many branches of applied science (physical, engineering, etc.).

We use the generalized chord method

$$F'(x_0)x_{n+1} + \partial\varphi(x_{n+1}) \ni F'(x_0)(x_n) - F(x_n) \quad (x_0 \in D) \tag{7.4.2}$$

and the generalized NK method

$$F'(x_n)(x_{n+1}) + \partial\varphi(x_{n+1}) \ni F'(x_n)(x_n) - F(x_n) \quad (x_0 \in D) \tag{7.4.3}$$

to approximate x^*.

We assume that φ is proper in the sense that

$$D(\varphi) = \{\varphi \in H : \varphi(x) < \infty\} \neq \emptyset.$$

For any $x \in H$, we denote by $\partial\varphi(x)$ the subgradient of φ at x, given by

$$\partial\varphi(x) = \{y \in H : \varphi(x) - \varphi(z) \leq \langle y, x - z \rangle \text{ for all } y \in D(\varphi)\}.$$

Semilocal convergence theorems for solving (7.4.1) using (7.4.2) or (7.4.3) are given here using hypotheses on the second Gâteaux derivative of F. We also show that our results compare favorably with relevant earlier ones [149]–[151].

Lemma 7.4.1. *Let $a > 0$, $b \geq 0$, $\eta \geq 0$ and $c > 0$ be constants. Define the polynomial p by*

$$p(r) = \frac{a}{6c}r^3 + \frac{b}{2c}r^2 - r + \frac{\eta}{c}. \tag{7.4.4}$$

The polynomial p has two positive zeros r_1, r_2 ($r_1 \leq r_2$) if and only if

$$p(q) \leq 0, \tag{7.4.5}$$

where q is the positive zero of p'.

Proof. Denote by q_0 the negative zero of p'. Clearly, p has a maximum at $r = q_0$ and a minimum at $r = q$. Hence a necessary and sufficient condition for p to have positive zeros is given by (7.4.5).

Lemma 7.4.2. *Let $a \geq 0$, $b \geq 0$, $r > 0$ be constants and $z \in H$ be fixed. Assume:*

$$\|F''(x) - F''(z)\| \leq a\|x - z\| \tag{7.4.6}$$

and

$$\|F''(z)\| \leq b \tag{7.4.7}$$

for all $x \in \bar{U}(z, r)$. Then the following estimate holds for all $x, y \in U(z, r)$:

$$\|F(x) - F(y) - F'(z)(x - y)\| \leq \frac{a}{2} \int_0^1 [(1-t)\|y - z\| + t\|x - z\|]^2 \|x - y\| dt$$

$$+ \frac{b}{2} [\|y - z\| + \|x - z\|] \|x - y\|. \tag{7.4.8}$$

We can show the following semilocal convergence theorem using twice Gâteaux-differentiable operators and the generalized method of chord:

Theorem 7.4.3. *Let F be a twice Gâteaux-differentiable operator defined on an open convex subset D of a Hilbert space H with values in H. Assume:*

(a) there exists $x_0 \in H$ satisfying

$$\partial\varphi(x_0) \ni 0, \tag{7.4.9}$$

where φ is a convex function;
(b) there exist constants η, c such that

$$\|F(x_0)\| \le \eta, \tag{7.4.10}$$

$$c\|y\|^2 \le \langle F'(x_0)(y), y \rangle \quad \text{for all } y \in D; \tag{7.4.11}$$

(c) operator F satisfies (7.4.6) and (7.4.7) for $z = x_0$;
(d) condition (7.4.5) holds for $a > 0$; and
(e) $U(x_0, r_2) \subseteq D$, where r_2 is given in Lemma 7.4.1.

Then, sequence $\{x_n\}$ ($n \ge 0$) generated by the generalized method of chord (7.4.2) is well defined, remains in $U(x_0, r_1)$, and converges to a solution x^* of (7.4.1), which is unique in $U(x_0, r_2)$.

Proof. The coercivity condition (7.4.11) and the Lions–Stampacchia Theorem imply that for any $x \in U(x_0, r_2)$ the operator g given by the variational inequality

$$F'(x_0)g(x) + \partial\varphi(g(x)) \ni F'(x_0)x - F(x) \tag{7.4.12}$$

is well defined. By (7.4.9), (7.4.12), and the monotonicity of $\partial\varphi$, we have in turn

$$\langle F(x) - F'(x_0)(x - g(x)), g(x) - x_0 \rangle \le 0,$$

or

$$\langle F'(x_0)(g(x) - x_0), g(x) - x_0 \rangle \le \langle F'(x_0)(x - x_0) - F(x), g(x) - x_0 \rangle,$$

and using (7.4.4)–(7.4.11), we get

$$
\begin{aligned}
\|g(x) - x_0\| &\le \frac{1}{c} \|F(x_0) - F'(x_0)(x - x_0)\| \\
&= \frac{1}{c} \|F(x_0) + F(x) - F(x_0) - F'(x_0)(x - x_0)\| \\
&\le \frac{1}{c} \|F(x_0)\| + \frac{1}{c} \|F(x) - F(x_0) - F'(x_0)(x - x_0)\| \\
&\le \frac{\eta}{c} + \frac{1}{6c} \|x - x_0\|^3 + \frac{b}{2c} \|x - x_0\|^2 \\
&\le \frac{\eta}{6} + \frac{a}{6c} r^3 + \frac{b}{2c} r^2 \le r
\end{aligned} \tag{7.4.13}
$$

for $r \in [r_1, r_2]$. That is, g maps $U(x_0, r)$ $(r_1 \le r \le r_2)$ into itself. Moreover (7.4.2) and the monotonicity of $\partial\varphi$ imply that for any $x, y \in U(x_0, r)$ we can have in turn

$$\langle F'(x_0)(g(x) - g(y)) + F(x) - F(y) - F'(x_0)(x - y), g(x) - g(y)\rangle \le 0,$$

$$\langle F'(x_0)(g(x) - g(y)), g(x) - g(y)\rangle \le \langle F'(x_0)(x - y) - F(x) + F(y), g(x) - g(y)\rangle,$$

and by (7.4.5)–(7.4.11), we get

$$\|g(x) - g(y)\| \le \frac{1}{c} \|F(x) - F(y) - F'(x_0)(x - y)\|$$

$$\le \frac{1}{c} \left\{ \frac{a}{2} \int_0^1 [(1 - t)\|y - x_0\| + t\|x - x_0\|]^2 \, dt \right.$$

$$\left. + \frac{b}{2} [\|y - x_0\| + \|x - x_0\|] \right\} \|x - y\|$$

$$\le \frac{1}{c} \left[\frac{a}{2} r + b \right] r \|x - y\|. \tag{7.4.14}$$

It follows from Lemma 7.4.1 and (7.4.14) that g is a contraction on $U(x_0, r)$ $(r_1 \le r \le r_2)$. The rest of the theorem follows from the Banach fixed point theorem and the observation that (7.4.2) is given by $x_{n+1} = g(x_n)$ $(n \ge 0)$.

The convergence in Theorem 7.4.3 is only linear. It can become quadratic if we use the set of conditions given in the following result:

Theorem 7.4.4. *Let F be a twice Gâteaux-differentiable operator defined on an open convex subset D of a Hilbert space H with values in H. Assume:*

(a) there exists $x_0 \in H$ satisfying

$$\partial\varphi(x_0) \ni 0,$$

   where φ is a convex function;
(b) there exist constants η, c such that

$$\|F(x_0)\| \le \eta,$$

$$c\|y\|^2 \le \langle F'(x)(y), y\rangle \quad \text{for all} \quad y \in H, \ x \in D; \tag{7.4.15}$$

(c) operator F satisfies (7.4.6) and (7.4.7) for $z = x_0$;
(d)

$$d = \alpha^{-1} \frac{\eta}{c} < 1, \tag{7.4.16}$$

   where,

$$\alpha^{-1} = \frac{1}{2c} \left(\frac{a\eta}{3c} + b \right) \tag{7.4.17}$$

   and

$$\alpha \sum_{i=0}^{\infty} d^{2^i} < r_0 \tag{7.4.18}$$

for some positive parameter r_0.

Then, sequence $\{x_n\}$ $(n \geq 0)$ generated by the generalized Newton's method (7.4.3) is well defined, remains in $U(x_0, r_0)$ for all $n \geq 0$ and converges to a unique solution x^* of (7.4.1) in $U(x_0, r_0)$. Moreover the following estimates hold for all $n \geq 0$

$$\|x_n - x^*\| \leq \alpha \sum_{i=n}^{\infty} d^{2^i}. \tag{7.4.19}$$

Proof. As in Theorem 7.4.3 the solution x_{n+1} of (7.4.3) exist for all $n \geq 0$. Using (7.4.3) for $n = 1$ as in Theorem 7.4.3 we obtain in turn

$$\langle F(x_0) - F'(x_0)(x_0 - x_1), x_1 - x_0 \rangle \leq 0,$$

or

$$\langle F'(x_0)(x_1 - x_0), x_1 - x_0 \rangle \leq \langle F(x_0), x_1 - x_0 \rangle,$$

and by (7.4.9) and (7.4.15), we get

$$\|x_1 - x_0\| \leq \frac{1}{c}\|F(x_0)\| \leq \frac{\eta}{c}. \tag{7.4.20}$$

Moreover by (7.4.3) and the monotonicity of $\partial \varphi$, we get

$$\langle F'(x_n)(x_{n+1} - x_n), x_{n+1} - x_n \rangle$$
$$\leq \langle F(x_n) - F(x_{n-1}) - F'(x_{n-1})(x_n - x_{n-1}), x_{n+1} - x_n \rangle,$$

and by (7.4.6), (7.4.7), (7.4.15), we obtain

$$\|x_{n+1} - x_n\| \leq \frac{1}{c}\|F(x_n) - F(x_{n-1}) - F'(x_{n-1})(x_n - x_{n-1})\|$$
$$\leq \frac{1}{c}\left[\frac{a}{6}\|x_n - x_{n-1}\| + \frac{b}{2}\right]\|x_n - x_{n-1}\|^2$$
$$\leq \alpha^{-1}\|x_n - x_{n-1}\|^2 \leq \alpha d^{2^n},$$

which leads to (7.4.19).

7.5 Results on generalized equations

In this section we are concerned with the problem of approximating a locally unique solution x^* of the problem

$$f(x) + g(x) \ni 0, \tag{7.5.1}$$

where f is a twice Gâteaux-differentiable operator defined on a Hilbert space H with values in H, and g is a multivalued (possible) operator from H into H.

We use the famous generalized Newton's method

$$x_{n+1} = (f'(x_n) + g)^{-1}(f'(x_n)(x_n) - f(x_n)) \quad (n \geq 0) \tag{7.5.2}$$

to approximate x^*.

In the special case, when $g = 0$, we obtain the classic NK method. Method (7.5.2) can be used where the classic method cannot by assuming several regularity assumptions for g.

Local and semilocal convergence theorems were given in [43], [149]–[151]. Especially in [151], Lipschitz hypotheses were used on the first Gâteaux derivative $f'(x)$ of $f(x)$. Here we use Lipschitz hypotheses on the second Gâteaux derivative $f''(x)$ of $f(x)$. This way our convergence conditions differ from earlier ones unless if the Lipschitz constant is zero. We complete this study with a numerical example to show that our results apply where corresponding earlier results [151] do not.

By a multivalued operator g from H into H being monotone, we mean

$$y_1 \in g(x_1), \ y_2 \in g(x_2) \Rightarrow \langle y_1 - y_2, x_1 - x_2 \rangle \geq 0.$$

Moreover, g is maximal if whenever g_0 is another multivalued monotone operator from H into H such that $y \in g(x) \Rightarrow y \in g_0(x)$, then $g = g_0$.

We can show the following local convergence theorem for the generalized NK method.

Theorem 7.5.1. *Let g be a maximal monotone multivalued operator from a Hilbert space H into itself and f be a Gâteaux-differentiable operator from H into H. Assume: there exist parameters $a > 0$, $b > 0$, $c > 0$ and a solution x^* of (7.5.1) such that:*

$$\|F''(x) - F''(y)\| \leq a\|x - y\|, \tag{7.5.3}$$
$$\|F''(x^*)\| \leq b, \tag{7.5.4}$$

and

$$\langle f'(y)(x), x \rangle \geq c\|x\|^2 \quad \text{for all } x, y \in H. \tag{7.5.5}$$

Then, generalized NK method $\{x_n\}$ $(n \geq 0)$ generated by (7.5.2) is well defined, remains in $U(x^*, \alpha)$ for all $n \geq 0$, and converges to x^* provided that $x_0 \in U(x^*, \alpha)$, and

$$\|x_n - x^*\| \leq \alpha d^{2^n} \quad (n \geq 0), \tag{7.5.6}$$

where, α is the positive zero of the equation

$$\frac{a}{3c}r^2 + \frac{b}{2c}r - 1 = 0, \tag{7.5.7}$$

and

$$d = \alpha^{-1}\|x_0 - x^*\|. \tag{7.5.8}$$

Proof. It follows from the choice of α that $d \in (0, 1)$. We note that all inverses $(f'(x_n) + g)^{-1}$ $(n \geq 0)$ exist, as g is a maximal monotone operator and $\langle f'(x_n)(x), x \rangle / \|x\| \to \infty$ as $\|x\| \to \infty$. Hence, generalized Newton's method $\{x_n\}$ $(n \geq 0)$ generated by (7.5.2) is well defined for all $n \geq 0$. By (7.5.5), (7.5.2), the monotonicity of g and $f'(x_n)(x^*) + g(x^*) \Rightarrow f'(x_n)(x^*) - f(x^*)$, we obtain in turn

$$c\|x_{n+1} - x^*\|^2 \leq$$
$$\leq \langle f'(x_n)(x_{n+1}) - f'(x_n)(x^*), x_{n+1} - x^* \rangle$$
$$\leq \langle f'(x_n)(x_n - x^*), x_{n+1} - x^* \rangle + \langle f(x^*) - f(x_n), x_{n+1} - x^* \rangle$$
$$\leq \langle f(x^*) - f(x_n) - f'(x_n)(x^* - x_n), x_{n+1} - x^* \rangle. \tag{7.5.9}$$

We showed (see, e.g., [43]) that under (7.5.3) and (7.5.4)

$$\langle f(x^*) - f(x_n) - f'(x_n)(x^* - x_n) \rangle$$
$$\leq \left[\frac{a}{6} \|x_n - x^*\| + \frac{b}{2} \right] \|x_n - x^*\|^2 \cdot \|x_{n+1} - x^*\|. \tag{7.5.10}$$

Using (7.5.8)–(7.5.10), we get

$$\|x_{n+1} - x^*\| \leq \alpha^{-1} \|x_n - x^*\|^2 \leq \alpha d^{2^n} \quad (n \geq 0).$$

The result follows by induction on the integer $n \geq 0$.

We show the semilocal convergence theorem for the generalized NK method.

Theorem 7.5.2. *Let g be a maximal, and continuous single-valued operator from a Hilbert space H into itself and f be a Gâteaux-differentiable operator from H into H.*

(a) there exist constants $a > 0$, $b > 0$, $\eta \geq 0$ such that (7.5.3) (with $y = x_0$), (7.5.4) and

$$\|f(x_0) + g(x_0)\| \leq \eta \tag{7.5.11}$$

hold;
(b) there exist $r_0 > 0$ such that

$$d_0 = \frac{1}{2c^2} \left[\frac{1}{3} a r_0 + b \right] \eta < 1 \tag{7.5.12}$$

and

$$\frac{\eta}{c} \sum_{i=0}^{\infty} d_0^{2^i - 1} < r_0. \tag{7.5.13}$$

Then, generalized NK method $\{x_n\}$ $(n \geq 0)$ generated by (7.5.2) is well defined, remains in $U(x_0, r_0)$ for all $n \geq 0$ and converges to a solution x^* of (7.5.1), so that

$$\|x_n - x^*\| \leq \sum_{i=n}^{\infty} d_0^{2^i - 1} \quad (n \geq 0). \tag{7.5.14}$$

Proof. As in Theorem 7.5.1, iterates $\{x_n\}$ $(n \geq 0)$ are well defined for all $n \geq 0$. Using (7.5.2), (7.5.5), and (7.5.11), we get

$$c\|x_1 - x_0\|^2 \leq \langle f'(x_0)(x_1 - x_0) + g(x_1) - g(x_0), x_1 - x_0 \rangle$$
$$= -\langle f(x_0) + g(x_0), x_1 - x_0 \rangle$$

or

$$\|x_1 - x_0\| \leq \frac{1}{4}\|f(x_0) + g(x_0)\| \leq \frac{\eta}{c} < r_0.$$

We will show

$$\|x_i - x_{i-1}\| \leq \frac{\eta}{c}d_0^{2^{i-1}-1} \tag{7.5.15}$$

and

$$\|x_i - x_0\| \leq r_0. \tag{7.5.16}$$

By (7.5.2) and (7.5.11), (7.5.15)–(7.5.16) hold for $i = 0$. Assume (7.5.15) and (7.5.16) hold for all integer $i \leq n$. Using (7.5.2), (7.5.5), (7.5.10), (7.5.11), and (7.5.12) we get

$$
\begin{aligned}
c\|x_{n+1} - x_n\|^2 &\leq \langle f'(x_n)(x_{n+1} - x_n) + g(x_{n+1}) - g(x_n), x_{n+1} - x_n\rangle \\
&= -\langle f(x_n) + g(x_n), x_{n+1} - x_n\rangle \\
&\leq \|f(x_n) + g(x_n)\| \|x_{n+1} - x_n\| \\
&= \|f(x_n) - f(x_{n-1}) - f'(x_{n-1})(x_n - x_{n-1})\| \|x_{n+1} - x_n\| \\
&\leq \frac{1}{2}\left[\frac{a}{3}\|x_n - x_{n-1}\| + b\right]\|x_n - x_{n-1}\|^2\|x_{n+1} - x_n\| \quad (7.5.17)
\end{aligned}
$$

or,

$$\|x_{n+1} - x_n\| \leq \frac{\eta}{c}d_0\|x_n - x_{n-1}\|^2 \leq \frac{\eta}{c}d_0^{2^n-1}$$

and

$$\|x_{n+1} - x_0\| \leq \sum_{i=0}^{n}\|x_{i+1} - x_i\| \leq \frac{\eta}{c}\sum_{i=0}^{n}d_0^{2^n-1} < r_0.$$

The induction is now complete.

For any integers n, m, we have

$$\|x_{n+m} - x_n\| \leq \sum_{i=m}^{n+m-1}\|x_{i+1} - x_i\| \leq \frac{\eta}{c}\sum_{i=n}^{n+m-1}d_0^{2^i-1} < r_0. \tag{7.5.18}$$

It follows from (7.5.18) that sequence $\{x_n\}$ $(n \geq 0)$ is Cauchy in a Hilbert space H and as such it converges to some $x^* \in \overline{U}(x_0, r_0)$ (as $\overline{U}(x_0, r_0)$ is a closed set). By letting $m \to \infty$ in (7.5.18), we obtain (7.5.14). Finally, because $f + g$ is continuous and (7.5.17) holds, we deduce $f(x^*) + g(x^*) = 0$.

The proof of the following local convergence theorem is omitted as it is identical to the one given in Theorem 7.5.1.

Theorem 7.5.3. *Let g be a maximal monotone multivalues operator from a Hilbert space H into itself, and f be a Gâteaux-differentiable operator from H into H. Assume:*

(a) there exist constants $a > 0$, $b > 0$, $c > 0$ such that (7.5.3)–(7.5.5) hold;
(b) (7.5.1) has a solution x^*.

Then, generalized Newton's sequence $\{x_n\}$ $(n \geq 0)$ *generated by* (7.5.2) *is well defined, remains in* $U(x^*, r_1)$ *for all* $n \geq 0$, *and converges to* x^*, *so that*

$$\|x_n - x^*\| \leq r_1 d_2^n,$$

where,

$$d_2 = \left(\frac{1}{3}ar_1 + b\right)\frac{r_1}{c} < 1,$$

and r_1 *is the positive zero of equation*

$$\frac{a}{3c}r^2 + \frac{br}{c} - 1 = 0.$$

Remark 7.5.4. (a) Theorems 7.5.1–7.5.3 become Theorems 1–3 in [203] respectively if $a = 0$.

(b) Theorems 7.5.1–7.5.3 weaken Theorems 1–3 in [149] (there g is taken to be the subgradient of a convex function).

(c) If $a = 0$, and $g = 0$ in Theorem 7.5.2, then we obtain Mysovskii's theorem [125].

(d) The radius r_0 in Theorem 7.5.2 can be determined from the solution of inequalities (7.5.12) and

$$\frac{\eta}{c}\frac{1}{1 - d_0} < r_0.$$

7.6 Semilocal convergence for quasivariational inequalities

Here, we use the contraction mapping principle to approximate a locally unique solution of a strongly nonlinear variational inequality on a Hilbert space under weak assumptions. Earlier results can be obtained as special cases of our locally convergent theorem.

Let C, $\langle \; ; \; , \; \rangle$, $\| \; \|$ denote a convex subset, inner product and norm of a Hilbert space H, respectively. Let $\varphi: H \to (-\infty, \infty]$ be a lower semicontinuous function, such that $D(\varphi) = \{y \in H \mid \varphi(y) < \infty\}$ is nonempty. Denote by $\partial\varphi(y)$ the set

$$\partial\varphi(y) = \{x \in H \mid \varphi(y) - \varphi(z) \leq \langle x, v - z\rangle \quad \text{for all} \quad z \in D(\varphi)\} \qquad (7.6.1)$$

the subgradient of φ at y.

Let A, T, g and f be operators from H into itself. We are concerned with the problem of finding $x \in H$ such that:

$$g(x) - f(x) \in D(\varphi) \qquad (7.6.2)$$

and

$$A(x) - T(x) + \partial\varphi(g(x)) - f(x) \ni 0, \qquad (7.6.3)$$

where,

$$\varphi(y) = \begin{cases} 0, & y \in C \\ \infty, & \text{otherwise} \end{cases} \qquad (7.6.4)$$

In this case, (7.6.3) reduces to finding $x \in H$ such that

$$g(x) \in C(x) \tag{7.6.5}$$

and

$$\langle A(x) - T(x), g(x) - y \rangle \le 0 \quad \text{for all } y \in C(x), \tag{7.6.6}$$

where,

$$C(x) = C + f(x). \tag{7.6.7}$$

Inequality (7.6.6) is called a general strongly nonlinear quasivariational inequality.

If $f = 0$ and $g(x) = x$, we obtain the general quasifunctional inequality studied in [178]. Moreover if $f = 0$ and $T = 0$, we get the general variational inequality considered in [149], [151].

The inverse $P_\mu = (I + \mu \partial \varphi)^{-1}$ exists as a single-valued function satisfying: for $\|P_\mu(x) - P_\mu(y)\| \le \|x - y\| \, \mu > 0$, for all $x, y \in H$, and is monotone: $f_1 \in \partial \varphi(x_1)$, $f_2 \in \partial y(x_2) \Rightarrow \langle f_1 - f_2, x_1 - x_2 \rangle \ge 0$. It is known that x^* is a solution of (7.6.5) if and only if

$$x^* = x^* - \lambda g(x^*) + \lambda f(x^*) + \lambda P_\mu \left[g(x^*) - f(x^*) - \mu A(x^*) + \mu T(x^*) \right] \tag{7.6.8}$$

for $\lambda > 0$, $\mu > 0$.

That is, solving (7.6.3) reduces to finding fixed points of the operator

$$P_{\mu,\lambda}(x) = x - \lambda g(x) + \lambda P_\mu \left[g(x) - f(x) - \mu A(x) + \mu T(x) \right]. \tag{7.6.9}$$

We assume: there exist a convex subset $H_0 \subseteq H$, $x_0 \in H_0$ and nonnegative constants $a_i, b_i, i = 1, 2, \ldots, 6$ such that:

$$\|A(x) - A(y)\| \le a_1 \|x - y\| \tag{7.6.10}$$

$$\langle A(x) - A(y), x - y \rangle \ge a_2 \|x - y\|^2, \tag{7.6.11}$$

$$\|g(x) - g(x)\| \le a_3 \|x - y\|, \tag{7.6.12}$$

$$\langle g(x) - g(y), x - y \rangle \ge a_4 \|x - y\|^2, \tag{7.6.13}$$

$$\|T(x) - T(y)\| \le a_5 \|x - y\|, \tag{7.6.14}$$

$$\|f(x) - f(y)\| \le a_6 \|x - y\|, \tag{7.6.15}$$

$$\|A(x_0) - A(y)\| \le b_1 \|x_0 - y\|, \tag{7.6.16}$$

$$\langle A(x_0) - A(y), x_0 - y \rangle \ge b_2 \|x_0 - y\|^2, \tag{7.6.17}$$

$$\|g(x_0) - g(y), x_0 - y\| \le b_3 \|x_0 - y\|, \tag{7.6.18}$$

$$\langle g(x_0) - g(y), x_0 - y \rangle \ge b_4 \|x_0 - y\|^2, \tag{7.6.19}$$

$$\|T(x_0) - T(y)\| \le b_5 \|x_0 - y\|, \tag{7.6.20}$$

and

$$\|f(x_0) - f(y)\| \le b_6 \|x_0 - y\| \tag{7.6.21}$$

for all $x, y \in H_0$.

Define the parameters c, d by

$$c = c_{\lambda,\mu} = 2\lambda a_6 + 2\sqrt{1 - 2a_4\lambda + \lambda^2 a_3^2} + \sqrt{1 - 2\lambda\mu a_2 + \mu^2\lambda^2 a_1^2} + \lambda\mu a_5, \quad (7.6.22)$$

$$d = \|P_{\mu,\lambda}(x_0) - x_0\|, \quad (7.6.23)$$

scalar function h on $[0, +\infty)$ by

$$h(r) = c_0 r + d, \quad (7.6.24)$$

where, c_0 is defined as above but a_i are replaced by b_i, $i = 1, 2, \ldots, 6$, and, ball H_1 by

$$H_1 = U(x_0, r_0), \quad (7.6.25)$$

where,

$$r_0 \geq \frac{d}{1-c_0} \quad \text{if } c_0 \neq 1. \quad (7.6.26)$$

We can now show the local fixed point result for (7.6.3).

Theorem 7.6.1. *Assume:*

(i) $c_0 \in [0, 1)$, $c \in [0, 1)$, $d \in (0, 1)$, $H_1 \subseteq H_0$;
(ii) conditions (7.6.10)–(7.6.21) hold for all $x, y \in H_1$.

Then (7.6.3) has a unique solution x^* in H_1. Moreover, x^* can be obtained as the limit of the sequence

$$x_{n+1} = P_{\mu,\lambda}(x_n) \quad (n \geq 0). \quad (7.6.27)$$

Proof. The result will follow from Banach's contraction mapping principle if we show:

(1) Operator $P_{\mu,\lambda}$ is a c-contraction on H_1;
(2) $P_{\mu,\lambda}$ maps H_1 into itself.

To show (7.6.1), let us choose $x, y \in H_1$. Using (7.6.9)–(7.6.16) and (7.6.22), we obtain in turn:

$$\|P_{\mu,\lambda}(x) - P_{\mu,\lambda}(y)\| \leq \|x - y - \lambda g(x) + \lambda g(y)\| + \lambda \|f(x) - f(y)\|$$
$$+ \lambda\|g(x) - g(y) - f(x) + f(y) - \mu A(x) + \mu A(y) + \mu T(x) - \mu T(y)\|$$

$$\leq 2\|x - y - \lambda g(x) + \lambda g(y)\| + 2\lambda\|f(x) - f(y)\|$$
$$+ \|x - y - \mu\lambda A(x) + \mu\lambda A(y)\| + \mu\lambda \|T(x) - T(y)\|$$

$$\leq \left[2\lambda a_6 + 2\sqrt{1 - 2a_4\lambda + \lambda^2 a_3^2} + \sqrt{1 - 2\lambda\mu a_2 + \mu^2\lambda^2 a_1^2} + \lambda\mu a_5 \right] \|x - y\|$$

$$= c \|x - y\|,$$

which shows that $P_{\mu,\lambda}$ is a c-contraction on H_1 as by hypothesis (i), $c \in [0, 1)$. Moreover, to show $P_{\mu,\lambda}$ maps H_1 into itself, let $y \in H_1$. Using (7.6.9), (7.6.12)–(7.6.21), (7.6.23), (7.6.24), and (7.6.26), we obtain

$$\|P_{\mu,\lambda}(y) - x_0\| \leq \|P_{\mu,\lambda}(y) - P_{\mu,\lambda}(x_0)\| + \|P_{\mu,\lambda}(x_0) - x_0\|$$
$$\leq c_0\|y - x_0\| + d \leq c_0 r_0 + d = h(r_0) \leq r_0,$$

by the choice of r_0, c_0 and d.

That completes the proof of the theorem.

Remark 7.6.2. The conditions on c_0, c and d can be dropped. Let us consider one such case: let $a_i = b_i$, $i = 1, 2, \ldots, 6$, define parameters:

$$\lambda_0 = \frac{a_4}{a_3^2} - \frac{1}{a_3} \sqrt{\frac{a_4^2}{a_3^2} - \frac{a_4^2 - a_6^2}{a_3^2 - a_6^2}},$$

$$\mu_0 = \frac{1}{\lambda_0} \left[\frac{a_2}{a_1^2} - \frac{1}{a_1} \sqrt{\frac{a_2^2}{a_1^2} - \frac{a_2^2 - a_5^2}{a_1^2 - a_5^2}} \right],$$

$$\alpha = 2a_6 \left[\frac{a_4}{a_3^2} - \frac{1}{a_3} \sqrt{\frac{a_4^2}{a_3^2} - \frac{a_4^2 - a_6^2}{a_3^2 - a_6^2}} \right]$$

$$+ a_5 \left[\frac{a_2}{a_1^2} - \frac{1}{a_1} \sqrt{\frac{a_2^2}{a_1^2} - \frac{a_2^2 - a_5^2}{a_1^2 - a_5^2}} \right] + \sqrt{\frac{a_1^2 - a_2^2}{a_1^2 - a_5^2}} + 2 \sqrt{\frac{a_3^2 - a_4^2}{a_3^2 - a_6^2}}.$$

Assume: $a_1 > a_5 > 0$, $a_3 > a_6 > 0$, and $\alpha < 1$. Then it is simple algebra to show that $c_{\lambda,\mu}$ given by (7.6.22) is minimized at (λ_0, μ_0) and the minimum value is given by α. The point r_0 is then given by (7.6.26) for $c_0 = \alpha$.

Note that if $H_0 = H_1 = H$, then x^* is the unique fixed point of $P_{\mu,\lambda}$ in H. However, our theorem is more useful than relevant earlier ones [149]–[151], [195] as conditions (7.6.10)–(7.6.21) rarely hold on the whole Hilbert space H.

7.7 Generalized equations in Hilbert space

In this study, we are concerned with the problem of approximating a locally unique solution x^* of the generalized equation

$$F(x) + g(x) \ni 0, \tag{7.7.1}$$

where F is a Fréchet-differentiable operator defined on a closed convex subset D of a Hilbert space H with values in H, and Dg is a nonempty subset of $H \times H$. Throughout this study we consider the expressions $[x, y] \in g$, $g(x) \ni y$, $-y + g(x) \ni 0$, and $y \in g(x)$ to be equivalent.

We use the generalized chord method

$$F'(x_0)x_{n+1} + g(x_{n+1}) \ni F'(x_0)x_n - F(x_n) \quad (n \ge 0), \ (x_0 \in D) \tag{7.7.2}$$

or, the generalized NK method

$$F'(x_n)x_{n+1} + g(x_{n+1}) \ni F'(x_n)(x_n) - F(x_n) \quad (n \ge 0), \ (x_0 \in D) \tag{7.7.3}$$

to approximate x^*.

Earlier results have used Lipschitz-type hypotheses on the first Fréchet derivative of F [195]. Here we use hypotheses on the second Fréchet derivative of F. It has already been shown that for regular equations on a Banach space, our new approach can solve problems not possible before and can also improve on the error bounds of the distances involved. Our convergence theorems for methods (7.7.2) and (7.7.3) to a solution x^* of (7.7.1) reduce to the ones in [195] if the Lipschitz constant of the second Fréchet derivatives is zero.

We will need the following Lemmas:

Lemma 7.7.1. *Let $a > 0$, $b \geq 0$, $\eta \geq 0$ and $\bar{c} > 0$ be constants. Define the polynomial p by*

$$p(r) = \frac{a}{6\bar{c}}r^3 + \frac{b}{2\bar{c}}r^2 - r + \frac{\eta}{c}. \qquad (7.7.4)$$

The polynomial p has two positive zeros r_1, r_2 $(r_1 \leq r_2)$ if and only if

$$p(q) \leq 0, \qquad (7.7.5)$$

where q is the positive zero of p'.

Proof. Denote by q_0 the negative zero of p'. Clearly, p has a maximum at $r = q_0$ and a minimum at $r = q$. Hence, a necessary and sufficient condition for p to have positive zeros is given by (7.7.5).

Lemma 7.7.2. *Let $a \geq 0$, $b \geq 0$, $r > 0$ be constants and $z \in H$ be fixed. Assume:*

$$\|F''(x) - F''(z)\| \leq a\|x - z\| \qquad (7.7.6)$$

and

$$\|F''(z)\| \leq b \qquad (7.7.7)$$

for all $x \in U(z, r)$. Then the following estimate holds for all $x, y \in U(z, r)$:

$$\|F(x) - F(y) - F'(z)(x - y)\| \leq \frac{a}{2}\int_0^1 [(1 - t)\|y - z\| + t\|x - z\|]^2 \|x - y\|dt$$

$$+ \frac{b}{2}[\|y - z\| + \|x - z\|]\|x - y\|. \qquad (7.7.8)$$

The following result gives sufficient conditions for generalized equation (7.7.1) to have a unique solution. The proof can be found in [195, p. 256].

Lemma 7.7.3. *Let g be a multivalued maximal monotone operator from H to H, in the sense that g is a nonempty subset of $H \times H$ and there exists $\alpha \geq 0$ such that:*

$$[v_1, w_1] \in g \quad and \quad [v_2, w_2] \in g \Rightarrow \langle w_2 - w_1, v_2 - v_1 \rangle \geq \alpha\|v_1 - v_2\|^2, \qquad (7.7.9)$$

and is not contained in any larger monotone subset of $H \times H$; L be a bounded linear operator from H into H. Assume there exists $c > -\alpha$, and

$$\langle L(x), x \rangle \geq c\|x\|^2 \quad for all \ x \in H. \qquad (7.7.10)$$

Then, for any $y \in H$, there exists $z \in H$ satisfying the generalized equation

$$L(z) + g(z) \ni y. \qquad (7.7.11)$$

We can show the following local result for method (7.7.3) to solve equation (7.7.1).

Theorem 7.7.4. *Let F be a twice Fréchet-differentiable operator defined on an open convex subset D of a Hilbert space H with values in H; g be a multivalued maximal monotone operator on H. Assume:*

(a) generalized equation (7.7.1) has a solution $x^* \in D$;
(b) there exist parameters $a \geq 0$, $b \geq 0$ such that

$$\|F''(x) - F''(x^*)\| \leq a\|x - x^*\|, \tag{7.7.12}$$

and

$$\|F''(x^*)\| \leq b \tag{7.7.13}$$

for all $x \in D$;
(c) there exists $c > -\alpha$ such that

$$\langle F'(z)(x), x \rangle \geq c\|x\|^2 \quad \text{for all } x \in H, \, z \in D \tag{7.7.14}$$

(d) $U(x^*, r_0) \subseteq D$, where r_0 is the positive zero of equation

$$p(r) = \frac{a}{6(c+\alpha)} r^2 + \frac{b}{2(c+\alpha)} r - 1 = 0. \tag{7.7.15}$$

Then, generalized NK sequence $\{x_n\}$ $(n \geq 0)$ generated by (7.7.3) is well defined, remains in $U(x^*, r_0)$ for all $n \geq 0$, and converges to x^* provided that $x_0 \in U(x^*, r_0)$. Moreover the following error bounds hold for all $n \geq 0$

$$\|x_n - x^*\| \leq \beta d^{2^n}, \tag{7.7.16}$$

where,

$$\beta^{-1} = \frac{1}{2(c+\alpha)} \left[\frac{a}{3} q + b \right] \quad \text{and} \quad d = \beta^{-1} q. \tag{7.7.17}$$

Proof. The existence of solutions to (7.7.3) follows from Lemma 7.7.3 and (7.7.14). Using (7.7.1), (7.7.9), and (7.7.3), we get

$$\alpha\|x_{n+1} - x^*\|^2 \leq \langle F(x^*) - F(x_n) - F'(x_n)(x_{n+1} - x_n), x_{n+1} - x^* \rangle$$

or

$$c\|x_{n+1} - x^*\|^2 + \langle F'(x_n)(x_{n+1} - x^*), x_{n+1} - x^* \rangle$$
$$\leq \langle F(x^*) - F(x_n) - F'(x_n)(x^* - x_n), x_{n+1} - x^* \rangle,$$

and by (7.7.8), (7.7.12)–(7.7.14), we have by induction on $n \geq 0$

$$\|x_{n+1} - x^*\| \leq \frac{1}{c+\alpha} \left[\frac{a}{6} \|x_n - x^*\| + \frac{b}{2} \right] \|x_n - x^*\|^2 \leq \beta d^{2^n},$$

and $\lim_{n \to \infty} x_n = x^*$ (as $d \in [0, 1)$).

We can show the following semilocal result for method (7.7.3) to solve equation (7.7.1).

Theorem 7.7.5. *Let F, g, D, (c) be as in Theorem 7.7.4. Moreover assume: there exist parameters $a \geq 0$, $b \geq 0$, $x_0 \in D$ such that*

$$\|F''(x) - F''(x_0)\| \leq a\|x - x_0\|, \tag{7.7.18}$$

$$\|F''(x_0)\| \leq b \tag{7.7.19}$$

$$d = \beta^{-1}\|x_1 - x_0\| < 1, \tag{7.7.20}$$

$$\beta^{-1} = \frac{1}{2(c+\alpha)}\left[\frac{a}{3}\|x_1 - x_0\| + b\right], \tag{7.7.21}$$

and

$$\overline{U}(x_0, r^*) \subseteq D, \tag{7.7.22}$$

$$r^* = \beta^{-1}\sum_{i=0}^{\infty} d^{2^i}. \tag{7.7.23}$$

Then, generalized NK iterates $\{x_n\}$ $(n \geq 0)$ generated by (7.7.3) are well defined, remain in $U(x_0, r^*)$ for all $n \geq 0$ and converge to a solution x^* of (7.7.1) in $\overline{U}(x_0, r^*)$ so that:

$$\|x_n - x^*\| \leq \beta\sum_{i=n}^{\infty} d^{2^i} \quad (n \geq 0). \tag{7.7.24}$$

Proof. The existence of solutions x_{n+1} to (7.7.3) follows from (7.7.14) and Lemma 7.7.3. Using (7.7.3) and (7.7.9), we get

$$\alpha\|x_{n+1} - x_n\|^2 + \langle F'(x_n)(x_{n+1} - x_n), x_{n+1} - x_n\rangle$$
$$\leq \langle F(x_n) - F(x_{n-1}) - F'(x_{n-1})(x_n - x_{n-1}), x_n - x_{n+1}\rangle,$$

and by (7.7.14), (7.7.8), we obtain by induction on $n \geq 0$

$$\|x_{n+1} - x_n\| \leq \frac{1}{c+\alpha}\left[\frac{a}{6}\|x_n - x_{n-1}\| + \frac{b}{2}\right]\|x_n - x_{n-1}\|^2 \leq \beta d^{2^n}.$$

Hence, we get

$$\|x_{n+1} - x_0\| \leq \|x_{n+1} - x_n\| + \|x_n - x_{n-1}\| + \cdots + \|x_1 - x_0\|$$
$$\leq \sum_{i=0}^{n}\|x_{i+1} - x_i\| \leq r^*,$$

which shows $x_{n+1} \in \overline{U}(x_0, r^*)$ $(n \geq 0)$. Let $m \geq 0$ then

$$\|x_{n+m} - x_n\| \leq \sum_{i=n}^{n+m-1}\|x_{i+1} - x_i\| \leq \beta\sum_{i=n}^{n+m-1} d^{2^i}. \tag{7.7.25}$$

That is, sequence $\{x_n\}$ $(n \geq 0)$ is Cauchy in H and as such it converges to $x^* \in \overline{U}(x_0, r^*)$ (as $\overline{U}(x_0, r^*)$ is a closed set). By letting $m \to \infty$ in (7.7.25), we obtain (7.7.24). Finally it follows from (7.7.3) and (7.7.9) that x^* is a solution of (7.7.1).

We show how to approximate solutions of equation (7.7.1) using method (7.7.2).

Theorem 7.7.6. *Let F, g, D, (7.7.18), and (7.7.19) be as in Theorem 7.7.5. Moreover, assume:*

(a) there exists $y_0 \in H$ such that

$$g(x_0) \ni y_0, \tag{7.7.26}$$

and

$$\|F(x_0) + y_0\| \le b_0 \quad \text{for some } b_0 > 0; \tag{7.7.27}$$

(b) there exists $c_0 > -\alpha$ such that

$$\langle F'(x_0)(x), x \rangle \ge c_0 \|x\|^2 \quad \text{for all } x \in H; \tag{7.7.28}$$

(c)

$$p(q) \le 0, \tag{7.7.29}$$

where, q is the positive zero of p' and

$$p(r) = \frac{a}{6(c_0 + \alpha)} r^3 + \frac{b}{2(c_0 + \alpha)} r^2 - r + \frac{b_0}{c_0 + \alpha} \tag{7.7.30}$$

Denote by r_0 and R_0 ($r_0 \le R_0$) the positive zeros of p guaranteed to exist by Lemma 7.7.1;
(d) $\overline{U}(x_0, r_0) \subseteq D$.

Then, the generalized chord iterates $\{x_n\}$ ($n \ge 0$) are well defined, remain in $U(x_0, r_0)$ for all $n \ge 0$, and converge to a solution x^* of (7.7.1) in $\overline{U}(x_0, r_0)$, which is unique in $U(x_0, R_0) \cap D$. Moreover, the following error bounds hold for all $n \ge 0$

$$\|x_n - x^*\| \le \gamma^n r_0, \tag{7.7.31}$$

where,

$$\gamma = \frac{1}{c_0 + \alpha} \left(\frac{a}{2} r_0 + b \right) r_0. \tag{7.7.32}$$

Proof. By Lemma 7.7.3, operator $w(x)$ is uniquely determined by

$$F'(x_0)w(x) + g(w(x)) \ni F'(x_0)(x) - F(x)$$

for all $x \in \overline{U}(x_0, r_0)$. Using (7.7.9), we can write

$$\alpha \|w(x) - x_0\|^2 \le \langle y_0 + F(x) - F'(x_0)(x - w(x)), x_0 - w(x) \rangle,$$

or

$$\alpha \|w(x) - x_0\|^2 + \langle F'(x_0)(w(x) - x_0), w(x) - x_0 \rangle$$
$$\le \langle F'(x_0)(x - x_0) - F(x) - y_0, w(x) - x_0 \rangle,$$

and by (7.7.18), (7.7.19), (7.7.26)–(7.7.30), we get

$$(c_0 + \alpha)\|w(x) - x_0\| \leq \|y_0 + F(x) - F'(x_0)(x - x_0)\|$$
$$= \|y_0 + F(x_0) + F(x) - F(x_0) + F'(x_0)(x - x_0)\|$$
$$\leq \|F(x_0) + y_0\| + \|F(x) - F(x_0) + F'(x_0)(x - x_0)\|$$
$$\leq b_0 + \tfrac{a}{6}\|x - x_0\|^3 + \tfrac{b}{2}\|x - x_0\|^2,$$

or

$$\|w(x) - x_0\| \leq \frac{1}{c_0 + \alpha}\left[b_0 + \frac{a}{6}\|x - x_0\|^3 + \frac{b}{2}\|x - x_0\|^2\right]$$
$$\leq \frac{1}{c_0 + \alpha}\left[b_0 + \frac{a}{6}r_0^3 + \frac{b}{2}r_0^2\right] = r_0. \qquad (7.7.33)$$

Hence, w maps $\overline{U}(x_0, r_0)$ into itself. Let $x, y \in \overline{U}(x_0, r_0)$. Then by (7.7.9), and the definition of w, we get in turn

$$\alpha\|w(x) - w(y)\|^2 \leq \langle F'(x_0)(w(x) - w(y)) + F(x) - F(y)$$
$$- F'(x_0)(x - y), w(y) - w(x)\rangle,$$

or

$$\alpha\|w(x) - w(y)\|^2 + \langle F'(x_0)(w(x) - w(y)), w(x) - w(y)\rangle$$
$$\leq \langle F'(x_0)(x - y) - F(x) + F(y), w(x) - w(y)\rangle,$$

and by (7.7.8), (7.7.29), we obtain

$$\|w(x) - w(y)\| \leq$$
$$\leq \frac{1}{c_0 + \alpha}\|F(x) - F(y) - F'(x_0)(x - y)\|$$
$$\leq \frac{1}{c_0 + \alpha}\left\{\frac{a}{2}\int_0^1 [(1 - t)\|y - x_0\| + t\|x - x_0\|]^2\right.$$
$$\left. + \frac{b}{2}[\|x - x_0\| + \|y - x_0\|]\right\}\|x - y\|$$
$$\leq \frac{1}{c_0 + \alpha}\left[\frac{a}{2}r_0^2 + br_0\right]\|x - y\| = \gamma\|x - y\|,$$

where γ is given by (7.7.32). It follows that w is a contraction on $\overline{U}(x_0, r_0)$, as by the definition of r_0, $\gamma \in [0, 1)$. Moreover, we can write $x_{n+1} = w(x_n)$ $(n \geq 0)$. The Banach contraction mapping principle guarantees the existence of a unique element x^* of $\overline{U}(x_0, r_0)$ satisfying $x^* = w(x^*)$ or equivalently (7.7.1). Moreover, we can write

$$\|x_n - x^*\| = \|w(x_{n-1}) - w(x^*)\| \leq \gamma^n\|x_0 - x^*\| \leq \gamma^n r_0,$$

which shows (7.7.31) for all $n \geq 0$. Finally to show uniqueness, let y^* be a solution of (7.7.1). As in (7.7.33), we get

$$(c_0 + \alpha)\|y^* - x_0\| \leq b_0 + \tfrac{a}{6}\|y^* - x_0\|^3 + \tfrac{b}{2}\|y^* - x_0\|^2,$$

which gives $\|y^* - x_0\| \geq R_0$ or $\|y^* - x_0\| \leq r_0$. If $y^* \in U(x_0, R_0)$ then $y^* \in \overline{U}(x_0, r_0)$, which implies $y^* = x^*$.

If iteration (7.7.2) is replaced by the generalized Newton method (7.7.3), then following the proofs of Theorems 7.7.4–7.7.6 and Theorem 2.11 in [195] we can show:

Theorem 7.7.7. *Let all hypotheses of Theorem 7.7.6 hold. Then, generalized NK method $\{x_n\}$ ($n \geq 0$), generated by (7.7.3) is well defined, remains in $\overline{U}(x_0, r_0)$ for all $n \geq 0$ and converges to a unique solution of (7.7.1) in $U(x_0, R_0) \cap D$. Moreover the following estimates hold for all $n \geq 0$:*

$$\|x_n - x^*\| \leq \beta \sum_{i=n}^{\infty} d^{2^i} \quad (n \geq 0),$$

where β, d are given by (7.7.21), (7.7.20), respectively.

Remark 7.7.8. Our results reduce to the ones in [195] if $a = 0$ in (7.7.12) and (7.7.18). The advantages of using second instead of second Fréchet derivative have been shown in for regular equations and for generalized equations. In particular, our error bounds can be finer and our convergence conditions hold whereas the corresponding ones in [195] do not.

7.8 Exercises

7.8.1. Consider the problem of approximating a locally unique solution of the variational inequality

$$F(x) + \partial\varphi(x) \ni 0, \tag{7.8.1}$$

where F is a Gâteaux-differentiable operator defined on a Hilbert space H with values in H; $\varphi: H \to (-\infty, \infty]$ is a lower semicontinuous convex function. We approximate solutions x^* of (7.8.1) using the generalized NK method in the form

$$F'(x_n)(x_{n+1}) + \partial\varphi(x_{n+1}) \ni F'(x_n)(x_n) - F(x_n) \tag{7.8.2}$$

to generate a sequence $\{x_n\}$ ($n \geq 0$) converging to x^*.
Define: the set

$$D(\varphi) = \{x \in H : \varphi(x) < \infty\} \quad \text{and assume } D(\varphi) \neq \phi;$$

the subgradient

$$\partial\varphi(x) = \{z \in H : \varphi(x) - \varphi(y) \leq \langle z, x - y \rangle, \, y \in D(\varphi)\};$$

and the set

$$D(\partial\varphi) = \{x \in D(\varphi): \partial\varphi(x) \neq 0\}.$$

Function $\partial\varphi$ is multivalued and for any $\lambda > 0$, $(1 + \lambda\partial\varphi)^{-1}$ exists (as a single-valued function) and satisfies

$$\|(1 + \lambda\partial\varphi)^{-1}(x) - (I + \lambda\partial\varphi)^{-1}(y)\| \leq \|x - y\| \quad (x, y \in H).$$

Moreover $\partial\varphi$ is monotone:

$$f_1 \in \partial\varphi(x_1), \ f_2 \in \partial\varphi(x_2) \Rightarrow \langle f_1 - f_2, x_1 - x_2 \rangle \geq 0.$$

Furthermore, we want $D(\bar{\varphi}) = D(\bar{\partial}\varphi)$, so that $D(\partial\varphi)$ is sufficient for our purposes.

We present the following local result for variational inequalities and twice Gâteaux-differentiable operators:

(a) Let $F: H \to H$ be a twice Gâteaux-differentiable function. Assume:
   (1) variational inequality (7.8.1) has a solution x^*;
   (2) there exist parameters $a \geq 0$, $b > 0$, $c > 0$ such that

$$\|F''(x) - F''(y)\| \leq a\|x - y\|,$$
$$\|F''(x^*)\| \leq b,$$

   and

$$c\|y - z\|^2 \leq \langle F'(x)(y - z), y - z \rangle$$

   for all $x, y, z \in H$;
   (3) $x_0 \in D(\varphi)$ and $x_0 \in U(x^*, r)$, where

$$r = 4c \left[b + \sqrt{b^2 + \tfrac{16ac}{3}} \right]^{-1}.$$

Then show: generalized NK method (7.8.2) is well defined, remains in $U(x^*, r)$, and converges to x^* with

$$\|x_n - x^*\| \leq p \cdot d^{2^n}, \quad (n \geq 0)$$

where,

$$p^{-1} = \tfrac{1}{c}\left[\tfrac{1}{3}a\|x^* - x_0\| + \tfrac{b}{2} \right] \quad \text{and} \quad d = p^{-1}\|x^* - x_0\|.$$

(b) We will approximate x^* using the generalized NK method in the form

$$f''(x_n)(x_{n+1}) + \partial\varphi(x_{n+1}) \ni f''(x_n)(x_n) - \nabla f(x_n). \qquad (7.8.3)$$

We present the following semilocal convergence result for variational inequalities involving twice Gâteaux-differentiable operators. Let $f: H \to R$ be twice Gâteaux-differentiable. Assume:

(1) for $x_0 \in D(\varphi)$ there exist parameters $\alpha \geq 0, \beta > 0, c > 0$ such that

$$\langle (f'''(x) - f'''(x_0))(y, y), z \rangle \leq \alpha \|x - x_0\| \, \|y\|^2 \|z\|,$$

$$\|f'''(x_0)\| \leq \beta$$

and

$$c\|y - z\|^2 \leq \langle f''(x)(y - z), y - z \rangle$$

for all $x, y, z \in H$;

(2) the first two terms of (7.8.3), x_0 and x_1, are such that for

$$\eta \geq \|x_1 - x_0\|$$

$$\eta \leq \begin{cases} c\left[\beta + 2\sqrt{\alpha c}\right]^{-1}, & \beta^2 - 4\alpha c \neq 0 \\ \\ c(2\beta)^{-1}, & \beta^2 - 4\alpha c = 0. \end{cases}$$

Then show: generalized NK method (7.8.3) is well defined, remains in $U(x_0, r_0)$ for all $\eta \geq 0$, where c_0 is the small zero of function δ,

$$\delta(r) = \alpha \eta r^2 - (c - \beta \eta)r + c\eta,$$

and converges to a unique solution x^* of inclusion $\nabla f(x) + \partial \varphi(x) \ni 0$. In particular $x^* \in \bar{U}(x_0, r_0)$. Moreover, the following error bounds hold for all $n \geq 0$

$$\|x_n - x^*\| \leq \gamma d^{2^n},$$

where,

$$\gamma^{-1} = \frac{\alpha r_0 + \beta}{c} \quad \text{and} \quad d = \eta \gamma^{-1}.$$

7.8.2. Let M, $\langle \cdot, \cdot \rangle$, $\| \cdot \|$ denote the dual, inner product and norm of a Hilbert space H, respectively. Let C be a closed convex set in H. Consider an operator $a \colon H \times H \to [0, +\infty)$. If a is continuous bilinear and satisfies

$$a(x, y) \geq c_0 \|y\|^2, \quad y \in H, \tag{7.8.4}$$

and

$$a(x, y) \leq c_1 \|x\| \cdot \|y\|, \quad x, y \in H, \tag{7.8.5}$$

for some constants $c_0 > 0$, $c_1 > 0$ then a is called a coercive operator. Given $z \in M$, there exists a unique solution $x \in C$ such that:

$$a(x, x - y) \geq \langle z, x - y \rangle, \quad y \in C. \tag{7.8.6}$$

Inequality (7.8.6) is called variational. It is well-known that x^* can be obtained by the iterative procedure

$$x_{n+1} = P_C(x_n - \rho F(G(x_n) - z)), \tag{7.8.7}$$

where P_C is a projection of H into C, $\rho > 0$ is a constant, F is a canonical isomorphism from M onto H, defined by

$$\langle z, y \rangle = \langle F(z), y \rangle, \quad y \in H, \ z \in M, \tag{7.8.8}$$

and

$$a(x, y) = \langle G(x), y \rangle, \quad y \in H. \tag{7.8.9}$$

Given a point-to-set operator C from H into M we define the quasivariational inequality problem to be: find $x \in C(x)$ such that:

$$a(x, y - x) \geq \langle z, y - x \rangle \quad y \in C(x). \tag{7.8.10}$$

Here, we consider $C(x)$ to be of the form

$$C(x) = f(x) + C, \tag{7.8.11}$$

where f is a point-to-point operator satisfying

$$\| f(x^*) - f(y) \| \leq c_2 \| x^* - y \|^\lambda \tag{7.8.12}$$

for some constants $c_2 \geq 0$, $\lambda \geq 1$, all $y \in H$ and x^* a solution of (7.8.10). We will extend (7.8.7) to compute the approximate solution to (7.8.10).

(a) Show: For fixed $z \in H$, $x \in C$ satisfies

$$\langle x - z, y - x \rangle \geq 0 \quad y \in C \tag{7.8.13}$$

$$\Leftrightarrow x = P_C(z), \tag{7.8.14}$$

where P_C is the projection of H into C.

(b) P_C given by (7.8.14) is nonexpansive, that is

$$\| P_C(x) - P_C(y) \| \leq \| x - y \|, \quad x, y \in H. \tag{7.8.15}$$

(c) For C given by (7.8.11), $x \in C(x)$ satisfies (7.8.10) $\Leftrightarrow$

$$x = f(x) + P_C(x - \rho F(G(x) - z)). \tag{7.8.16}$$

Result (c) suggests the iterative procedure

$$x_{n+1} = f(x_n) + P_C\big(x_n - \rho F(G_n) - z) - f(x_n)\big) \tag{7.8.17}$$

for approximating solutions of (7.8.10).
Let us define the expression

$$\theta = \theta(\lambda, \rho) = 2c_2 \| x_0 - x \|^{\lambda - 1} + \sqrt{1 + \rho^2 c_1^2 - 2c_0 \rho}.$$

It is simple algebra to show that $\theta \in [0, 1)$ in the following cases:

(1) $\lambda = 1$, $c_2 \leq \frac{1}{2}$, $c_0 \geq 2c_1 \sqrt{c_2(1 - c_2)}$, $0 < \rho < \dfrac{c_0 + \sqrt{c_0^2 - 4c_1^2 c_2(1 - c_2)}}{c_1^2}$;

(2) $\lambda > 1, c_0 \leq c_1, \|x_0 - x\| < \left[\frac{1}{2c_2}\left(1 + \sqrt{1 - \left(\frac{c_0}{c_1}\right)^2}\right)\right]^{1/\lambda - 1} = c_3,$

$$0 < \rho < \frac{c_0 + \sqrt{c_0^2 + qc_1^2}}{c_1^2}, \quad q = \left(1 - 2c_2\|x_0 - x\|^{\lambda - 1}\right)^2 - 1;$$

(3) $\lambda > 1, c_0 > c_1, \|x_0 - x\| \leq \left(\frac{1}{2c_2}\right)^{1/\lambda - 1} = c_4, 0 < \rho < \frac{c_0 + \sqrt{c_0^2 + qc_1^2}}{c_1^2}.$

Denote by H_0, H_1 the sets

$$H_0 = \{y \in H \mid \|y - x^*\| \leq c_3\} \quad \text{and} \quad H_1 = \{y \in H \mid \|y - x^*\| \leq c_4\}.$$

(d) Let operator f satisfy (7.8.12) and C be a nonempty closed convex subset of H. If $a(x, y)$ is a coercive, continuous bilinear operator on H, x^* and x_{n+1} are solutions of (7.8.10) and (7.8.17), respectively, then x_{n+1} converges to x^* strongly in H if (7.8.1) or (7.8.2) or (7.8.3) above hold.

It follows from (d) that a solution x^* of (7.8.10) can be approximated by the iterative procedure

(1) $x^* \in C(x^*)$ is given,

(2) $x_{n+1} = f(x_n) + P_C\big(x_n - \rho F(G(x_n) - z) - f(x_n)\big),$

where ρ, x_0 are as in (7.8.1) or (7.8.2) or (7.8.3).

If $\lambda = 1$ our result (d) reduces to Theorem 3.2 in [150] (provided that (7.8.12) is replaced by $\|f(x) - f(y)\| \leq c_2^1\|x - y\|$ for all $x, y \in H$). Note also that as $c_2^1 \geq c_2$, in general our error bounds on the distances $\|x_n - x^*\|$ $(n \geq 0)$ are smaller. Moreover, if $C(x)$ is independent of x, then $f = 0$ and $c_2 = 0$, in which case (c) and (d) reduce to the ones in [149].

7.8.3. Let $x_0 \in D$ and $R > 0$ be such that $D \equiv U(x_0, R)$. Suppose that f is m-times Fréchet-differentiable on D, and its mth derivative $f^{(m)}$ is in a certain sense uniformly continuous:

$$\|f^{(m)}(x) - f^{(m)}(x_0)\| \leq w(\|x - x_0\|), \quad \text{for all } x \in D, \qquad (7.8.18)$$

for some monotonically increasing positive function w satisfying

$$\lim_{t \to \infty} w(r) = 0, \qquad (7.8.19)$$

or, even more generally, that

$$\|f^{(m)}(x) - f^{(m)}(x_0)\| \leq w(r, \|x - x_0\|), \quad \text{for all } x \in \bar{D}, r \in (0, R), \quad (7.8.20)$$

for some monotonically increasing in both variables positive function w satisfying

$$\lim_{t \to 0} w(r, t) = 0, \quad r \in [0, R]. \qquad (7.8.21)$$

Let us define function θ on $[0, R]$ by

$$\theta(r) = \frac{1}{c+\alpha} \left[\eta + \frac{\alpha_m}{m!} r^m + \cdots + \frac{\alpha_2}{2!} r^2 \right.$$
$$\left. + \int_0^{v_{m-2}} \cdots \int_0^{v_1} w(v_{m-1})(r - v_1)dv_1 \cdots dv_{m-1} \right] - r \quad (7.8.22)$$

for some constants $\alpha, c, \eta, \alpha_i, i = 2, \ldots, m$; the equation,

$$\theta(r) = 0; \qquad (7.8.23)$$

and the scalar iteration $\{r_n\}$ $(n \geq 0)$ by

$$r_0 = 0, \quad r_{n+1} = r_n - \frac{\theta(r_n)}{\theta'(r_n)}. \qquad (7.8.24)$$

Let g be a maximal monotone operator satisfying $L(z) + g(z) \ni y$, and suppose: (7.8.18) holds, there exist α_i $(i = 2, \ldots, m)$ such that

$$\|F^{(i)}(x_0)\| \leq \alpha_i, \qquad (7.8.25)$$

and equation (7.8.23) has a unique $r^* \in [0, R]$ and $\theta(R) \leq 0$.
Then show: the generalized NK method $\{x_n\}$ $(n \geq 0)$ generated by

$$f'(x_n)x_{n+1} + g(x_{n+1}) \ni f'(x_n)(x_n) - f(x_n) \quad (n \geq 0), \quad (x_0 \in D)$$

is well defined, remains in $V(x_0, r^*)$ for all $n \geq 0$, and converges to a solution x^* of

$$f(x) + g(x) \ni x.$$

Moreover, the following error bounds hold for all $n \geq 0$:

$$\|x_{n+1} - x_n\| \leq r_{n+1} - r_n, \qquad (7.8.26)$$

and

$$\|x_n - x^*\| \leq r^* - r_n, \quad r^* = \lim_{n \to \infty} r_n. \qquad (7.8.27)$$

7.8.4. Let $x_0 \in D$ and $R > 0$ be such that $D \equiv U(x_0, R)$. Suppose that f is Fréchet-differentiable on D, and its derivative f' is in a certain sense uniformly continuous as an operator from D into $L(H, H)$; the space of linear operators from H into H. In particular we assume:

$$\|f'(x) - f'(y)\| \leq w(\|x - y\|), \quad x, y \in D, \qquad (7.8.28)$$

for some monotonically increasing positive function w satisfying

$$\lim_{t \to \infty} w(r) = 0,$$

or, even more generally, that

$$\|f'(x) - f'(y)\| \leq w(r, \|x - y\|), \quad x, y \in \bar{D}, \ r \in (0, R), \qquad (7.8.29)$$

for some monotonically increasing in both variables positive function w satisfying

$$\lim_{t \to 0} w(r, t) = 0, \quad r \in [0, R].$$

Conditions of this type have been studied in the special cases $w(t) = dt^\lambda$, $w(r, t) = d(r)t^\lambda$, $\lambda \in [0, 1]$ for regular equations; and for $w(t) = dt$ for generalized equations of the form $f(x) + g(x) \ni x$. The advantages of using (7.8.28) or (7.8.29) have been explained in great detail in the excellent paper [6]. It is useful to pass from the function w to

$$\bar{w}(r) = \sup\{w(t) + w(s) : t + s = r\}.$$

The function may be calculated explicitly in some cases. For example, if $w(r) = dr^\lambda$ ($0 < \lambda \leq 1$), then $\bar{w}(r) = 2^{1-\lambda}dr^\lambda$. More generally, if w is a concave function on $[0, R]$, then $\bar{w}(r) = 2w\left(\frac{r}{2}\right)$, and $\bar{w}$ is increasing, convex, and $\bar{w}(r) \geq w(r), r \in [0, R]$.

Let us define the functions $\theta, \bar{\theta}$ on $[0, R]$ by

$$\theta(r) = \frac{1}{c + \alpha}\left[\eta + \int_0^r w(t)dt\right] - r, \quad \text{for some } \alpha > 0, \eta > 0, c > 0$$

$$\bar{\theta}(r) = \frac{1}{c + \alpha}\left[\eta + \int_0^r \bar{w}(t)dt\right] - r$$

and the equations

$$\theta(r) = 0,$$
$$\bar{\theta}(r) = 0.$$

Let g be a maximal monotone operator satisfying

there exists $c > -\alpha$ such that $\langle f'(z)(x), x \rangle \geq c\|x\|^2$, for all $x \in H, z \in D$,

and suppose: (7.8.28) holds and equation $\bar{\theta}(r) = 0$ has a unique solution $r^* \in [0, R]$.

Then, show: the generalized NK method $\{x_n\}$ ($n \geq 0$) generated by

$$f'(x_n)x_{n+1} + g(x_{n+1}) \ni f'(x_n)(x_n) - f(x_n) \quad (n \geq 0), \ (x_0 \in D)$$

is well defined, remains in $U(x_0, r^*)$ for all $n \geq 0$, and converges to a solution x^* of $f(x) + g(x) \ni x$. Moreover, the following estimates hold for all $n \geq 0$:

$$\|x_{n+1} - x_n\| \leq r_{n+1} - r_n$$
$$\|x_n - x^*\| \leq r^* - r_n,$$

where,

$$r_0 = 0, \ r_{n+1} = r_n - \frac{\bar{\theta}(r_n)}{\theta'(r_n)},$$

and

$$\lim_{n \to \infty} r_n = r^*.$$

8

Convergence Involving Operators with Outer or Generalized Inverses

Local and semilocal convergence of iterative methods using outer or generalized inverses under weaker conditions than before are examined hence.

8.1 Convergence with no Lipschitz conditions

In this study, we are concerned with the problem of approximating a solution x^* of the equation

$$F'(x_0)^\# F(x) = 0, \qquad (8.1.1)$$

where F is an m-times Fréchet-differentiable operator ($m \geq 2$ an integer) defined on an open convex subset of a Banach space X with values in a Banach space Y, and $x_0 \in D$. Operator $F'(x)^\#$ ($x \in D$) denotes an outer inverse of $F'(x)$ ($x \in D$). Many authors have provided local and semilocal results for the convergence of NK method to x^* using hypotheses on the Fréchet derivative (see earlier Chapters 2–6).

Here we provide local convergence theorems for NK method using outer or generalized inverses given by

$$x_{n+1} = x_n - F'(x_n)^\# F(x_n) \quad (n \geq 0) \quad (x_0 \in D). \qquad (8.1.2)$$

Our Newton-Kantorovich-type convergence hypothesis is different from the corresponding famous condition used in the above-mentioned works (see Remark 8.1.10 (b)). Hence, our results have theoretical and practical value. In fact, we show using a simple numerical example that our convergence ball contains earlier ones. This way, we have a wider choice of initial guesses than before. Our results can be used to solve undetermined systems, nonlinear least squares problems, and ill-posed nonlinear operator equations [59], [60].

In this section, we restate some of the definitions and lemmas given in the elegant paper [59].

Let $A \in L(X, Y)$. A linear operator $B: Y \to X$ is called an inner inverse of A if $ABA = A$. A linear operator B is an outer inverse of A if $BAB = B$. If B is both an inner and an outer inverse of A, then B is called a generalized inverse of A. There

I.K. Argyros, *Convergence and Applications of Newton-type Iterations*,
DOI: 10.1007/978-0-387-72743-1_8, © Springer Science+Business Media, LLC 2008

exists a unique generalized inverse $B = A^\dagger_{P,Q}$ satisfying $ABA = A$, $BAB = B$, $BA = I - P$, and $AB = Q$, where P is a given projector on X onto $N(A)$ (the null set of A) and Q is a given projector of Y onto $R(A)$ (the range of A). In particular, if X and Y are Hilbert spaces, and P, Q are orthogonal projectors, then $A^\dagger_{P,Q}$ is called the Moore-Penrose inverse of A.

We will need five lemmas of Banach type and perturbation bounds for outer inverses and for generalized inverses in Banach spaces. The Lemmas 8.1.1–8.1.5 stated here correspond with Lemmas 2.2–2.6 in [59] respectively.

Lemma 8.1.1. *Let $A \in L(X, Y)$ and $A^\# \in L(Y, X)$ be an outer inverse of A. Let $B \in L(X, Y)$ be such that $\|A^\#(B - A)\| < 1$. Then $B^\# = (I + A^\#(B - A))^{-1}A^\#$ is a bounded outer inverse of B with $N(B^\#) = N(A^\#)$ and $R(B^\#) = R(A^\#)$. Moreover, the following perturbation bounds hold:*

$$\|B^\# - A^\#\| \leq \frac{\|A^\#(B - A)A^\#\|}{1 - \|A^\#(B - A)\|} \leq \frac{\|A^\#(B - A)\| \, \|A^\#\|}{1 - \|A^\#(B - A)\|}$$

and

$$\|B^\# A\| \leq (1 - \|A^\#(B - A)\|)^{-1}.$$

Lemma 8.1.2. *Let $A, B \in L(X, Y)$ and $A^\#$, $B^\# \in L(Y, X)$ be outer inverses of A and B, respectively. Then $B^\#(I - AA^\#) = 0$ if and only if $N(A^\#) \subseteq N(B^\#)$.*

Lemma 8.1.3. *Let $A \in L(X, Y)$ and suppose X and Y admit the topological decompositions $X = N(A) \oplus M$, $Y = R(A) \oplus S$. Let $A^\dagger (= A^\dagger_{M,S})$ denote the generalized inverse of A relative to these decompositions. Let $B \in L(X, Y)$ satisfy*

$$\|A^\dagger(B - A)\| \leq 1$$

and

$$(I + (B - A)A^\dagger)^{-1}B \quad maps \ N(A) \ into \ R(A).$$

Then $B^\dagger = B^\dagger_{R(A^\dagger),N(A^\dagger)}$ exists and is equal to

$$B^\dagger = A^\dagger(I + TA^\dagger)^{-1} = (I + A^\dagger T)^{-1}A^\dagger,$$

where $T = B - A$. Moreover, $R(B^\dagger) = R(A^\dagger)$, $N(B^\dagger) = N(A^\dagger)$ and $\|B^\dagger A\| \leq (1 - \|A^\dagger(B - A)\|)^{-1}$.

Lemma 8.1.4. *Let $A \in L(X, Y)$ and $A^\dagger$ be the generalized inverse of Lemma 8.1.3. Let $B \in L(X, Y)$ satisfy the conditions $\|A^\dagger(B - A)\| < 1$ and $R(B) \subseteq R(A)$. Then the conclusion of Lemma 8.1.3 holds and $R(B) = R(A)$.*

Lemma 8.1.5. *Let $A \in L(X, Y)$ and $A^\dagger$ be a bounded generalized inverse of A. Let $B \in L(X, Y)$ satisfy the condition $\|A^\dagger(B - A)\| < 1$. Define $B^\# = (I + A^\dagger(B - A))^{-1}A^\dagger$. Then $B^\#$ is a generalized inverse of B if and only if $\dim N(B) = \dim N(A)$ and $\operatorname{codim} R(B) = \operatorname{codim} R(A)$.*

Let $A \in L(X, Y)$ be fixed. Then, we will denote the set on nonzero outer inverses of A by

$$\Delta(A) = \{B \in L(Y, X) : BAB = B, B \neq 0\}.$$

In [18], [27] we showed the following semilocal convergence theorem for NK method (8.1.2) using outer inverses for m-Fréchet-differentiable operators ($m \geq 2$ an integer).

Theorem 8.1.6. *Let $F: D \subseteq X \to Y$ be an m-times Fréchet-differentiable operator ($m \geq 2$ an integer). Assume:*

(a) there exist an open convex subset D_0 of D, $x_0 \in D_0$, a bounded outer inverse $F'(x_0)^\#$ of $F'(x_0)$, and constants α_i, $\eta \geq 0$ such that for all $x, y \in D_0$ the following conditions hold:

$$\| F'(x_0)^\# (F^{(m)}(x) - F^{(m)}(y)) \| \leq q, \ q > 0, \quad \forall x \in U(x_0, \delta_0), \delta_0 > 0,$$
$$\tag{8.1.3}$$

$$\| F'(x_0)^\# F(x_0) \| \leq \eta, \tag{8.1.4}$$

$$\| F'(x_0)^\# F^{(i)}(x_0) \| \leq \alpha_i, \quad i = 2, 3, \ldots, m; \tag{8.1.5}$$

the positive zero s of $p'(s) = 0$ is such that:

$$p(s) \leq 0, \tag{8.1.6}$$

where

$$f(t) = \eta - t + \frac{\alpha_2}{2!} t^2 + \cdots + \frac{\alpha_m + q}{m!} t^m. \tag{8.1.7}$$

Then polynomial p has only two positive zeros denoted by t^*, t^{**} ($t^* \leq t^{**}$).

(b)

$$\bar{U}(x_0, \delta) \subseteq D_0, \quad \delta = \max\{\delta_0, t^*, t^{**}\}. \tag{8.1.8}$$

(c) $\delta_0 \in \left[t^*, t^{**} \right]$ or $\delta_0 > t^{**}$.

Then

(i) NK method $\{x_n\}$ ($n \geq 0$) generated by (8.1.2) with

$$F'(x_n)^\# = \left[I + F'(x_0)^\# (F'(x_n) - F'(x_0)) \right]^{-1} F'(x_0)^\# \quad (n \geq 0)$$

is well defined, remains in $U(x_0, t^*)$, and converges to a solution $x^* \in \bar{U}(x_0, t^*)$ of equation $F'(x_0)^\# F(x) = 0$;

(ii) the following estimates hold for all $n \geq 0$

$$\| x_{n+1} - x_n \| \leq t_{n+1} - t_n, \tag{8.1.9}$$

and

$$\| x_n - x^* \| \leq t^* - t_n, \tag{8.1.10}$$

where $\{t_n\}$ ($n \geq 0$) is a monotonically increasing sequence generated by

$$t_0 = 0, \quad t_{n+1} = t_n - \frac{f(t_n)}{f'(t_n)}; \tag{8.1.11}$$

(iii) equation $F'(x_0)^{\#}$ has a unique solution in $\tilde{U} \cap \{R(F'(x_0)^{\#}) + x_0\}$, where

$$\tilde{U} = \begin{cases} \bar{U}(x_0, t^*) \cap D_0 & \text{if } \delta_0 \in [t^*, t^{**}] \\ U(x_0, t^{**}) \cap D_0 & \text{if } \delta_0 > t^{**}, \end{cases} \qquad (8.1.12)$$

and

$$R(F'(x_0)^{\#}) + x_0 := \{x + x_0 : x \in R(F'(x_0)^{\#})\}.$$

We provide a local convergence theorem for NK method $\{x_n\}$ $(n \geq 0)$ generated by (8.1.2) for m-Fréchet-differentiable operators.

Theorem 8.1.7. *Let $F: D \subseteq X \to Y$ be an m-times Fréchet-differentiable operator $(m \geq 2$ an integer). Assume:*

(a) $F^{(i)}(x)$, $i = 2, 3, \ldots, m$ satisfies

$$\|F^{(m)}(x) - F^{(m)}(y)\| \leq q_0,$$
$$\|F^{(i)}(x) - F^{(i)}(y)\| \leq b^i \|x - y\|, \quad \text{for all } x, y \in D; \qquad (8.1.13)$$

(b) there exists $x^* \in D$ such that $F(x^*) = 0$ and

$$\|F^{(i)}(x^*)\| \leq b_i, \quad i = 2, 3, \ldots, m; \qquad (8.1.14)$$

(c) let r_0 be the positive zero of equation $g'(t) = 0$, where

$$g(t) = p\left[\frac{b_m + q_0}{m!}t^m + \cdots + \frac{b_2}{2!}t^2\right] - t + b_0, \quad \text{for any } b_0, p > 0, \quad (8.1.15)$$

and such that $U(x^*, r_0) \subseteq D$;
(d) there exists an $F'(x^*)^{\#} \in \Delta(F'(x^*))$ such that

$$\|F'(x^*)^{\#}\| \leq p, \qquad (8.1.16)$$

and for any $x \in U(x^*, r_1)$, for given $\varepsilon_0 > 1$, r_1 is the positive zero of equation $g_1(t) = 0$, where

$$g_1(t) = p\varepsilon_0\left[\frac{b_m + q_0}{(m-1)!}t^m + \cdots + b_2 t\right] + (1 - \varepsilon_0), \qquad (8.1.17)$$

the set $\Delta(F'(x))$ contains an element of minimal mean.

Then, there exists $U(x^*, r) \subseteq D$ with $r \in (0, r_1)$ such that for any $x_0 \in U(x^*, r)$, NK method $\{x_n\}$ $(n \geq 0)$ generated by (8.1.2) for

$$F'(x_0)^{\#} \in \text{argmin}\{\|B\|: B \in \Delta(F'(x_0))\}$$

with $F'(x_n)^{\#} = \left[I + F'(x_0)^{\#}(F'(x_n) - F'(x_0))\right]^{-1} F'(x_0)^{\#}$, converges to $y \in U(x_0, r_0) \cap \{R(F'(x_0)^{\#}) + x_0\}$ such that $F'(x_0)^{\#} F(y) = 0$. Here, we denote

$$R(F'(x_0)^{\#}) + x_0 = \{x + x_0 : x \in R(F'(x_0)^{\#})\}.$$

Proof. (i) We first define parameter ε by

$$\varepsilon \in (0, \min\{\varepsilon_1, \varepsilon_2\}],$$

where,

$$\varepsilon_1 = \frac{1}{p\varepsilon_0}\left[s - \frac{\alpha_2}{2!}s^2 - \cdots - \frac{\alpha_m + q}{m!}s^m\right], \tag{8.1.18}$$

$$\varepsilon_2 = \frac{1}{p\varepsilon_0}\left[(r_0 - r_1) - \frac{\alpha_2}{2!}(r_0 - r_1)^2 - \cdots - \frac{\alpha_m + q}{m!}(r_0 - r_1)^m\right], \tag{8.1.19}$$

and α_i, $i = 2, 3, \ldots, m + 1$ are given by

$$\alpha_m = p\varepsilon_0\left[b_m + (\alpha_m + q_0)r_1\right], \quad q = p\varepsilon_0(\alpha_m + q_0)$$

and

$$\alpha_i = p\varepsilon_0(b_i + b^i r_1), \quad i = 2, 3, \ldots, m - 1.$$

We will use Theorem 8.1.6. Operator F is continuous at x^*. Hence, there exists $U(x^*, r) \subseteq D, r \in (0, r_1)$, such that

$$\|F(x)\| \leq \varepsilon \quad \text{for all } x \in U(x^*, r_1). \tag{8.1.20}$$

Using the identity,

$$F'(x) - F'(x^*) - F''(x^*)(x - x^*) + F''(x^*)(x - x^*) =$$

$$= \int_0^1 \left[F''(x^* + \theta_1 \varepsilon) - F''(x^*)\right]\varepsilon d\theta_1 + \int_0^1 F''(x^*)\varepsilon d\theta_1$$

$$= \int_0^1 \int_0^1 F''(\beta_2)(\beta_1 - x^*)\varepsilon d\theta_2 d\theta_1 + \int_0^1 F''(x^*)\varepsilon d\theta_1$$

$$= \cdots = \int_0^1 \cdots \int_0^1 F^{(m)}(\beta_{m-1})(\beta_{m-2} - x^*)\cdots(\beta_1 - x^*)\varepsilon d\theta_{m-1}\cdots d\theta_1$$

$$+ \int_0^1 \cdots \int_0^1 F^{(m-1)}(\beta_{m-2})(\beta_{m-3} - x^*)\cdots(\beta_1 - x^*)\varepsilon d\theta_{m-2}\cdots d\theta_2 d\theta_1$$

$$+ \cdots + \int_0^1 F''(x^*)\varepsilon d\theta_1$$

$$= \int_0^1 \cdots \int_0^1 \left[F^{(m)}(\beta_{m-1}) - F^{(m)}(x^*)\right](\beta_{m-2} - x^*)\cdots(\beta_1 - x^*)\varepsilon d\theta_{m-1}\cdots d\theta_1$$

$$+ \int_0^1 \cdots \int_0^1 F^{(m)}(x^*)(\beta_{m-2} - x^*)\cdots(\beta_1 - x^*)\varepsilon d\theta_{m-1}\cdots d\theta_1$$

$$+ \int_0^1 \cdots \int_0^1 F^{(m-1)}(x^*)(\beta_{m-3} - x^*)\cdots(\beta_1 - x^*)\varepsilon d\theta_{m-2}$$

$$\cdots d\theta_1 + \cdots + \int_0^1 F''(x^*)\varepsilon d\theta_1, \tag{8.1.21}$$

where we used $\varepsilon = x - x^*$, $\beta_1 = x^* + \theta_1\varepsilon$, $\beta_i = x^* + \theta_i(\beta_{i-1} - x^*)$, $\theta_i \in [0, 1]$, $i = 1, 2, 3, \ldots, m - 1$, conditions (8.1.13), (8.1.14), (8.1.15), and (8.1.16), we get

$$\|F'(x^*)^\#(F'(x) - F'(x^*))\| \le p\left[\frac{b_m + q_0}{(m-1)!}r_0^m + \cdots + b_2r_0\right] < 1,$$

by the choice of r_0.

It follows from Lemma 8.1.1 that

$$F'(x)^\# = \left[I + F'(x^*)^\#(F'(x) - F'(x^*))\right]^{-1}F'(x^*)^\#, \tag{8.1.22}$$

is an outer inverse $F'(x)$, and

$$\|F'(x)^\#\| \le \frac{\|F'(x^*)^\#\|}{1 - p\left[\frac{b_m+q_0}{(m-1)!}r_1^{m-1} + \cdots + b_2r_1\right]} \le p\varepsilon_0, \tag{8.1.23}$$

by the choice of r_1 and ε_0. That is, for any $x_0 \in U(x^*, r)$, the outer inverse

$$F'(x_0)^\# \in \operatorname{argmin}\{\|B\| : B \in \Delta(F'(x_0))\} \quad \text{and} \quad \|F'(x_0)^\#\| \le p\varepsilon_0.$$

We can then obtain for all $x, y \in D$

$$\|F'(x_0)^\#(F^{(m)}(x) - F^{(m)}(y))\| \le p\varepsilon_0\|F^{(m)}(x) - F^{(m)}(y)\| \le p\varepsilon_0q_0 = q,$$
$$\|F'(x_0)^\# F^{(m)}(x_0)\| \le p\varepsilon_0\|F^{(m)}(x_0)\| \le p\varepsilon_0\left[b_m + b_{m+1}r_1\right] = \alpha_m$$
$$\text{(by (8.1.13) and (8.1.14))},$$

and

$$\|F'(x_0)^\# F^{(i)}(x_0)\| \le p\varepsilon_0(b_i + b^ir_1) = \alpha_i, \quad i = 2, 3, \ldots, m - 1,$$
$$\eta \le \|F'(x_0)^\# F(x_0)\| \le p\varepsilon_0 \le s - \frac{\alpha_2}{2!}s^2 - \cdots - \frac{\alpha_m + q}{m!}s^m, \tag{8.1.24}$$

by the choice of ε and ε_1. Hence, there exists a minimum positive zero $t^* < r_1$ of polynomial f given by (8.1.7). It also follows from (8.1.15), (8.1.17), and the choice of ε_2 that $f(r_0 - r_1) \le 0$. That is,

$$r_1 + t^* \le r_0. \tag{8.1.25}$$

Hence, for any $x \in U(x_0, t^*)$ we have

$$\|x^* - x\| \le \|x_0 - x^*\| + \|x_0 - x\| \le r_1 + t^* \le r_0. \tag{8.1.26}$$

It follows from (8.1.26) that $U(x_0, t^*) \subseteq U(x^*, r_0) \subseteq D$. The hypotheses of Theorem 8.1.6 hold at x_0. Consequently, NK method $\{x_n\}$ ($n \ge 0$) stays in $U(x_0, t^*)$ for all $n \ge 0$ and converges to a solution y of equation $F'(x_0)^\# F(x) = 0$.

In the next theorem, we examine the order of convergence of NK method $\{x_n\}$ ($n \ge 0$).

Theorem 8.1.8. *Under the hypotheses of Theorem 8.1.7, if $F'(x_0)^\# F(y) = 0$, then*

$$\|y - x_{n+1}\| \le \frac{\frac{\alpha_m + q}{m!}\|x_n - y\|^{m-2} + \cdots + \frac{\alpha_2}{2!}}{1 - \alpha_2\|x_n - y\| - \cdots - \frac{\alpha_m + q}{(m-1)!}\|x_n - y\|^m}\|y - x_n\|^2, \quad \text{for all } n \ge 0, \quad (8.1.27)$$

and, if $y \in U(x_0, r_2)$, where r_2 is the positive zero of equation $g_2(t) = 0$,

$$g_2(t) = \frac{(\alpha_m + q)(m+1)}{m!}t^{m-1} + \cdots + \frac{3\alpha_2}{2!}t - 1, \quad (8.1.28)$$

then, sequence $\{x_n\}$ $(n \ge 0)$ converges to y quadratically.

Proof. We first note that $r_2 < r_0$. By Lemma 8.1.1 we get $R(F'(x_0)^\#) = R(F'(x_n)^\#)$ $(n \ge 0)$. We have

$$x_{n+1} - x_n = F'(x_n)^\# F(x_n) \in R(F'(x_n)^\#) \quad (n \ge 0),$$

from which it follows

$$x_{n+1} \in R(F'(x_n)^\#) + x_n = R(F'(x_{n-1})^\#) + x_n = R(F'(x_0)^\#) + x_0,$$

and $y \in R(F'(x_n)^\#) + x_{n+1}$ $(n \ge 0)$. That is, we conclude that

$$y \in R(F'(x_0)^\#) + x_0 = R(F'(x_n)^\#) + x_0,$$

and

$$F'(x_n)^\# F'(x_n)(y - x_{n+1}) = F'(x_n)^\# F'(x_n)(y - x_0) - F'(x_n)^\# F'(x_n)(x_{n+1} - x_0)$$
$$= y - x_{n+1}.$$

We also have by Lemma 8.1.2 $F'(x_n)^\# = F'(x_n)^\# F'(x_0) F'(x_0)^\#$. By $F'(x_0)^\# F(y) = 0$ and $N(F'(x_0)^\#) = N(F'(x_n)^\#)$, we get $F'(x_n)^\# F(y) = 0$. Using the estimate

$$\|y - x_{n+1}\| =$$
$$= \|F'(x_n)^\# F'(x_n)(y - x_{n+1})\|$$
$$= \left\|F'(x_n)^\# F'(x_n)\left[y - x_n + F'(x_n)^\#(F(x_n) - F(y))\right]\right\|$$
$$\le \|F'(x_n)^\# F'(x_0)\|$$
$$\cdot \left\|F'(x_0)^\# \left\{\int_0^1 \left[F''[x_n + t(y - x_n)] - F''(x^*)\right](1 - t)dt(y - x_n)^2\right.\right.$$
$$\left.\left. + \tfrac{1}{2}F''(x^*)(y - x_n)^2\right\}\right\|$$
$$\le \frac{\frac{\alpha_m + q}{m!}\|x_n - y\|^{m-2} + \cdots + \frac{\alpha_2}{2!}}{1 - \alpha_2\|x_n - y\| - \cdots - \frac{\alpha_m + 1}{m!}\|x_n - y\|^m}\|y - x_n\|^2 \quad (n \ge 0),$$

which shows (8.1.27) for all $n \ge 0$. By the choice of r_2 and (8.1.27) there exists $\alpha \in [0, 1)$ such that $\|y - x_{n+1}\| \le \alpha\|y - x_n\|$ $(n \ge 0)$, which together with (8.1.27) show that $x_n \to y$ quadratically as $n \to \infty$.

We provide a result corresponding with Theorem 8.1.7 but involving generalized instead of outer inverses.

Theorem 8.1.9. *Let F satisfy the hypotheses of Theorems 8.1.7 and 8.1.8 except (d) which is replaced by*

(d)' the generalized inverse $F'(x^*)$ exists, $\|F'(x^*)^{\dagger}\| \leq p$,

$$\dim N(F'(x)) = \dim N(F'(x^*)) \tag{8.1.29}$$

and

$$\operatorname{codim} R(F'(x)) = \operatorname{codim} R(F'(x^*)) \tag{8.1.30}$$

for all $x \in U(x^*, r_1)$.

Then, the conclusions of Theorems 8.1.7 and 8.1.8 hold with

$$F'(x_0)^{\#} \in \left\{ B : B \in \Delta(F'(x_0)), \|B\| \leq \left\|F'(x_0)^{\dagger}\right\| \right\}. \tag{8.1.31}$$

Proof. In Theorem 8.1.7 we showed that the outer inverse $F'(x)^{\#} \in \operatorname{argmin}\{\|B\| : B \in \Delta(F'(x))\}$ for all $x \in U(x^*, r)$, $r \in (0, r_1)$ and $\|F'(x)^{\#}\| \leq p\varepsilon_0$. We must show that under (d)' the outer inverse

$$F'(x)^{\#} \in \left\{ B : B \in \Delta(F'(x)), \|B\| \leq \left\|F'(x)^{\dagger}\right\| \right\}$$

satisfies $\|F'(x)^{\#}\| \leq p\varepsilon_0$. As in (8.1.21), we get

$$\left\|F'(x^*)^{\dagger}(F'(x) - F'(x^*))\right\| \leq p\left[\frac{b_m + q_0}{(m-1)!}r_0^{m-1} + \cdots + b_2 r_0\right] < 1.$$

Moreover, by Lemma 8.1.5

$$F'(x)^{\dagger} = \left[I + F'(x^*)^{\dagger}(F'(x) - F'(x^*))\right]^{-1} F'(x^*)^{\dagger} \tag{8.1.32}$$

is the generalized inverse of $F'(x)$. Furthermore, by Lemma 8.1.1 as in (8.1.23) $\|F'(x)^{\dagger}\| \leq p\varepsilon_0$. That is, the outer inverse

$$F'(x_0)^{\#} \in \left\{ B : B \in \Delta(F'(x_0)), \|B\| \leq \left\|F'(x_0)^{\dagger}\right\| \right\}$$

satisfies $\|F'(x_0)^{\#}\| \leq p\varepsilon_0$, provided that $x_0 \in U(x^*, r)$.
The rest follows exactly as in Theorems 8.1.7 and 8.1.8.

Remark 8.1.10. (a) We note that Theorem 8.1.6 was proved in [43] with the weaker condition

$$\left\|F'(x_0)^{\#}\left(F^{(m)}(x) - F^{(m)}(x_0)\right)\right\| \leq \bar{\alpha}_{m+1}\|x - x_0\|$$

replacing (8.1.3).

(b) Our conditions (8.1.3)–(8.1.7) differ from the corresponding ones in [52] (see, for example, Theorem 3.1) unless if $\alpha_i = 0$, $i = 2, 3, \ldots, m$, $q = 0$, in which case our condition (8.1.6) becomes the Newton-Kantorovich hypothesis (3.3) in [60, p. 450]:

$$K\eta \leq \tfrac{1}{2}, \tag{8.1.33}$$

where K is such that

$$\left\| F'(x_0)^{\#} \left(F'(x) - F'(y) \right) \right\| \leq K \|x - y\| \tag{8.1.34}$$

for all $x, y \in D$. Similarly (if $\alpha_i = 0$, $i = 2, 3, \ldots, m$), our r_0 equals the radius of convergence in Theorem 3.2 [60, p. 450].

(c) In Theorem 3.2 [60], the condition

$$\left\| F'(x) - F'(y) \right\| \leq c_0 \|x - y\| \quad \text{for all } x, y \in D \tag{8.1.35}$$

was used instead of (8.1.34). The ball used there is $U(x^*, r^*)$, (corresponding with $U(x^*, r_0)$) where

$$r^* = \frac{1}{c_0 p}. \tag{8.1.36}$$

Finally, for convergence $x_0 \in U(x^*, r_1^*)$, where

$$r_1^* = \tfrac{2}{3} r^*. \tag{8.1.37}$$

Below we consider such a case. For simplicity we have taken $F'(x)^{\#} = F'(x)^{-1}$ ($x \in D$) and $m = 2$.

Remark 8.1.11. Methods/routines of how to construct the required outer generalized inverses of the derivative can be found at a great variety in Exercise 8.2.3 at the end of this chapter.

Example 8.1.12. Let us consider the system of equations

$$F(x, y) = 0,$$

where $F: \mathbb{R}^2 \to \mathbb{R}^2$,

$$F(x, y) = (xy - 1, xy + x - 2y).$$

Then, we get

$$F'(x, y) = \begin{bmatrix} y & x \\ y+1 & x-2 \end{bmatrix},$$

and

$$F'(x, y)^{-1} = \frac{1}{x+2y} \begin{bmatrix} 2-x & x \\ y+1 & -y \end{bmatrix},$$

provided that (x, y) does not belong on the straight line $x + 2y = 0$. The second derivative is a bilinear operator on $\mathbb{R}^2$ given by the following matrix

$$F''(x, y) = \begin{bmatrix} 0 & 1 \\ 1 & 0 \\ - & - \\ 0 & 1 \\ 1 & 0 \end{bmatrix}.$$

We consider the max-norm in $\mathbb{R}^2$. Moreover in $L(\mathbb{R}^2, \mathbb{R}^2)$ we use for

$$A = \begin{bmatrix} a_{11} & a_{12} \\ a_{21} & a_{22} \end{bmatrix}$$

the norm,

$$\|A\| = \max\{|a_{11}| + |a_{12}|, |a_{21}| + |a_{22}|\}.$$

As in [6], we define the norm of a bilinear operator B on $\mathbb{R}^2$ by

$$\|B\| = \sup_{\|z\|=1} \max_i \sum_{j=1}^{2} \left| \sum_{k=1}^{2} b_i^{jk} z_k \right|,$$

where,

$$z = (z_1, z_2) \quad \text{and} \quad B = \begin{bmatrix} b_1^{11} & b_1^{12} \\ b_1^{21} & b_1^{22} \\ - & - \\ b_2^{11} & b_2^{12} \\ b_2^{21} & b_2^{22} \end{bmatrix}.$$

For $m = 2$ and $(x^*, y^*) = (1, 1)$, we get $c_0 = \frac{4}{3}$, $r_1^* = .5$, $\alpha_2 = 1$. We can set $q = .001$ to obtain $r_2 = .666444519$. Note that $r_2 > r_1^*$.

8.2 Exercises

8.2.1 (a) Assume there exist nonnegative parameters $K, M, L, \ell, \mu, \eta, \delta \in [0, 1]$ such that:

$$L \leq K, \tag{8.2.1}$$

$$\ell + 2\mu < 1, \tag{8.2.2}$$

and

$$h_\delta \equiv \left(K + L\delta + \frac{4M}{2-\delta} \right) \eta + \delta\ell + 2\mu \leq \delta. \tag{8.2.3}$$

Show: iteration $\{t_n\}$ $(n \geq 0)$ given by

$$t_0 = 0, \quad t_1 = \eta, \quad t_{n+2} = t_{n+1} + \frac{K(t_{n+1}-t_n)+2(Mt_n+\mu)}{2(1-\ell-Lt_{n+1})}(t_{n+1} - t_n) \quad (n \geq 0)$$

is nondecreasing, bounded above by t^{**}, and converges to some t^* such that

$$0 \leq t^* \leq \frac{2\eta}{2-\delta} \equiv t^{**}.$$

Moreover, the following estimates hold for all $n \geq 0$

$$t_{n+2} - t_{n+1} \leq \tfrac{8}{2}(t_{n+1} - t_n) \leq \left(\tfrac{8}{2}\right)^{n+1} \eta.$$

(b) Let $F: D \subseteq X \to Y$ be a Fréchet-differentiable operator. Assume: there exist an approximation $A(x) \in L(X, Y)$ of $F'(x)$, an open convex subset D_0 of D, $x_0 \in D_0$, a bounded outer inverse $A^\#$ of $A(x_0)$, and parameters $\eta > 0$, $K > 0$, $M \geq 0$, $L \geq 0$, $\mu \geq 0$, $\ell \geq 0$ such that (8.2.1)–(8.2.3) hold

$$\|A^\# F(x_0)\| \leq \eta,$$

$$\|A^\# \left[F'(x) - F'(y)\right]\| \leq K\|x - y\|,$$

$$\|A^\# \left[F'(x) - A(x)\right]\| \leq M\|x - x_0\| + \mu,$$

and

$$\|A^\# \left[A(x) - A(x_0)\right]\| \leq L\|x - x_0\| + \ell$$

for all $x, y \in D_0$, and

$$\bar{U}(x_0, t^*) \subseteq D_0$$

Show: sequence $\{x_n\}$ ($n \geq 0$) generated by Newton-like method with

$$A(x_n)^\# = \left[I + A^\#(A(x_n) - A(x_0))\right]^{-1} A^\#$$

is well defined, remains in $U(x_0, s^*)$ for all $n \geq 0$, and converges to a unique solution x^* of equation $A^\# F(x) = 0$, $\bar{U}(x_0, t^*) \cap D_0$
Moreover, the following estimates hold for all $n \geq 0$

$$\|x_{n+1} - x_n\| \leq t_{n+1} - t_n,$$

and

$$\|x_n - x^*\| \leq t^* - t_n.$$

(c) Assume:
—there exist an approximation $A(x) \in L(X, Y)$ of $F'(x)$, a simple solution $x^* \in D$ of equation $F(x) = 0$, a bounded outer inverse $A_1^\#$ of $A(x^*)$, and nonnegative parameters $\bar{K}, \bar{L}, \bar{M}, \bar{\mu}, \bar{\ell}$, such that:

$$\|A_1^\# \left[F'(x) - F'(y)\right]\| \leq \bar{K}\|x - y\|,$$

$$\|A_1^\# \left[F'(x) - A(x)\right]\| \leq \bar{M}\|x - x^*\| + \bar{\mu},$$

and

$$\|A_1^\# \left[A(x) - A(x^*)\right]\| \leq \bar{L}\|x - x^*\| + \bar{\ell}$$

for all $x, y \in D$;
—equation

$$\left(\tfrac{\bar{K}}{2} + \bar{M} + \bar{L}\right) r + \bar{\mu} + \bar{\ell} - 1 = 0$$

has a minimal nonnegative zero r^* satisfying

$$\bar{L}r + \bar{\ell} < 1,$$

and

$$U(x^*, r^*) \subseteq D.$$

Show: sequence $\{x_n\}$ $(n \geq 0)$ generated by Newton-like method is well defined, remains in $U(x^*, r^*)$ for all $n \geq 0$, and converges to x^* provided that $x_0 \in U(x^*, r^*)$. Moreover, the following error bounds hold for all $n \geq 0$:

$$\|x^* - x_{n+1}\| \leq$$

$$\leq \frac{1}{1 - \bar{L}\|x^* - x_n\| - \bar{\ell}} \left[\frac{\bar{K}}{2}\|x^* - x_n\| + (\bar{M}\|x^* - x_n\| + \bar{\mu}) \right] \|x^* - x_n\|$$

$$< \frac{\left(\frac{\bar{K}}{2} + \bar{M} \right) r^* + \bar{\mu}}{1 - \bar{L}r^* - \bar{\ell}} \|x^* - x_n\|.$$

8.2.2 (a) Let $F : D \subseteq X \to Y$ be an m-times Fréchet-differentiable operator ($m \geq 2$ integer).
Assume:

(a_1) there exist an open convex subset D_0 of D, $x_0 \in D_0$, a bounded outer inverse $F'(x_0)^{\#}$ of $F'(x_0)$, and constants $\eta > 0$, $\alpha_i \geq 0$, $i = 2, ..., m + 1$ such that for all $x, y \in D_0$ the following conditions hold:

$$\|F'(x_0)^{\#}(F^{(m)}(x) - F^{(m)}(x_0))\| \leq \varepsilon, \quad \varepsilon > 0, \tag{8.2.4}$$

$$\text{for all } x \in U(x_0, \delta_0) \text{ and some } \delta_0 > 0.$$

$$\|F'(x_0)^{\#} F(x_0)\| \leq \eta,$$

$$\|F'(x_0)^{\#} F^{(i)}(x_0)\| \leq \alpha_i$$

the positive zeros s of p' is such that

$$p(s) \leq 0,$$

where,

$$p(t) = \eta - t + \frac{\alpha_2 t^2}{2!} + \cdots + \frac{\alpha_m + \varepsilon}{m!} t^m.$$

Show: polynomial p has only two positive zeros denoted by t^*, t^{**} ($t^* \leq t^{**}$).

(a_2)

$$\bar{U}(x_0, \delta) \subseteq D_0, \quad \delta = \max\{\delta_0, t^*, t^{**}\}.$$

(a_3) $\delta_0 \in [t^*, t^{**}]$ or $\delta_0 > t^{**}$.

Moreover show: sequence $\{x_n\}$ $(n \geq 0)$ generated by NK method with $F'(x_n)^{\#} = [I + F'(x_0)^{\#}(F'(x_n) - F'(x_0))]^{-1} F'(x_0)^{\#}$ $(n \geq 0)$ is well defined, remains in $U(x_0, t^*)$, and converges to a solution $x^* \in \bar{U}(x_0, t^*)$ of equation $F'(x_0)^{\#} F(x) = 0$;

—the following estimates hold for all $n \geq 0$

$$\|x_{n+1} - x_n\| \leq t_{n+1} - t_n$$

and

$$\|x_n - x^*\| \leq t^* - t_n,$$

where $\{t_n\}$ $(n \geq 0)$ is a monotonically increasing sequence converging to t^* and generated by

$$t_0 = 0, \quad t_{n+1} = t_n - \frac{p(t_n)}{p'(t_n)}.$$

(b) Let $F: D \subseteq X \to Y$ be an m-times Fréchet-differentiable operator ($m \geq 2$ an integer). Assume:

(b$_1$) condition (8.2.4) holds;

(b$_2$) there exists an open convex subset D_0 of D, $x_0 \in D_0$, and constants $\alpha, \beta, \eta \geq 0$ such that for any $x \in D_0$, there exists an outer inverse $F'(x)^{\#}$ of $F'(x)$ satisfying $N(F'(x)^{\#}) = N(F'(x_0)^{\#})$ and

$$\left\| F'(x_0)^{\#} F(x_0) \right\| \leq \eta,$$

$$\left\| F'(y)^{\#} \int_0^1 F''[x + t(y - x)] (1 - t) dt (y - x)^2 \right\| \leq$$

$$\leq \left[\frac{\alpha_m + \varepsilon}{m!} \|y - x\|^{m-2} + \cdots + \frac{\alpha_2}{2!} \right] \|y - x\|^2,$$

for all $x, y \in D_0$,

$$\left[\frac{\alpha_m + \varepsilon}{m!} \eta^{m-2} + \cdots + \frac{\alpha_2}{2!} \right] \eta < 1,$$

and

$$\bar{U}(x_0, r) \subseteq D_0 \quad \text{with } r = \min \left\{ \frac{\eta}{1 - r_0}, \delta_0 \right\},$$

where,

$$r_0 = \left[\frac{\alpha_m + \varepsilon}{m!} \eta^{m-2} + \cdots + \frac{\alpha_2}{2!} \right] \eta.$$

Show: sequence $\{x_n\}$ $(n \geq 0)$ generated by NK method is well defined, remains in $\bar{U}(x_0, r)$ for all $n \geq 0$, and converges to a solution x^* of $F'(x_0)^{\#} F(x) = 0$ with the iterates satisfying $N(F'(x_n)^{\#}) = N(F'(x_0)^{\#})$ $(n \geq 0)$. Moreover, the following estimates hold for all $n \geq 0$

$$\|x_{n+1} - x_n\| \leq r_0^n \|x_1 - x_0\|,$$

$$\|x^* - x_n\| \leq \frac{r_0^n}{1 - r_0} \|x_1 - x_0\|,$$

and

$$\|x_n - x_0\| \leq \frac{1 - r_0^n}{1 - r_0} \|x_1 - x_0\| \leq \frac{1 - r_0^n}{1 - r_0} \eta \leq r.$$

8.2.3 Let X and Y be Banach spaces, and let L be a bounded linear operator on X into Y. A linear operator $M: Y \to X$ is said to be an inner inverse of L if $LMA = L$. A linear operator $M: Y \to X$ is an outer inverse of L if $MLM = M$. Let L be an $m \times n$ matrix, with $m > n$. Any outer inverse M of L will be an $n \times m$ matrix. Show:

(a) If rank $(L) = n$, then L can be written as

$$L = A \begin{bmatrix} I \\ 0 \end{bmatrix},$$

where I is the $n \times n$ identity matrix, and A is an $m \times m$ invertible matrix. The $n \times m$ matrix

$$M = \begin{bmatrix} I & B \end{bmatrix} A^{-1}$$

is an outer inverse of L for any $n \times (m - n)$ matrix B.

(b) If rank $(L) = r < n$, then L can be written as

$$L = A \begin{bmatrix} I & 0 \\ 0 & 0 \end{bmatrix} C,$$

where A is an $m \times m$ invertible matrix, I is the $r \times r$ identity matrix, and C is an $n \times n$ invertible matrix. If E is an outer (inner) inverse of the matrix $\begin{bmatrix} I & 0 \\ 0 & 0 \end{bmatrix}$, then the $n \times m$ matrix

$$M = C^{-1} E A^{-1}$$

is an outer (inner) inverse of L.

(c) E is both an inner and an outer inverse of $\begin{bmatrix} I & 0 \\ 0 & 0 \end{bmatrix}$ if and only if E can be written in the form

$$E = \begin{bmatrix} I & M \\ C & CM \end{bmatrix} = \begin{bmatrix} I \\ C \end{bmatrix} \begin{bmatrix} I & M \end{bmatrix}.$$

(d) For any $(n - r) \times r$ matrix T, the matrix $E = \begin{bmatrix} I & 0 \\ T & 0 \end{bmatrix}$ is an outer inverse of $\begin{bmatrix} I & 0 \\ 0 & 0 \end{bmatrix}$.

8.2.4. Let $F: D \subseteq X \to Y$ be a Fréchet-differentiable operator between two Banach spaces X and Y, $A(x) \in L(X, Y)$ $(x \in D)$ be an approximation to $F'(x)$. Assume that there exist an open convex subset D_0 of D, $x_0 \in D_0$, a bounded outer inverse $A^{\#}$ of $A (= A(x_0))$, and constants $\eta, k > 0$, $M, L, \mu, l \geq 0$ such that for all $x, y \in D_0$ the following conditions hold:

$$\| A^{\#} F(x_0) \| \leq \eta, \quad \| A^{\#} \left(F'(x) - F'(y) \right) \| \leq k \| x - y \|,$$

$$\| A^{\#} \left(F'(x) - A(x) \right) \| \leq M \| x - x_0 \| + \mu,$$

$$\| A^{\#} (A(x) - A) \| \leq L \| x - x_0 \| + l, \quad b := \mu + l < 1.$$

Assume $h = \sigma \eta \leq \frac{1}{2}(1 - b)^2$, $\sigma := \max(k, M + L)$, and $\bar{U} = \bar{U}(x_0, t^*) \subseteq D_0$, $t^* = \frac{1 - b - \sqrt{(1-b)^2 - 2h}}{\sigma}$. Then show

(i) sequence $\{x_n\}$ $(n \geq 0)$ generated by $x_{n+1} = x_n - A(x_n)^{\#} F(x_n)$ $(n \geq 0)$
with $A(x_n)^{\#} = [I + A^{\#}(A(x_n) - A)]^{-1} A^{\#}$ remains in U and converges to
a solution $x^* \in \bar{U}$ of equation $A^{\#} F(x) = 0$.

(ii) equation $A^{\#} F(x) = 0$ has a unique solution in $\bar{U} \cap \{R(A^{\#}) + x_0\}$, where

$$\tilde{U} = \begin{cases} (x_0, t^*) \cap D_0 & \text{if } h = \frac{1}{2}(1 - b)^2 \\ U(x_0, t^{**}) \cap D_0 & \text{if } h < \frac{1}{2}(1 - b)^2, \end{cases}$$

$$R(A^{\#}) + x_0 := \left\{ x + x_0 : x \in R(A^{\#}) \right\},$$

and

$$t^* = \frac{1 - b + \sqrt{(1-b)^2 - 2h}}{\sigma}$$

(iii) $\|x_{n+1} - x_n\| \leq t_{n+1} - t_n$, $\|x^* - x_n\| \leq t^* - t_n$, where $t_0 = 0$, $t_{n+1} = t_n + \frac{f(t_n)}{g(t_n)}$, $f(t) = \frac{\sigma}{2} t^2 - (1 - b) t + \eta$, and $g(t) = 1 - Lt - \ell$.

8.2.5. Let $F: D \subseteq X \to Y$ be a Fréchet-differentiable operator between two Banach
spaces X and Y and let $A(x) \in L(X, Y)$ be an approximation of $F'(x)$. Assume
that there exist an open convex subset D_0 of D, a point $x_0 \in D_0$, and constants
$\eta, k > 0$ such that for any $x \in D_0$, there exists an outer inverse $A(x)^{\#}$ of $A(x)$
satisfying $N(A(x)^{\#}) = N(A^{\#})$, where $A = A(x_0)$ and $A^{\#}$ is a bounded outer
inverse of A, and for this outer inverse the following conditions hold:

$$\|A^{\#} F(x_0)\| \leq \eta,$$

$$\|A(y)^{\#}(F'(x + t(y - x)) - F'(x))\| \leq kt \|x - y\|$$

for all $x, y \in D_0$ and $t \in [0, 1]$, $h = \frac{1}{2} k \eta < 1$ and $\bar{U}(x_0, r) \subseteq D_0$
with $r = \frac{\eta}{1-h}$. Then show sequence $\{x_n\}$ $(n \geq 0)$ generated by $x_{n+1} = x_n - A(x_n)^{\#} F(x_n)$ $(n \geq 0)$ with $A(x_n)^{\#}$ satisfying $N(A(x_n)^{\#}) = N(A^{\#})$ remains in $\bar{U}(x_0, r)$ and converges to a solution x^* of equation $A^{\#} F(x) = 0$.

8.2.6. Show that NK method with outer inverses $x_{n+1} = x_n - F'(x_n)^{\#}$ $(n \geq 0)$
converges quadratically to a solution $x^* \in \tilde{U} \cap \{R(F'(x_0)^{\#}) + x_0\}$ of equation $F'(x_0)^{\#} F(x) = 0$ under the conditions of Exercise 8.2.4 with $A(x) = F'(x)$ $(x \in D_0)$.

8.2.7. Let $F: D \subseteq X \to Y$ be Fréchet-differentiable and assume that $F'(x)$ satisfies
a Lipschitz condition

$$\|F'(x) - F'(y)\| \leq L \|x - y\|, \quad x, y \in D.$$

Assume $x^* \in D$ exists with $F(x^*) = 0$. Let $a > 0$ such that $U\left(x^*, \frac{1}{a}\right) \subseteq D$.
Suppose there is an

$$F'(x^*)^{\#} \in \Omega(F'(x^*)) = \{B \in L(Y, X) : BF'(x^*) B = B, \ B \neq 0\}$$

such that $\|F'(x^*)^{\#}\| \leq a$ and for any $x \in U\left(x^*, \frac{1}{3La}\right)$, the set $\Omega(F'(x))$ con-
tains an element of minimum norm. Then show there exists a ball $U(x^*, r) \subseteq D$

with $cr < \frac{1}{3aL}$ such that for any $x_0 \in U(x^*, r)$ the sequence $\{x_n\}$ $(n \geq 0)$
$x_{n+1} = x_n - F'(x_n)^\# F(x_n)$ $(n \geq 0)$ with

$$F'(x_0)^\# \in \text{argmin}\left\{\|B\| \,\big|\, B \in \Omega\left(F'(x_0)\right)\right\}$$

and with $F'(x_n)^\# = (I + F'(x_0)^\# (F'(x_n) - F'(x_0)))^{-1} F'(x_0)^\#$ converges
quadratically to $\bar{x}^* \in U\left(x_0, \frac{1}{La}\right) \cap \{R(F'(x_0)^\#) + x_0\}$, which is a solution
of equation $F'(x_0)^\# F(x) = 0$. Here, $R(F'(x_0)^\#) + x_0 = \{x + x_0 : x \in R(F'(x_0)^\#)\}$.

9

Convergence on Generalized Banach Spaces: Improving Error Bounds and Weakening of Convergence Conditions

The local and semilocal convergence of iterative methods in generalized spaces with a convergence structure as well as in K-normed spaces under weak conditions is examined in this chapter.

9.1 K-normed spaces

In this section, we are concerned with the problem of approximating a solution x^* of equation

$$F(x) + G(x) = 0, \tag{9.1.1}$$

where F, G are operators between two Banach spaces X, Y defined on a closed ball centered at some point $x_0 \in X$ and of radius $R > 0$. Operator F is differentiable, whereas the differentiability of G is not assumed.

We propose the Newton-Kantorovich method

$$x_{n+1} = x_n - F'(x_n)^{-1}(F(x_n) + G(x_n)) \quad (n \geq 0) \tag{9.1.2}$$

to generate a sequence approximating x^*.

This study is motivated by the elegant work in [53], where X is a real Banach space ordered by a closed convex cone K. We note that passing from scalar majorants to vector majorants enlarges the range of applications, as the latter uses the spectral radius, which is usually smaller than its norm used by the former.

Here using finer vector majorants than before (see [53]), we show under the same hypotheses:

(a) sufficient convergence conditions can be obtained that are always weaker than before.

(b) finer estimates on the distances involved and an at least as precise information on the location of the solution x^* are provided.

I.K. Argyros, *Convergence and Applications of Newton-type Iterations*,
DOI: 10.1007/978-0-387-72743-1_9, © Springer Science+Business Media, LLC 2008

Several applications are provided. In particular, we show as a special case that the famous Newton-Kantorovich hypothesis is weakened. Finally, we study the local convergence of method (9.1.2).

In order to make the study as self-contained as possible, we need to introduce some concepts involving K-normed spaces.

Let X be a real Banach space ordered by a closed convex cone K. We say that cone K is regular if every increasing sequence $\gamma_1 \leq \gamma_2 \leq \cdots \leq \gamma_n \leq \cdots$ that is bounded above converges in norm. If $\gamma_n^0 \leq \gamma_n \leq \gamma_n^1$ and $\lim_{n\to\infty} \gamma_n^0 = \lim_{n\to\infty} \gamma_n^1 = \gamma^*$, the regularity of K implies $\lim_{n\to\infty} \gamma_n = \gamma^*$.

Let $\alpha, \beta \in X$, the conic segment $\langle \alpha, \beta \rangle = \{\gamma \mid \alpha \leq \gamma \leq \beta\}$. An operator Q in X is called positive if $Q(\gamma) \in K$ for all $\gamma \in K$. Denote by $L(X, X)$ the space of all bounded linear operators in X, and $L_{\text{sym}}(X^2, X)$ the space of bilinear, symmetric, bounded operators from X^2 to X. By the standard linear isometry between $L(X^2, X)$, and $L(X, L(X, X))$, we consider the former embedded into the latter.

Let D be a linearly connected subset of K, and φ be a continuous operator from D into $L(X, X)$ or $L(X, L(X, X))$. We say that the line integral of φ is independent of the path if for every polygonal line L in D, the line integral depends only on the initial and final point of L. We define

$$\int_{r_0}^{r} \varphi(t)dt = \int_0^1 \varphi\left[(1-s)r_0 + sr\right](r - r_0)ds. \tag{9.1.3}$$

We need the definition of K-normed space:

Definition 9.1.1. *Let X be a real linear space. Then X is said to be K-normed if operator $]\cdot[: X \to Y$ satisfies:*

$$]x[\geq 0 \quad (x \in X);$$
$$]x[= 0 \Leftrightarrow x = 0;$$
$$]rx[= |r|]x[\quad (x \in X, r \in \mathbf{R});$$

and

$$]x + y[\leq]x[+]y[\quad (x, y \in X).$$

Let $x_0 \in X$ and $r \in K$. Then we denote

$$\overline{U}(x_0, r) = \{x \in X \mid]x - x_0[\leq r\}. \tag{9.1.4}$$

Using K-norm, we can define convergence on X. A sequence $\{y_n\}$ $(n \geq 0)$ in X is said to be

(1) convergent to a limit $y \in X$ if

$$\lim_{n\to\infty}]y_n - y[= 0 \quad \text{in } X$$

and we write

$$(X) - \lim_{n\to\infty} y_n = y;$$

(2) a Cauchy sequence if

$$\lim_{m,n\to\infty}]y_m - y_n[\ = 0.$$

The space X is complete if every Cauchy sequence is convergent.

We use the following conditions:

F is differentiable on the K-ball $U(x_0, R)$, and for every $r \in S = \langle 0, R \rangle$ there exist positive operators $w_0(r), \overline{w}(r) \in L_{\text{sym}}(X^2, X)$ such that for all $z \in X$

$$](F'(x) - F'(x_0))(z)[\ \leq w_0(r)(]x - x_0[,]z[) \tag{9.1.5}$$

for all $x \in \overline{U}(x_0, r)$,

$$](F'(x) - F'(y))(z)[\ \leq \overline{w}(r)(]x - y[,]z[) \tag{9.1.6}$$

for all $x, y \in \overline{U}(x_0, r)$, where operators $w_0, \overline{w}: S \to L_{\text{sym}}(X^2, X)$ are increasing. Moreover, the line integral of $\overline{w}$ (similarly for w_0) is independent of the path, and the same is true for the operator $w: S \to L(X, X)$ given by

$$w(r) = \int_0^r \overline{w}(t)dt. \tag{9.1.7}$$

Note that in general

$$w_0(r) \leq \overline{w}(r) \quad \text{for all } r \in S. \tag{9.1.8}$$

The Newton-Leibniz formula holds for F on $\overline{U}(x_0, R)$:

$$F(x) - F(y) = \int_x^t F'(z)dz, \tag{9.1.9}$$

for all segments $[x, y] \in \overline{U}(x_0, R)$; for every $r \in S$ there exists a positive operator $w_1(r) \in L(X, X)$ such that

$$]G(x) - G(y)[\ \leq w_1(r)(]x - y[) \quad \text{for all } x, y \in \overline{U}(x_0, r), \tag{9.1.10}$$

where $w_1: S \to L(X, X)$ is increasing and the line integral of w_1 is independent of the path;

Operator $F'(x_0)$ is invertible and satisfies:

$$]F'(x_0)(y)[\ \leq b]y[\quad \text{for all } y \in Y \tag{9.1.11}$$

for some positive operator $b \in L(X, X)$.

Let

$$\eta =]F'(x_0)^{-1}(F(x_0) + G(x_0))[, \tag{9.1.12}$$

and define operator $f: S \to X$ by letting

$$f(r) = \eta + b \int_0^r w(t)dt + b \int_0^r w_1(t)dt. \tag{9.1.13}$$

By the monotonicity of operators w, w_1, we see that f is order convex, i.e., for all $r, \bar{r} \in S$ with $r \leq \bar{r}$,

$$f((1-s)r + s\bar{r}) \leq (1-s)f(r) + sf(\bar{r}) \quad \text{for all } s \in [0,1]. \qquad (9.1.14)$$

We will use the following results whose proofs can be found in [53]:

Lemma 9.1.2. *(a) If Lipschitz condition (9.1.6) holds, then*

$$](F'(x+y) - F'(x))(z)[\; \leq (w(r +]y[) - w(r))(]z[) \qquad (9.1.15)$$

for all $r, r +]y[\in S$, $x \in \overline{U}(x_0, r)$, $z \in X$;
(b) If Lipschitz condition (9.1.10) holds, then

$$]G(x+y) - G(x)[\; \leq \int_r^{r+]y[} w_1(t)dt \quad \text{for all } r, r +]y[\in S, \; x \in \overline{U}(x_0, r).$$
$$(9.1.16)$$

Denote by $\text{Fix}(f)$ the set of all fixed points of the operator f.

Lemma 9.1.3. *Assume:*

$$\text{Fix}(f) \neq \emptyset. \qquad (9.1.17)$$

Then there is a minimal element r^* in $\text{Fix}(f)$ that can be found by applying the method of successive approximations

$$r = f(r) \qquad (9.1.18)$$

with 0 as the starting point.
The set

$$B(f, r^*) = \left\{ r \in S: \lim_{n \to \infty} f^n(r) = r^* \right\} \qquad (9.1.19)$$

is the attracting zone of r^*.

Remark 9.1.4. [53] *Let $r \in S$. If*

$$f(r) \leq r, \qquad (9.1.20)$$

and

$$\langle 0, r \rangle \cap \text{Fix}(f) = \{r^*\} \qquad (9.1.21)$$

then,

$$\langle 0, r \rangle \subseteq B(f, r^*). \qquad (9.1.22)$$

Note that the successive approximations

$$\varepsilon_{n+r} = \delta(\varepsilon_n) \quad (\varepsilon_0 = r) \; (n \in N) \qquad (9.1.23)$$

converges to a fixed point ε^* of f, satisfying $0 \leq s^* \leq r$. Hence, we conclude $s^* = r^*$, which implies $r \in B(f, r^*)$.

In particular, Remark 9.1.4 implies

$$\langle 0, (1 - s)r^* + sr \rangle \subseteq B(f, r^*) \tag{9.1.24}$$

for every $r \in \mathrm{Fix}(f)$ with $\langle 0, r \rangle \cap \mathrm{Fix}(f) = \{r^*, r\}$, and for all $\lambda \in [0, 1)$.

In the scalar case $X = \mathbf{R}$, we have

$$B(f, r^*) = [0, r^*] \cup \{r \in S : r^* < r, f(q) < q, (r^* < q \leq r)\}. \tag{9.1.25}$$

We will also use the notation

$$E(r^*) = \bigcup_{r \in B(f, r^*)} \overline{U}(x_0, r). \tag{9.1.26}$$

Returning back to method (9.1.2), we consider the sequences of approximations

$$r_{n+1} = r_n - (bw_0(r_n) - I)^{-1}(f(r_n) - r_n) \quad (r_0 = 0, \ n \geq 0) \tag{9.1.27}$$

and

$$\overline{r}_{n+1} = \overline{r}_n - (bw(\overline{r}_n) - I)^{-1}(f(\overline{r}_n) - \overline{r}_n) \quad (\overline{r}_0 = 0, \ n \geq 0) \tag{9.1.28}$$

for the majorant equation (9.1.18).

Lemma 9.1.5. *If operators*

$$I - bw_0(r) \quad r \in [0, r^*) \tag{9.1.29}$$

are invertible with positive inverses, then sequence $\{r_n\}$ $(n \geq 0)$ given by (9.1.27) is well defined for all $n \geq 0$, monotonically increasing, and convergent to r^*.

Proof. We first show that if $a_1 \leq a_2$, $a_2 \neq r^*$, then

$$(I - bw_0(a_1))^{-1}\theta \leq (I - bw_0(a_2))^{-1}\theta \quad \text{for all} \ \theta \in K. \tag{9.1.30}$$

We have

$$\theta + bw_0(a_2)\theta + \cdots + (bw_0(a_2))^n\theta$$
$$= (I - bw_0(a_2))^{-1}\theta - (bw_0(a_2))^{n+1}(I - bw_0(a_2))^{-1}\theta \leq \theta_2, \tag{9.1.31}$$

where,

$$\theta_2 = (I - bw_0(a_2))^{-1}\theta. \tag{9.1.32}$$

Using the monotonicity of w we get

$$\theta + bw_0(a_1)\theta + \cdots + (bw_0(a_1))^n\theta$$
$$\leq \theta + bw_0(a_1)\theta + \cdots + (bw_0(a_1))^n\theta \leq \theta_2, \tag{9.1.33}$$

which implies increasing sequence $\{(bw_0(a_1))^n\theta\}$ $(n \geq 0)$ is bounded in a regular cone K, and as such it converges to some

$$\theta_1 = (I - bw_0(a_1))^{-1}\theta \qquad (9.1.34)$$

with

$$\theta_1 \leq \theta_2.$$

We also need to show that the operator

$$g(r) = r - (bw_0(r) - I)^{-1}(f(r) - r) \qquad (9.1.35)$$

is increasing on $\langle 0, r^* \rangle - \{r^*\}$. Let $b_1 \leq b_2$ with $b_1, b_2 \in [0, r^*]$, then using (9.1.8) and (9.1.35), we obtain in turn

$$g(b_2) - g(b_1)$$
$$= b_2 - b_1 - (bw_0(b_2) - I)^{-1}(f(b_2) - b_2) + (bw_0(b_1) - I)^{-1}(f(b_1) - b_1)$$
$$= b_2 - b_1 - (bw_0(b_1) - I)^{-1}[(f(b_2) - b_2) - (f(b_1) - b_1)]$$
$$\quad + \left[(bw_0(b_1) - I)^{-1} - (bw_0(b_2) - I)^{-1}\right](f(b_2) - b_2)$$
$$= (I - bw_0(b_1))^{-1}[f(b_2) - f(b_1) - bw_0(b_1)(b_2 - b_1)] \qquad (9.1.36)$$
$$\quad + (I - bw_0(b_2))^{-1}(bw_0(b_2) - bw_0(b_1))(I - bw_0(b_1))^{-1}(f(b_2) - b_2) \geq 0,$$

as all terms in the right hand of equality (9.1.36) are in the cone.

Moreover, g leaves $\langle 0, r^* \rangle - \{r^*\}$ invariant, since

$$0 = g(0) \leq g(r) \leq g(r^*) = r^*. \qquad (9.1.37)$$

Hence sequence

$$r_{n+1} = g(r_n) \quad (r_0 = 0, \; n \geq 0) \qquad (9.1.38)$$

is well defined for all $n \geq 0$, lies in the set $\langle 0, r^* \rangle - \{r^*\}$, and is increasing. Therefore the limit of this sequence exists. Let us call it r_1^*. The point r_1^* is a fixed point of f in $\langle 0, r^* \rangle$. But r^* is the unique fixed point of f in $\langle 0, r^* \rangle$. Hence, we deduce

$$r_1^* = r^*. \qquad (9.1.39)$$

Remark 9.1.6. If equality holds in (9.1.8) then sequence $\{\bar{r}_n\}$ becomes $\{r_n\}$ $(n \geq 0)$ and Lemma 9.1.5 reduces to Lemma 5 in [53, p. 555]. Moreover as it can easily be seen using induction on n

$$r_{n+1} - r_n \leq \bar{r}_{n+1} - \bar{r}_n, \qquad (9.1.40)$$

and

$$r_n \leq \bar{r}_n \qquad (9.1.41)$$

for all $n \geq 0$. Furthermore, if strict inequality holds in (9.1.8), so it does in (9.1.40) and (9.1.41). If $\{r_n\}$ $(n \geq 0)$ is a majorizing sequence for method (9.1.2), then (9.1.40) shows that the error bounds on the distances $\|x_{n+1} - x_n\|$ are improved. It turns out that this is indeed the case.

We can show the semilocal convergence theorem for method (9.1.2).

Theorem 9.1.7. *Assume:*
hypotheses (9.1.6), (9.1.7), (9.1.9), (9.1.10), (9.1.11), (9.1.17) hold, and operators (9.1.29) are invertible with positive inverses.
Then sequence $\{x_n\}$ $(n \geq 0)$ generated by Newton-Kantorovich method (9.1.2) is well defined, remains in the K-ball $U(x_0, r^*)$ for all $n \geq 0$, and converges to a solution x^* of equation $F(x) + G(x) = 0$ in $E(r^*)$, where $E(r^*)$ is given by (9.1.26). Moreover, the following error bounds hold for all $n \geq 0$:

$$]x_{n+1} - x_n[\,\leq r_{n+1} - r_n, \tag{9.1.42}$$

and

$$]x^* - x_n[\,\leq r^* - r_n, \tag{9.1.43}$$

where sequence $\{r_n\}$ is given by (9.1.27).

Proof. We first show (9.1.42) using induction on $n \geq 0$ (by (9.1.12)). For $n = 0$;

$$]x_1 - x_0[\,= \,]F'(x_0)^{-1}(F(x_0) + G(x_0))[\,= \eta = r_1 - r_0. \tag{9.1.44}$$

Assume:

$$]x_k - x_{k-1}[\,\leq r_k - r_{k-1} \tag{9.1.45}$$

for $k = 1, 2, \ldots, n$.
Using (9.1.45) we get

$$]x_n - x_0[\,\leq \sum_{k=1}^{n}]x_k - x_{k-1}[\,\leq \sum_{k=1}^{n}(r_k - r_{k-1}) = r_n. \tag{9.1.46}$$

Define operators $Q_n \colon X \to X$ by

$$Q_n = -F'(x_0)^{-1}\left[F'(x_n) - F'(x_0)\right]. \tag{9.1.47}$$

By (9.1.5) and (9.1.11) we get

$$
\begin{aligned}
]Q_n(z)[\,&= \,]F'(x_0)^{-1}(F'(x_n) - F'(x_0))(z)[\\
&\leq b\,](F'(x_n) - F'(x_0))(z)[\\
&\leq bw_0(r_n)(]z[),
\end{aligned} \tag{9.1.48}
$$

and

$$]Q_n^i(z)[\,\leq (bw_0(r_n))^i\,(]z[) \quad (i \geq 1). \tag{9.1.49}$$

Hence, we have:

$$\sum_{i=0}^{\infty}]Q_n^i(z)[\,\leq \sum_{j=0}^{\infty}(bw_0(r_n))^i\,(]z[). \tag{9.1.50}$$

That is, series $\sum_{i=0}^{\infty} Q_n^i(z)$ is convergent in X. Hence operator $I - Q_n$ is invertible, and

$$](I - Q_n)^{-1}(z)[\,\leq (I - bw_0(r_n))^{-1}(]z[). \tag{9.1.51}$$

Operator $F'(x_n)$ is invertible for all $n \geq 0$, as $F'(x_n) = F'(x_0)(I - Q_n)$, and for all $x \in Y$ we have:

$$]F'(x_n)^{-1}(x)[\,= \,](I - Q_n)^{-1} F'(x_0)^{-1}(x)[$$
$$\leq (I - bw_0(r_n))^{-1} \left( ]F'(x_0)^{-1}(x)[\right)$$
$$\leq (I - bw_0(r_n))^{-1}(b\,]x[). \tag{9.1.52}$$

Using (9.1.3) we obtain the approximation

$$]x_{n+1} - x_n[$$
$$= \,]F'(x_n)^{-1}(F(x_n) + G(x_n)) - F'(x_n)^{-1}(F'(x_{n-1})(x_n - x_{n-1})$$
$$+ F(x_{n-1}) + G(x_{n-1}))[. \tag{9.1.53}$$

It now follows from (9.1.5)–(9.1.11), (9.1.13), (9.1.27), and (9.1.53)

$$]x_{n+1} - x_n[\,\leq$$
$$\leq \,]F'(x_n)^{-1}(F(x_n) - F(x_{n-1}) - F'(x_{n-1})(x_n - x_{n-1})[$$
$$+ \,]F'(x_n)^{-1}(G(x_n) - G(x_{n-1}))$$
$$\leq (I - bw_0(r_n))^{-1} \left\{ b \,] \int_0^1 (F'((1 - \lambda)x_{n-1} + \lambda x_n) \right.$$
$$\left. - F'(x_{n-1}))(x_n - x_{n-1})d\lambda [\right\}$$
$$+ (I - bw_0(r_n))^{-1}(b\,]G(x_n) - G(x_{n-1})[)$$
$$\leq (I - bw_0(r_n))^{-1} \left\{ b \int_0^1 (w((1 - \lambda)r_{n-1} + \lambda r_n) - w(r_{n-1}))(r_n - r_{n-1})d\lambda \right\}$$
$$+ (I - bw_0(r_n))^{-1} \left(b \int_{r_{n-1}}^{r_n} w_1(t)dt \right)$$
$$= (I - bw_0(r_n))^{-1} \left\{ b \int_{r_{n-1}}^{r_n} w(t)dt - bw(r_{n-1})(r_n - r_{n-1}) \right.$$
$$\left. + b \int_{r_{n-1}}^{r_n} w_1(t)dt \right\} \tag{9.1.54}$$

$$= (I - bw_0(r_n))^{-1}(f(r_n) - f(r_{n-1}) - bw(r_{n-1})(r_n - r_{n-1}))$$
$$= (I - bw_0(r_n))^{-1}((f(r_n) - r_n) - (f(r_{n-1}) - r_{n-1})$$
$$\quad - (bw(r_{n-1}) - I)(r_n - r_{n-1}))$$
$$= (I - bw_0(r_n))^{-1}((f(r_n) - r_n) - (f(r_{n-1}) - r_{n-1})$$
$$\quad - (bw(r_{n-1}) - I)(r_n - r_{n-1}))$$
$$\leq (I - bw_0(r_n))^{-1}((f(r_n) - r_n) - (f(r_{n-1}) - r_{n-1})$$
$$\quad - (bw_0(r_{n-1}) - I)(r_n - r_{n-1})) \tag{9.1.55}$$
$$= (I - bw_0(r_n))^{-1}(f(r_n) - r_n) = r_{n+1} - r_n. \tag{9.1.56}$$

By Lemma 9.1.5, sequence $\{r_n\}$ $(n \geq 0)$ converges to r^*. Hence $\{x_n\}$ is a convergent sequence, and its limit is a solution of equation (9.1.1). Therefore, x_n converges to x^*.

Finally, (9.1.43) follows from (9.1.42) by using standard majorization techniques. The uniqueness part is omitted as it follows exactly as in Theorem 2 in [53]. That completes the proof of the theorem.

Remark 9.1.8. It follows immediately from (9.1.54) and (9.1.55) that sequence

$$t_0 = t_0,$$
$$t_1 = \eta,$$
$$t_{n+1} - t_n = (I - bw_0(t_n))^{-1}\left\{ b \int_{t_{n-1}}^{t_n} w(t)dt - bw(t_{n-1})(t_n - t_{n-1}) \right.$$
$$\left. + b \int_{t_{n-1}}^{t_n} w_1(t)dt \right\} \quad (n \geq 1) \tag{9.1.57}$$

is also a finer majorizing sequence of $\{x_n\}$ $(n \geq 0)$ and converges to some t^* in $\langle 0, r^* \rangle$. Moreover, the following estimates hold for all $n \geq 0$

$$]x_1 - x_0[\leq t_1 - t_0 = r_1 - r_0, \tag{9.1.58}$$
$$]x_{n+1} - x_n[\leq t_{n+1} - t_n \leq r_{n+1} - r_n, \tag{9.1.59}$$
$$]x^* - x_n[\leq t^* - t_n \leq r^* - r_n, \tag{9.1.60}$$
$$t_n \leq r_n, \tag{9.1.61}$$

and

$$t^* \leq r^*. \tag{9.1.62}$$

That is, $\{t_n\}$ is a finer majorizing sequence than $\{r_n\}$ and the information on the location of the solution x^* is more precise. Therefore, we wonder if studying the convergence of $\{t_n\}$ without assuming (9.1.17) can lead to weaker sufficient convergence conditions for method (9.1.2). In Theorem 9.1.10, we answer this question.

But first we need the following lemma on majorizing sequences for method (9.1.2).

Lemma 9.1.9. *If:*
there exist parameters $\eta \geq 0$, $\delta \in [0, 2)$ such that: Operators

$$I - bw_0 \left[2(2I - \delta I)^{-1} \left(I - \left(\tfrac{\delta I}{2} \right)^{n+1} \right) \eta \right] \tag{9.1.63}$$

be positive, invertible, and with positive inverses for all $n \geq 0$;

$$2(I - bw_0(\eta))^{-1} \left[bw_1(\eta) + b \int_0^1 w(s\eta) ds - bw(0) \right] \leq \delta I, \tag{9.1.64}$$

and

$$2b \int_0^1 w \left[2(2I - \delta I)^{-1} \left(I - \left(\tfrac{\delta I}{2} \right)^{n+1} \right) \eta + s \left(\tfrac{\delta I}{2} \right)^{n+1} \eta \right] ds$$

$$- 2bw \left[2(2I - \delta I)^{-1} \left(I - \left(\tfrac{\delta I}{2} \right)^{n+1} \right) \eta \right]$$

$$+ 2bw_1 \left[2(2I - \delta I)^{-1} \left(I - \left(\tfrac{\delta I}{2} \right)^{n+1} \right) \eta \right]$$

$$+ \delta bw_0 \left[2(2I - \delta I)^{-1} \left(I - \left(\tfrac{\delta I}{2} \right)^{n+1} \right) \eta \right]$$

$$\leq \delta I, \quad \text{for all } n \geq 0. \tag{9.1.65}$$

Then iteration $\{t_n\}$ ($n \geq 0$) given by (9.1.57) is nondecreasing, bounded above by

$$t^{**} = 2(2I - \delta I)^{-1} \eta, \tag{9.1.66}$$

and converges to some t^* such that:

$$0 \leq t^* \leq t^{**}. \tag{9.1.67}$$

Moreover, the following error bounds hold for all $n \geq 0$:

$$0 \leq t_{n+2} - t_{n+1} \leq \tfrac{\delta I}{2} (t_{n+1} - t_n) \leq \left(\tfrac{\delta I}{2} \right)^{n+1} \eta. \tag{9.1.68}$$

Proof. We must show:

$$2(I - bw_0(t_{k+1}))^{-1} \left[b \int_0^1 w(t_k + s(t_{k+1} - t_k)) ds - bw(t_k) + bw_1(t_{k+1}) \right] \leq \delta I, \tag{9.1.69}$$

and operators

$$0 \leq t_{k+1} - t_k, \tag{9.1.70}$$

$$I - bw_0(t_{k+1}), \tag{9.1.71}$$

positive, invertible, and with positive inverses.

Estimate (9.1.69) can then follow immediately from (9.1.70)–(9.1.72). Using induction on the integer k, we get for $k = 0$

$$2(I - bw_0(t_1))^{-1}\left[b\int_0^1 w\left[t_0 + s(t_1 - t_0)\right]ds - bw(t_1) + w_1(t_1)\right] \le \delta I,$$

$$bw_0(t_1) < I, \qquad\qquad (9.1.72)$$

by the initial conditions. But (9.1.57) then gives

$$0 \le t_2 - t_1 \le \tfrac{\delta I}{2}(t_1 - t_0). \qquad\qquad (9.1.73)$$

Assume (9.1.70)–(9.1.72) hold for all $k \le n + 1$. Using (9.1.63)–(9.1.66), we obtain in turn:

$$2b\int_0^1 w\left[t_{k+1} - s(t_{k+2} - t_{k+1})\right]ds - 2bw(t_{k+1}) + 2bw_1(t_{k+1}) + \delta bw_0(t_{k+1})$$

$$\le 2b\int_0^1 w\left[2(2I - \delta I)^{-1}\left(I - \left(\tfrac{\delta I}{2}\right)^{k+1}\right)\eta + s\left(\tfrac{\delta I}{2}\right)^{k+1}\eta\right]ds$$

$$- 2bw\left[2(2I - \delta I)^{-1}\left(I - \left(\tfrac{\delta I}{2}\right)^{k+1}\right)\eta\right]$$

$$+ 2bw_1\left[2(2I - \delta I)^{-1}\left(I - \left(\tfrac{\delta I}{2}\right)^{k+1}\eta\right)\right]$$

$$+ \delta bw_0\left[2(2I - \delta I)^{-1}\left(I - \left(\tfrac{\delta I}{2}\right)^{k+1}\right)\eta\right]$$

$$\le \delta I. \qquad\qquad (9.1.74)$$

Moreover, we show:

$$t_k \le t^{**}. \qquad\qquad (9.1.75)$$

We have:

$$t_0 = \eta \le t^{**}, \ t_1 = \eta \le t^{**}, \ t_2 \le \eta + \tfrac{\delta I}{2}\eta = \tfrac{2I + \delta I}{2}\eta \le t^{**}.$$

Assume (9.1.75) holds for all $k \le n + 1$. It follows from (9.1.57), (9.1.70)–(9.1.72):

$$t_{k+2} \le t_{k+1} + \tfrac{\delta I}{2}(t_{k+1} - t_k) \le t_k + \tfrac{\delta I}{2}(t_k - t_{k-1}) + \tfrac{\delta I}{2}(t_{k+1} - t_k)$$

$$\le \cdots \le \eta + \tfrac{\delta I}{2}\eta + \left(\tfrac{\delta I}{2}\right)^2\eta + \cdots + \left(\tfrac{\delta I}{2}\right)^{k+1}\eta$$

$$= 2(2I - \delta I)^{-1}\left[I - \left(\tfrac{\delta I}{2}\right)^{k+2}\right]\eta \le 2(2I - \delta I)^{-1}\eta = t^{**}. \qquad (9.1.76)$$

Hence, sequence $\{t_n\}$ ($n \ge 0$) converges to some t^* satisfying (9.1.68). That completes the proof of Lemma 9.1.9.

We can show the main semilocal convergence theorem for method (9.1.2).

Theorem 9.1.10. *Assume:*
hypotheses (9.1.5)–(9.1.7), (9.1.9)–(9.1.11), (9.1.63)–(9.1.66) hold, and

$$t^{**} \leq R, \tag{9.1.77}$$

where t^{**} is given by (9.1.67).
Then sequence $\{x_n\}$ $(n \geq 0)$ generated by Newton-Kantorovich method (9.1.2) is well defined, remains in the K-ball $U(x_0, t^*)$ for all $n \geq 0$, and converges to a solution x^* of equation $F(x) + G(x) = 0$, which is unique in $E(t^*)$. Moreover, the following error bounds hold for all $n \geq 0$:

$$]x_{n+1} - x_n[\, \leq t_{n+1} - t_n \tag{9.1.78}$$

and

$$]x^* - x_n[\, \leq t^* - t_n, \tag{9.1.79}$$

where sequence $\{t_n\}$ $(n \geq 0)$ and t^* are given by (9.1.57) and (9.1.68), respectively.

Proof. The proof is identical to Theorem 9.1.7 with sequence t_n replacing r_n until the derivation of (9.1.54). But then the right-hand side of (9.1.54) with these changes becomes $t_{n+1} - t_n$. By Lemma 9.1.9, $\{t_n\}$ converges to t^*. Hence $\{x_n\}$ is a convergent sequence, its limit converges to a solution of equation $F(x) + G(x) = 0$. Therefore $\{x_n\}$ converges to x^*. Estimate (9.1.79) follows from (9.1.78) by using standard majorization techniques. The uniqueness part is omitted as it follows exactly as in Theorem 2 in [53].

That completes the proof of Theorem 9.1.10.

Remark 9.1.11. Conditions (9.1.63), (9.1.66) can be replaced by the stronger but easier to check

$$I - bw_0 \left[2(2I - \delta I)^{-1}\eta \right] \tag{9.1.80}$$

and

$$2b \int_0^1 w \left[2(2I - \delta I)^{-1}\eta + s \left(\tfrac{\delta I}{2} \right) \eta \right] s - 2bw \left[2(2I - \delta I)^{-1}\eta \right]$$
$$+ \, 2bw_1 \left[2(2I - \delta I)^{-1}\eta \right] + \delta bw_0 \left[2(2I - \delta I)^{-1}\eta \right]$$
$$\leq \delta I, \tag{9.1.81}$$

respectively.

Application 9.1.12. *Assume operator $][$ is given by a norm $\| \cdot \|$ and set $G(x) = 0$ for all $x \in \overline{U}(x_0, R)$. Choose for all $r \in S$, $b = 1$ for simplicity,*

$$\overline{w}(r) = \ell r, \tag{9.1.82}$$
$$w_0(r) = \ell_0 r \tag{9.1.83}$$

and

$$w_1(r) = 0. \tag{9.1.84}$$

With these choices, our conditions reduce to the ones in Section 2.3 that have already been compared favorably with the Newton-Kantorovich theorem.

Remark 9.1.13. The results obtained here hold under even weaker conditions. Indeed, because (9.1.6) is not "directly" used in the proofs above, it can be replaced by the weaker condition (9.1.15) throughout this study.

The local convergence for method (9.1.2) was not examined in [53]. Let x^* be a simple solution of equation (9.1.1), and assume $F(x^*)^{-1} \in L(Y, X)$. Moreover, assume with x^* replacing x_0 that hypotheses (9.1.5), (9.1.6), (9.1.7), (9.1.9), (9.1.10), (9.1.11) hold. Then exactly as in (9.1.54) but using the local conditions, and the approximation

$$x_{n+1} - x^* = \tag{9.1.85}$$

$$= \left[F'(x_n)^{-1} F'(x^*) \right] F'(x^*)^{-1}$$

$$\times \left\{ \int_0^1 \left[F'(x_n + t(x^* - x_n)) - F'(x_n) \right] (x^* - x_n) dt + (G(x^*) - G(x_n)) \right\} \tag{9.1.86}$$

we can show the following local result for method (9.1.87).

Theorem 9.1.14. *Assume there exists a minimal solution $r^* \in S$ of equation*

$$p(r) = 0, \tag{9.1.87}$$

where,

$$p(r) = b \int_0^1 [w((1+s)r) - w(r)] ds + bw_0(r) + bw_1(r) - 1. \tag{9.1.88}$$

Then, sequence $\{x_n\}$ $(n \geq 0)$ generated by Newton-Kantorovich method (9.1.2) is well defined, remains in $\overline{U}(x^*, r^*)$ for all $n \geq 0$, and converges to x^* provided that $x_0 \in U(x^*, r^*)$.
Moreover the following estimates hold for all $n \geq 0$

$$]x^* - x_{n+1}[\leq \varepsilon_{n+1}, \tag{9.1.89}$$

where,

$$\varepsilon_{n+1} = \frac{b \int_0^1 (w((1+s)]x_n - x^*[) - w(]x_n - x^*[)) ds]x_n - x^*[+ \int_0^{]x_n - x^*[} w_1(s) ds}{1 - bw_0(]x_n - x^*[)} \tag{9.1.90}$$

$(n \geq 0)$.

Application 9.1.15. *Returning back to the choices of Application 9.1.12 and using (9.1.87) we get*

$$r^* = \frac{1}{2\ell_0 + \ell}. \tag{9.1.91}$$

See also Section 2.4.

9.2 Generalized Banach spaces

In this section, we are concerned with the problem of approximating a locally unique solution x^* of equation

$$G(x) = 0, \qquad (9.2.1)$$

where G is a Fréchet-differentiable operator defined on an open subset D of a Banach space X with values in a Banach space Y. The results will be stated for an operator

$$F = L_0 G, \qquad (9.2.2)$$

where, $L_0 \in L(Y, X)$ is an approximate inverse of $G'(x_0)$ ($x_0 \in D$).

Using the concept of a generalized norm that is an operator from a linear space into a partially ordered Banach space, sufficient semilocal convergence conditions for NK method were given in [17], [22], [43], [139]–[141]. This way, convergence results and error estimates are improved compared with the real norm theory. Several examples for the benefits of this approach can be found in [141].

Here we use Lipschitz as well as center-Lipschitz conditions on F. It turns out that this way under the same information, we obtain finer error bounds under in general weaker sufficient convergence conditions in the semilocal convergence case. In the local case not covered in [139]–[141], we also show that our radius of convergence is larger than before.

We complete our study with an example where our results compare favorably with earlier ones using the same information.

We first need some definitions on ordered spaces:

Definition 9.2.1. *By a generalized Banach space we mean a triplet $(X, E, / \cdot /)$ such that:*

(i) X is a linear space over $\mathbf{R}(\mathbf{C})$;
(ii) $E = (E, K, // \cdot //)$ is a partially ordered Banach space in the sense:
(ii)$_1$ $(E, // \cdot //)$ is a real Banach space,
(ii)$_2$ E is partially ordered by a closed convex cone K,
(ii)$_3$ the norm $// \cdot //$ is monotone on K;
(iii) operator $/ \cdot /: X \to K$ is such that

$$/x/ = 0 \Leftrightarrow x = 0,$$
$$/sx/ = /s/ /x/,$$
$$/x + y/ \le /x/ + /y/;$$

(iv) X is a Banach space with the induced norm

$$// \cdot //_i := // \cdot // \bullet / \bullet /.$$

The operator $/ \cdot /$ is called a generalized norm. All topological terms are understood with respect to this norm.

If X, Y are partially ordered, $L_+(X^n, Y)$ is the subset of monotone operators W such that

$$0 \le u_i \le v_i \Rightarrow W(u_1, \ldots, u_n) \le W(v_1, \ldots, v_n).$$

Definition 9.2.2. *The set of bounds for an operator $L \in L(X, X)$ on $(X, E, / \cdot /)$ is given by:*

$$B(L) = \{W \in L_+(E, E) \mid /L(x)/ \leq W/x/ \text{ for } x \in X\}.$$

For $x_0 \in D \subseteq X$, $J: D \to D$, we use the notation

$$x_{n+1} = J(x_n) = J^{n+1}(x_0), \tag{9.2.3}$$

and in the case of convergence

$$J^{\infty}(x_0) = \lim_{n \to \infty} (J^n(x_0)) = \lim_{n \to \infty} \{x_n\}. \tag{9.2.4}$$

The Newton iterates are determined through a fixed point approach:

$$x_{n+1} = x_n + y_n, \quad F'(x_n)(y_n) + F(x_n) = 0, \tag{9.2.5}$$
$$\Leftrightarrow \quad y_n = J_n(y_n) = (I - F'(x_n))(y_n) - F(x_n). \tag{9.2.6}$$

In case of convergence, we can write NK method in the form:

$$x_{n+1} = x_n + J_n^{\infty}(0) \quad (n \geq 0). \tag{9.2.7}$$

Proposition 9.2.3. *Let $(E, K, // \cdot //)$ be a partially ordered Banach space, $\delta \in K$, $M \in L_+(E, E)$, $N \in L_+(E^2, E)$ be given operators.*
Assume there exist:

(a) $c \in K$ such that

$$R(c) = M(c) + Nc^2 + \delta \leq c \text{ and } (M + 2N(c))^i(c) \to 0 \text{ as } i \to \infty. \tag{9.2.8}$$

> *Then $\delta_0 = R^{\infty}(0)$ is well defined, solves $\delta_0 = R(\delta_0)$, and is the smaller solution of inequality $R(\delta_1) \leq \delta_1$.*

(b) $\delta_2 \in K$, $\lambda \in (0, 1)$ such that $R(\delta_2) \leq \lambda \delta_2$. Then there exists $c \leq \delta_2$ satisfying (9.2.8).

Proposition 9.2.4. *Let $(X, K, // \cdot //), / \cdot /)$ be a generalized Banach space and $W \in B(L)$ be a bound for $L \in L(X, X)$. If for $y \in X$, there exists $\eta \in K$ such that*

$$W(\eta) + |y| \leq \eta \text{ and } W^i(\eta) \to 0 \text{ as } i \to \infty, \tag{9.2.9}$$

then

$$z = J^{\infty}(0), \quad J(x) = L(x) + y \tag{9.2.10}$$

is well defined, and satisfies:

$$z = L(z) + y, \quad \text{and} \quad /z/ \leq W/z/ + /y/ \leq \eta. \tag{9.2.11}$$

We can show the following semilocal result for NK method (9.2.7) on generalized Banach spaces:

Theorem 9.2.5. *Let $(X, (E, K, // \cdot //, / \cdot /), Y$ be generalized Banach spaces, D an open subset of X, $G: D \to Y$ a Fréchet-differentiable operator, and point $x_0 \in D$ be given. Assume there exist:*

(a) operators $M \in B(I - F'(x_0))$, $N_0, N \in L_+(E^2, E)$ such that:

$$N_0 \leq N, \tag{9.2.12}$$
$$/F'(w)(z) - F'(v)(z)/ \leq 2N(/w - v/, /z/), \tag{9.2.13}$$
$$/F'(w)(z) - F'(x_0)(z)/ \leq 2N_0(/w - x_0/, /z/) \tag{9.2.14}$$

for all $v, w \in D$, $z \in X$;
(b) a solution $r \in K$ of

$$R_0(q) = M(q) + Nq^2 + /F(x_0)/ \leq q \tag{9.2.15}$$

satisfying

$$(M + 2N(r))^i(r) \to 0 \quad as \quad i \to \infty, \tag{9.2.16}$$

and

$$U(x_0, r) = \{x \in X /x - x_0/ \leq r\} \subseteq D. \tag{9.2.17}$$

Then sequence $\{x_n\}$ ($n \geq 0$) generated by NK method (9.2.7) is well defined, remains in $U(x_0, r)$ for all $n \geq 0$, and converges to a unique zero x^* of F in $U(x_0, r)$. Moreover, a priori estimates are given by the sequence $\{r_n\}$ ($n \geq 0$):

$$r_0 = r, \quad r_n = P_n^\infty(0) \quad (n \geq 0) \tag{9.2.18}$$

where,

$$P_n(q) = M(q) + 2N_0(r - r_{n-1})(q) + \overline{N}r_{n-1}^2 \tag{9.2.19}$$
$$\overline{N} = N_0 \quad if \quad n = 1, \quad \overline{N} = N \quad if \quad n > 1, \tag{9.2.20}$$

and

$$\lim_{n \to \infty} r_n = 0. \tag{9.2.21}$$

Furthermore, a posteriori estimates are given by sequence $\{c_n\}$ ($n \geq 0$):

$$c_n = R_n^\infty(0), \tag{9.2.22}$$

where

$$R_n(q) = M(q) + 2N_0(b_n)(q) + Nq^2 + Na_{n-1}^2, \tag{9.2.23}$$
$$a_{n-1} = /x_n - x_{n-1}/ \tag{9.2.24}$$

and

$$b_n = /x_n - x_0/. \tag{9.2.25}$$

Proof. We use induction on the integer n to show the claim:

(I_k) $(x_k, r_k) \in (X, k)$ are well defined and

$$r_k + a_{k-1} \leq r_{k-1}. \tag{9.2.26}$$

The claim holds for $k = 1$. Indeed by (9.2.8), (9.2.15), and (9.2.16), there exists q_1 such that:

$$q_1 \leq r, \quad M(q_1) + /F(x_0)/ = q_1 \text{ and } M^i(q_1) \leq M^i(r) \to 0 \text{ as } i \to \infty. \tag{9.2.27}$$

It follows from (9.2.9) that x_1 is well defined

$$a_0 \leq q_1.$$

Using (9.2.8) and the estimate

$$\begin{aligned}
P_1(r - q_1) &= M(r - q_1) + 2N_0(r - r_0)(r - q_1) + N_0 r_0^2 \\
&\leq M(r - q_1) + 2N(r - r_0)(r - q_1) + N r_0^2 \\
&= R_0(r) - q_1 \leq r - q_1,
\end{aligned} \tag{9.2.28}$$

we deduce c_1 is well defined and

$$r_1 + a_0 \leq r - q_1 + q_1 = r_0. \tag{9.2.29}$$

That is, (9.2.26) holds for $k = 1$. Assume (I_k) holds for all $k \leq n$. We can have

$$M(r_k) + 2N_0(r - r_k)(r_k) + N(r_{k-1} - r_k)^2$$
$$\leq M(r_k) + 2N(r - r_k)(r_k) + N(r_{k-1} - r_k)^2 = P_k(r_k) - N r_k^2 \leq r_k. \tag{9.2.30}$$

It follows by (9.2.8) there exists $q_k \leq c_k$ such that

$$q_k = M(q_k) + 2N(r - r_k)(q_k) + N(r_{k-1} - r_k)^2, \tag{9.2.31}$$

and

$$(M + 2N(r - r_k))^i q_k \to 0 \quad \text{as} \quad i \to \infty. \tag{9.2.32}$$

By the induction hypothesis

$$b_k = /x_k - x_0/ \leq \sum_{j=0}^{k-1} a_j \leq \sum_{j=0}^{k-1} (r_j - r_{j+1}) = r - r_k \leq r, \tag{9.2.33}$$

which implies $x_k \in U(x_0, r)$. We must find a bound for operator $I - F'(x_k)$. Using (9.2.14) we get from

$$I - F'(x_k) = (I - F'(x_0)) + (F'(x_0) - F'(x_k))$$

that

$$/I - F'(x_k)/ \leq /1 - F'(x_0)/ + /F'(x_0) - F'(x_k)/$$
$$\leq M + 2N_0(/x_k - x_0/) \leq M + 2N_0(r - r_k). \tag{9.2.34}$$

Moreover by (9.2.5) and (9.2.13), we get

$$/F(x_k)/ = /F(x_k) - F'(x_{k-1}) - F'(x_{k-1})(x_k - x_{k-1})/$$

$$= / \int_0^1 \left[F'(x_{k-1} + t(x_k - x_{k-1})) - F'(x_{k-1}) \right] (x_k - x_{k-1}) dt /$$

$$\leq N a_{k-1}^2 \leq N(r_{k-1} - r_k)^2. \tag{9.2.35}$$

By (9.2.34) and (9.2.35), we get

$$M(q_k) + 2N(r - r_k)(q_k) + /F(x_k)/ \leq q_k. \tag{9.2.36}$$

That is, x_{k+1} is well defined, and

$$a_k \leq q_k \leq r_k. \tag{9.2.37}$$

To show the existence of r_{k+1}, we note

$$P_{k+1}(r_k - q_k) = P_k(r_k) - q_k = r_k - q_k, \tag{9.2.38}$$

which implies the existence of r_{k+1}, and

$$r_{k+1} + a_k \leq r_k - q_k + q_k = r_k. \tag{9.2.39}$$

The induction for (I_k) is now complete.

We can obtain the estimates

$$/x_{m+1} - x_k/ \leq \sum_{j=k}^m a_j \leq \sum_{j=k}^m (r_j - r_{j+1}) = r_k - r_{m+1} \leq r_k, \tag{9.2.40}$$

and

$$r_{k+1} = P_{k+1}(r_{k+1}) \leq P_{k+1}(r_k) \leq (M + 2N(r))r_k \leq (M + 2N(r))r_k$$
$$\leq \cdots \leq (M + 2N(r))^{k+1}(r) \to \infty \tag{9.2.41}$$

as $k \to \infty$. Hence $\{x_n\}$ is a Cauchy sequence and as such it converges to some $x^* \in X$. By letting $m \to \infty$ in (9.2.40), we deduce $x^* \in U(x_k, r_k)$, whereas by letting $k \to \infty$ in (9.2.35), we get $F(x^*) = 0$. The proof of the uniqueness of x^* in $U(x_0, r)$ is omitted as identical to [140, Theorem 4.1]. We note

$$R_k(r_k) \leq P_k(r_k) \leq r_k. \tag{9.2.42}$$

Hence a posteriori estimates (9.2.22) are well defined by (9.2.8). That is, $c_k \leq r_k$ in general. The hypotheses of the theorem hold if x_0 is replaced by x_k and M becomes $M + 2N_0(b_k)$. Using (9.2.35), we get c_k is a solution of (9.2.15). By (9.2.40), hypotheses of the theorem hold. It follows from (9.2.7) that $x^* \in U(x_k, c_k)$ proving (9.2.22).

Remark 9.2.6. Our Theorem 9.2.5 reduces to Theorem 2.1 in [141, p. 251] if $N_0 = N$. However in general (9.2.12) holds. It follows from the proof of the theorem that if strict inequality holds in (9.2.12) then the error bounds r_n, a_n, b_n are finer than the corresponding ones $\bar{r}_n, \bar{a}_n, \bar{b}_n$ in [141] under the same information. That is, for all $n \geq 1$:

$$r_n < \bar{r}_n, \tag{9.2.43}$$

$$a_n < \bar{a}_n, \tag{9.2.44}$$

$$b_n < \bar{b}_n, \tag{9.2.45}$$

and

$$r \leq r^*. \tag{9.2.46}$$

Remark 9.2.7. It turns out that (9.2.15) and (9.2.16) can be weakened. Indeed, assume:

(b') there exist $\bar{r} \in K$ satisfying

$$Q_0(q) = M(q) + N_0 q^2 + /F(x_0)/ \leq q; \tag{9.2.47}$$

$r : r \geq \bar{r}$ such that (9.2.15), and (9.2.16) hold with N_0 replacing N;   (9.2.48)

$r_n : r_n \leq r$ $(n \geq 1)$ solving   (9.2.49)

$$M(r_n) + 2N_0(r - r_n)(r_n) + N(r_{n-1} - r_n)^2 \leq r_n. \tag{9.2.50}$$

It follows from the proof of Theorem 9.2.5 that if (b'), Q replace (b), R then the conclusions also hold. Moreover if (9.2.12) is a strict inequality, then error bounds (9.2.43)–(9.2.46) hold.

To show a local result for NK method, assume x^* is a zero of F. Then, we can easily see from (9.2.5) that we must solve the equation

$$
\begin{aligned}
x_{n+1} &- x^* \\
&= \left[I - F'(x_n)\right](x_{n+1} - x^*) \\
&\quad + \left[-\int_0^1 \left[F'(x^* + t(x_n - x^*)) - F'(x_n)\right](x_n - x^*)dt\right]
\end{aligned}
\tag{9.2.51}
$$

for each $n \geq 0$.

Using identity (9.2.51), we can show the following local result for NK method (9.2.7):

Theorem 9.2.8. *Let* $(X, (E, K, // \cdot //, / \cdot /), Y$ *be generalized Banach spaces, D an open subset of X, and $G: D: D \to Y$ a Fréchet-differentiable operator. Assume there exist:*

(a) a zero $x^* \in D$ of operator F;

(b) operators $A \in B(I - F'(x^*))$, H_0, $H \in L_+(E^2, E)$ such that:

$$H_0 \leq H, \tag{9.2.52}$$

$$/F'(w)(z) - F'(v)(z)/ \leq 2H(/w - v/, /z/), \tag{9.2.53}$$

$$/F'(w)(z) - F'(x^*)(z)/ \leq 2H_0(/w - x^*/, /z/) \tag{9.2.54}$$

for all $v, w \in D, z \in X$;
(c) a solution $\gamma \in K$ of

$$T_0(q) = A(q) + 2H_0 q^2 + H q^2 \leq q \tag{9.2.55}$$

satisfying

$$(A + 4H_0(\gamma) + 2H(\gamma))^i(\gamma) \to 0 \quad as \quad i \to \infty, \tag{9.2.56}$$

and

$$U(x^*, \gamma) \subseteq D; \tag{9.2.57}$$

(d) $x_0 \in D$ such that:

$$0 < /x_0 - x^*/ = \gamma. \tag{9.2.58}$$

Then sequence $\{x_n\}$ $(n \geq 0)$ generated by NK method (9.2.7) is well defined, remains in $\overline{U}(x^*, \gamma)$ for all $n \geq 0$, and converges to x^*. Moreover, the following estimates hold $n \geq 1$

$$\beta_{n-1} = /x_{n-1} - x^*/ \leq \gamma_{n-1} - \gamma_n \tag{9.2.59}$$

where,

$$\gamma_0 = \gamma, \quad \gamma_n = \Gamma_n^{\infty}(0), \quad \Gamma_n(q) = A(q) + 2H_0(\gamma - \gamma_{n-1})(q) + H\gamma_{n-1}^2, \tag{9.2.60}$$

and

$$\lim_{n \to \infty} \gamma_n = 0. \tag{9.2.61}$$

Proof. As in Theorem 9.2.5, using induction on the integer n we show
(II_n) $x_n \in X$ and $\gamma_n \in K$ are well defined and satisfy (9.2.39).
By (9.2.8), (9.2.55), and (9.2.56), there exists α_1 such that

$$\alpha_1 \leq \gamma, \quad A(\alpha_1) + 2H_0(\gamma)(\alpha_1) + H\gamma^2 = \alpha_1,$$

$$(A + 2H_0(\gamma))^i(\alpha_1) \to 0. \tag{9.2.62}$$

It follows from (9.2.9) x_1 is well defined, and

$$\beta_0 \leq \alpha_1. \tag{9.2.63}$$

Using (9.2.60) we get in turn

$$\Gamma_1(\gamma - \alpha_1) = A(\gamma - \alpha_1) + 2H_0(\gamma - \gamma_0)(\gamma - \alpha_1) + H\gamma_0^2$$

$$= A(\gamma - \alpha_1) + H\gamma^2$$

$$= A(\gamma) + H\gamma^2 - A(\alpha_1)$$

$$= A(\gamma) + H\gamma^2 + 2H_0(\gamma)(\gamma) - A(\alpha_1) - 2H_0(\gamma)(\gamma)$$

$$\leq T_0(\gamma) - A(\alpha_1) - 2H_0(\gamma)(\alpha_1) \tag{9.2.64}$$

$$= T_0(\gamma) - H\gamma^2 - \alpha_1 \tag{9.2.65}$$

$$\leq T_0(\gamma) - \alpha_1 \leq \gamma - \alpha_1. \tag{9.2.66}$$

By (9.2.8), γ_1 is well defined, and

$$\gamma_1 + \beta_0 \leq \gamma - \alpha_1 + \alpha_1 = \gamma_0. \tag{9.2.67}$$

Hence we showed (II_1) holds. Suppose $(II_1), \ldots (II_k)$ hold for all $k \leq n$. We must show the existence of x_{k+1} and find a bound α_k for b_k.

We can have

$$A(\gamma_k) + 2H_0(\gamma - \gamma_k)(\gamma_k) + H(\gamma_{k-1} - \gamma_k)^2 \leq \Gamma_k(\gamma_k) = \gamma_k. \tag{9.2.68}$$

That is by (9.2.8), there exists α_k such that

$$\alpha_k \leq \gamma_k, \tag{9.2.69}$$

$$\alpha_k = A(\alpha_k) + 2H_0(\gamma - \gamma_k)(\alpha_k) + H(\gamma_{k-1} - \gamma_k)^2, \tag{9.2.70}$$

and

$$[A + 2H_0(\gamma - \gamma_k)]^i (\alpha_k) \to 0. \tag{9.2.71}$$

It follows from (9.2.67) that $x_k \in U(x^*, \gamma)$. We must find a bound for $I - F'(x_k)$. Using (9.2.6), and (9.2.52), we get

$$/I - F'(x_k)/ = /(1 - F'(x_k)) + (F'(x^*) - F'(x_k))/ \leq A + 2H_0(/x_k - x^*/). \tag{9.2.72}$$

By (9.2.51), (9.2.68), and (9.2.70), we deduce

$$A(\alpha_k) + 2H_0(\gamma - \gamma_k)(\alpha_k)$$

$$+ /- \int_0^1 [F'(x^* + t(x_k - x^*)) - F'(x_k)] (x_k - x^*)dt / \leq \alpha_k. \tag{9.2.73}$$

Hence by (9.2.9), x_{k+1} is well defined, and

$$\beta_k \leq \alpha_k \leq \gamma_k. \tag{9.2.74}$$

Moreover, we can have in turn

$$\Gamma_{k+1}(\gamma_k - \alpha_k) =$$
$$= A(\gamma_k - \alpha_k) + 2H_0(\gamma - \gamma_k)(\gamma_k - \alpha_k) + H\gamma_k^2$$
$$= \Gamma_k(\gamma_k) - (A(\gamma_k) + 2H_0(\gamma - \gamma_{k-1})(\gamma_k) + H\gamma_{k-1}^2)$$
$$\quad + A(\gamma_k) - A(\alpha_k) + 2H_0(\gamma - \gamma_k)(\gamma_k - \alpha_k) + H\gamma_k^2$$
$$= \Gamma_k(\gamma_k) - \alpha_k + 2H_0(\gamma - \gamma_k)(\alpha_k) + H(\gamma_{k-1} + \gamma_k)^2 - 2H_0(\gamma - \gamma_{k-1})(\gamma_k)$$
$$\quad - H\gamma_{k-1}^2 + 2H_0(\gamma - \gamma_k)(\gamma_k - \alpha_k) + H\gamma_k^2$$
$$= \Gamma_k(\gamma_k) - \alpha_k + 2H_0(\gamma - \gamma_k)(\gamma_k) - 2H_0(\gamma - \gamma_{k-1})(\gamma_k)$$
$$\quad + H(\gamma_{k-1} - \gamma_k)^2 - H\gamma_{k-1}^2 + H\gamma_k^2$$
$$= \Gamma_k(\gamma_k) - \alpha_k + 2H_0(\gamma_{k-1} - \gamma_k)(\gamma_k) + H(\gamma_{k-1} - \gamma_k)^2 - H\gamma_{k-1}^2 + H\gamma_k^2$$
$$\leq \Gamma_k(\gamma_k) - \alpha_k + 2H(\gamma_{k-1} - \gamma_k)(\gamma_k) + H(\gamma_{k-1} - \gamma_k)^2 - H\gamma_{k-1}^2 + H\gamma_k^2$$
$$\leq \Gamma_k(\gamma_k) - \alpha_k = \gamma_k - \alpha_k. \tag{9.2.75}$$

That is by (9.2.8) and (9.2.73), γ_{k+1} is well defined and:

$$\gamma_{k+1} + \beta_k \leq \gamma_k - \alpha_k + \alpha_k = \gamma_k, \tag{9.2.76}$$

which completes the induction, and shows (9.2.59).

As in (9.2.41), we show

$$\lim_{n \to \infty} \gamma_n = 0. \tag{9.2.77}$$

Finally by letting $n \to \infty$, we deduce

$$\lim_{n \to \infty} x_n = x^*. \tag{9.2.78}$$

Remark 9.2.9. Local results were not given in earlier studies [139]–[141]. However, from Theorem 9.2.8 for $H_0 = H$, such results can immediately be obtained. Therefore we can only compare Theorem 9.2.8 with earlier ones in the case of a real-normed space (i.e., $E = \mathbf{R}$). Assume for simplicity that

$$F'(x^*) = I, \tag{9.2.79}$$

and there exist ℓ, ℓ_0 such that

$$//F'(x) - F'(x^*)// \leq \ell_0 //x - x^*// \tag{9.2.80}$$
$$//F'(x) - F'(y)// \leq \ell //x - y// \tag{9.2.81}$$

for all $x, y \in D$. Choose:

$$A = 0, \quad \ell_0 = 2H_0 \quad \text{and} \quad \ell = 2H. \tag{9.2.82}$$

Then the convergence radius γ solving (9.2.55) is given by

$$\gamma = \frac{2}{2\ell_0 + \ell}, \tag{9.2.83}$$

and the corresponding error bounds by

$$\beta_{n+1} \leq \frac{\ell//x^* - x_n//^2}{2[1 - \ell_0//x_n - x^*//]} \quad (n \geq 0). \tag{9.2.84}$$

Rheinboldt [177] using only (9.2.81) showed:

$$\overline{\gamma} = \frac{2}{3\ell}, \tag{9.2.85}$$

and

$$\overline{\beta}_{n+1} \leq \frac{\ell//x^* - x_n//^2}{2[1 - \ell//x_n - x^*//]} \quad (n \geq 0). \tag{9.2.86}$$

In general we have:

$$\ell_0 \leq \ell. \tag{9.2.87}$$

Hence we get

$$\overline{\gamma} \leq \gamma, \tag{9.2.88}$$

and

$$\beta_{n+1} \leq \overline{\beta}_{n+1} \quad (n \geq 0). \tag{9.2.89}$$

Moreover, if strict inequality holds in (9.2.85), so it does in (9.2.86) and (9.2.87). Hence the convergence radius is enlarged and the error bounds are finer using the same information as before.

Condition (9.2.56) needed for the computation of the inverses can be dropped in this case as by the Banach Lemma on invertible operators $F'(x_n)^{-1}$ exist and

$$//F'(x_n)^{-1}// \leq [1 - \ell_0//x_n - x^*//]^{-1} \quad (n \geq 0) \tag{9.2.90}$$

(see also Section 2.2).

9.3 Inexact Newton-like methods on Banach spaces with a convergence structure

In this section, we are concerned with approximating a solution x^* of the nonlinear operator equation

$$F(x) + Q(x) = 0, \tag{9.3.1}$$

where F is a Fréchet-differentiable operator defined on a convex subset D of a Banach space X with values in X, and Q is a nondifferentiable nonlinear operator with the same domain and values in X.

We introduce the inexact Newton-like method

$$x_{n+1} = x_n + A(x_n)^* [-(F(x_n) + Q(x_n))] - z_n, \quad x_0 = 0 \quad (n \geq 0) \tag{9.3.2}$$

to approximate a solution x^* of equation (9.3.1). Here $A(x_n)^*$ $(n \geq 0)$ denotes a linear operator that is an approximation for $F'(x_n)^{-1}$ $(n \geq 0)$. For $A(x_n) = F'(x_n)$ $(n \geq 0)$, we obtain the inexact Newton's method. The residual points $z_n \in D$ $(n \geq 0)$, depend on x_n, $F(x_n) + Q(x_n)$ $(n \geq 0)$ and are such that $\lim_{n \to \infty} z_n = 0$. Some special choices of points z_n $(n \geq 0)$ are given in the Remark 9.3.10 (see (c), (d), and (f)) after Proposition 9.3.9 and in [43].

The importance of studying inexact Newton-like methods comes from the fact that many commonly used variants of Newton's method can be considered procedures of this type. Indeed, approximation (9.3.2) characterizes any iterative process in which the corrections are taken as approximate solutions of the Newton equations. Moreover, we note that if for example an equation on the real line is solved $F(x_n) + Q(x_n) \geq 0$ $(n \geq 0)$ and $A(x_n)^*$ $(n \geq 0)$ overestimates the derivative, $x_n + A(x_n)^* [-(F(x_n) + Q(x_n))]$ $(n \geq 0)$ is always larger than the corresponding Newton-iterate. In such cases, a positive correction term is appropriate.

The notion of a Banach space with a convergence structure was used in the elegant paper [141] (see also [140]) to solve equation (9.3.1), when $A(x) = F'(x)$, $Q(x) = 0$ for all $x \in D$ and $z_n = 0$ for all $n \geq 0$. However, there are many interesting real-life applications already in the literature, where equation (9.3.1) contains a nondifferentiable term. See for example the applications at the end of this study. The case when $A(x) = F'(x)$, $Q(x) = 0$ for all $x \in D$ has already been considered but on a Banach space without generalized structure [43], [140], [141].

By imposing very general Lipschitz-like conditions on the operators involved, on the one hand, we cover a wider range of problems, and on the other hand, by choosing our operators appropriately we can find sharper error bounds on the distances involved than before.

As in [141], we provide semilocal results of Kantorovich-type and global results based on monotonicity considerations from the same general theorem. Moreover, we show that our results can be reduced to the one obtained in [141], when $A(x) = F'(x)$, $Q(x) = 0$ $(x \in D)$ and $z_n = 0$ $(n \geq 0)$, and furthermore to the ones obtained in [140] by further relaxing the requirements on X.

Finally, our results apply to solve a nonlinear integral equation involving a nondifferentiable term that cannot be solved with existing methods.

We will need the definitions:

Definition 9.3.1. *The triple* (X, V, E) *is a Banach space with a convergence structure if*

(C_1) $(X, \| \cdot \|)$ *is a real Banach space;*

(C_2) $(V, C, \| \cdot \|_V)$ *is a real Banach space that is partially ordered by the closed convex cone* C*; the norm* $\| \cdot \|_V$ *is assumed to be monotone on* C*;*

(C_3) E *is a closed convex cone in* $X \times V$ *satisfying* $\{0\} \times C \subseteq E \subseteq X \times C$*;*

(C_4) *the operator* $| \cdot |: D_0 \to C$ *is well defined:*

$$|x| = \inf \{q \in C \mid (x, q) \in E\}$$

for

$$x \in D_0 = \{x \in X \mid \exists q \in C : (x, q) \in E\};$$

and

(C_5) *for all* $x \in D_0$ $\|x\| \leq \| |x| \|_V$.

The set

$$U(a) = \{x \in X \mid (x, a) \in E\}$$

defines a sort of generalized neighborhood of zero.

Let us give some motivational examples for $X =: \mathbb{R}^m$ with the maximum-norm:

(a) $V := \mathbb{R}, E := \{(x, e) \in \mathbb{R}^m \times \mathbb{R} \mid \|x\|_\infty \le e\}$.
(b) $V := \mathbb{R}^m, E := \{(x, e) \in \mathbb{R}^m \times \mathbb{R}^m \mid |x| \le e\}$
- (componentwise absolute value).
(c) $V := \mathbb{R}^m, E := \{(x, e) \in \mathbb{R}^m \times \mathbb{R}^m \mid 0 \le x \le e\}$.

Case (a) involves classic convergence analysis in a Banach space, (b) allows componentwise analysis and error estimates, and (c) is used for monotone convergence analysis.

The convergence analysis will be based on monotonicity considerations in the space $X \times V$. Let (x_n, e_n) be an increasing sequence in $E^{\mathbb{N}}$, then

$$(x_n, e_n) \le (x_{n+k}, e_{n+k}) \Rightarrow 0 \le (x_{n+k} - x_n, e_{n+k} - e_n).$$

If $e_n \to e$, we obtain: $0 \le (x_{n+k} - x_n, e - e_n)$ and hence by (C$_5$)

$$\|x_{n+k} - x_n\| \le \|e - e_n\|_V \to 0, \quad \text{as } n \to \infty.$$

Hence $\{x_n\}$ ($n \ge 0$) is a Cauchy sequence. When deriving error estimates, we shall as well use sequences $e_n = w_0 - w_n$ with a decreasing sequence $\{w_n\}$ ($n \ge 0$) in $C^{\mathbb{N}}$ to obtain the estimate

$$0 \le (x_{n+k} - x_n, w_n - w_{n+k}) \le (x_{n+k} - x_n, w_n).$$

If $x_n \to x^*$, as $n \to \infty$, this implies the estimate $|x^* - x_n| \le w_n$ ($n \ge 0$). Moreover, if $(x, e) \in E$, then $x \in D_0$ and by (C$_4$) we deduce $|x| \le e$.

Definition 9.3.2. *An operator $L \in C^1(V_1 \to V)$ defined on an open subset V_1 of an ordered Banach space V is order convex on $[a, b] \subseteq V_1$ if*

$$c, d \in [a, b], \ c \le d \Rightarrow L'(d) - L'(c) \in L_+(V),$$

where for $m \ge 0$

$$L_+(V^m) = \{L \in L(V^m) \mid 0 \le x_i \Rightarrow 0 \le L(x_1, x_2, \ldots, x_m)\}$$

and $L(V^m)$ denotes the space of m-linear, symmetric, bounded operators on V.

Definition 9.3.3. *The set of bounds for an operator $H \in L(X^m)$ is defined to be*

$$B(H) = \{L \in L_+(V^m) \mid (x_i, q_i) \in E \Rightarrow (H(x_1, \ldots, x_m), L(q_1, \ldots, q_m)) \in E\}.$$

Definition 9.3.4. *Let $H \in L(X)$ and $y \in X$ be given, then*

$$H^*(y) = z \Leftrightarrow z = T^\infty(0) = \lim_{n \to \infty} T^n(0),$$

$$T(x) = (I - H)(x) + y \Leftrightarrow z = \sum_{i=0}^{\infty} (I - H)^i y,$$

if this limit exists.

We will also need the Lemmas [43], [141]:

Lemma 9.3.5. *Let $L \in L_+(V)$ and $a, q \in C$ be given such that:*

$$L(q) + a \leq q \quad \text{and} \quad L^n(q) \to 0 \quad \text{as } n \to \infty.$$

Then the operator

$$(I - L)^*: [0, a] \to [0, a]$$

is well defined and continuous.

The following is a generalization of Banach's lemma [43], [141] (see also Chapter 1).

Lemma 9.3.6. *Let $H \in L(X)$, $L \in B(H)$, $y \in D_0$ and $q \in C$ be such that*

$$L(q) + |y| \leq q \quad \text{and} \quad L^n(q) \to 0 \quad \text{as } n \to \infty.$$

Then the point $x = (I - H)^*(y)$ is well defined, $x \in S$ and

$$|x| \leq (I - L)^* |y| \leq q.$$

Moreover, the sequence

$$b_{n+1} = L(b_n) + |y|, \quad b_0 = 0$$

is well defined,
and

$$b_{n+1} \leq q, \quad \lim_{n \to \infty} b_n = b = (I - L)^* |y| \leq q.$$

Lemma 9.3.7. *Let $H_1: [0, 1] \to L(X^m)$ and $H_2: [0, 1] \to L_+(V^m)$ be continuous operators, then for all $t \in [0, 1]$:*

$$H_2(t) \in B(H_1(t)) \Rightarrow \int_0^1 H_2(t)dt \in B\left(\int_0^1 H_1(t)dt\right)$$

which will be used for the remainder of Taylor's formula [141] (see also Chapter 1).

We can now provide a convergence analysis for the Newton-like method (9.3.2).

Let $a \in C$, operators K_1, K_2, M, M_1, $K_3(w) \in C(V_1 \to C)$, $V_1 \subseteq V$, $w \in [0, a]$, and points $x_n \in D$ ($n \geq 0$). It is convenient to define the sequences c_n, d_n, a_n, b_n ($n \geq 0$) by

$$c_{n+1} = |x_{n+1} - x_n| \quad (n \geq 0), \tag{9.3.3}$$
$$d_{n+1} = (K_1 + K_2 + M + M_1)(d_n) + K_3(|x_n|)c_n, \quad d_0 = 0 \ (n \geq 0), \tag{9.3.4}$$
$$a_n = (K_1 + K_2 + M + M_1)^n(0), \tag{9.3.5}$$
$$b_n = (K_1 + K_2 + M + M_1)^n(0), \tag{9.3.6}$$

and the point b by

$$b = (K_1 + K_2 + M + M_1)^\infty(0). \tag{9.3.7}$$

We can now state and prove the main result of this section:

Theorem 9.3.8. *Let X be a Banach space with convergence structure (X, V, E) with $V = (V, C, \| \cdot \|_V)$, an operator $F \in C^1(D \to X)$ $(D \subseteq X)$, an operator $Q \in C(D \to X)$, an operator $A(x) \in L(x)$ $(x \in D)$, a point $a \in C$, operators K_1, $K(w)$, $K_3(w)$, $M_1 \in L_+(V)$ $(w \in [0, a])$, an operator $M_0 = M_0(v, w) \in C(V_1 \times V_1 \to V)$ $(V_1 \subseteq V)$, operators M, $K_2 \in C(V_1 \to C)$, continuous operator K_t such that for each $v, w \in V_1$, $K_t(v, w): [0, 1] \to L_+(V)$, and a null sequence $\{z_n\} \in D$ $(n \geq 0)$ such that the following conditions are satisfied:*

(C_6) $U(a) \subseteq D$, $[0, a] \subseteq V_1$, $K_3(0) \in B(I - A(0))$, $(-(F(0) + Q(0) + A(0)(z_0), (K_1 + K_2 + M + M_1)(0)) \in E$;

(C_7) *the operator K is increasing in both variables and*

$$K_1 + K(|x| + t|y|) - K(|x|) \geq K_t(|x| + t|y|, |x|) \in B(A(x) - F'(x + ty))$$

for all $t \in [0, 1]$, $x, y \in U(a)$ with $|x| + |y| \leq a$;

(C_8) $0 \leq (Q(x) - Q(x + y), M_0(|x|, |y|)) \in E$ *and* $M_0(v, w) \leq M(v + w) - M(v)$ *for all $v, w \in [0, a]$, $x, y \in U(a)$ with $|x| + |y| \leq a$;*

(C_9) $0 \leq (A(x_n)(z_n) - A(x_{n-1})(z_{n-1}), M_1(c_{n-1})) \in E$ $(n \geq 1)$;

(C_{10}) $K_3(|x|) - K_3(0) \in B(A(0) - A(x))$ *and* $K_3(|x|) \leq K_1 + K_2$ $(x \in U(a))$;

(C_{11}) $R(a) := (K_1 + K_2 + M + M_1)(0) \leq a$;

(C_{12}) $(K_1 + K_2 + M + M_1)^n a \to 0$ *as* $n \to \infty$.

(C_{13}) $\int_0^1 K(w + t(v - w))(v - w)dt \leq K_2(v) - K_2(w)$ *for all $v, w \in [0, a]$ with $w \leq v$;*

(C_{14}) $M_2(w) \geq 0$ *and* $0 \leq M_2(w_1 + w_2) - M_2(w_1) \leq M_2(w_3 + w_4) - M_2(w_3)$ *for all $w, w_1, w_2, w_3, w_4 \in [0, a]$ with $w_1 \leq w_3$ and $w_2 \leq w_4$, where M_2 is M or K_2; and*

(C_{15}) $0 \leq K(w) \leq K(v)$, $K(w) \leq K_2(w)$, $K_3(w) \leq K(w)$, *and* $0 \leq K_3(w) \leq K_3(v)$ *for all $v, w \in [0, a]$ with $w \leq v$.*

Then

(i) *the sequences (x_n, d_n), $(x_n, b_n) \in (X \times V)^{\mathbb{N}}$ are: well defined, remain in $E^{\mathbb{N}}$, monotone and satisfy*

$$b_n \le d_n \le b, \quad b_n \le a_n \quad and \quad \lim_{n \to \infty} b_n = \lim_{n \to \infty} d_n = b.$$

(ii) *Iteration $\{x_n\}$ $(n \ge 0)$ generated by (9.3.2) is: well defined, remains in $U(b)$ and converges to a solution $x^* \in U(b)$ of equation $F(x) + Q(x) = 0$, where b is the unique fixed point of R on $[0, a]$. Moreover if $z_n = 0$ $(n \ge 0)$, x^* is unique in $U(a)$.*

(iii) *Furthermore, the following error bounds are true:*

$$|x_{n+1} - x_n| \le d_{n+1} - d_n,$$

$$|x_n - x^*| \le b - d_n$$

and

$$|x_n - x^*| \le a_n - b_n \quad if \ z_n = 0 \ (n \ge 0),$$

where d_n, a_n and b_n are given by (9.3.4), (9.3.5), and (9.3.6), respectively.

Proof. We first note that b replacing a also satisfies the conditions of the theorem. Using condition (C_6) and (C_{10}), we obtain

$$
\begin{aligned}
|I - A(0)|(b) + | - (F(0) + Q(0) + A(0)(z_0))| &\le \\
\le K_3(0)(b) + (K_1 + K_2 + M + M_1)(0) &\\
\le (K_1 + K_2)(b - 0) + (K_1 + K_2 + M + M_1)(0) &\\
\le (K_1 + K_2 + M + M_1)(b - 0) + (K_1 + K_2 + M + M_1)(0) &\\
= (K_1 + K_2 + M + M_1)(b) &\\
= R(b) \le b \ \ (\text{by } (9.3.12)). &
\end{aligned}
$$

Hence, by Lemma 9.3.6, x_1 is well defined and $(x_1, b) \in E$. We also get

$$x_2 = (I - A(0))(x_2) + (-(F(0) + Q(0) + A(0)(z_0)))$$

$\Rightarrow$

$$
\begin{aligned}
|x_2| &\le K_3(0)|x_1| + (K_1 + K_2 + M + M_1)(0) \\
&\le (K_1 + K_2)|x_1| + (K_1 + K_2 + M + M_1)(0) = d_1,
\end{aligned}
$$

and by the order convexity of L

$$
\begin{aligned}
d_1 &= (K_1 + K_2)|x_1| + (K_1 + K_2 + M + M_1)(0) \\
&\le (K_1 + K_2)|x_1| + (K_1 + K_2 + M + M_1)(0) \\
&\le (K_1 + K_2)(b - 0) + (K_1 + K_2 + M + M_1)(0) \\
&\le (K_1 + K_2 + M + M_1)(b - 0) + (K_1 + K_2 + M + M_1)(0) \\
&= R(b) = b.
\end{aligned}
$$

That is, we get $|x_1 - x_0| \leq d_1 - d_0$ or $0 \leq (x_0, d_0) \leq (x_1, d_1)$.

We assume that

$$0 \leq (x_{k-1}, d_{k-1}) \leq (x_k, d_k), \quad \text{and} \quad d_k \leq b \quad \text{for } k = 1, 2, \ldots, n.$$

We need to find a bound for $I - A(x_n)$ $(n \geq 0)$. We will show that $K_3(|x_n|) \in B(I - A(x_n))$. This fact follows from (C_6), (C_{10}), and the estimate

$$|I - A(x_n)| \leq |I - A(0)| + |A(0) - A(x_n)| \leq K_3(0) + K_3(|x_n|) - K_3(0) = K_3(|x_n|).$$

Using (9.3.2), we obtain the approximation

$$\begin{aligned}
&- [F(x_n) + Q(x_n) + A(x_n)(z_n)] = \qquad\qquad\qquad (9.3.8)\\
&= -F(x_n) - Q(x_n) - A(x_n)(z_n) + A(x_{n-1})(x_n - x_{n-1})\\
&\quad + F(x_{n-1}) + Q(x_{n-1}) + A(x_{n-1})(z_{n-1}).
\end{aligned}$$

By (C_7)–(C_9), (C_{13})–(C_{15}), Lemma 9.3.7, and the induction hypotheses, we obtain in turn

$$\begin{aligned}
&|- F(x_n) + F(x_{n-1}) + A(x_{n-1})(x_n - x_{n-1})| + |Q(x_{n-1}) - Q(x_n)|\\
&\quad + |A(x_n)(z_n) - A(x_{n-1})(z_{n-1})| \leq\\
&\leq \int_0^1 K_t(|x_{n-1}| + t|x_n - x_{n-1}|, |x_{n-1}|)c_{n-1}dt + M_0(|x_{n-1}|, |x_n|) + M_1 c_{n-1}\\
&\leq \int_0^1 \Big[K(|x_{n-1}| + tc_{n-1}) - K(|x_{n-1}|) + K_1\Big]c_{n-1}dt + M(|x_{n-1}| + c_{n-1})\\
&\quad - M(|x_{n-1}|) + M_1 c_{n-1}\\
&\leq \int_0^1 \Big[K(d_{n-1} + t(d_n - d_{n-1}))(d_n - d_{n-1})\, dt - K(|x_{n-1}|)c_{n-1} + K_1 c_{n-1}\\
&\quad + M(d_{n-1} + d_n - d_{n-1}) - M(d_{n-1}) + M_1(d_n) - M_1(d_{n-1})\\
&\leq K_2(d_n) - K_2(d_{n-1}) - K(|x_{n-1}|)c_{n-1} + K_1 c_{n-1} + M(d_n) - M(d_{n-1})\\
&\quad + M_1(d_n) - M_1(d_{n-1}) \leq
\end{aligned}$$

$$\leq (K_1 + K_2 + M + M_1)(d_n) - d_n.$$

We can now obtain that

$$\begin{aligned}
K_3(|x_n|)(b - d_n) + |- (F(x_n) &+ Q(x_n) + A(x_n)(z_n))| + d_n \leq\\
&\leq (K_1 + K_2 + M + M_1)(b - d_n) + (K_1 + K_2 + M + M_1)(d_n) \qquad (9.3.9)\\
&= R(b) = b.
\end{aligned}$$

That is, x_{n+1} is also well defined by Lemma 9.3.6 and $c_n \leq b - d_n$. Hence, d_{n+1} is well defined too and as in (9.3.9), we obtain that:

$$d_{n+1} \leq R(b) \leq b.$$

The monotonicity $(x_n, d_n) \leq (x_{n+1}, d_{n+1})$ can be derived from

$$c_n + d_n \leq K_3(|x_n|)c_n + |-(F(x_n) + Q(x_n) + A(x_n)(z_n))| + d_n$$
$$\leq K_3(|x_n|)c_n + (M + M_1 + K_1 + K_2)d_n \leq d_{n+1}.$$

The induction has now been completed. We need to show that

$$b_n \leq d_n \quad \text{for all } n \geq 1.$$

For $n = 1$ and from the definitions of b_n, d_n

$$b_1 = (K_1 + K_2 + M + M_1)^1(0) \leq d_1.$$

Assume that

$$b_k \leq d_k \quad \text{for } k = 1, 2, \ldots, n.$$

Then, we obtain in turn

$$b_{n+1} = (K_1 + K_2 + M + M_1)^{n+1}(0)$$
$$= (K_1 + K_2 + M + M_1)(K_1 + K_2 + M + M_1)^n(0)$$
$$\leq (K_1 + K_2 + M + M_1)(d_n) \leq d_n \leq d_{n+1}.$$

Because $d_n \leq b$, we have $b_n \leq d_n \leq b$. By (9.3.5), and (9.3.6) it follows that

$$0 \leq a_n - b_n \leq (K_1 + K_2 + M + M_1)^n(a) \quad (n \geq 1).$$

By condition (C$_{13}$) and the above, we deduce that the sequence $\{b_n\}$ $(n \geq 0)$ is Cauchy in a Banach space C, and as such it converges to some $b = (K_1 + K_2 + M + M_1)^\infty(0)$. From $(K_1 + K_2 + M + M_1)(b) = (K_1 + K_2 + M + M_1)(\lim_{n\to\infty}(K_1 + K_2 + M + M_1)^n(0)) = \lim_{n\to\infty}(K_1 + K_2 + M + M_1)^{n+1}(0) = b$, we obtain

$$(K_1 + K_2 + M + M_1)(b) = b \leq a,$$

which makes b smaller than any solution of the inequality

$$(K_1 + K_2 + M + M_1)(p) \leq p.$$

It also follows that the sequence $\{x_n\}$ $(n \geq 0)$ is Cauchy in X, and as such it converges to some $x^* \in U(b)$. By letting $n \to \infty$ in (9.3.8) and using the hypotheses that $\lim_{n\to\infty} z_n = 0$, we deduce that x^* is a solution of the equation $F(x) + Q(x) = 0$.

To show uniqueness, let us assume that there exists another solution y^* of the equation $F(x) + Q(x) = 0$ in $U(a)$. Then, exactly as in [43], [140], by considering the modified Newton-process

$$x_{n+1} = x_n - (F(x_n) + Q(x_n)),$$

we can show that this sequence converges, under the hypotheses of the theorem.

Moreover, as above, we can easily show (see also [43], [141]) that

$$|y^* - x_n| \le a_n - b_n \quad \text{for } z_n = 0 \ (n \ge 0),$$

from which follows that $x_n \to y^*$ as $n \to \infty$. Finally, the estimates (iii) are obtained by using standard majorization techniques.

That completes the proof of the theorem.

We will now introduce results on a posteriori estimates. It is convenient to define the operator

$$R_n(q) = (I - K_3(|x_n|))^* S_n(q) + c_n$$

where

$$S_n(q) = (K_1 + K_2 + M + M_1)(|x_n| + q)$$
$$- (K_1 + K_2 + M + M_1)(|x_n|) - K_3(|x_n|)(q)$$

and the interval

$$I_n = [0, a - |x_n|].$$

It can easily be seen that the operators S_n are monotone on I_n. Moreover, the operators $R_n : [0, a - d_n] \to [0, a - d_n]$ are well defined and monotone. This fact follows from Lemma 9.3.5 and the scheme

$$d_n + c_n \le d_{n+1} \Rightarrow R(a) - d_{n+1} \le a - d_n - c_n$$
$$\Rightarrow S_n(a - d_n) + K_3(|x_n|)(a - d_n - c_n) \le a - d_n - c_n \ (n \ge 0).$$

Then, exactly as in [141], we can show:

Proposition 9.3.9. *The following implications are true:*

(i) *if $q \in I_n$ satisfy $R_n(q) \le q$, then*

$$c_n \le R_n(q) = p \le q,$$

and

$$R_{n+1}(p - c_n) \le p - c_n \quad \text{for all } n \ge 0;$$

(ii) *under the hypotheses of Theorem 9.3.8, let $q_n \in I_n$ be a solution of $R_n(q) \le q$, then*

$$|x^* - x_m| \le a_m \quad (m \ge n)$$

where

$$a_n = q_n \quad \text{and} \quad a_{m+1} = R_m(a_m) - c_m;$$

and

(iii) *under the hypotheses of Theorem 9.3.8, any solution $q \in I_n$ of $R_n(q) \le q$ is such that*

$$|x^* - x_n| \le R_n^\infty(0) \le q.$$

Remark 9.3.10. (a) The results obtained in Theorem 9.3.8 and the Proposition reduce immediately to the corresponding ones in [141, Theorem 5 and Lemmas 10–12], when $A(x) = F'(x)^{-1}$, $Q(x) = 0$ $(x \in D)$, $z_n = 0$ $(n \geq 0)$, $t = 1$, $K_1 = 0$, $K_2 = L$, where L is order convex on $[0, a]$, $K = L'$ and $K_3(0) = L'(0)$. On the one hand, using our conditions, we cover a wider range of problems, and on the other hand, it is because it may be possible to choose K_t, K, K_1 so that $K_t(p+tq, p) \leq K_1 + K(p+tq, p) \leq L'(p+tq) - L'(p)$ for all $p, q \in K_1, t \in [0, 1]$. Then it can easily be seen that our estimates on the distances $|x_{n+1} - x_n|$ and $|x^* - x_n|$ $(n \geq 0)$ will be sharper. One such choice for K_t could be

$$K_t(p + tq, p) = \sup_{\substack{|x|+|y|\leq a,\, t\in[0,1] \\ |x|\leq p,\, |y|\leq q}} |A(x) - F'(x + ty)|$$

for all $x, y \in U(a)$, $p, q \in [0, a]$.

(b) As in [43], [141], we can show that if conditions (C_6)–(C_{10}), (C_{13})–(C_{15}) are satisfied and there exists $t \in (0, 1)$ such that $(K_1 + K_2 + M + M_1)(a) \leq ta$, then there exists $a_1 \in [0, ta]$ satisfying conditions (C_6)–(C_{15}). The solution $x^* \in U(a_1)$ is unique in $U(a)$ (when $z_n = 0$ $(n \geq 0)$).

(c) From the approximation

$$A(x_n)(z_n) - A(x_{n-1})(z_{n-1}) =$$
$$= (A(x_n)(z_n) - z_n) + (A(x_{n-1})(z_{n-1}) - z_{n-1}) + I(z_n - z_{n-1}),$$

we observe that (C_9) will be true if $M_1 = 2K_3(b) + I$ and points z_n $(n \geq 0)$ are such that $|z_n| + |z_{n-1}| + |z_n - z_{n-1}| \leq c_{n-1}$ $(n \geq 1)$.

(d) Another choice for M_1, z_n can be $M_1 = |\varepsilon| I$, $z_n = z_{n-1} + \varepsilon_n (x_n - x_{n-1})$ $(n \geq 1)$ with $|\varepsilon_n| \leq |\varepsilon|$ $(n \geq 0)$, where e, e_n $(n \geq 0)$ are numbers or operators in $L_+(V)$ and provided that $F'(x) = I$ $(x \in D)$. It can then easily be seen that (C_9) is satisfied. The sequence $\{z_n\}$ $(n \geq 0)$ must still be chosen to be null. At the end of this paper, in part V, we have given examples for this case. Several other choices are also possible.

(e) Condition (C_7) can be replaced by the set of conditions

$$K_4 + K_5(|x| + t|y|) - K_5(|x|) \in B(F'(x) - F'(x + ty))$$

and

$$K_6(|x|) - K_6(|y|) \in B(A(x) - F'(x))$$

for all $t \in [0, 1]$, $x, y \in U(a)$ with $|x| + |y| \leq a$.

(f) Define the residuals $r_n = -A(x_n)(z_n)$ $(n \geq 0)$ and set $\delta_n = x_{n+1} - x_n$ $(n \geq 0)$. Then from the approximation

$$r_n = [(I - A(x_n)) - I](z_n)$$

we obtain

$$|r_n| \leq (K_3(|x_n|) + I)|z_n|$$

which shows that $r_n \to 0$ if $z_n \to \infty$ as $n \to \infty$. Consequently, the results obtained in Theorem 9.3.8 and Proposition 9.3.9 remain true for the system

$$x_{n+1} - x_n = \delta_n, \quad A(x_n)\delta_n = -(F(x_n) + Q(x_n)) + r_n \quad (n \geq 0).$$

(g) It can easily be seen from the proof of Theorem 9.3.8 that the results obtained in this Theorem remain valid if (C_9) is replaced by the condition

$(C_9)'$ $(A(x_n)(z_n) - A(x_{n-1})(z_{n-1}), M_2(d_n^* - d_{n-1}^*)) \in E$

with $d_n^* = d_n$ or $d_n^* = b_n$ $(n \geq 0)$, for some $M_2 \in L_+(V)$. This is equivalent to the condition $(A(x_n)(z_n), M_2(d_n^*)) \in E^{\mathbb{N}}$ $(n \geq 1)$ is an increasing sequence.

We now examine the monotone case.

Let $J \in L(X \to X)$ be a given operator. Define the operators $P, T(D \to X)$ by

$$P(x) = JT(x + u), \quad T(x) = G(x) + G_1(x), \quad P(x) = F(x) + Q(x),$$
$$F(x) = JG(x + u) \quad \text{and} \quad Q(x) = JG_1(x + u),$$

where G, G_1 are as F, Q, respectively. We deduce immediately that under the hypotheses of Theorem 9.3.8, the zero x^* of P is a zero of JT also, if $u = 0$.

We will now provide a monotonicity result to find a zero x^* of JT. The space X is assumed to be partially ordered and satisfies the conditions for V given in (C_1)–(C_5). Moreover, we set $X = V$, $D = C^2$ so that $|\cdot|$ turns out to be I.

Theorem 9.3.11. *Let V be a partially ordered Banach space satisfying conditions (C_1)–(C_5), Y be a Banach space, G, G_1 as $F, Q, D \subseteq V, J \in L(V \to V), K_t, M, M_1; K, K_1, K_2, K_3$ as in Theorem 9.3.8 and $u, v \in V$ such that*

(C_{16}) $[u, v] \subseteq D$;
(C_{17}) *sequence $\{z_n\}$ $(n \geq 0)$, and iteration*

$$y_0 = u, \quad y_{n+1} = y_n + [JA(y_n)]^* (-JT(y_n)) - z_n \quad (n \geq 0) \qquad (9.3.10)$$

are such that

$$y_n + [JA(y_n)]^* (-JT(y_n)) - v \leq z_n \leq [JA(y_n)]^* (-JT(y_n)),$$

$$z_n \in [u, v] \ (n \geq 0).$$

(C_{18}) *conditions (C_6)–(C_{15}) are satisfied for $a = v - u$. Then iteration (9.3.10) is well defined for all $n \geq 0$, monotone and converges to a zero x^* of JT in $[u, v]$. Moreover x^* is unique in $[u, v]$ if $z_n = 0$ $(n \geq 0)$.*

Proof. It then follows immediately from Theorem 9.3.8 by setting $a = v - u$.

We will complete this study with two applications that show how to choose the terms introduced in Theorem 9.3.8, in practical cases. From now on, we choose $t = 1$, $A(x) = F'(x)^{-1}$ $(x \in D)$, $K_1 = 0$, $K = L'$ (order convex), $K_2 = L$ and $K_3(0) = L'(0)$. It can then easily be seen from the proof of Theorem 9.3.8 that conditions (C_{12}) and (C_{13}) can be replaced by $(L + M + M_1)(a) \leq a$ and $(L'(a) + M + M_1)^n(a) \to 0$ as $n \to \infty$ respectively (see also Remark 9.3.10 (a)).

Application 9.3.12. *We discuss the case of a real Banach space with norm* $\| \; \|$*. Assume that* $F'(0) = I$*, there exists a monotone operator*

$$f \colon [0, a] \to \mathbb{R}$$

such that

$$\|F''(x)\| \leq f(\|x\|) \quad \text{for all } x \in U(a)$$

and a continuous nondecreasing function g on $[0, r]$*,* $r \leq a$ *such that*

$$\|Q(x) - Q(y)\| \leq g(r)\|x - y\| \tag{9.3.11}$$

for all $x, y \in U\left(\frac{r}{2}\right)$*.*

We showed in [43], (see also [147]) that (9.3.11) implies that

$$\|Q(x + l) - Q(x)\| \leq h(r + \|l\|) - h(r), \quad x \in U(a), \quad \|l\| \leq a - r, \tag{9.3.12}$$

where,

$$h(r) = \int_0^r g(t)\,dt.$$

Conversely, it is not hard to see that we may assume, without loss of generality, that the function h and all functions $h(r + t) - h(r)$ *are monotone in r. Hence, we may assume that* $h(r)$ *is convex and hence differentiable from the right. Then, as in [43], we show that (9.3.12) implies (9.3.11) and* $g(r) = h'(r + 0)$*. Hence, we can set*

$$L(q) = \|F(0) + Q(0)\| + \int_0^q ds \int_0^s f(t)\,dt \tag{9.3.13}$$

and

$$M(q) = \int_0^q g(t)\,dt. \tag{9.3.14}$$

In Remark 9.3.10 (c) and (d), we have already provided some choices for M_1, z_n*. Here, however, for simplicity let us choose* $z_n = 0$ $(n \geq 0)$ *and* $M_1 = 0$*.*

Then condition (C_{11}) *will be true if*

$$\tfrac{1}{2}f(a)a^2 - (1 - g(a))a + \|F(0) + Q(0)\| \leq 0. \tag{9.3.15}$$

If we set $Q = 0$ *and* $g = 0$*, (9.3.15) is true if* $\|F(0)\|f(a) \leq \frac{1}{2}$*, which is a well-known condition due to Kantorovich. If* $Q \neq 0$*, condition (9.3.15) is the same condition with the one found in [147] for the Zincenko iteration.*

In the application that follows, we show that our results can apply to solve nonlinear integral equations involving a nondifferentiable term, whereas the results obtained in [140] (or [141]) cannot apply.

Application 9.3.13. *Let* $X = V = C[0, 1]$*, and consider the integral equation*

$$x(t) = \int_0^1 k(t, s, x(s))\,ds \quad \text{on } X, \tag{9.3.16}$$

where the kernel $k(t, s, x(s))$ with $(t, s) \in [0, 1] \times [0, 1]$ is a nondifferentiable operator on X. Consider (9.3.16) in the form (9.3.1) where $F, Q: X \to X$ are given by

$$F(x)(t) = Ix(t) \quad and \quad Q(x)(t) = -\int_0^1 k(t, s, x(s))\, ds.$$

The operator $| \; |$ is defined by considering the sup-norm. We assume that V is equipped with natural partial ordering, and there exists $\alpha, a \in [0, +\infty)$, and a real function $\alpha(t, s)$ such that

$$\|k(t, s, x) - k(t, s, y)\| \leq \alpha(t, s)\|x - y\|$$

for all $t, s \in [0, 1]$, $x, y \in U\left(\frac{a}{2}\right)$, and

$$\alpha \geq \sup_{t \in [0,1]} \int_0^1 \alpha(t, s)\, ds.$$

Define the real functions h, f, g on $[0, a]$ by $h(r) = \alpha r$, $f(r) = 0$ and $g(r) = \alpha$ for all $r \in [0, a]$. By choosing L, M, M_1 as in (9.3.13), and (9.3.14), and Remark 9.3.10 (c), respectively, we can easily see that the conditions (C_1)–(C_{10}), (C_{13})–(C_{15}) of Theorem 9.3.8 are satisfied. In particular, condition (C_{12}) becomes

$$(1 - \alpha - |\varepsilon|)a - \|Q(0)\| \geq 0 \tag{9.3.17}$$

which is true in the following cases: if $0 \leq \alpha < 1 - |\varepsilon|$, choose $a \geq \beta = \frac{\|Q(0)\|}{1 - \alpha - |\varepsilon|}$; if $\alpha = 1 - |\varepsilon|$ and $Q(0) = 0$, choose $a \geq 0$; if $\alpha > 1 - |\varepsilon|$ and $Q(0) = 0$, choose $a = 0$. If in (9.3.17) strict inequality is valid, then there exists a solution a^* of equation (9.3.17) satisfying condition (C_{13}). Note that if we choose $\alpha \in (0, 1 - |\varepsilon|)$, $a \in (\beta, +\infty)$ and $\varepsilon \in (-1, 1)$, condition (9.3.17) is valid as a strict inequality. Finally, we remark that the results obtained in [140], [141] cannot apply here to solve equation (9.3.16), because Q is nondifferentiable on X and the z_n's are not necessarily zero. This example is useful, especially when the z_n's are not necessarily all zero. Otherwise, results on (9.3.2) with general convergence structure have already been found (see, e.g., [43], [140], [141] and the references there).

In the remaining of this section we show how to control the residuals in the Netwon-like method (9.3.2).

We generate a sequence $\{x_n\}$ $(n \geq 0)$ using the perturbed Newton-like method scheme given by

$$x_{n+1} = x_n + \delta_n \quad (n \geq 0) \tag{9.3.18}$$

where the correction δ_n satisfies

$$A(x_n)\delta_n = -(F(x_n) + Q(x_n)) + r_n \quad (n \geq 0) \tag{9.3.19}$$

Here we derive sufficient conditions for controlling the residuals r_n in such a way that the convergence of the sequence $\{x_n\}$ $n \geq 0$ to a solution of equation $F(x) = 0$ is ensured.

We also refer the reader to [43], [140], [141] and the references there for relevant work, which however is valid on a Banach space X without a convergence structure. The advantages of working on a Banach space with a convergence structure have been explained in some detail [43], [140], [141].

We will need the following basic result:

Lemma 9.3.14. *Let V be a regular partially ordered topological space, an operator $L \in C^1 (V_1 \to V)$ with $[0, a] \subseteq V_1 \subseteq V$ for some $a \in V$, an operator $M \in C (V_1 \to V)$, operator $R < I$, $B > I$, T, $K \in L_+ (V)$, a point $c \in V$ with $c > 0$ and a point $p \in [0, a]$.*

Assume:

(a) The equation

$$g(q) = BT \left[L(p+q) - L(p) - L'(p)q + M(p+q) \right.$$
$$\left. - M(p) + K(p+q) - K(p) \right] - (I - r)q + c = 0 \quad (9.3.20)$$

has solutions in the interval $[0, a]$ and denote by q^* the least of them.

(b) Let $G \in L_+ (V)$ be given and $R_+ \in L_+ (V)$, $c_+, p_+ \in V$ be such that the following conditions are satisfied:

$$R_+ < \min \left\{ (G - 2I) BTL'(p) + BTGL'(p_+) + G(R - I) + I, I \right\} = \alpha \quad (9.3.21)$$

$$0 < c_+$$
$$\leq BTG (L(p_+) + M(p_+) + K(p_+)) + GBT (L(p) + M(p) + K(p))$$
$$- 2BT (L(p) + M(p) + K(p)) + \left(R_+ + G - I - BTGL'(p_+) \right) c = \beta \quad (9.3.22)$$

and

$$0 \leq p_+ \leq p + c, \quad (9.3.23)$$

where α and β are functions of the operators and points involved.

(c) The following estimate is true

$$M(p) \leq M(p+q) \quad (9.3.24)$$

for all $p, q \in [0, a]$.

Then the equation

$$g_+ (q) =$$
$$= BGT \left[L(p_+ + q) - L(p_+) - L'(p_+)q + M(p_+ + q) - M(p_+) \right.$$
$$\left. + K(p_+ + q) - K(p_+) \right] - (I - R_+)q + c_+ = 0 \quad (9.3.25)$$

has nonnegative solutions and the least of them, denoted by q^*_+, lies in the interval $[c_+, q^* - c]$.

Proof. Using the hypotheses $g(q^*) = 0$ and $R < I$, we deduce from (9.3.20) that $c \le q^*$. We will show that

$$g_+(q^* - c) \le 0. \tag{9.3.26}$$

From equation (9.3.25), and using (9.3.22), we obtain in turn

$$g_+(q^* - c) \le$$
$$\le \Big[BGTL(p + c + q^* - c) - L(p_+) - L'(p_+)(q^* - c)$$
$$\quad + M(p + c + q^* - c) - M(p_+) + K(p + c + q^* - c) - K(p_+) \Big]$$
$$\quad - (I - R_+)(q^* - c) + c_+$$
$$= g(q^*)$$
$$\quad - BT\Big[L(p + q^*) - L(p) - L'(p)q^* + M(p + q^*) - M(p)$$
$$\qquad + K(p + q^*) - K(p) \Big]$$
$$\quad + BTG\Big[L(p + q^*) - L(p_+) - L'(p_+)(q^* - c) + M(p + q^*) - M(p_+)$$
$$\qquad + K(p + q^*) - K(p_+) \Big]$$
$$\quad + (I - R)q^* - c - (I - R_+)(q^* - c) + c_+$$
$$= \Big[BTL'(p) - BTGL'(p_+) + (I - R) - (I - R_+) \Big] q^*$$
$$\quad + (G - I)\Big\{ (I - R)q^* - c - BT\big[L(p) + M(p) + K(p) + L'(p)q^* \big] \Big\}$$
$$\quad + BT\Big[L(p) + M(p) + K(p) - GL(p_+) - GM(p_+) \Big]$$
$$\quad + BTGL'(p_+)c + (I - R_+)c + c_+ - c$$
$$= \Big[(2I - G)BTL'(p) - BTGL'(p_+) + R_+ + G(I - R) - I \Big] q^*$$
$$\quad + 2BT(l(p) + M(p) + K(p)) - BTG(L(p_+) + M(p_+) + K(p_+))$$
$$\quad - GBT(L(p) + M(p) + K(p)) + \big(BTGL'(p_+) - G + I - R_+ \big)c + c_+$$
$$\le 0,$$

because (9.3.21) and (9.3.22) are satisfied.

Moreover from (9.3.25) for $q = c_+$, we obtain

$$g_+(c_+) \ge 0. \tag{9.3.27}$$

By inequalities (9.3.26), (9.3.27), the fact that g is continuous and isotone on $[c_+, q^* - c]$, and as V is a regular partially ordered topological space, from the proposition, we deduce that there exists a point q_+^* with

$$g_+(q_+^*) = 0 \tag{9.3.28}$$

and

$$c_+ \leq q_+^* \leq q^* - c. \tag{9.3.29}$$

We can assume that q_+^* denotes the least of the solutions of equation (9.3.25). That completes the proof of the lemma.

The following result is a consequence of the above lemma.

Theorem 9.3.15. *Let $\{c_n\} \in V$, $\{T_n\}$, $\{R_n\}$, $\{G_n\} \in L_+(V)$ $(n \geq 0)$ be sequences and V as in the above lemma. Assume:*

(a) There exists a sequence $\{p_n\} \in [0, a] \subseteq V_1 \subseteq V$ for some $a \in V$ with $p_0 = 0$, and

$$p_{n+1} \leq \sum_{j=0,1,\dots,n} c_j \text{ for } n \geq 0. \tag{9.3.30}$$

(b) $R_0 < I$ and the function

$$g_0(q) = BT_0\Big[L(p_0 + q) - L(p_0) - L'(p_0)q + M(p_0 + q) \tag{9.3.31}$$
$$- M(p_0) + K(p_0 + q) - K(p_0)\Big] - (I - R)q + c_0 = 0$$

has root on $[0, a]$, where B, L, M, K are as in the above lemma. Denote by q_0^* the least of them.

(c) The following conditions are satisfied for all $n \geq 0$

$$R_{n+1} < \alpha_{n+1}, \tag{9.3.32}$$
$$0 < c_{n+1} \leq \beta_{n+1}, \tag{9.3.33}$$

and

$$0 \leq p_{n+1} \leq p_n + c_n. \tag{9.3.34}$$

(d) The linear operators T_n are boundedly invertible for all $n \geq 0$, and set $G_n = T_{n+1}T_n^{-1}$ $(n \geq 0)$.

(e) Condition (9.3.24) is satisfied.

Then, the equation

$$g_n(q) = BG_nT_n\Big[L(p_n + q) - L(p_n) - L'(p_n)q + M(p_n + q) \tag{9.3.35}$$
$$- M(p_n) + K(p_n + q) - K(p_n)\Big] - (I - R_n)q + c_n = 0$$

has solution in $[0, a]$ for every $n \geq 0$ and denoting by q_n^* the least of them, we have

$$\sum_{j=n,\dots,\infty} c_j \leq q_n^* \quad (n \geq 0). \tag{9.3.36}$$

Proof. Let us assume that for some nonnegative integer n, $I - R_n > 0$, $g_n(q)$ has roots on $[0, a]$ and denote by q_n^* the least of them. We use introduction on n. We also observe that this is true by hypothesis (b) for $n = 0$. Using (9.3.30), (9.3.32),

(9.3.33), (9.3.34), the lemma, and setting $c = c_n$, $c_+ = c_{n+1}$, $R = R_n$, $R_+ = R_{n+1}$, $G = G_n$ we deduce that q_{n+1}^* exists, and

$$c_{n+1} \leq q_{n+1}^* \leq q_n^* - c_n. \tag{9.3.37}$$

The induction is now complete and (9.3.36) follows immediately from (9.3.37).

That completes the proof of the theorem. From now on we assume that X is a Banach space with a convergence structure in the sense of [141].

The following result is an immediate consequence of Theorem 9.3.15.

Theorem 9.3.16. *Assume:*

(a) the hypotheses of Theorem 9.3.15 are satisfied;

(b) there exists a sequence $\{x_n\}$ $(n \geq 0)$ in a Banach space X with a convergence structure such that $|x_{n+1} - x_n| \leq c_n$.

Then,

(i) the sequence $\{x_n\}$ $(n \geq 0)$ converges to some point x^*;

(ii) moreover the following error estimates hold

$$\left| x^* - x_n \right| \leq q_n^*, \tag{9.3.38}$$

and

$$\left| x^* - x_{n+1} \right| \leq q_n^* - c_n, \quad \text{for all } n \geq 0. \tag{9.3.39}$$

We can introduce the main result:

Theorem 9.3.17. *Let X be a Banach space with convergence structure (X, V, E) with $V = (V, C, \|\cdot\|_v)$, an operator $F \in C^1 (D \to X)$ with $D \subseteq X$, an operator $Q \in C (D \to X)$, an operator $A(x) \in C (X \to D)$, an operator $L \in C^1 (V_1 \to V)$ with $V_1 \subseteq V$, an operator $M \in C (V_1 \to V)$, an operator $K \in L_+ (V)$, and a point $a \in C$ such that the following conditions are satisfied:*

(a) the inclusions $U(a) \subseteq D$, and $[0, a] \subseteq V_1$ are true;

(b) L is order-convex on $[0, a]$, and satisfies

$$K + L' |x| + |y| - L' (|x|) \in B \left(A(x) - F'(x + y) \right) \tag{9.3.40}$$

for all $x, y \in U(a)$ with $|x| + |y| \leq a$;

(c) M satisfies the condition

$$M(|x| + |y|) - M(|x|) \in B (Q(x) - Q(x + y)), \quad M(0) = 0 \tag{9.3.41}$$

for all $x, y \in U(a)$ with $|x| + |y| \leq a$;

(d) for the sequences $\{c_n\}$, $\{T_n\}$, $\{R_n\}$, $\{G_n\}$, $\{p_n\}$ $(n \geq 0)$ the hypotheses (9.3.30), (b), (9.3.32), (9.3.34) and (d) of Theorem 9.3.15 are satisfied;

(e) the following conditions are also satisfied

$$|\delta_n| \le c_n \le T_n \,|-(F(x_n) + Q(x_n))| \le \gamma_n \le \beta_n \quad (n \ge 1), \qquad (9.3.42)$$

$$|-r_n| \le T_n^{-1} R_n c_n, \qquad (9.3.43)$$

where,

$$\gamma_n = T_n \Big[L(p_n + c_n) - L(p_n) - L'(p_n) c_n + M(p_n + c_n) - M(p_n)$$

$$+ K(p_n + c_n) - K(p_n) \Big] + R_n c_n \quad (n \ge 1). \qquad (9.3.44)$$

Then,

(i) *the sequence $\{x_n\}$ $(n \ge 0)$ generated by*

$$x_{n+1} = x_n + \delta_n, \quad \text{with } x_0 = 0$$

remains in $\bar{U}(x_0, t_0^*)$ and converges to a solution x^* of equation $F(x) = 0$;

(ii) *moreover, the error estimates (9.3.38) and (9.3.39) are true where q_n^* is the least root in $[0, \alpha]$ of the function $g_n(q)$ defined in (9.3.35), with $p_n = \|x_n - x_0\|$ $(n \ge 0)$.*

Proof. Let us assume that $x_n, x_{n+1} \in U(x_0, q_0^*)$, where the existence of q_0^* is guaranteed from hypotheses (d). We note that $|\delta_0| \le c_0$. Using the approximation

$$- (F(x_{n+1}) + Q(x_{n+1}))$$

$$= (F(x_n) - F(x_{n+1}) + A(x_n)(x_{n+1} - x_n))$$

$$+ (Q(x_n) - Q(x_{n+1})) - r_n, \qquad (9.3.45)$$

(9.3.40), (9.3.41), (9.3.43) and setting $p_n = |x_n - x_0|$, we obtain in turn

$$|-(F(x_{n+1}) + Q(x_{n+1}))|$$

$$\le |F(x_n) - F(x_{n+1}) + A(x_n)(x_{n+1} - x_n)|$$

$$+ |Q(x_n) - Q(x_{n+1})| + |-r_n|$$

$$\le \int_0^1 \Big[L'(p_n + t\,|x_{n+1} - x_n|) - L'(p_n) + K \Big] |x_{n+1} - x_n|\, dt$$

$$+ M(p_n + |x_{n+1} - x_n|) - M(p_n) + |-r_n|$$

$$\le L(p_n + c_n) - L(p_n) - L'(p_n) c_n + K c_n$$

$$+ M(p_n + c_n) - M(p_n) + T_n^{-1} R_n c_n.$$

Hence, by (9.3.42) we get

$$c_{n+1} \le T_{n+1} \,|-(F(x_{n+1}) + Q(x_{n+1}))| \le \gamma_n \le \beta_n \quad (n \ge 1),$$

which shows (9.3.33).

It can easily be seen that by using induction on n, the hypotheses of Theorem 9.3.15 are satisfied. Hence, by (9.3.40) and (9.3.43) the iteration $\{x_n\}$ $(n \geq 0)$ remains in $U\left(x_0, q_0^*\right)$ and converges to x^* so that (9.3.38), and (9.3.39) satisfied. Moreover, from the estimate

$$|-(F(x_n) + Q(x_n))| \leq |A(x_n) - F'(x_n)| c_n + |F'(x_n)| c_n + |-r_n|,$$

(9.3.40), (9.3.43), the continuity of F, F', A, T_n, R_n, and $c_n \to 0$ as $n \to \infty$, we deduce that

$$F(x^*) + Q(x^*) = 0.$$

That completes the proof of the theorem.

We complete this section with an application.

Application 9.3.18. *Returning back to Application 9.3.12, define the functions f_1, f_2, f_3 on $[0, a]$ by*

$$f_1(q) = L(q) - q, \quad f_2(q) = f_1(q) + h(q) \text{ and } f_3(q) = \int_0^q f(t)\,dt.$$

Choose $B = T_0 = 1$, $R_0 = K = 0$ and $p_0 = 0$. Then by (9.3.31) we get

$$g_0(q) = f_2(q).$$

Example 9.3.19. It can easily be seen that with the above choices of L and M, conditions (9.3.40) and (9.3.41) are satisfied.

Suppose that the function g_0 has a unique zero q_0^* in $[0, a]$ and $g_0(a) \leq 0$. It is then known [43], [141] that there exists a solution x^* in $U(q_0^*)$, this solution is unique in $U(a)$, and the iteration $\{x_n\}$ $(n \geq 0)$ given by (9.3.18) is well defined, remains in $U(q_0^*)$ for all $n \geq 0$, and converges to x^*. By applying the Banach Lemma on invertible operators, we can show that $A(x_n)$ is invertible and $\|A(x_n)^{-1}\| \leq -f_1'(\|x_n\|)^{-1} = T_n$ $(n \geq 0)$.

Assume that instead of conditions (9.3.32) and (9.3.33), the weaker condition (9.3.26) is satisfied. Using the approximation

$$r_n = [(F'(x_n) - F'(x_0)) + F'(x_0)]\delta_n + F(x_n) + Q(x_n),$$

we obtain

$$\|r_n\| \leq (f_3(\|x_n\|) + 1) c_n + T_n^{-1}\gamma_n \quad (n \geq 1).$$

The above estimate provides us with a possible (but not the only) choice for the R_n's. Indeed, (9.3.43) will be true if the R_n's $(n \geq 0)$ can be chosen so that

$$(f_3(\|x_n\|) + 1) c_n + T_n^{-1}\gamma_n \leq T_n^{-1} R_n c_n \quad (n \geq 1).$$

Using the above choice and Theorem 9.3.17, it finally also follows that estimates (9.3.38) and (9.3.39) are satisfied for all $n \geq 0$.

9.4 Exercises

9.4.1. Let L, M, M_1 be operators such that $L \in C^1 (V_1 \to V)$, $M_1 \in L_+ (V)$, $M \in C (V_1 \to V)$, and x_n be points in D. It is convenient for us to define the sequences c_n, d_n, a_n, b_n ($n \geq 0$) by

$$d_{n+1} = (L + M + M_1) (d_n) + L' (|x_n|) c_n, \quad d_0 = 0,$$
$$c_n = |x_{n+1} - x_n|,$$
$$a_n = (L + M + M_1)^n (a) \quad \text{for some } a \in C,$$
$$b_n = (L + M + M_1)^n (0),$$

and the point b by

$$b = (L + M + M_1)^\infty (0).$$

Prove the result:

Let X be a Banach space with convergence structure (X, V, E) with $V = (V, C, \|\cdot\|_v)$, an operator $F \in C^1 (D \to X)$ with $D \subseteq X$, an operator $Q \in C (D \to X)$, an operator $L \in C^1 (V_1 \to V)$ with $V_1 \subseteq V$, an operator $M \in C (V_1 \to V)$, an operator $M_1 \in L_+ (V)$, a point $a \in C$, and a null sequence $\{z_n\} \in D$ such that the following conditions are satisfied:

(C6) the inclusions $U (a) \subseteq D$ and $[0, a] \subseteq V1$ are true;

(C7) L is order-convex on $[0, a]$, and satisfies

$$L' (|x| + |y|) - L' (|x|) \in B \left(F' (x) - F' (x + y) \right)$$

for all $x, y \in U (a)$ with $|x| + |y| \leq a$;

(C8) M satisfies the conditions

$$0 \leq (Q (x) - Q (x + y)), \; M (|x| + |y|) - M (|x|) \in E$$

for all $x, y \in U (a)$ with $|x| + |y| \leq a$, and

$$M (w_1) - M (w_2) \leq M (w_3) - M (w_4) \quad \text{and } M (w) \geq 0$$

for all $w, w_1, w_2, w_3, w_4 \in [0, a]$ with $w_1 \leq w_3$, $w_2 \leq w_4$, $w_2 \leq w_1$, $w_4 \leq w_3$;

(C9) M_1, x_n, z_n satisfy the inequality

$$0 \leq \left(F' (x_n) (z_n) - F' (x_{n-1}) (z_{n-1}), \; M_1 (d_n - d_{n-1}) \right) \in E$$

for all $n \geq 1$;

(C10) $L' (0) \in B \left(I - F' (0) \right)$, and
$$\left(- \left(F (0) + Q (0) + f' (0) (z_0) \right), \; L (0) + M (0) + M_1 (0) \right) \in E;$$

(C11) $(L + M + M_1) (a) \leq a$ with $0 \leq L + M + M_1$; and

(C12) $\left(M + M_1 + L' (a) \right)^n a \to 0$ as $n \to \infty$.

Then,

(i) The sequence $(x_n, d_n) \in (X \times V)^N$ is well defined, remains in E^N, is monotone, and satisfies for all $n \geq 0$

$$d_n \leq b$$

where b is the smallest fixed point of $L + M + M_1$ in $[0, a]$.

(ii) Moreover, the iteration $\{x_n\}$ $(n \geq 0)$ generated by

$$x_{n+1} = x_n + F'(x_n)^*[-(F(x_n) + Q(x_n))] - z_n, \ z_0 = 0$$

converges to a solution $x^* \in U(b)$ of the equation $F(x) + Q(x) = 0$, which is unique in $U(a)$.

(iii) Furthermore, the following estimates are true for all $n \geq 0$:

$$b_n \leq d_n \leq b,$$
$$b_n \leq a_n,$$
$$|x_{n+1} - x_n| \leq d_{n+1} - d_n,$$
$$\left| x_n - x^* \right| \leq b - d_n,$$

and

$$\left| x_n - x^* \right| \leq a_n - b_n, \quad \text{for } M_1 = 0, \text{ and } z_n = 0 \ (n \geq 0).$$

9.4.2. We will now introduce results on a posteriori estimates for the iteration introduced in Exercise 9.4.1. It is convenient to define the operator

$$R_n(q) = \left(I - L'(|x_n|) \right)^* S_n(q) + c_n$$

where,

$$S_n(q) = (L + M + M_1)(|x_n| + q) - (L + M + M_1)(|x_n|) - L'(|x_n|)(q),$$

and the interval

$$I_n = [0, a - |x_n|].$$

Show:

(a) operators S_n are monotone on I_n;

(b) operators $R_n: [0, a - d_n] \to [0, a - d_n]$ are well defined, and monotone.
    Hint: Verify the scheme:

$$d_n + c_n \leq d_{n+1} \implies R(a) - d_{n+1} \leq a - d_n - c_n$$
$$\implies S_n(a - d_n) + L'(|x_n|)(a - d_n - c_n) \leq a - d_n - c_n \ (n \geq 0);$$

(c) if $q \in I_n$ satisfy $R_n(q) \leq q$, then

$$c_n \leq R_n(q) = p \leq q,$$

and

$$R_{n+1}(p - c_n) \leq p - c_n \quad \text{for all } n \geq 0;$$

(d) under the hypotheses of Exercise 9.4.1, let $q_n \in I_n$ be a solution of $R_n(q) \leq q$, then

$$\left|x^* - x_m\right| \leq a_m \quad (m \geq n),$$

where

$$a_n = q_n \text{ and } a_{m+1} = R_m(a_m) - c_m;$$

and

(e) under the hypotheses of Exercise 9.4.1, any solution $q \in I_n$ of $R_n(q) \leq q$ is such that

$$\left|x^* - x_n\right| s < R_n^\infty(0) \leq q.$$

9.4.3. Let $A \in L(X \to X)$ be a given operator. Define the operators $P, T(D \to X)$ by

$$P(x) = AT(x + u),$$
$$T(x) = G(x) + R(x), \quad P(x) = F(x) + Q(x),$$

and

$$F(x) = AG(x + u), \quad Q(x) = AR(x + u),$$

where $A \in L(X \to X)$ G, R are as F, Q, respectively. We deduce immediately that under the hypotheses of Exercise 9.4.1, the zero x^* of P is also a zero of AT, if $u = 0$.

We will now provide a monotonicity result to find a zero x^* of AT. The space X is assumed to be partially ordered and satisfies the conditions for V given in (C_1)–(C_5). Moreover, we set $X = V, D = C^2$ so that $|\cdot|$ turns out to be I.

Prove the result:

Let V be a partially ordered Banach space satisfying conditions (C_1)–(C_5), Y be a Banach space, $G \in C^1(D \to Y)$, $R \in C(D \to Y)$ with $D \subseteq V$, $A \in L(X \to V)$, $M \in C(D \to V)$, $M_1 \in L_+(V)$ and $u, v \in V$ such that:

(C_{13}) $[u, v] \subseteq D$;

(C_{14}) $I - AG'(u) + M + M_1 \in L_+(V)$;

(C_{15}) for all $w_1, w_2 \in [u, v]: w_1 \leq w_2 \implies AG'(w_1) \geq AG'(w_2)$;

(C_{16}) $AT(u) + AG'(u)(z_0) \leq 0$, $AT(v) + AG'(v)(z_0) \geq 0$ and $AT(v) - M_1(v - u) \geq 0$;

(C_{17}) condition (C_8) is satisfied, and $M(v - u) \leq -Q(v - u)$;

(C_{18}) condition (C_9) is satisfied;

(C_{19}) the following initial condition is satisfied

$$-\left(Q(0) + AG'(u)(z_0)\right) \leq (M + M_1)(0);$$

and

(C_{20}) $\left(I - AG'(v) + m + M_1\right)^n(v - u) \to 0$ as $n \to \infty$.

Then the NK sequence

$$y_0 = u, \quad y_{n+1} = y_n + \left(AG'(y_n)\right)^*[-AT(y_n)] - z_n \quad (n \geq 0)$$

is well defined for all $n \geq 0$, monotone, and converges to a unique zero x^* of AT in $[u, v]$.

9.4.4. Let X be a Banach space with convergence structure (X, V, E) with $V = (V, C, \|\cdot\|_v)$, an operator $F \in C^1 (X_F \to X)$ with $X_F \subseteq X$, an operator $L \in C^1 (V_L \to V)$ with $V_L \subseteq V$, and a point of C such that
(a) $U(a) \subseteq X_F$, $[0, a] \subseteq V_L$;
(b) L is order convex on $[0, a]$ and satisfies for $x, y \in U(a)$ with $|x| + |y| \leq a$;

$$L'(|x| + |y|) - L'(|x|) \in B (F'(x) - F'(x + y));$$

(c) $L'(0) \in B (I - F'(0))$, $(-F(0), L(0)) \in E$;
(d) $L(a) \leq a$;
(e) $L'(a)^n \to 0$ as $n \to \infty$.
Then show NK sequence $x_0 := 0$, $x_{n+1} = x_n + F'(x_n)^* (-F(x_n))$ is well defined, and converges to the unique zero z of F in $U(a)$.

9.4.5. Under the hypotheses of Exercise 9.4.4 consider the case of a Banach space with a real norm $\|\cdot\|$. Let $F'(0) = I$ and define a monotone operator

$$k: [0, a] \to \mathbb{R} \; \forall x \in U(a): \|F''(x)\| \leq k(\|x\|)$$

and

$$L(t) = \|F(0)\| + \int_0^t ds \int_0^s d\theta k(\theta).$$

Show (d) above is equivalent to $\|F(0)\| + .5k(a) a^2 \leq a$. Under what conditions is this inequality true.
If conditions (a)–(c) of Exercise 9.4.4 are satisfied, and

$$\exists t \in (0, 1): L(a) \leq ta,$$

then show there exists $a' \in [0, ta]$ satisfying (a)–(e). The zero $z \in U(a')$ is unique in $U(a)$.

9.4.6. Let $L \in L_+ (V)$ and $a, e \in C$ be given such that: Let $a \leq e$ and $L^n e \to 0$ as $n \to \infty$. Then show: operator

$$(I - L)^* : [0, a] \to [0, a],$$

is well defined and continuous.

9.4.7. Let $A \in L(X)$, $L \in B(A)$, $y \in D$, and $e \in C$ such that

$$Le + |y| \leq e \text{ and } L^n e \to 0 \text{ as } n \to \infty.$$

Then show $x := (I - A)^* y$ is well defined, $x \in D$, and $|x| \leq (I - L)^* |y| \leq e$.

9.4.8. Let V be a partially ordered Banach space, Y a Banach space, $G \in C^1 (V_G \to Y)$, $A \in L(X \to Y)$ and $u, v \in V$ such that
(a) $[u, v] \subseteq V_G$;
(b) $I - AG'(u) \in L_+ (V)$;
(c) $\forall w_1, w_2 \in [u, v] : w_1 \leq w_2 \Longrightarrow AG'(w_1) \geq AG'(w_2)$;
(d) $AG(u) \leq 0$ and $AG'(v) \geq 0$;
(e) $(I - AG'(v))^n (v - u) \to 0$ as $n \to \infty$.

Then show: the NK sequence

$$u_0 = u, u_{n+1} := u_+ \left[AG'(u_n)\right]^* \left[-AG(u_n)\right] \quad (n \geq 0)$$

is well defined, monotone, and converges to the unique zero z of AG in $[u, v]$.

9.4.9. Consider the two boundary value problem

$$-x''(s) = 4\sin(x(s)) + f(s), \quad x(0) = x(1) = 0$$

as a possible application of Exercises 9.4.1–9.4.8 in the space $X = C[0, 1]$ and $V = X$ with natural partial ordering; the operator $/\cdot/$ is defined by taking absolute values. Let $G \in L_+(C[0, 1])$ be given by

$$Gx(s) = \frac{1}{2\sin(2)} \left\{ \int_0^s \sin(2t)\sin(2 - 2s)x(t)\,dt \right.$$
$$\left. + \int_0^s \sin(2 - 2t)\sin(2s)x(t)\,dt \right\}$$

satisfying $Gx = y \iff -y'' - 4y = x, y(0) = y(1) = 0$. Define the operator

$$F: C[0, 1] \to C[0, 1], \quad F(x) := x - G(4\sin(x) - 4x + f).$$

Let

$$L: C[0, 1] \to C[0, 1], \quad L(e) = 4G(e - \sin(e)) + |Gf|.$$

For $x, y, w \in C[0, 1]$

$$|x|, |x| + |y| \leq .5\pi \Rightarrow$$
$$\left| \left[F'(x) - F'(x + y)\right] w \right| \leq \left[L'(|x| + |y|) - L'L'(|x|)\right] |w|.$$

Further we have $L(0) = |Gf|$ and $L'(0) = 0$. We have to determine $a \in C_+[0, 1]$ with $|a| \leq .5\pi$ and $s \in (0, 1)$ such that $L(a) = 4G(a - \sin(a)) + |Gf| \leq sa$. We seek a constant function as a solution. For $e_0(s) = 1$, we compute

$$p: \|Ge_0\|_\infty = .25 \left(\frac{1}{\cos(e)} - 1 \right).$$

Show that $a = te_0$ will be a suitable solution if

$$4p(t - \sin(t)) + \|Gf\|_\infty < t.$$

9.4.10. (a) Assume: given a Banach space X with a convergence structure (X, V, E) with $V = (V, C, \|\cdot\|_V)$, an operator $F \in C^1(X_0 \subseteq X \to X)$, operators M, L_0, $L \in L^1(V_0 \subseteq V \to V)$, and a point $p \in C$ such that the following conditions hold:

$$U(p) \subseteq X_0, \quad [0, p] \subseteq V_0;$$

M, L_0, L are order convex on $[0, p]$, and such that for $x, y, z \in U(p)$ with $|x| \leq p, |y| + |x| \leq p$

$$L_0' \left(|x|\right) - L_0' \left(0\right) \in B \left(F' \left(0\right) - F' \left(x\right)\right),$$
$$L' \left(|y| + |x|\right) - L' \left(|y|\right) \in B \left(F' \left(y\right) - F' \left(y + x\right)\right),$$
$$M \left(p\right) \le p$$
$$L_0(p_0) \le M(p_0) \quad \text{for all} \quad p_0 \in [0, p],$$
$$L_0'(p_0) \le M'(p_0) \quad \text{for all} \quad p_0 \in [0, p],$$
$$L_0' \in B(I - F' \left(0\right)), \quad (-F(0), L_0(0)) \in E,$$
$$M' \left(p\right)^n \left(p\right) \to 0 \quad \text{as} \quad n \to \infty,$$
$$M' \left(d_n\right) \left(b - d_n\right) + L \left(d_n\right) \le M \left(b\right) \quad \text{for all} \quad n \ge 0,$$

where

$$d_0 = 0, \, d_{n+1} = L \left(d_n\right) + L_0' \left(|x_n|\right) \left(c_n\right), \quad c_n = |x_{n+1} - x_n| \quad (n \ge 0)$$

and

$$b = M^\infty \left(0\right).$$

Show: sequence $\{x_n\}$ $(n \ge 0)$ generated by NK method is well defined, remains in E^N, is monotone, and converges to a unique zero x^* in $U(b)$, where b is the smallest fixed point of M in $[0, p]$. Moreover, the following bounds hold for all $n \ge 0$

$$d_n \le b,$$
$$\left|x^* - x_n\right| \le b - d_n,$$
$$\left|x^* - x_n\right| \le M^n \left(p\right) - M^n \left(0\right),$$
$$c_n + d_n \le d_{n+1},$$

and

$$c_n = |x_{n+1} - x_n|.$$

(b) If $r \in [0, p - |x_n|]$ satisfies $R_n(r) \le r$ then show: the following holds for all $n \ge 0$:

$$c_n \le R_n(r) = p \le r, \tag{9.4.1}$$

and

$$R_{\{n+1\}}(p - c_n) \le p - c_n$$

(c) Assume hypotheses of (a) hold and let $r_n \in [0, p - |x_n|]$ be a solution (9.4.1). Then show the following a posteriori estimates hold for all $n \ge 0$

$$\left|x^* - x_m\right| < q_m,$$

where

$$q_n = r_n, \quad q_{m+1} = R_m(q_m) - c_m \quad (m \ge n).$$

(d) Under hypotheses of (a), show any solution $r \in [0, p - |x_n|]$ of $R_n(r) \le r$ yields the a posteriori estimate

$$|x^* - x_n| \le R_n^\infty (0) \le r \quad (n \ge 0).$$

Let X be partially ordered and set

$$X = V, \quad D = C^2 \text{ and } |\cdot| = I.$$

(e) Let X be partially ordered, Y a Banach space, $G \in C^1 (X_0 \subseteq X \to Y)$, $A \in L (Y \to X)$ and $x, y \in X$ such that:

1. $[x, y] \in X_0$.
2. $I - AG' (x) \in L_+ (X)$,
3. $z_1 \le z_2 \implies AG' (z_1) \le AG' (z_2)$ for all $z_1, z_2 \in [x, y]$,
4. $AG (x) \le 0, \quad AG (y) \ge 0$,
5. $(I - AG' (y))^n (y - x) \to 0$ as $n \to \infty$.

Show: sequence $\{y_n\}$ $(n \ge 0)$ generated by

$$y_0 = x, \quad y_{n+1} = y_n + AG' (y_n)^* [-AG (y_n)]$$

is well defined for all $n \ge 0$, monotone, and converges to a unique zero y^* of AG in $[x, y]$.

9.4.11. Let there be given a Banach space X with convergence structure (X, V, E) where $V = (V, C, \|\cdot\|_V)$, and operator $F \in C'' (X_0 \to X)$ with $X_0 \subseteq X$, and operator $M \in C' (V_0 \to V)$ with $V_0 \subseteq V$, and a point $p \in C$ satisfying:

$$U (p) \subseteq X_0, \quad [0, p] \subseteq V_0;$$

M is order-convex on $[0, p]$ and for all $x, y \in U (p)$ with $|x| + |y| \le p$

$$M' (|x + y|) - M' (x) \in B \left(\left[F'' (x) - F'' (x + y) \right] (y) \right)$$
$$M' (0) \in B \left(I - F'(0) \right), \quad (-F(0), M (0)) \in E;$$
$$M (p) \le p;$$

and

$$M' (p)^n p \to 0 \text{ as } n \to \infty.$$

Then, show sequence $(x_n, t_n) \in (X \times V)^{\mathbb{N}}$, where $\{x_n\}$ is generated by Newton's method and $\{t_n\}$ $(n \ge 0)$ is given by

$$t_0 = 0, \quad t_{n+1} = M (t_n) + M' (|x_n|) (a_n), \quad a_n = |x_{n+1} - x_n|$$

is well defined for all $n \ge 0$, belongs in $E^{\mathbb{N}}$, and is monotone. Moreover, the following hold for all $n \ge 0$

$$t_n \ge b,$$

where,

$$b = M^\infty (0),$$

is the smallest fixed point of M in $[0, p]$.

Show corresponding results as in Exercises 9.4.10 (b)–(c).

9.4.12. Assume hypotheses of Exercise 9.4.10 hold for $M = L$.

Show: (a) conclusions (a)–(e) hold (under the revised hypothesis) and the last hypothesis in (a) can be dropped;

(b) error bounds $|x^* - x_n|$ $(n \geq 0)$ obtained in this setting are finer and the information on the location of the solution more precise than the corresponding ones in Exercise 9.4.4 provided that $L_0(p_0) < L(p_0)$ or $L_0'(p_0) < L'(p_0)$ for all $p \in [0, p]$.

As in Exercise 9.4.5, assume there exists a monotone operator $k_0: [0, p] \to R$ such that

$$\|F'(x) - F'(0)\| \leq k_0(\|x\|) \|x\|, \quad \text{for all } x \in U(p),$$

and define operator L_0 by

$$L_0(t) = \|F(0)\| + \int_0^t ds \int_0^s d\theta k_0(\theta).$$

Sequence $\{d_n\}$ given by $d_0 = 0$, $d_{n+1} = L(d_n) + L_0'(|x_n|) c_n$ converges to some $p^* \in [0, p]$ provided that

$$\left(k_0(p) + \frac{k(p)}{2}\right) \|F(0)\| \leq 1.$$

Conclude that the above semilocal convergence condition is weaker than (d) in Exercise 9.4.4 or equivalently (for $p = a$)

$$2k(p) \|F(0)\| \leq 1.$$

Finally, conclude that in the setting of Exercise 9.4.12 and under the same computational cost, we always obtain and under weaker conditions: finer error bounds on the distances $|x_n - x^*|$ $(n \geq 0)$ and a better information on the location of the solution x^* than in Exercise 9.4.5 (i.e., [140], see also [43], [141]).

10

Point-to-Set-Mappings

This chapter gives an outline of general iterative procedures and their convergence under general hypotheses.

10.1 Algorithmic models

Let X denote an abstract set, and introduce the following notation:

$$X^1 = X, \quad X^2 = X \times X, ..., X^m = X^{m-1} \times X, \quad k \geq 2.$$

Assume that k is a positive integer, and for all $m \geq k - 1$, the point-to-set mapping F_m are defined on X^{m+1}, furthermore for all $\left(x^{(1)}, ..., x^{(m+1)}\right) \in X^{m+1}$ and $x \in F_m\left(x^{(1)}, ..., x^{(m+1)}\right)$, $x \in X$. For the sake of brevity, we will use the notation $F_m : X^{m+1} \to 2^X$, where 2^X denotes the set of all subsets of X.

Definition 10.1.1. *Select* $x_0, x_1, ..., x_{k-1} \in X$ *arbitrarily, and construct the sequence*

$$x_{m+1} \in F_m(x_0, x_1, ..., x_m) \quad (m \geq k - 1), \tag{10.1.1}$$

where arbitrary point from the set $F_m(x_0, x_1, ..., x_m)$ *can be accepted as the successor of* x_m. *Recursion (10.1.1) is called the general algorithmic model.*

Remark 10.1.2. Because the domain of F_m is X^{m+1} and $F_m(x_0, x_1, ..., x_m) \subseteq X$, the recursion is well defined for all $m \geq k - 1$. Points $x_0, ..., x_{k-1}$ are called initial approximations, and the maps F_m are called iteration mappings.

Definition 10.1.3. *The algorithmic model (10.1.1) is called a k-step process if for all* $m \geq k - 1$, F_m *does not depend explicitly on* $x_0, x_1, ..., x_{m-1}$, *that is, if algorithm (10.1.1) has the special form*

$$x_{m+1} \in F_m(x_{m-k+1}, ..., x_{m-1}, x_m). \tag{10.1.2}$$

I.K. Argyros, *Convergence and Applications of Newton-type Iterations*,
DOI: 10.1007/978-0-387-72743-1_10, © Springer Science+Business Media, LLC 2008

It is easy to show that any k-step process is equivalent to a certain single-step process defined on X^k. For $m \geq 0$, introduce vectors

$$y_m = \left(y_m^{(1)}, y_m^{(2)}, ..., y_m^{(k)} \right).$$

Starting from the initial approximation

$$y_0 = (x_0, x_1, ..., x_{k-1}),$$

consider the single-step algorithmic model:

$$
\begin{aligned}
y_{m+1}^{(1)} &= y_m^{(2)} \\
y_{m+1}^{(2)} &= y_m^{(3)} \\
&\vdots \\
y_{m+1}^{(k-1)} &= y_m^{(k)} \\
y_{m+1}^{(k)} &= F_m \left(y_m^{(k)}, ..., y_m^{(k)} \right).
\end{aligned}
\tag{10.1.3}
$$

This iteration algorithm is a single-step process, and obviously it is equivalent to the algorithmic model (10.1.2), as for all $m \geq 0$,

$$y_m^{(1)} = x_m, \, y_m^{(2)} = x_{m+1}, ..., y_m^{(k)} = x_{m+k-1}.$$

This equivalence is the main reason why only single-step iteration methods are discussed in most publications.

Definition 10.1.4. *A k-step process is called stationary, if mappings Fm do not depend on m. Otherwise the process is called nonstationary.*

Iteration models in the most general form (10.1.1) have a great importance in certain optimization methods. For example, in using cutting plane algorithms, very early cuts can still remain in the latter stages of the process by assuming that they are not dominated by later cuts. Hence the optimization problem of each step may depend on the solutions of very early problems. Multistep processes are also used in many other fields of applied mathematics. As an example, we mention that the secant method for solving nonlinear equations is a special two-step method. Nonstationary methods have a great practical importance in analyzing the global asymptotical stability of dynamic economic systems, when the state transition relation is time-dependent.

In this chapter, the most general algorithmic model (10.1.1) will be first considered, and then, special cases will be derived from our general convergence theorem. In order to establish any kind of convergence, X should have some topology.

Assume now that X is a Hausdorff topological space that satisfies the first axiom of countability [189]. Let $S \subset X$ be the set of desirable points, which are considered as the solutions to the problems being solved by the algorithm. For example, in the case of an optimization problem, X can be selected as the feasible set and P as the set of the optimal solutions. If a linear or nonlinear fixed point problem is solved, then X is the domain of the mapping and P is the set of all fixed points. In analyzing the global asymptotic stability of a discrete dynamic system, set X is the state space and P is the set of equilibrium points.

Definition 10.1.5. *An algorithmic model is said to be convergent, if the accumulation points of any iteration sequence $\{x_n\}$ constructed by the algorithm are in P.*

Note that the convergence of an algorithm model does not imply that the iteration sequence is convergent.

We now impose convergence criteria for algorithmic models.

Assume that for $m \geq 0$ there exist functions $g_k: X \to \mathbf{R}^1$ with the following properties:

(A_1) For large m, functions $\{g_k\}$ are uniformly locally bounded below on $X \setminus P$. That is, there is a nonnegative integer N_1 such that for all $x \in X \setminus P$ there is a neighborhood U of x and a $b \in \mathbf{R}^1$ (which may depend on x) such that for all $m \geq N_1$ and $x' \in U$,

$$g_k\left(x'\right) \geq b; \tag{10.1.4}$$

(A_2) If $m \geq N_1$, $x' \in F_m\left(y^{(1)}, ..., y^{(m)}, x\right) (x, y^{(i)} \in X \; i = 1, ..., m)$, then

$$g_{k+1}\left(x'\right) \leq g_k\left(x\right); \tag{10.1.5}$$

(A_3) For each $y \in X \setminus P$ if $\{y_i\} \subset \setminus \subseteq X$ is any sequence such that $y_i \to y$ and $\{m_i\}$ is any strictly increasing sequence of nonnegative integers such that $g_k\left(h_i\right) \to g^*$, then for all iteration sequences $\{x_i\}$ such that $x_{m_i} = y_i$ $(i \geq 0)$ there exists an integer N_2 such that $m_{N_2} \geq N_1 - 1$ and

$$g_{k_{N_2}+1}\left(y\right) < g^* \text{ for all } y \in F_{m_{N_2}}\left(x_0, x_1, ..., x_{m_{N_2}}\right). \tag{10.1.6}$$

Theorem 10.1.6. *If conditions (A_1), (A_2) and (A_3) hold, then the algorithmic model (10.1.1) is convergent.*

Proof. Let x^* be an accumulation point of the iteration sequence $\{x_m\}$ constructed by the algorithmic model (10.1.1), and assume that $x^* \in X \setminus P$. Let $\{m_i\}$ denote the index set such that $\{x_{m_i}\}$ is a subsequence of $\{x_m\}$ converging to x^*. Assumption (A_1) implies that for large m, $\{g_k(x_m)\}$ is decreasing, and from assumption (A_2), we conclude $\{g_{k_i}\left(x_{m_i}\right)\}$ is convergent. Therefore the entire sequence $\{g_k(x_m)\}$ converges to a $g^* \in \mathbf{R}^1$. From (10.1.5) we know that for $m \geq N_1$,

$$g_k\left(x_m\right) \geq g^*. \tag{10.1.7}$$

Use subsequence $\{x_{m_i}\}$ as sequence $\{y_i\}$ in condition (A_3) to see that there exists an N_2 such that $m N_2 \geq N_1 - 1$ and with the notation $M = m N_2 + 1$,

$$g_M\left(x_M\right) < g^*,$$

which contradicts relation (10.1.7) and completes the proof.

Remark 10.1.7. Note first that in the special case when (10.1.1) is single-step non-stationary process and g_k does not depend on m, this theorem generalized Theorem 4.3 of Tishyadhigama et al. [189]. If the process is stationary, then this result further specializes to Theorem 3.5 of the same paper.

Remark 10.1.8. The conditions of the theorem do not imply that sequence $\{x_m\}$ has an accumulation point, as the next example shows.

Example 10.1.9. Select $X = \mathbf{R}^1$, $P = \{0\}$, and consider the single-step process with $F_m(x) \equiv F(x) = x - 1$, and choose $g_k(x) = x$ for all $x \in X$.

Because functions g_k are continuous and $F_m(x) < x$ for all x, condition (A_1) obviously holds, and because functions F_m are strictly decreasing and continuous, assumptions (A_2) and (A_3) also hold. However, for arbitrary $x_0 \in X$, the iteration sequence is strictly decreasing and divergent. (Infinite limit is not considered here as limit point from X.)

Remark 10.1.10. Even in cases when the iteration sequence has an accumulation point, the sequence does not need to converge as the following example shows.

Example 10.1.11. Select $X = \mathbf{R}^1$, $P = \{0, 1\}$, and consider the single-step iteration algorithm with function

$$F_m(x) \equiv F(x) = \begin{cases} 1 & \text{if } x = 0 \\ 0 & \text{if } x = 1 \\ x - 1 & \text{if } x \notin P. \end{cases}$$

Choose

$$g_k(x) \equiv g(x) = \begin{cases} 0 & \text{if } x \in P \\ x & \text{otherwise.} \end{cases}$$

On $X \setminus P$, function g is continuous, hence assumption (A_1) is satisfied. If $x \notin P$, then $F(x) < x$, which implies that $g(F(x)) < g(x)$. If $x \in P$, then $F(x) \in P$. Therefore in this case $g(F(x)) = g(x)$. Hence condition (A_2) also holds. Assumption (A_3) follows from the definition of functions g_k and from the fact that $F(x) < x$ on $X \setminus P$. If x_0 is selected as a nonnegative integer, then the iteration sequence has two accumulation points: 0 and 1. If x_0 is selected otherwise, then no accumulation point exists.

Note that Definition 10.1.5 is considered as the definition of global convergence on X, because the initial approximations $x_0, x_1, \ldots, x_{k-1}$ are arbitrary elements of X. Local convergence of algorithmic models can be defined in the following way:

Definition 10.1.12. *An algorithmic model is said to be locally convergent, if there is a subset X_1 of X such that the accumulation points of any iteration sequence $\{x_m\}$ constructed by the algorithm starting with initial approximations $x_0, x_1, \ldots, x_{k-1}$ from X_1 are in P.*

Theorem 10.1.6 can be modified as a local convergence theorem by substituting X and P by X_1 and $X_1 \cap P$, respectively.

10.2 A general convergence theorem

Consider again the algorithmic model

$$x_{m+1} \in F_m(x_0, x_1, ..., x_m) \quad (m \geq k - 1), \tag{10.2.1}$$

where for $m \geq k - 1$, $F_m: X^{m+1} \to 2^X$. Here we assume again that X is a Hausdorf topological space that satisfies the first axiom of countability, and k is a given positive integer, furthermore in relation (10.2.1), any point from the set $F_m(x_0, x_1, ..., x_m)$ can be accepted as the successor of x_m. Assume furthermore that the set P of desirable points has only one element p^*.

Assume that:

(B$_1$) There is a compact set $C \subseteq X$ such that for all m, $x_m \in C$;
(B$_2$) conditions (A$_1$), (A$_2$), and (A$_3$) of Theorem 10.1.6 are satisfied.

The main result of this section is given as

Theorem 10.2.1. *Under assumptions (B$_2$) and (B$_2$), $x_m \to p^*$ as $m \to \infty$ with arbitrary points $x_0, x_1, ..., x_{k-1} \in X$.*

Proof. Because C is compact, sequence $\{x_m\}$ has a convergent subsequence. From Theorem 10.1.6 we also know that all the limit points of this iteration sequence belong to P, which has only one point p^*. Hence the iteration sequence has only one limit point p^*, which implies that it converges to p^*.

Remark 10.2.2. The theorem in this formulation can be interpreted as a global convergence result. However, if the conditions of the theorem hold only in a neighborhood X_1 of p^* such that $F_m(x^{(1)}, x^{(2)}, ..., x^{(m+1)}) \subseteq X_1$ for all $m \geq k - 1$ and $x^{(i)} \in X_1$ ($i = 1, 2, ..., m + 1$), then local convergence results are obtained.

The speed of convergence of algorithm (10.2.1) can be estimated as follows. Assume that:

(B$_3$) X is a metric space with distance $d: X \times X \to \mathbf{R}^1$;
(B$_4$) There exist nonnegative constants a_{mi} ($m \geq k - 1, 0 \leq i \leq m$) such that if $m \geq k - 1$ and $x \in F_m(x^{(0)}, x^{(1)}, ..., x^{(m)})$, then

$$d(x, p^*) \leq \sum_{i=0}^{m} a_{mi} d(x^{(i)}, p^*).$$

From (10.2.1) we have

$$\varepsilon_{m+1} \leq \sum_{i=0}^{m} a_{mi} \varepsilon_i,$$

where $\varepsilon_i = d(x_i, p^*)$ for all $i \geq 0$.

Starting from initial value $\delta_i = \varepsilon_i$ ($i = 0, 1, ..., k - 1$), consider the nonstationary difference equation

$$\delta_{m+1} = \sum_{i=0}^{m} a_{mi} \delta_i. \tag{10.2.2}$$

Obviously, for all $m \geq 0$, $\varepsilon_m \leq \delta_m$. In order to obtain a direct expression for δ_m, and therefore the same for the error bound of x_m ($m \geq k - 1$), introduce the following additional notation:

$$d_m = (\delta_0, \delta_1, ..., \delta_m)^T,$$

$$A_m = \begin{bmatrix} 1 & & & \\ & 1 & & \\ & & \ddots & \\ & & & 1 \\ a_{m0} & \cdots & & a_{mm} \end{bmatrix}, \quad \text{and } a_m^T = (a_{m0}, a_{m1}, ..., a_{mm}).$$

Then from (10.2.2)

$$d_{m+1} = A_m d_m,$$

and hence, finite induction shows that for all $m \geq 1$,

$$d_m = A_{m-1} A_{m-2} ... A_{k-1} d_{k-1}.$$

Note that the components of d_{k-1} are the error of the initial approximations $x_0, x_1, ..., x_{k-1}$. From (10.2.2) we have

$$\delta_{m+1} = a_m^T d_m = \left(a_m^T A_{m-1} A_{m-2} ... A_{k-1} \right) d_{k-1} = b_m^T d_{k-1}$$

with

$$b_m^T = a_m^T A_{m-1} A_{m-2} ... A_{k-1}$$

being a one-dimensional row vector. Introducing finally the notation

$$b_m^T = \left(b_{m0}, b_{m1}, ..., b_{m,k-1} \right),$$

the definition of the numbers δ_m and relation (10.2.2) imply the following result:

Theorem 10.2.3. *Under assumptions (B$_3$)-(B$_4$),*

$$d \left(x_{m+1}, p^* \right) \leq \sum_{i=0}^{k-1} b_{mi} d \left(x_i, p^* \right), \quad (m \geq k - 1). \tag{10.2.3}$$

Corollary 10.2.4. *If for all $i = 0, 1, ..., k - 1$, $b_{mi} \to 0$ as $m \to \infty$, then the iteration sequence $\{x_m\}$ generated by algorithm (10.2.1) converges to p^*. Hence, in this case conditions (B$_1$) and (B$_2$) are not needed to establish convergence.*

The conditions of Theorem 10.1.6 are usually difficult to be verified in practical cases. Therefore in the next section, we will relax these conditions in order to derive sufficient convergence conditions that can be easily verified.

10.3 Convergence of k-step methods

In this section, k-step iteration processes of the form

$$x_{m+1} \in F_m\, (x_{m-k+1}, x_{m-k+2}, ..., x_m) \qquad (10.3.1)$$

are discussed, where $k \geq 1$ is a given integer, and for all m, $F_m\colon X^1 \to 2^X$. Assume again that the set P of desirable points has only one element p^*.

Definition 10.3.1. *A function* $V\colon X^k \to \mathbf{R}_+^k$ *is called the Liapunov function of process (10.3.1), if for arbitrary*

$$x^{(i)} \in X \left(i = 1, 2, ..., k, x^{(k)} \neq p^* \right)$$

and

$$y \in F_m \left(x^{(1)}, x^{(2)}, ..., x^{(k)} \right)\ (m \geq k - 1),$$

$$V \left(x^{(2)}, ..., x^{(k)}, y \right) < V \left(x^{(1)}, x^{(2)}, ..., x^{(k)} \right). \qquad (10.3.2)$$

Definition 10.3.2. *The Liapunov function* V *is called closed, if it is defined on* $\overline{X}^k$, *where* $\overline{X}$ *is the closure of* X, *furthermore if* $m_i \to \infty$, $x_i^{(j)} \to x^{(j)*}$ *as* $i \to \infty$ $(x_i^{(j)} \in X$ *for* $i \geq 0$ *and* $j = 1, 2, ..., k$ *such that* $x^{(k)*} \neq p^*)$ *and* $y_i \in F_{m_i} \left(x_i^{(1)}, ..., x_i^{(k)} \right)$ $(i \geq 0)$ *such that* $y_i \to y^*$ *as* $i \to \infty$, *then*

$$V \left(x^{(2)*}, ..., x^{(k)*}, y \right) < V \left(x^{(1)*}, ..., x^{(k)*} \right). \qquad (10.3.3)$$

Assume now that the following conditions hold:

(C_1) For all $m \geq k - 1$,

$$F_m \left(x^{(1)}, ..., x^{(k-1)}, p^* \right) = \{ p^* \}$$

with arbitrary $x^{(1)}, ..., x^{(k-1)} \in X$;

(C_2) Process (10.3.1) has a continuous, closed Liapunov function;
(C_3) X is a compact.

Theorem 10.3.3. *Under assumptions* (C_1), (C_2) *and* (C_3), $x_m \to p^*$ *as* $m \to \infty$.

Proof. Note first that this process is equivalent to the single-step method (10.1.3), where set X is replaced by $\hat{X} = X^k$, and the new set of desirable points is now $\hat{P} = P^k$. Select function g as the Liapunov function V.

We can now easily verify that the conditions of Theorem 10.2.1 are satisfied, which implies the convergence of the iteration sequence $\{x_m\}$.

Assumption (A_1) follows from (C_3) and the continuity of V. Condition (C_1) and the monotonicity of V imply assumption (A_2). And finally, assumption (A_3) is the consequence of condition (C_2) and relation (10.3.3).

Remark 10.3.4. Assumption (C₃) can be weakened as follows:

(C′₃) For all $x \in X \setminus P$, there is a compact neighborhood $U \subseteq X$ of x.

In this case, we have to assume that $p^* \in \overline{X}$, and condition (C₁) is required only if $p^* \in X$.

Remark 10.3.5. Assumption $p^* \in \overline{X}$ is needed in order to obtain p^* as the limit of sequences from X. Assumption (C₁) guarantees that if at any iteration step the solution p^* is obtained, then the process remains at the solution. We may also show that the existence of the Liapunov function is not a too strong assumption. Assume that X is a metric space, and consider the special iteration process $x_{m+1} = F(x_m)$ and assume that starting from arbitrary initial point, $\{x_m\}$ converges to the solution p^* of equation $x = F(x)$. Let $V: X \to \mathbf{R}^1$ be constructed as follows. With selecting $x_0 = x$, consider sequence $x_{m+1} = F(x_m)$, $(m \geq 0)$, and define

$$V(x) = \begin{cases} & \text{if } x = p^* \\ \max d(x_m, p^*), & m \geq 0 \end{cases}$$

where d is the distance. Obviously, $V(F(x)) \leq V(x)$ for all $x \in X$. The continuity-type assumptions in (C₂) are also natural, because without certain continuity assumptions no convergence can be established. Assumption (C₃) says that the entire sequence $\{x_m\}$ is contained in a compact set. This condition is necessarily satisfied for example, if X is in a finite-dimensional Euclidean space, and is bounded or if for every $K > 0$ there exists a $Q > 0$ such that $t^{(1)}, ..., t^{(k)} \in X$ and $\|t^{(j)}\| > Q$ (for at least one index j) imply that

$$V\left(t^{(1)}, ..., t^{(k)}\right) > K.$$

In the case of one-step processes (that is, if $k = 1$), this last condition can be reformulated as $V(x) \to \infty$ as $\|x\| \to \infty, x \in X$.

Assume next that the iteration process is stationary, that is, mappings F_m do not depend on m. Replace condition (C₂) by the following pair of conditions:

(C′₂) The process has a continuous Liapunov function;
(C″₂) Mapping F is closed on X, that is, if $x_i^{(j)} \to x^{(j)^*}$ as $i \to \infty$ $(j = 1, 2, ..., k)$ and $y_i \to F\left(x_i^{(k)}, ..., x_i^{(k)}\right)$ such that $y_i \to y^*$, then $y^* \in F\left(x^{(1)^*}, ..., x^{(k)^*}\right)$.

Theorem 10.3.6. *If process (10.2.1) is stationary and conditions* (C₁), (C′₂), (C″₂) *and* (C₃) *hold, then* $x_m \to p^*$ *as* $m \to \infty$.

Remark 10.3.7. This result in the special case of $k = 1$ can be considered as the discrete-time counterpart of the famous stability theorem of Uzwa [198].

Remark 10.3.8. Assume that for all $m \geq k - 1$, mapping F_m is closed, and the iteration sequence converges to p^*. Then for all $m \geq k - 1$, $p^* \in F_m(p^*, ..., p^*)$. Hence, p^* is a common fixed point of mappings F_m.

The speed of convergence of process (10.3.1) is next examined. Two results will be introduced. The first one is based on Theorem 10.2.1, and the second one is based on special properties of the Liapunov function.

Note first that in the case of a k-step process assumption (B$_4$) is modified as

(C$_4$) There exist nonnegative constants a_{mi} ($m \geq k - 1, m - k + 1 \leq i \leq m$), such that for all $m \geq k$ and $x \in F_m \left(x^{(k)}, ..., x^{(k)} \right)$, $\left(x^{(k)}, ..., x^{(k)} \right) \in X$ are arbitrary),

$$d\left(x, p^*\right) \leq \sum_{i=1}^{k} a_{m,m-k+i} d\left(x^{(i)}, p^*\right).$$

Then Theorem 10.2.3 remains valid with the specification that $a_{mi} = 0$ for all $i \leq m - k$.

In the case of a stationary process, constants $a_{m,m-k+1}$ do not depend on m. If we introduce the notation $\bar{a}_i = a_{m,m-k+1}$, then (10.2.2) reduces to

$$\delta_{m+1} = \sum_{i=1}^{k} \bar{a}_i \delta_{m+k-i}. \tag{10.3.4}$$

Observe that sequence $\{\delta_m\}$ is the solution of this kth order linear difference equation. Note first that the characteristic polynomial of this equation is as follows:

$$\varphi(\lambda) = \lambda^k - \bar{a}_1 \lambda^{k-1} - \bar{a}_2 \lambda^{k-2} - \cdots - \bar{a}_{k-1}\lambda - \bar{a}_k.$$

Assume that the roots of φ are $\lambda_1, \lambda_2, ..., \lambda_R$ with multiplicities $m_1, m_2, ..., m_R$, then the general solution of Equation (10.3.4) is given as

$$\delta_{m+1} = \sum_{r=1}^{R} \sum_{s=0}^{m_r-1} g_{rs} m^s \lambda_r^m,$$

where the coefficients g_{rp} are obtained by solving the initial-value equations

$$\sum_{r=1}^{R} \sum_{s=0}^{m_r-1} g_{rs} i^s \lambda_r^i = d\left(x_{i-1}, p^*\right) \quad (i = 1, 2, ..., k).$$

Hence, we proved the following:

Theorem 10.3.9. *Under assumption (C$_4$),*

$$d\left(x_{m+1}, p^*\right) \leq \sum_{r=1}^{R} \sum_{s=0}^{m_r-1} g_{rs} m^s \lambda_r^m \quad (m \geq k - 1). \tag{10.3.5}$$

Corollary 10.3.10. *If for all $r, r = 1, 2, ..., R, |\lambda_r| < 1$, then $x_m \to p^*$ as $m \to \infty$.*

Hence, in this case the conditions of Theorem 10.3.6 are not needed to establish convergence.

In the rest of the section the speed of convergence of process (10.3.1) is estimated based on some properties of the Liapunov function.

Assume now that

(C5) There exist constants a_i, b_i $(i = 1, 2, ..., k, a_k > 0)$ such that

$$\sum_{i=1}^{k} a_i, d\left(x^{(i)}, p^*\right) \leq V\left(x^{(k)}\right) \leq \sum_{i=1}^{k} b_i d\left(x^{(i)}, p^*\right)$$

for all $x^{(i)} \in X$ $(i = 1, 2, ..., k)$.

The following result holds.

Theorem 10.3.11. *Assume that process (10.3.1) has a Liapunov function V, which satisfies condition (C5). Then for $m \geq k - 1$,*

$$d\left(x_{m+1}, p^*\right) \leq a_k^{-1} \sum_{i=1}^{k} (b_i - a_{i-1}) d\left(x_{m-k+i}, p^*\right) (a_0 = 0). \qquad (10.3.6)$$

Proof. If $x_m = p^*$, then $x_{m+1} = p^*$, and therefore (10.3.6) obviously holds, as the left-hand side equals zero. If $x_m \neq p^*$, then condition (C5) implies that

$$\sum_{i=1}^{k} a_i d\left(x_{m-k+i+1}, p^*\right) \leq V\left(x_{m+2-k}, ..., x_{m+1}\right)$$

$$\leq V\left(x_{m+k-1}, ..., x_m\right) \leq \sum_{i=1}^{k} b_i d\left(x_{m-k+i}, p^*\right).$$

The assertion is a simple consequence of this inequality.

Corollary 10.3.12. *Introduce next the notation $\bar{a}_i = (b_i - a_{i-k})/a_k$ $(i = 1, 2, ..., k)$, and let sequence $\{\delta_m\}$ denote now the solution of difference equation (10.3.4) with initial conditions $\delta_i = d(x_{i-1}, p^*)$ $(i = 1, 2, ..., k)$. Then obviously, $d(x_m, p^*) \leq \delta_m$ for all $m \geq k - 1$, and with the above coefficients $\bar{a}_i$, Theorem 10.3.9 remains true.*

10.4 Convergence of single-step methods

In this section, single-step processes generated by point-to-step mappings are first examined. For the sake of simplicity we assume that X is a subset of a Banach space B and contains the origin. The iteration process now has the form

$$x_{m+1} \in F_m(x_m) \quad (m \geq 0), \qquad (10.4.1)$$

where $F_m: X \to 2^X$. It is also assumed that O is in X and $p^* = O$. We may have this last assumption without losing generality, as any solution p^* can be transformed into zero by introducing the transformed mappings

$$g_k(x) = \{y - p^* \,/\, y \in F_m(x + p^*)\}.$$

It is also assumed that for all m, $F_m(O) = \{O\}$.

We start our analysis with the following useful result.

Theorem 10.4.1. *Assume that X is compact, and there is a real valued continuous function $\alpha: X \setminus \{O\} \to [0, 1)$ such that*

$$\|y\| \le \alpha(x)\|x\| \tag{10.4.2}$$

for all $m \ge 0$, $x \ne 0$, and $y \in F_m(x)$.
Then the iteration sequence (10.4.1) converges to O as $m \to \infty$.

Proof. we now verify that all conditions of Theorem 10.3.3 are satisfied with the Liapunov function $V(x) = \|x\|$ and $p^* = O$. Note that (C_1) and (C_3) obviously hold, and condition (C_2) is implied by the facts that a and the norm are continuous, and $\alpha(x) < 1$ for $x \ne 0$.

Remark 10.4.2. If (10.4.2) is replaced by the weaker assumption that

$$\|y\| < \|x\|$$

for all $m \ge 0$, $x \ne 0$, and $y \in F_m(x)$, then the result may not hold, as the following example shows.

Example 10.4.3. Select $B = \mathbf{R}^1$, $X = [0, 2]$, and for $m \ge 0$,

$$F_m(x) = \left[(m+1)^2 - 1\right](m+1)^{-2} x.$$

If the initial points is chosen as $x_0 = 2$, then finite induction shows that

$$x_m = 1 + (k+1)^{-1} \to 1 \ne 0 \text{ as } m \to \infty.$$

Furthermore for all $m \ge 0$ and $x \ne 0$,

$$|F_m(x)| < |x|.$$

Corollary 10.4.4. *Recursion (10.4.1) and inequality (10.4.2) imply that for $m \ge 0$,*

$$\|x_{m+1}\| \le \alpha(x_m)\|x_m\|,$$

and therefore finite induction shows that

$$\|x_{m+1}\| \le \alpha(x_m)\,\alpha(x_{m-1}) \cdots \alpha(x_0)\|x_0\|. \tag{10.4.3}$$

As a special case assume that $\alpha(x) \leq q < a$ for all $o \neq x \in X$. Then for all $m \geq 0$,

$$\|x_{m+1}\| \leq q^{m+1} \|x_0\|, \tag{10.4.4}$$

which shows the linear convergence of the process in this special case.

Relation (10.4.4) serves as the error formula of the algorithm. In addition, it has the following consequence: Assume that (10.4.2) holds for all $O \neq x \in X$, furthermore $\alpha(x_m) \alpha(x_{m-1}) \cdots \alpha(x_0) \to O$ as $m \to \infty$. Then $x_m \to O$ for $m \to \infty$. Hence in this case we may drop the assumptions that $\alpha(x) \in [0, 1)$ ($O \neq x \in X$) and X is compact.

An alternative approach to Theorem 10.4.1 is based on the assumption that there exists a function $h \colon (0, \infty) \to \mathbf{R}$ such that

$$\|y\| \leq h(r) \|x\| \tag{10.4.5}$$

for all $m \geq 0, r > 0, \|x\| \leq r, x \in X$ and $y \in F_m(x)$.

In this case it is easy to verify that for all m,

$$\|x_m\| \leq q_m,$$

where q_m is the solution of the nonlinear difference equation

$$q_{m+1} = h(q_m) q_m, \quad q_0 = \|x_0\|.$$

Hence, the convergence analysis of iteration algorithms defined in a Banach space is reduced to the examination of the solution of a special scalar nonlinear difference equation.

We will use the following special result to derive further practical convergence conditions.

Lemma 10.4.5. *Assume that X is convex, and function $h \colon X \to X$ satisfies the following condition:*

$$\left\| h(x) - h(x') \right\| \leq \alpha(\xi) \left\| x - x' \right\| \tag{10.4.6}$$

for all $x, x' \in X$, where ξ is a point on the linear segment between x and x', furthermore $\alpha \colon X \to \mathbf{R}^1$ is a real-valued function such that for all fixed x and $x' \in X, \alpha(x' + t(x - x'))$ as the function of the parameter t is Riemann integrable on $[0, 1]$. Then for all x and $x' \in X$,

$$\left\| h(x) - h(x') \right\| \leq \int_0^1 \alpha(x' + t(x - x')) \, dt \, \left\| x - x' \right\|. \tag{10.4.7}$$

Proof. Let $x, x' \in X$ and define $t_i = i/N$ ($i = 0, 1, 2, \ldots, N$), where N is a positive integer. Then from (10.4.6),

$$\left\| h(x) - h(x') \right\| \leq \sum_{i=1}^{N} \left\| h(x' + t_i(x - x')) - h(x' + t_{i-1}(x - x')) \right\|$$

$$\leq \sum_{i=1}^{N} \alpha(x' + \tau_i(x - x')) \left\| (t_i - t_{i-1})(x - x') \right\|,$$

where $\tau_i \in [t_{i-1}, t_i]$, which implies that

$$\|h(x) - h(x')\| \le \left\{ \sum_{i=1}^{N} \alpha \left(x' + \tau_i \left(x - x'\right)\right) (t_i - t_{i-1}) \right\} \|x - x'\|.$$

Observe that the first factor is a Riemann-sum of the integral $\int_0^1 \alpha \left(x' + t \left(x - x'\right)\right) dt$ which converges to the integral. Let $N \to \infty$ in the above inequality to obtain the result.

Remark 10.4.6. If function α is continuous, then $\alpha \left(x' + t \left(x - x'\right)\right)$ is continuous in t, therefore it is Riemann integrable.

Assume next that maps F_m are point-to-point and process (10.4.1) satisfies the following conditions:

(D_1) $F_m(O) = O$ for $m \ge 0$;
(D_2) for all $m \ge 0$,

$$\|F_m(x) - F_m(x')\| \le \alpha(\xi_m) \|x - x'\| \tag{10.4.8}$$

for all $x, x' \in X$, where $\alpha: X \to \mathbf{R}^1$ is a continuous function, and ξ_m is a point on the linear segment connecting x and x'.
(D_3) $\alpha(x) \in [0, 1)$ for all $O \ne x \in X$;
(D_4) X is compact and convex.

Theorem 10.4.7. *Under the above conditions, $x_m \to O$ as $m \to \infty$.*

Proof. Let $O \ne x \in X$, then relation (10.4.7) implies that for all m,

$$\|F_m(x)\| \le \int_0^1 \alpha(tx) \, dt \, \|x\|, \tag{10.4.9}$$

where we have selected $x' = O$. Break the integral into two parts to obtain

$$\|F_m(x)\| \le \left\{ \int_0^\delta \alpha(tx) \, dt + \int_\delta^1 \alpha(tx) \, dt \right\} \|x\|.$$

Because α is continuous, $\alpha(O) \le 1$, and because the interval $[\delta, 1]$ is compact, $\alpha(tx) \le \beta_\delta(x) < 1$ for all $\delta \le t \le 1$, where $\beta_\delta: X \setminus \{O\} \to \mathbf{R}^1$ is the real-valued function defined as

$$\beta_\delta(x) = \max_{\delta \le t \le 1} \{\alpha(tx)\}.$$

Therefore,

$$\|F_m(x)\| \le \{\delta + (1 - \delta) \beta_\delta(x)\} \|x\| = \gamma_\delta(x) \|x\|,$$

where $\gamma_\delta: X \setminus \{O\} \to \mathbf{R}^1$ is a continuous function such that for all $x \ne O$, $\gamma_\delta(x) \in [0, 1)$.

Hence the conditions of Theorem 10.4.1 are satisfied with $\alpha = \gamma_\delta$, which implies the assertion.

Remark 10.4.8. Replace (10.4.8) by the following weaker condition: Assume that for all $m \geq 0$ and $x, x' \in X$,

$$\left\| F_m(x) - F_m(x') \right\| \leq \alpha_m(\xi_m) \left\| x - x' \right\|, \tag{10.4.10}$$

where for $m \geq 0$, $\alpha_m : X \to \mathbf{R}^1$ is a continuous function, ξ_m is a point on the linear segment connecting x and x', and $\alpha_m(x) \in [0, 1)$ for all $m \geq 0$ and $O \neq x \in X$.

Then the assertion of the theorem may not hold, as it is illustrated in the case of Example 10.4.3.

Corollary 10.4.9. *Recursion (10.4.1) and inequality (10.4.9) imply that for $m \geq 0$,*

$$\| x_{m+1} \| = \| F_m(x_m) \| \leq \overline{\alpha}(x_m) \| x_m \|,$$

where

$$\overline{\alpha}(x) = \int_0^1 \alpha(tx)\, dt.$$

Hence, by replacing $\alpha(x)$ by $\overline{\alpha}(x)$, Corollary of Theorem 10.4.1 remains valid.

In the previous results no differentiability of functions F_m was assumed. In the special case of Fréchet-differentiable functions F_m, the above theorem can be reduced to very practical convergence conditions. These results are presented in the next section.

10.5 Convergence of single-step methods with differentiable iteration functions

Assume now that B is a Banach space, $X \subseteq B$, and functions $F_m : X \to X$ are continuously differentiable on X. It is also assumed that X is compact and convex, $O \in X$, furthermore O is a common fixed point of functions F_m. In this special case, the following result holds.

Theorem 10.5.1. *Let $F'_m(x)$ denote the Fréchet derivative of F_m at x. Assume that for all $m \geq 0$,*

$$\left\| F'_m(x) \right\| \leq \beta(x), \tag{10.5.1}$$

where $\beta : X \to \mathbf{R}^1_+$ is a continuous function such that for $x \neq O$, $\beta(x) \in [0, 1)$.
Then $x_m \to O$ as $m \to \infty$.

Proof. Select

$$X_0 = \{ x \setminus x \in X \text{ and } \| x \| \leq \| x_0 \| \},$$

then X_0 is compact. Select furthermore $\alpha = \beta$. We can easily verify that all conditions of Theorem 10.4.7 are satisfied with X_0 replacing X, which implies the assertion. Assumptions (D_1) and (D_3) are obviously satisfied. Assumption (D_2) follows from the mean value theorem of derivatives and that the linear segment between x and x' is compact and function α is continuous. In order to verify assumption (D_4), we have to show that $x_m \in X_0$ for all $m \geq 0$. From the beginning of the proof of Theorem 10.3.6 we conclude that for $O \neq x \in X$, $\|F_m(x)\| < \|x\|$. Then finite induction implies that for all $m \geq 0$, $\|x_m\| \leq \|x_0\|$. Hence $x_m \in X_0$ for all $m \geq 0$, which completes the proof.

Remark 10.5.2. If (10.5.1) is replaced by the weaker assumption that for all $m \geq 0$ and $x \in O$,

$$\|F_m'(x)\| < 1,$$

the result may not hold, as the case of Example 10.4.3 illustrates. However, if F_m does not depend on m, that is, when $F_m = F$, the condition

$$\|F'(x)\| < 11 \text{ for all } x \neq O$$

implies that $x_m \to O$ as $k \to \infty$. To see this assertion, select $\beta(x) = \|F'(x)\|$. Note that this special result was first introduced by Wu and Brown [208].

Corollary 10.5.3. *Note that the corollary of Theorem 10.4.7 remains valid with* $\alpha(x) = \beta(x)$.

Remark 10.5.4. Assume that no assumption is made on the derivatives at the fixed point O.

Consider next the special case, when $B = \mathbf{R}^N$. Obviously the above results are still valid. However this further specialization enables us to derive even stronger conditions for the convergence of the iteration process

$$x_{m+1} = F_m(x_m),$$

where $F_m: B \to B$.

Theorem 10.5.5. *Let U be an open neighborhood of O. Assume that for all m, F_m is differentiable, and there exists a continuous function $\alpha: \mathbf{R}^N \to \mathbf{R}^1$ such that $\alpha(x) \in [0, 1)$ for $x \neq O$, furthermore*

(E_1) $\|F_m(x)\| \leq \alpha(x) \|x\|$ *for all m and $O \neq x \in U$;*
(E_2) *If $x \notin U$ and $\|F_m(x)\| = \alpha(x)\|x\|$ with some m, then $\|F_m'(x)x\| \leq \alpha(x)\|x\|$.*

Under these assumptions, $x_m \to O$ as $m \to \infty$.

Proof. We will prove that for all $m \geq 0$ and $x \notin O$, relation (10.4.2) holds, which implies the assertion.

Assume that for some m, (10.4.2) does not hold in the entire set $\mathbf{R}^N \setminus \{O\}$, then

$$r^* = \inf \{\|x\| \setminus O \text{ and } (10.4.2) \text{ does not hold for } m\}$$

exists and is positive. If for all vectors satisfying $\|x\| = r^*$, $\|F_m(x)\| > \alpha(x)\|x\|$, then the continuity of functions F_m and α implies that r^* can be reduced, which contradicts the definition of r^*. Therefore there is at least one x^* such that

$$\|x^*\| = r^* \quad \text{and} \quad \|F_m(x^*)\| = \alpha(x)\|x^*\|. \tag{10.5.2}$$

Because F_m is differentiable, we know that for any $\varepsilon > 0$, and sufficiently large $\lambda \in (0, 1)$,

$$\|F_m((1-\lambda)x^*) - F_m(x^*) - \lambda F_m'(x^*)x^*\| < \varepsilon\lambda\|x^*\|,$$

which together with (E₂) implies that

$$\|F_m((1-\lambda)x^*) - F_m(x^*)\| < \lambda\left[\|F_m'(x^*)x^*\| + \varepsilon\|x^*\|\right]$$
$$= \lambda\left[\beta(x^*) + \varepsilon\right]\|x^*\|,$$

where

$$\beta(x^*) = \|F_m'(x^*)x^*\|\,\|x^*\|^{-1} < \alpha(x^*).$$

From this and equality (10.5.1) we conclude that

$$\|F_m((1-\lambda)x^*)\| > \|F_m(x^*)\| - \lambda\left[\beta(x^*) + \varepsilon\right]\|x^*\|$$
$$= (\alpha(x^*) - \lambda\beta(x^*) - \lambda\varepsilon)\|x^*\|$$
$$\geq \|x^*\|\,\alpha(x^*)(1-\lambda)$$
$$= \|(1-\lambda)x^*\|\,\alpha(x^*),$$

when ε is selected small enough. Because α is continuous, this inequality contradicts again the definition of r^*, which completes the proof.

Corollary 10.5.6. *Note that the corollary of Theorem 10.4.1 can be applied again for estimating the convergence speed under the assumption of the theorem.*

Corollary 10.5.7. *Consider the special case, when $F_m = F$. The assertion of the theorem remains valid, if conditions (E₁) and (E₂) are substituted by the following assumptions:*
There exists an $\varepsilon > 0$ and a $0 \leq q < 1$ such that

(E₁') For all $x \neq O$ and $\|x\| < \varepsilon$,

$$\|F(x)\| \leq q\|x\|;$$

(E₂') If $\|x\| \geq \varepsilon$ and $\|F(x)\| = \|x\|$, then

$$\|F'(x)x\| < \|x\|.$$

Proof. Define

$$r_m = \max \left\{ \left\| F'(x) x \right\| \left\| x \right\|^{-1} / \left\| F(x) \right\| = \left\| x \right\|, m\varepsilon \leq \left\| x \right\| \leq (m+1)\varepsilon \right\}$$

for $m = 1, 2, \ldots$.

Obviously $r_m < 1$. Introduce constants

$$R_m = \max \left\{ q; r_1; r_2; \ldots; r_m \right\},$$

and the piecewise linear function $p(t)$ with vertices $(0, q)$, (ε, R_1), $(2\varepsilon, R_2)$, $(3\varepsilon, R_3)$, ... Then all conditions of the theorem are satisfied with $U = \{x \setminus \|x\| < \varepsilon\}$ and $\alpha(x) = p(\|x\|)$.

Remark 10.5.8. Then mean value theorem of derivatives implies that if $\left\| F'(O) \right\| < 1$, then there exist $\varepsilon > 0$ and $0 \leq q < 1$ that satisfy condition (E'_1). Assume furthermore that if $x \neq O$ and $\| F(x) \| = \|x\|$, then $\left\| F'(x) x \right\| < \|x\|$. In this case condition (E'_2) is also satisfied. Hence the iteration sequence $\{x_m\}$ converges to O. This special result was first introduced by Fujimoto [96], [97].

Assume again that $X \subseteq B$, where B is Banach space, furthermore for all $m \geq 0$, F_m is Fréchet-differentiable at O, and $\left\| F'_m(O) \right\| \leq 1$. As the following example shows, these conditions do not imply even the local convergence of the algorithm.

Example 10.5.9. Select $X = \mathbf{R}^1$, and for $m \geq 0$ let

$$F_m(x) = \frac{(m+1)(m+4)x}{(m+2)(m+3)}.$$

It is easy to verify that for all $m \geq 0$,

$$0 \leq F'_m(0) = \frac{(m+1)(m+4)}{(m+2)(m+3)} < 1.$$

If $x_0 \neq 0$ is any initial approximation, then finite induction shows that

$$x_m = \frac{m+3}{3(m+1)} x_0 \rightarrow \frac{1}{3} x_0 \neq 0 \text{ as } m \rightarrow \infty.$$

However if the process is stationary, then the following result holds:

Theorem 10.5.10. *Assume that $F_m = F$ $(m \geq 0)$, O is in the interior of X, and F is Fréchet-differentiable at O, furthermore $\left\| F'(O) \right\| < 1$. Then there is a neighborhood U of O such that $x_0 \in U$ implies that $x_m \rightarrow O$ as $m \rightarrow \infty$.*

Proof. Because F is differentiable at O, we can write $F(x) = L(x) + R(x)$, where L is a bounded linear mapping of X into itself and $\lim \| R(x) \| \|x\|^{-1} = O$ as $x \rightarrow O$. By assumption $\|L\| < 1$. Select a number $b > 0$ such that $\|L\| < b < 1$. There exists a $d > 0$ such that

$$\| R(x) \| < (1 - b) \|x\| \text{ if } \|x\| < d.$$

Let $U = \{x \in X / \|x\| < d\}$. We shall now prove that U has the required properties. Using the triangle inequality, we can easily show that

$$\|F(x)\| < e\|x\|, \quad \text{if } x \in U,$$

where $e = \|L\| + 1 - b$. Because $0 < e < 1$, it follows that U is F-invariant. Consequently, if $x_0 \in U$, the entire sequence of iterates x_m is also contained in U, and using finite induction we get

$$\|x_m\| \leq e^m \|x\|.$$

Because $e^m \to O, x_m \to O$ as $m \to \infty$.

Remark 10.5.11. Assumption $\|F'(O)\| < 1$ can be weakened by assuming only that the spectral radius of $F'(O)$ is less than one. In this case, $\|F'(O)^N\| < 1$ with some $N > 1$, and then apply the theorem for the function

$$F^N(x) = (F \circ F \circ \cdots \circ F)(x).$$

Remark 10.5.12. Note that no differentiability is assumed for $x \neq O$.

Remark 10.5.13. When $X = \mathbf{R}^N$, our results can be reduced to the ones obtained by Ostrowskii [155] and Ortega-Rheinboldt [154].

In the previous results, the special Liapunov function $V(x) = \|x\|$ was used, where $\|\cdot\|$ is some vector norm. Select now the Liapunov function $V(x) = \|Px\|$, where P is an $n \times n$ constant nonsingular matrix. For the sake of simplicity, we assume that $F_m = F$ for all $m \geq 0$. Then in Theorem 10.4.1 and 10.5.10 conditions (10.4.2) and (10.5.1) can be substituted by the modified relations

$$\|PF(x)\| < \|Px\|$$

and

$$\|PF'(x)u\| < \|Pu\| \quad \text{(for all } u \neq O).$$

If one selects the Euclidean norm $\|x\| = x^T x$, then these conditions are equivalent to the relations

$$F^T(x) P^T P F(x) < x^T P^T P x \tag{10.5.3}$$

and

$$u^T F'(x)^T P^T P F'(x) u < u^T P^T P u. \tag{10.5.4}$$

Note that (10.5.3) holds for all $u \neq O$ if and only if matrix $F'(x)^T P^T P F'(x) - P^T P$ is negative definite. This condition has been derived in Fujimoto [97] and it is a generalization of Theorem 1.3.2.3 of Okuguchi [153]. The case of other Liapunov functions can be discussed in an analogous manner, the details are omitted.

In the rest of the section, we are concerned with the problem of approximating a locally unique common solution x^* of the sequence of nonlinear operator equations

$$F_n(x) = 0 \quad (n \geq 0), \tag{10.5.5}$$

where each F_n ($n \geq 0$) is defined on the same convex subset D of a Banach space X with values in a Banach space Y.

Let $x_0, y_0 \in D$ be fixed, and define the two-step Newton-like method for all $n \geq 0$ by

$$y_n = x_n - A_n(x_n)^{-1} F_n(x_n) \tag{10.5.6}$$

$$x_{n+1} = y_n - y_n. \tag{10.5.7}$$

Here, $A_n(x_n)$ denotes a linear operator that is a "conscious" approximation to the Fréchet derivative $F_n'(x_n)$ of F evaluated at $x = x_n$ for all $n \geq 0$. The points $y_n \in X$ for all $n \geq 0$ are to be determined in such a way that the iteration $\{x_n\}$ ($n \geq 0$) converges to a common solution x^* of Equations (10.5.5).

We will assume that 0 is in D and $x^* = 0$.

Here we provide convergence results for the iteration (10.5.6)–(10.5.7) as well as an error analysis in a Banach space setting. The monotone convergence of this iteration is also examined in a partially ordered topological space setting.

Finally, some application of our results are provided to the solution of nonlinear integral equations of Uryson-type.

Let $R > 0$ be fixed. We assume that the following conditions are satisfied

$$\left\| A_n(0)^{-1} (A_n(x) - A_n(0)) \right\| \leq C_0(\|h\|) \tag{10.5.8}$$

and

$$\left\| A_n(0)^{-1} (F_n'(x+h) - A_n(x)) \right\| \leq C(r, r + \|h\|) \tag{10.5.9}$$

for all $x \in U(0, r)$ and $\|h\| \leq R - r$.

Here C_0 is a nondecreasing function, and C is a nondecreasing function of two variables on $[0, R]$ and $[0, R] \times [0, R]$, respectively.

Let $x_0, y_0 \in D$, and $R > 0$ be fixed. We introduce the constants

$$\|x_0\| \leq t_0 \quad \text{and} \quad \ell_0 \geq \|y_0\|, \tag{10.5.10}$$

and the iterations for all $n \geq 0$

$$\bar{s}_n = \frac{1}{1 - C_0(\|x_n\|)} \int_{\|x_n\|}^{2\|x_n\|} C(\|x_n\|, q) \, dq, \tag{10.5.11}$$

$$s_n = \frac{1}{1 - C_0(t_n)} \int_{t_n}^{2t_n} C(t_n, q) \, dq \tag{10.5.12}$$

$$t_{n+1} = s_n + \ell_n, \tag{10.5.13}$$

for some given sequence $\{\ell_n\}$ ($n \geq 0$).

We can now prove the following result:

Theorem 10.5.14. *Let $F_n: D \subseteq X \to Y$ $(n \geq 0)$ be Fréchet-differentiable nonlinear operators.*

Assume:

(a) The point $x^* = 0$ is a common solution of the operators F_n for all $n \geq 0$;
(b) the conditions (10.5.8) and (10.5.9) are satisfied on $U(0, R)$ and $U(0, R) \subseteq D$;
(c) the following estimates are true:

$$\|z_n\| \leq \ell_n \leq \beta_n t_n \quad n \geq 0 \text{ for some } \beta_n \geq 0, \tag{10.5.14}$$

$$C_0(t_0) < 1, \tag{10.5.15}$$

$$0 \leq \beta_n + \delta_n \leq 1, \quad \prod_{n=0}^{\infty} (\beta_n + \delta_n) = 0 \tag{10.5.16}$$

where,

$$\delta_n = \frac{C(t_n, 2t_n)}{1 - C_0(t_n)},$$

$$\int_0^{t_0} C(0, q) \, dq < 1, \tag{10.5.17}$$

and

$$t_0 \leq R. \tag{10.5.18}$$

Then,

(i) The sequence $\{t_n\}$ $(n \geq 0)$ generated by (10.5.12)–(10.5.13) is monotonically decreasing to 0, and

$$t_n \leq \prod_{n=0}^{n} (\beta_n + \delta_n) t_0 \quad n \geq 0. \tag{10.5.19}$$

(ii) The iterates $\{x_n\}$ $(n \geq 0)$ generated by (10.5.6)–(10.5.7) are well defined, belong to $U(0, t_0)$ for all $n \geq 0$ and converge to 0 which is a unique common solution of the operators F_n $(n \geq 0)$ in $U(0, t_0)$. Moreover, we have

$$\|x_n\| \leq t_n \quad (n \geq 0). \tag{10.5.20}$$

Proof. (i) Inequality (10.5.19) is true for $n = 0$ as equality. Let us assume that (10.5.19) is true for $m = 0, 1, 2, ..., n$. Then by (10.5.12), we get

$$s_m = \frac{1}{1 - C_0(t_m)} \int_{t_m}^{2t_m} C(t_m, q) \, dq$$

$$\leq \frac{1}{1 - C_0(t_m)} C(t_m, 2t_m)(2t_m - t_m) \leq \delta_m t_m \leq t_m. \tag{10.5.21}$$

From relations (10.5.13), (10.5.14), (10.5.16), and (10.5.21), we get

$$t_{m+1} = s_m + \ell_m \leq \delta_m t_m + \beta_m t_m = (\delta_m + \beta_m) t_m \prod_{i=0}^{m+1} (\delta_i + \beta_i) t_0. \tag{10.5.22}$$

Hence, relation (10.5.22) shows that the sequence $\{t_n\}$ $(n \geq 0)$ is monotonically decreasing to 0.

(ii) By induction on n, we will show (10.5.20). For $n = 0$, relation (10.5.20) becomes $\|x_0\| \leq t_0$, which is true by (10.5.10). Suppose the relation (10.5.20) holds for $i < n$. Because

$$A_m(x_m) = A_m(x_m) - A_m(0) + A_m(0)$$
$$= A_m(0)\left[I + A_m(0)^{-1}(A_m(x_m) - A_m(0))\right],$$

from relations (10.5.8) and (10.5.20), we obtain

$$\left\| A_m(0)^{-1}(A_m(x_m) - A_m(0)) \right\| \leq C_0(\|x_m\|) \leq C_0(t_m) \leq C_0(t_0) < 1$$

(by (10.5.15)).

It now follows from the Banach Lemma on invertible operators that $A_m(x_m)$ $(m \geq 0)$ is invertible, and

$$\left\| A_m(x_m)^{-1} A_m(0) \right\| \leq \frac{1}{1 - C_0(\|x_m\|)} \leq \frac{1}{1 - C_0(t_m)} \leq \frac{1}{1 - C_0(t_0)} \quad (m \geq 0).$$
$$(10.5.23)$$

From relations (10.5.6) and (10.5.13), we now obtain the approximation

$$y_m = \left(A_m(x_m)^{-1} A_m(0)\right)\left[A_m(0)^{-1}\int_0^1 \left(F_m'(tx_m) - A_m(x_m)\right)x_m dt\right],$$

and using (10.5.9), (10.5.23), and (10.5.20), we get in turn

$$\|y_m\| \leq \left\| A_m(x_m)^{-1} A_m(0) \right\| \cdot \left\| \left[A_m(0)^{-1}\int_0^1 \left(F_m'(tx_m) - A_m(x_m)\right)x_m dt\right]\right\|$$

$$\leq \frac{1}{1 - C_0(\|x_m\|)}\int_0^1 C(\|x_m\|, \|x_m\| + (1-t)\|x_m\|)\|x_m\| dt$$

$$= \bar{s}_m \leq s_m \quad (m \geq 0). \qquad (10.5.24)$$

Using relations (10.5.14), (10.5.19), (10.5.22), and (10.5.24), we obtain

$$\|x_{m+1}\| \leq \|y_m\| + \|y_m\| \leq s_m + \ell_m = t_{m+1} \leq t_0,$$

which together with part (i) show (ii) except the uniqueness part.

To show uniqueness, let us assume that there exists a second common solution y^* of the operators F_n $(n \geq 0)$ in $U(0, t_0)$. Then we get the estimate

$$\int_0^1 \left\| A(0)^{-1}\left[F_m'(ty^*) - A(0)\right]\right\| \leq \int_0^1 C(0, 0 + t\|y^*\|) dt$$

$$\leq \int_0^1 C(0, tt_0) dt \leq \int_0^{t_0} C(0, q) dq < 1$$

by (10.5.17).

It now follows from the above estimate that the linear operator $\int_0^1 F_m' (ty^*) \, dt$ is invertible, and from the approximations

$$F_m (y^*) - F_m (0) = \int_0^1 F_k' (ty^*) \, dt \, y^* \quad (m \geq 0)$$

it follows that $y^* = 0$.

That completes proof of the Theorem.

Remark 10.5.15. If we set $A_m (x) = F_m' (x)$ for all $x \in D$, then the iterations (10.5.6)–(10.5.7) become

$$y_n = x_n - F_n' (x_n)^{-1} F_n (x_n) \qquad (10.5.25)$$

$$x_{n+1} = y_n - y_n \quad (n \geq 0). \qquad (10.5.26)$$

Let us also set $C (0, t) = C_0 (t)$ for all $t \in [0, +\infty)$. Then under the hypotheses of the theorem, the conclusions will also hold by for the NK iteration $\{x_n\}$ $(n \geq 0)$ generated by (10.5.25)–(10.5.26) for all $n \geq 0$.

Finally note that the sufficient convergence conditions as well as the error bounds can be further improved if center-Lipschitz conditions are also introduced along the lines of Section 2.2.

Example 10.5.16. We assume that $A_n (x) = F_n' (x)$ for all $n \geq 0$ and $x \in U (0, R)$. Let us assume that the following condition is satisfied

$$\left\| F_n' (0)^{-1} \left(F_n' (v) - F_n' (w) \right) \right\| \leq q (r) \|v - w\| \qquad (10.5.27)$$

for all $v, w \in U (0, R)$, $n \geq 0$, and for some nondecreasing function q on $[0, R]$. Then as in [35] we can show that by setting

$$C (r, r + \|h\|) = \int_r^{r+\|h\|} w (t) \, dt, \quad w (t) = \int_0^r q (t) \, dt \qquad (10.5.28)$$

and

$$C_0 (t) = C (0, t) \quad t \in [0, R] \qquad (10.5.29)$$

conditions (10.5.8) and (10.5.9) are satisfied.

Let assume that $X = Y = C = C [0, 1]$ the space of continuous functions on $[0, 1]$ equipped with the usual supremum norm. We consider Uryson-type nonlinear integral equations of the form

$$F (x) (t) = x (t) - \int_0^1 K (t, s, x (s)) \, ds. \qquad (10.5.30)$$

We make use of the following standard result whose proof can be found for example in [6], [43].

Theorem 10.5.17. *The Lipschitz condition (10.5.27) for the Fréchet derivative F' of the operator (10.5.30) holds if and only if the second derivative $K''_{uu}(t,s,u)$ exists for all t and almost all s and u, and*

$$\sup_{s\in[0,1]} \int_0^1 \sup_{|u|\leq r} |K''_{uu}(t,s,u)|\, ds < \infty \tag{10.5.31}$$

Moreover, the left-hand side in relation (10.5.31) is then the minimal Lipschitz constant $\frac{q(r)}{\alpha}$, $\alpha = \|F'(0)^{-1}\|$ in (10.5.27).

Moreover, the constant α is given by

$$\alpha = 1 + \sup_{t\in[0,1]} \int_0^1 |r(t,s)|\, ds, \tag{10.5.32}$$

where $r(t,s)$ is the resolvant kernel of the equation

$$h(t) - \int_0^1 K'_u(t,s,0)h(s)\, ds = -\int_0^1 K(t,s,0)\, ds. \tag{10.5.33}$$

Proof. Let us consider a simple example. Suppose that $K(t,s,u) = c_1(t)c_2(s)c_3(u)$ with two continuous functions c_1 and c_2, and $c_3 \in C^2$. We set

$$d_1 = \int_0^1 c_2(s)\, ds, \quad d_2 = \int_0^1 c_1(s)c_2(s)\, ds. \tag{10.5.34}$$

Then relation (10.5.33) becomes

$$h(t) = \left[c'_4 c'_3(0) - d_1 c_3(0)\right]c_1(t), \tag{10.5.35}$$

where

$$c'_4 = \int_0^1 c_2(s)h(s)\, ds. \tag{10.5.36}$$

substituting relation (10.5.35) into (10.5.36), one may calculate c'_4 and hence find the resolvent kernel $r(t,s)$ in case $d_2 c'_3(0) < 1$, to get

$$r(t,s) = \frac{c_1(t)c_2(t)c'_3(0)}{1-d_2 c'_3}. \tag{10.5.37}$$

Using relation (10.5.31), (10.5.32), we obtain

$$q(t) = \|c_1\|\, d_1 \sup_{\|u\|\leq r} |c''_3(u)|, \tag{10.5.38}$$

$$\eta = \frac{d_1 c_3(0)}{1-d_2 c'_3(0)}\, \|c_1\| \tag{10.5.39}$$

and

$$\beta = 1 + \frac{d_1 c_3(0)}{1-d_2 c'_3(0)}\, \|c_1\|. \tag{10.5.40}$$

Let us now consider the sequence of equations

$$F_n(x)(t) = 0 \qquad (10.5.41)$$

where

$$F_n(x)(t) = x(t) - \int_0^1 K_n(t, s, x(s))\, ds, \qquad (10.5.42)$$

and we choose for all $n \geq 0$

$$c_1^n(t) = \frac{n+1}{n+2}\frac{2}{3}t, \quad c_2^n(t) = \frac{n+1}{n+2}\frac{2}{10}s, \quad c_3^n(u) = \frac{n+2}{n+1}\left(\frac{1}{3}u^2 + \frac{1}{10}\right)u.$$

Then using relations (10.5.28), (10.5.29), (10.5.34), (10.5.37)–(10.5.40), we obtain for all $n \geq 0$

$$d_1^n = \frac{1}{10}\frac{n+1}{n+2}, \quad d_2^n = \frac{2}{100}\frac{n+1}{n+2}, \quad d_2^n c_3^{\prime n}(0) = 0 < 1,$$

$$r_n(t, s) = 0, \quad \alpha = 1, \quad q_n(t) = \frac{6}{100}\left(\frac{n+1}{n+2}\right)^2 r \leq \frac{6}{100}r = q(r),$$

$$w(r) = \frac{3}{100}r^2, \quad C(r, r + \|h\|) = \frac{1}{100}\left[(r + \|h\|)^3 - r^3\right],$$

and

$$C_0(t) = C(0, t) \quad \text{for all } t \in [0, R].$$

We select $z_n = 0$ and $\beta_n = 0$ for all $n \geq 0$ for simplicity. Then (10.5.15), (10.5.16), and (10.5.18) will be satisfied if the following conditions are satisfied, respectively

$$t_0 < \sqrt[3]{100}, \quad t_0 < \frac{1}{2}\sqrt[3]{100}, \quad \text{and} \quad t_0 < \sqrt[4]{400}.$$

By setting $x_0(t) = 2 = t_0$ for all t and $R = t_0$, with the above choices all conditions (10.5.14)–(10.5.16), (10.5.17), and (10.5.18) are satisfied. Hence the conclusions of Theorem 10.5.17 for equations (10.5.41) follow.

10.6 Monotone convergence

Let X be a linear space. We examine the monotone convergence of POTL-space (see Section 1.2).

Theorem 10.6.1. Let $F_n: D \subset X \to Y$ $(n \geq 0)$ where X is a regular POTL-space and Y is a POTL-space. Let $\bar{x}_0, x_0, x_{-1}$ be three points of D such that

$$\bar{x}_0 \leq x_0 \leq x_{-1}, \quad \langle \bar{x}_0, x_{-1}\rangle \subset D, \quad F_0(\bar{x}_0) \leq 0 \leq F_0(x_0), \qquad (10.6.1)$$

and denote

$$S_1 = \left\{(x, y) \in X^2 \mid \bar{x}_0 \leq x \leq y \leq x_0\right\}, \qquad (10.6.2)$$

$$S_2 = \left\{(u, x_{-1}) \in X^2 \mid \bar{x}_0 \leq u \leq x_0\right\}, \qquad (10.6.3)$$

$$S_3 = S_1 \cup S_2. \qquad (10.6.4)$$

Assume:

(a) The operators $A_n (\cdot, \cdot): S_3 \rightarrow LX (X, Y)$ are such that

$$F_n (y) - F_n (x) \leq A_n (w, z) (y - y) \qquad (10.6.5)$$

for all $n \geq 0$ (x, y), $(y, w) \in S_1$, $(w, y) \in S_3$.
(b) The linear operators $A_n (u, v)$, $(n \geq 0)$ have a continuous nonsingular nonneg-
ative subinverse.
(c) The following conditions hold:

$$F_n (x) \leq F_{n-1} (x) \text{ for all } x \in \langle x_0, y_0 \rangle \ (n \geq 1), \ F_{n-1} (x) \leq 0, \qquad (10.6.6)$$
$$F_n (y) \geq F_{n-1} (x) \text{ for all } x \in \langle x_0, y_0 \rangle \ (n \geq 1), \ F_{n-1} (y) \geq 0. \qquad (10.6.7)$$

(d) There exist sequences $\{y_n\}$, $\{y_n\}$, $\{\bar{y}_n\}$, $\{\bar{y}_n\}$ $(n \geq 0)$ satisfying

$$F_n (\bar{y}_n) - A_n (\bar{y}_n) \leq 0, \quad A_n = A_n (\bar{x}_n, x_n) \qquad (10.6.8)$$
$$F_n (y_n) - A_n (y_n) \geq 0, \qquad (10.6.9)$$
$$y_n - \bar{y}_n \geq y_n - \bar{y}_n, \qquad (10.6.10)$$
$$y_n \geq 0 \qquad (10.6.11)$$

and

$$\bar{y}_n \leq 0 \ \text{ for all } \ n \geq 0. \qquad (10.6.12)$$

Then there exist sequence $\{\bar{x}_n\}$, $\{x_n\}$ $(n \geq 0)$ and points $\bar{x}^*$, x^*, such that for all $n \geq 0$,

$$F_n (x_n) + A_n (y_n - x_n) = 0, \qquad (10.6.13)$$
$$y_n + x_{n+1} - y_n = 0, \qquad (10.6.14)$$
$$F_n (\bar{x}_n) + A_n (\bar{y}_n - \bar{x}_n) = 0, \qquad (10.6.15)$$
$$\bar{y}_n + \bar{x}_{n+1} - \bar{y}_n = 0, \qquad (10.6.16)$$
$$F_n (\bar{x}_n) \leq F_{n-1} (\bar{x}_n) \leq 0 \leq F_{n-1} (x_n) \leq F_n (x_n), \qquad (10.6.17)$$
$$\bar{x}_0 \leq \bar{y}_0 \leq \bar{x}_1 \leq \cdots \leq \bar{y}_n \leq \bar{x}_{n+1} \leq x_{n+1} \leq y_n \leq \cdots \leq x_1 \leq y_0 \leq x_0, \qquad (10.6.18)$$

and

$$\lim_{n \to \infty} \bar{x}_n = \bar{x}^* \leq x^* = \lim_{n \to \infty} x_n. \qquad (10.6.19)$$

Moreover, if the operators A_n are inverse nonnegative, then any solution u of the equations $F_n (x) = 0$ $(n \geq 0)$ in $\langle \bar{x}_0, x_0 \rangle$ in $\langle \bar{x}_0, x_0 \rangle$ belongs to $\langle \bar{x}^*, x^* \rangle$. Further-more, if $\bar{x}^* = x^*$, then we get $\bar{x}^* = u = x^*$.

Proof. Let L_0 be a continuous nonsingular nonnegative left subinverse of A_0, and consider the mapping $P: \langle 0, x_0 - \bar{x}_0 \rangle \rightarrow X$ defined by

$$P (x) = x - L_0 (F_0 (\bar{x}_0) + A_0 (x))$$

where $A_0 (x)$ denotes the image of x with respect to the mapping $A_0 = A_0 (\bar{x}_0, x_0)$. It is easy to see that P is isotone and continuous. We also have

$$P(0) = -L_0(F_0(\overline{x}_0)) \geq 0, \quad \text{(by (10.6.1))},$$
$$P(x_0 - \overline{x}_0) = x_0 - \overline{x}_0 + L_0(F(x_0) F_0(\overline{x}_0) - A_0(x_0 - \overline{x}_0))$$
$$\leq x_0 - \overline{x}_0 + L_0(F_0(\overline{x}_0) - A_0(x_0 - \overline{x}_0)) \leq x_0 - \overline{x}_0.$$

According to Kantorovich's Lemma (see Section 1.2), the operator P has a fixed point $w \in \langle 0, x_0 - \overline{x}_0 \rangle$. Taking $\overline{y}_0 = \overline{x}_0 + w$, we have

$$F_0(\overline{x}_0) + A_0(\overline{y}_0 - \overline{x}_0) = 0, \quad \overline{x}_0 \leq \overline{y}_0 \leq x_0. \tag{10.6.20}$$

Using (10.6.20) and the above approximation, we get $F_0(\overline{y}_0) = F_0(\overline{y}_0) - F_0(\overline{x}_0) - A_0(\overline{y}_0 - \overline{x}_0) \leq 0$. Hence, we obtain by (10.6.6), $F_1(\overline{y}_0) \leq F_0(\overline{y}_0) \leq 0$. From the approximation (10.6.16) and estimates (10.6.8), (10.6.12) we have that

$$\overline{x}_1 - \overline{y}_0 = -\overline{z}_0 \geq 0 \Longrightarrow \overline{x}_1 \leq \overline{y}_0, \tag{10.6.21}$$

$$F_0(\overline{x}_1) \leq F_0(\overline{x}_1) - F_0(\overline{y}_0) - A_0(\overline{x}_1 - \overline{y}_0) \leq 0, \tag{10.6.22}$$

and by (10.6.6)

$$F_1(\overline{x}_1) \leq F_0(\overline{x}_1) \leq 0. \tag{10.6.23}$$

Consider now the operator $G: \langle 0, x_0 - \overline{y}_0 \rangle \rightarrow E_1$ defined by $G(x) = x + L_0(F_0(x_0) - A_0(x))$. G is clearly continuous, isotone, and

$$G(0) = L_0(F_0(x_0)) \geq 0, \quad \text{(by (10.6.1))}$$
$$G(x_0 - \overline{y}_0) = x_0 - \overline{y}_0 + L_0(F_0(\overline{y}_0)) + L_0(F_0(x_0) - F_0(\overline{y}_0) - A_0(x_0 - \overline{y}_0))$$
$$\leq x_0 - \overline{y}_0 + L_0(F_0(x_0) - F_0(\overline{y}_0) - A_0(x_0 - \overline{y}_0))$$
$$\leq x_0 - \overline{y}_0, \quad \text{(by (10.6.5))}.$$

Applying the Kantorovich Lemma again, we deduce the existence of a point $v \in \langle 0, x_0 - \overline{y}_0 \rangle$ such that $G(v) = v$. Taking $y_0 = x_0 - v$,

$$F_0(x_0) + A_0(y_0 - x_0) = 0, \quad \overline{y}_0 \leq y_0 \leq x_0. \tag{10.6.24}$$

Using (10.6.5) and (10.6.24), we get

$$F_0(y_0) = F_0(y_0) - F_0(x_0) - A_0(y_0 - x_0) \geq 0.$$

Hence, by (10.6.7), we obtain

$$F_1(y_0) \geq F_0(y_0) \geq 0.$$

Using (10.6.11), and the approximation

$$x_1 - y_0 = -y_0 \leq 0 \Longrightarrow x_1 \leq y_0, \tag{10.6.25}$$

and by (10.6.9), we obtain

$$F_0(x_1) \geq F_0(x_1) - F_0(y_0) - A_0(x_1 - y_0) \geq 0.$$

Hence, we get by (10.6.7)

$$F_1(x_1) \geq F_0(x_1) \geq 0. \tag{10.6.26}$$

By induction, it easy to show that there exist sequence $\{\bar{x}_n\}$, $\{x_n\}$ $(n \geq 0)$ satisfying (10.6.13)–(10.6.18) in a regular space E_1 and as such there exist $\bar{x}^*, x^* \in \langle \bar{x}_0, x_0 \rangle$ satisfying (10.6.19).

If $\bar{x}_0 \leq u \leq x_0$ and $F_n(u) = 0$ $(n \geq 0)$, then we can obtain in turn

$$A_0(y_0 - u) = A_0(x_0 - u) - (F_0(x_0) - F_0(u)) \geq 0,$$

and

$$A_0(\bar{y}_0 - u) = A_0(\bar{x}_0 - u) - (F_0(x_0) - F_0(u)) \leq 0.$$

Because the operator A_0 is inverse nonnegative, we get $\bar{y}_0 \leq u \leq y_0$. Proceeding by induction, we deduce that $\bar{y}_n \leq u \leq y_n$, from which it follows that $\bar{y}_n \leq \bar{x}_n \leq \bar{y}_{n+1} \leq u \leq y_{n+1} \leq x_n \leq y_n$ for all $n \geq 0$. That is, we have $\bar{x}_n \leq u \leq x_n$ for all $n \geq 0$. Hence, we get $\bar{x}^* \leq u \leq x^*$. Furthermore, if $\bar{x}^* = x^*$, then we get

$$\bar{x}^* = u = x^*,$$

which completes the proof of the Theorem.

Remark 10.6.2. The linear operators A_n $(n \geq 0)$ are usually chosen as divided differences of order one related with the operators F_n for each $n \geq 0$. For example, we can set $A_n = [x_n, x_n]$ $(n \geq 0)$, or $A_n = [x_n, x_{n+1}]$ $(n \geq 0)$, etc., where each $[\cdot, \cdot]$ depends on the operators F_n $(n \geq 0)$. The hypotheses (10.6.8)–(10.6.12) will then be conditions on divided differences for each $n \geq 0$. It then turns out that for appropriate choices of the A_n's $(n \geq 0)$, relations (10.6.8)–(10.6.12) turn out to be standard natural conditions on divided differences.

10.7 Exercises

10.7.1. Maximize $F = 240x_1 + 104x_2 + 60x_3 + 10x_4$ subject to

$$20x_1 + 9x_2 + 6x_3 + x_4 \leq 20$$
$$10x_1 + 4x_2 + 2x_3 + x_4 \leq 10$$
$$x_i \geq 0, \quad i = 1, ..., 4.$$

10.7.2. Minimize $F = 3x_1 + 2x_2$ subject to

$$8x_1 - x_2 \geq 8$$
$$2x_1 - x_2 \geq 6$$
$$x_1 + 3x_2 \geq 6$$
$$x_1 + 6x_2 \geq 8$$
$$x_1 \geq 0, \quad x_2 \geq 0.$$

10.7.3. Find an interval $[a, b]$ containing a root x^* of the equation $x = \frac{1}{2} \cos x$ such that for every $x_0 \in [a, b]$, the iteration $x_{n+1} = \frac{1}{2} \cos x_n$ will converge to x^*. Solve the equation by using NK or the secant method.

10.7.4. Solve the following nonlinear equations by the method of your choice:
   (a) $\ln x = x - 4$
   (b) $xe^x = 7$
   (c) $e^x \ln x = 7$.

10.7.5. Find $(I - A)^{-1}$ (if it exists) for the matrix

$$A = \begin{pmatrix} 0.12 \ 0.04 \\ 0.01 \ 0.03 \end{pmatrix}.$$

Perform three steps.

10.7.6. Repeat Exercise 10.7.5 for the matrix

$$A = \begin{pmatrix} -1 \ 3 \\ 2 \ 4 \end{pmatrix}.$$

10.7.7. Solve problem

$$\dot{x} = x - y - 1 \quad x(0) = 0$$
$$\dot{y} = x + y \qquad y(0) = 1.$$

Perform three steps.

10.7.8. Solve the boundary-value problem

$$\ddot{x} = tx - \dot{x} - 1, \quad x(0) = x(1) = 1$$

by the discretization method. Select $h = 0.1$.

10.7.9. Solve

$$x(t) = \int_0^1 \frac{(t-s)^2}{10} x(s) \, ds - 7.$$

10.7.10. Solve

$$x(t) = \int_0^t (t + s) x(s) \, ds - 4.$$

10.7.11. Consider a continuous map $P : \mathbf{R}^n \to \mathbf{R}^n$ such that $P \in C^1$ and $P(0) = 0$. Set $S_1 = \{x \mid \|P(x)\| < \|x\|\}$, $S_2 = \{x \mid \|P(x)\| \geq \|x\|\}$. Assume that
   (a) S_1 is invariant under P, $P(S_1) \subseteq S_1$ and
   (b) For all $a \in S_2$, there exists a positive integer $i(a)$ such that $P^{i(a)}(a) \in S_1$.
Show that for all $x \in \mathbf{R}^n$, $P^m(x) \to 0$ as $m \to \infty$.

10.7.12. Assume that there exists a function $h : (0, \infty) \to \mathbf{R}$ such that $|y| \leq h(r) |x|$ for all $m \geq 0, r > 0, |x| \leq r, x \in X$ and $y \in F_m(x)$. Show that

$$|x_m| \leq q_m,$$

where

$$q_{m+1} = h(q_m) q_m, \quad q_0 = |x_0|.$$

Provide a convergence analysis of iteration (10.4.1) based on the above estimate.

10.7.13. To find a zero for $G(x) = 0$ by iteration, where G is a real function defined on $[a, b]$ rewrite the equation as,

$$x = x + c \cdot G(x) \equiv F(x)$$

for some constant $c \neq 0$. If x^* is a root of $G(x)$ and if $G'(x^*) \neq 0$, how should c be chosen in order that the sequence $x_{m+1} = F(x_m)$ converge to x^*?

10.7.14. Solve the initial value problem

$$\dot{x}(t) = 1 + \cos(x(t)), \quad x(0) = 0.$$

10.7.15. The predator-prey population models describe the iteration of a prey population X and a predator population Y. Assume that their interaction is modeled by the system of ordinary differential equations

$$\dot{x} = x - \tfrac{1}{4}x^2 - \tfrac{1}{10}xy + 1$$
$$\dot{y} = -\tfrac{1}{4}y + \tfrac{1}{7}xy + \tfrac{2}{3}.$$

(Assume $x(0) = y(0) = 0$.) Solve the system.

10.7.16. Assume that $F_m = F$ $(m \geq O)$, O is in the interior of X, and F is Fréchet-differentiable at O, furthermore the special radius of $F'(0)$ is less than 1. Then show that there is a neighborhood U of O such that $x_0 \in U$ implies that $x_m \to O$ as $m \to \infty$.

10.7.17. Let $F: \mathbf{R}^n \to \mathbf{R}^n$ be a function such that $F(0) = 0$, $F \in C^0$, and consider the difference equation $x(t+1) = F(x(t))$. If, for some norm, $\|F(x)\| \leq \|x\|$ for any $x \neq O$, then show that the origin is globally asymptotically stable equilibrium for the equation.

10.7.18. Assume that there exists a strictly increasing function $g: \mathbf{R} \to \mathbf{R}$ such that $g(O) = 0$, and a norm such that $g(\|F(x)\|) < g(\|x\|)$ for all $x \neq 0$. Then show that O is a globally asymptotically stable equilibrium for equation $x(t+1) = F(x(t))$, where $F(O) = 0$.

10.7.19. Consider the following equation in $\mathbf{R}^2$:

$$x_1(t+1) = 8\sin\left(x_1(t) + \tfrac{\pi}{\ell}\right) + .2x_2(t)$$
$$x_2(t+1) = 8x_1(t) + .1x_2(t).$$

10.7.20. Solve the Fredholm-type integral equation

$$x(t) = \int_0^1 \frac{t \cdot x(s)}{10} dx + 1.$$

10.7.21. Solve the Volterra-type integral equation

$$x(t) = \int_0^t \frac{t \cdot x(s)}{10} ds + 1.$$

10.7.22. Solve equation

$$x = \frac{\sin x}{2} + 1.$$

11

The Newton-Kantorovich Theorem and Mathematical Programming

The Newton-Kantorovich Theorem 2.2.4 with a few notable exceptions [173], [174], [165], has not been sufficiently utilized in the mathematical programming community. The purpose of this chapter is to provide a bridge between the two research communities by showing that the Newton-Kantorovich theorem can be used in analyzing LP and interior point methods and can be used for obtaining optimal error bounds along the way.

In Sections 11.1 and 11.2 we show how to improve on the elegant works of Renegar, Shub in [174] and Potra in [165]. To avoid repetitions, we simply refer the reader to the above excellent works for the development of these methods. We simply start from the point where the hypotheses made can be replaced by ours, which are weaker. The benefits of this approach have also been explained in the Introduction and in Section 2.2.

We also note that the work in [174] is motivated by a Theorem of Smale in combination with the Newton-Kantorovich theorem, whereas the work in [165] differs from the above as it applies the latter theorem directly.

11.1 Case 1: Interior point methods

It has already been shown in [165] that the Newton-Kantorovich theorem can be used to construct and analyze optimal-complexity path following algorithms for linear complementary problems. Potra has chosen to apply this theorem to linear complementary problems because such problems provide a convenient framework for analyzing primal-dual interior point methods. Theoretical and experimental work conducted over the past decade has shown that primal-dual path following algorithms are among the best solution methods for linear programming (LP), quadratic programming (QP), and linear complementary problems (LCP). Primal-dual path following algorithms are the basis of the best general-purpose practical methods, and they have important theoretical properties [207].

I.K. Argyros, *Convergence and Applications of Newton-type Iterations*,
DOI: 10.1007/978-0-387-72743-1_11, © Springer Science+Business Media, LLC 2008

Potra using Newton-Kantorovich Theorem 2.2.4 in particular showed how to construct path following algorithms for LCP that have $O\left(\sqrt{n}L\right)$ iteration complexity.

Given a point x that approximates a point $x\left(\tau\right)$ on the central path of the LCP with complementary gap τ, the algorithms compute a parameter $\theta \in (0, 1)$ so that x satisfies the Newton-Kantorovich hypothesis (2.2.17) for the equation defining $x\left((1 - \theta)\tau\right)$. It is proven that θ is bounded below by a multiple of $n^{-1/2}$. Because (2.2.17) is satisfied, the sequence generated by NK method (2.1.3) or by the MNK method (2.1.5) with starting x, will converge to $x\left((1 - \theta)\tau\right)$. He showed that the number of steps required to obtain an acceptable approximation of $x\left((1 - \theta)\tau\right)$ is bounded above by a number independent of n. Therefore, a point with complementarity gap less than ε can be obtained in at most $O\left(\sqrt{n}\log\left(\frac{\varepsilon_0}{\varepsilon}\right)\right)$ steps (for both methods), where ε_0 is the complementary gap of the starting point. For linear complementary problems with rational input data of bit length L, this implies that an exact solution can be obtained in at most $O\left(\sqrt{n}L\right)$ iterations plus a rounding procedure involving $O\left(n^3\right)$ arithmetic operations [207].

The differences between Potra's work and the earlier works by Renegar [173] and Renegar and Shub [174] have been stated in the introduction of this chapter and further analyzed in [165].

We also refer the reader to the excellent monograph of Nesterov and Nemirovskii [145] for an analysis of the construction of interior point methods for a larger class of problems than that considered in [165].

Below one can find our contribution.

Let $\| \cdot \|$ be a given norm on $\mathbf{R}^i$, i a natural integer, and x_0 be a point of D such that the closed ball of radius r centered at x_0,

$$\overline{U}(x_0, r) = \{x \in R^i : \|x - x_0\| \le r\} \tag{11.1.1}$$

is included in $D \subseteq \mathbf{R}^i$, i.e.,

$$\overline{U}(x_0, r) \subseteq D. \tag{11.1.2}$$

We assume that the Jacobian $F'(x_0)$ of $F: D \subseteq X \to X$ is nonsingular and that the Lipschitz condition (2.2.36) is satisfied.

The Newton-Kantorovich Theorem 2.2.4 states that if the quantity

$$k = \ell\eta \le \frac{1}{2}, \tag{11.1.3}$$

then there exists $x^* \in \overline{U}(x_0, r)$ with $F(x^*) = 0$. Moreover the sequences produced by NK method (2.1.3) and by the modified NK method

$$y_{n+1} = y_n - F'(y_0)^{-1}F(y_n), \quad y_0 = x_0 \quad (n \ge 0) \tag{11.1.4}$$

are well defined and converge to x^*.

Define

$$k^0 = \bar{\ell}\eta \le \frac{1}{2}, \tag{11.1.5}$$

where,

$$\bar{\ell} = \frac{\ell_0 + \ell}{2}.$$ (11.1.6)

where ℓ_0 is the constant appearing in the center-Lipschitz conditions (2.2.43).
Note that

$$k \leq \frac{1}{2} \Rightarrow k^0 \leq \frac{1}{2}$$ (11.1.7)

but not vice versa unless if $\ell_0 = \ell$.

Similarly by simply replacing ℓ with ℓ_0 (as (2.2.43) instead of (2.2.36) is actually needed in the proof) and condition (11.1.3) by the weaker

$$k^1 = \ell_0 \eta \leq \frac{1}{2}$$ (11.1.8)

in the proof of Theorem 1 in [165] we show that NK method (2.1.3) also converges to x^* and the improved bounds

$$\|y_n - x^*\| \leq \frac{2\beta_0 \lambda_0^2}{1 - \lambda_0^2} \xi_0^{n-1} \quad (n \geq 1)$$ (11.1.9)

where

$$\beta_0 = \frac{\sqrt{1 - 2k^1}}{\omega_0}, \quad \lambda_0 = \frac{1 - \sqrt{1 - 2k^1} - h^1}{k^1} \quad \text{and}$$ (11.1.10)

$$\xi_0 = 1 - \sqrt{1 - 2k^1},$$ (11.1.11)

hold. In case $\ell_0 = \ell$ (11.1.8) reduces to (11.1.3) used in [165]. Otherwise our error bounds are finer. Note also that

$$k \leq \frac{1}{2} \Rightarrow k^1 \leq \frac{1}{2}$$ (11.1.12)

but not vice versa unless if $\ell_0 = \ell$.

We can now describe the linear complementarity problem as follows: Given two matrices $Q, R \in \mathbf{R}^{n \times n}$ ($n \geq 2$) and a vector $b \in \mathbf{R}^n$, the horizontal linear complementarity problem (HLCP) consists of approximating a pair of vectors (w, s) such that

$$ws = 0$$
$$Q(w) + R(s) = b$$ (11.1.13)
$$w, s \geq 0.$$

The monotone linear complementarity problem (LCP) is obtained by taking $R = -I$ and Q positive semidefinite.

Moreover, the linear programming problem (LP) and the quadratic programming problem (QP) can be formulated as HLCPs. That is, HLCP is a suitable way for studying interior point methods.

We assume HLCP (11.1.13) is monotone in the sense that:

$$Q(u) + R(v) = 0 \text{ implies } u^t v \geq 0, \text{ for all } u, v \in \mathbf{R}^n. \tag{11.1.14}$$

Condition (11.1.14) holds if the HLCP is a reformulation of a QP. If the HLCP is a reformulation of a LP, then the following stronger condition holds:

$$Q(u) + R(v) = 0 \text{ implies } u^t v = 0, \text{ for all } u, v \in \mathbf{R}^n. \tag{11.1.15}$$

Then we say in this case that the HLCP is skew-symmetric.

If the HLCP has an interior point, i.e., there is $(w, s) \in \mathbf{R}^n_{++} \times \mathbf{R}^n_{++}$ satisfying $Q(w) + R(s) = b$, then for any parameter $\tau > 0$ the nonlinear system

$$ws = \tau e$$
$$Q(w) + R(s) = b \tag{11.1.16}$$
$$w, s \geq 0$$

has a unique positive solution $x(\tau) = \left[w(\tau)^t, s(\tau)^t \right]^t$.

The set of all such solutions defines the central path C of the HLCP. It can be proved that $(w(\tau), s(\tau))$ converges to a solution of the HLCP as $\tau \to 0$. Such an approach for solving the HLCP is called the path following algorithm.

At a basic step of a path following algorithm, an approximation (w, s) of $(w(\tau), s(\tau))$ has already been computed for some $\tau > 0$. The algorithm determines the smaller value of the central path parameter $\tau_+ = (1 - \theta) \tau$, where the value $\theta \in (0, 1)$ is computed in some unspecified way. The approximation (w^t, s^t) of $(w(\tau_+), s(\tau_+))$ is computed. The procedure is then repeated with (w^+, s^+, τ^+) in place of $(w, s.\tau)$.

In order for us to relate the path following algorithm and the Newton-Kantorovich theorem, we introduce the notations

$$x = \begin{bmatrix} w \\ s \end{bmatrix}, \quad x(\tau) = \begin{bmatrix} w(\tau) \\ s(\tau) \end{bmatrix},$$
$$x^+ = \begin{bmatrix} w^+ \\ s^+ \end{bmatrix}, \quad x(\tau_+) = \begin{bmatrix} w(\tau_+) \\ s(\tau_+) \end{bmatrix}, \quad \text{etc.}$$

Then for any $\theta > 0$ we define the nonlinear operator

$$F_\sigma(x) = \begin{bmatrix} ws - \sigma e \\ Q(w) + R(s) - b \end{bmatrix}. \tag{11.1.17}$$

Then system (11.1.16) defining $x(\tau)$ becomes

$$F_\sigma(x) = 0, \tag{11.1.18}$$

whereas the system defining $x(\tau_+)$ is given by

$$F_{(1-\theta)\tau}(x) = 0. \tag{11.1.19}$$

We assume that the initial guess x belongs in the interior of the feasible set of the HLCP

$$F^0 = \left\{ x = (w^t, s^t)^t \in \mathbf{R}^{2n}_{++} : Q(w) + R(s) = b \right\}. \tag{11.1.20}$$

In order to verify the Newton-Kantorovich hypothesis for equation (11.1.18), we introduce the quantity

$$\eta = \eta(x, \tau) = \left\| F'(x)^{-1} F_\tau(x) \right\|, \tag{11.1.21}$$

the measure of proximity

$$k = k(x, \tau) = \eta \ell, \ \ell = \ell(x) \tag{11.1.22}$$
$$k^0 = k^0(x, \tau) = \eta \bar{\ell}, \ \bar{\ell} = \bar{\ell}(x)$$
$$k^1 = k^1(x, \tau) = \eta \ell_0, \ \ell_0 = \ell_0(x)$$

and the normalized primal-dual gap

$$\mu = \mu(x) = \frac{w^t s}{\eta}. \tag{11.1.23}$$

If for a given interior point x and a given parameter τ we have $k^0(x, \tau) \leq .5$ for the Newton-Kantorovich method or $k^1(x, \tau) \leq .5$ for the modified Newton-Kantorovich method, then corresponding sequences starting from x will converge to the point $x(\tau)$ on the central path. We can now describe our algorithm, which is a weaker version of the one given in [208]:

Algorithm 11.1.1. *(using Newton-Kantorovich method).*
Given $0 < k_1^0 < k_2^0 < .5$, $\varepsilon > 0$, and $x_0 \in F^0$ satisfying $k^0(x_0, \mu(x_0)) \leq k_1^0$;
Set $k^0 \leftarrow 0$ and $\tau_0 \leftarrow \mu(x_0)$;
repeat (outer iteration)
Set $(x, \tau) \leftarrow (x_k, \tau_k)$, $\bar{x} \leftarrow x_k$;
Determine the largest $\theta \in (0, 1)$ such that $k^0(x, (1 - \theta)\tau) \leq k_2^0$;
Set $\tau \leftarrow (1 - \theta)\tau$;
repeat (inner iteration)

$$\text{Set } x \leftarrow x - F'(x)^{-1} F_\tau(x) \tag{11.1.24}$$

until $k^0(x, \mu) \leq k_1^0$;
Set $(x_{k+1}, \tau_{k+1}) \leftarrow (x, \tau)$;
Set $k \leftarrow k + 1$;
until $(w^k)^t s^k \leq \varepsilon$.

For the modified the Newton-Kantorovich algorithm k_1^0, k_2^0, k^0 should be replaced by k_1^1, k_2^1, k^1, and (11.1.24) by

$$\text{Set } x \leftarrow x - F'(\bar{x})^{-1} F_\tau(x)$$

respectively.

In order to obtain Algorithm 1 in [208], we need to replace k_1^0, k_2^0, k^0 by k_1, k_2, k, respectively.

The above suggest that all results on interior methods obtained in [208] using (11.1.3) can now be rewritten using only the weaker (11.1.5) (or (11.1.8)).

We only state those results for which we will provide applications.

Let us introduce the notation

$$
\Psi_i^a = \begin{cases} 1 + \theta_i^a + \sqrt{2\theta_i^a + r_i^a}, & \text{if HLCP is monotone,} \\[2mm] 1 + q_{ia} + \sqrt{2q_{ia} + q_{ia}^2}, & \text{if HLCP is skew-symmetric} \end{cases} \tag{11.1.25}
$$

where $\sqrt{r_i^a} = \theta_i^a$, $\sqrt{t_i^a} = k_i^a$, $a = 0, 1$,

$$
\theta_i^a = t_i \left[1 + \frac{t_i^a}{1 - t_i^a} \right], \quad q_{ia} = \frac{t_i^a}{2}, \quad i = 1, 2. \tag{11.1.26}
$$

Then by simply replacing k, k_1, k_2 by k^0, k_1^0, k_2^0, respectively, in the corresponding results in [208], we obtain the following improvements:

Theorem 11.1.2. *The parameter θ determined at each outer iteration of Algorithm 11.1.1 satisfies*

$$
\theta \geq \frac{\chi^a}{\sqrt{n}} = \lambda^a
$$

where

$$
\chi^a = \begin{cases} \dfrac{\sqrt{2}\left(k_2^a - k_1^a\right)}{\sqrt{2 + p^2 t_i}\sqrt{\psi_1^a}}, & \begin{array}{c}\text{if HLCP is skew-symmetric or if} \\ \text{no simplified Newton-Kantorovich} \\ \text{steps are performed,}\end{array} \\[6mm] \dfrac{\sqrt{2}\left(k_2^a - k_1^a\right)}{\left(\sqrt{2} + p k_1^a\right)\sqrt{\psi_1^a}}, & \text{otherwise} \end{cases} \tag{11.1.27}
$$

where

$$
p = \begin{cases} \sqrt{2}, & \text{if HLCP is monotone,} \\[2mm] 1, & \text{if HLCP is skew-symmetric.} \end{cases} \tag{11.1.28}
$$

Clearly, the lower bound on λ^a on θ is an improvement over the corresponding one in [208, Corollary 4].

In the next result, a bound on the number of steps of the inner iteration that depends only on k_1^0 and k_2^0 is provided.

Theorem 11.1.3. *If Newton-Kantorovich method is used in Algorithm 11.1.1 then each inner iteration terminates in at most $N^0\left(k_1^0, k_2^0\right)$ steps, where*

$$N^0\left(k_1^0, k_2^0\right) = \text{integer part}\left[\log_2\left(\frac{\log_2\left(x_{N^0}\right)}{\log_2\left[\left(1 - \sqrt{1 - 2k_2^0} - k_2^0\right)/k_2^0\right]}\right)\right]$$

(11.1.29)

and

$$x_{N^0} = \frac{\left(1 - \frac{pk_2^0}{\sqrt{2}}\right)\left[t_2^0 - \left(1 - \sqrt{1 - 2k_2^0}\right)^2\right]k_1^0}{2\sqrt{2}t_2^0\sqrt{1 - 2k_2^0}\left[\sqrt{\psi_2^0} + 1 - \sqrt{1 - 2k_2^0}\right]\left(1 + k_1^0\right)}.$$

(11.1.30)

If the modified the Newton-Kantorovich method is used in Algorithm 11.1.1 then each iteration terminates in at most $S^0\left(k_1, k_2\right)$ steps, where

$$S^1\left(k_1^1, k_2^1\right) = \text{integer part}\left[\frac{\log_2\left(x_{S^1}\right)}{\log_2\left(1 - \sqrt{1 - \sqrt{1 - 2k_2^1}}\right)}\right] + 1$$

(11.1.31)

and

$$x_{S^1} = \frac{\left(1 - \frac{pk_2^1}{\sqrt{2}}\right)\left[t_2^1 - \left(1 - \sqrt{1 - 2k_2^1} - k_2^1\right)^2\right]k_1^0}{2\sqrt{2}\sqrt{1 - 2k_2^1}\left(1 - \sqrt{1 - 2k_2^1} - k_2^1\right)^2\left(\sqrt{\psi_2^1} + 1 - \sqrt{1 - 2k_2^1}\right)\left(1 + k_1^1\right)}.$$

Clearly, if $k_1^1 = k_1^0 = k_1$, $k_2^1 = k_2^0 = k_2$, $k^1 = k^0 = k$, Theorem 11.1.2 reduces to the corresponding Theorem 2 in [208]. Otherwise the following improvement holds:

$$N^0\left(k_1^0, k_2^0\right) < N\left(k_1, k_2\right), \qquad\qquad N^0 < N,$$

$$S^1\left(k_1^1, k_2^1\right) < S\left(k_1, k_2\right), \qquad\qquad \text{and } S^1 < S.$$

Because $\frac{k_1}{k_1^0}, \frac{k_2}{k_2^0}, \frac{k_1}{k_1^1}, \frac{k_2}{k_2^1}$ can be arbitrarily large for a given triplet η, ℓ and ℓ_0, the choices

$$k_1^0 = k_1^1 = .12, \quad k_2^0 = k_2^1 = .24 \text{ when } k_1 = .21 \text{ and } k_2 = .42$$

and

$$k_1^0 = k_1^1 = .24, \quad k_2^0 = k_2^1 = .48 \text{ when } k_1 = .245 \text{ and } k_2 = .49$$

are possible. Then using formulas (11.1.28), (11.1.29), and (11.1.31) for our results and (9)–(11) in [208], we obtain the following tables:

(a) If the HLCP is monotone and only Newton directions are performed, then:

Potra	Argyros
$\chi(.21, .42) > .17$	$\chi^0(.12, .24) > .1$
$\chi(.245, .49) > .199$	$\chi^0(.24, .48) > .196$

Potra	Argyros
$N(.21, .42) = 2$	$N^0(.12, .24) = 1$
$N(.245, .49) = 4$	$N^0(.24, .48) = 3$

(b) If the HLCP is monotone and Modified Newton directions are performed:

Potra	Argyros
$\chi(.21, .42) > .149$	$\chi^1(.12, .24) > .098$
$\chi(.245, .49) > .164$	$\chi^1(.24, .48) > .162$

Potra	Argyros
$S(.21, .42) = 5$	$S^1(.12, .24) = 1$
$S(.245, .49) = 18$	$S^1(.24, .48) = 12$

All the above improvements are obtained under weaker hypotheses and the same computational cost (in the case of Newton's method) or less computational cost (in the case of the modified Newton method) as in practice the computation of ℓ requires that of ℓ_0 and in general the computation of ℓ_0 is less expensive than that of ℓ.

11.2 Case 2: LP methods

We are motivated by paper [174], where a unified complexity analysis for Newton LP methods was given. Here we show that it is possible under weaker hypotheses to provide a finer semilocal convergence analysis, which in turn can reduce the computational time for interior algorithms appearing in linear programming and convex quadratic programming.

Finally, an example is provided to show that our results apply where the ones in [174] cannot.

Let us assume that $F: D \subseteq X \to Y$ is an analytic operator and $F'(x)^{-1}$ exists at $x = x_0 \in D$. Define

$$\eta = \eta(x_0) = \|F'(x_0)^{-1} F(x_0)\| \tag{11.2.1}$$

and

$$\gamma = \gamma(x_0) = \sup_{k \geq 2} \left\| \frac{1}{k!} F'(x_0)^{-1} F^{(k)}(x) \right\|^{\frac{1}{k-1}}. \tag{11.2.2}$$

Note that $\eta(x_0)$ is the step length at x_0 when NK method is applied to approximate x^*.

Smale in [184] showed that if $D = X$, and

$$\gamma \eta \leq \frac{1}{8} = .125, \tag{11.2.3}$$

then Newton-Kantorovich method (2.1.3) converges to x^*, so that

$$\|x_{n+1} - x_n\| \leq 2 \left(\frac{1}{2} \right)^n \|x_1 - x_0\| \quad (n \geq 0). \tag{11.2.4}$$

Rheinboldt in [177] using the Newton-Kantorovich theorem showed convergence assuming

$$D \subseteq X \quad \text{and} \quad \gamma \eta \leq .11909565. \tag{11.2.5}$$

We need the following semilocal convergence result for Newton-Kantorovich method (2.1.3) and twice Fréchet-differentiable operators:

Theorem 11.2.1. *Let* $F: D \subseteq X \to Y$ *be a twice Fréchet-differentiable operator. Assume there exist a point* $x_0 \in D$ *and parameters* $\eta \geq 0$, $\ell_0 \geq 0$, $\ell \geq 0$, $\delta \in [0, 1]$ *such that:*

$$F'(x_0)^{-1} \in L(Y, X), \tag{11.2.6}$$

$$\|F'(x_0)^{-1} [F'(x) - F'(x_0)]\| \leq \ell_0 \|x - x_0\|, \tag{11.2.7}$$

$$\|F'(x_0)^{-1} F''(x)\| \leq \ell \tag{11.2.8}$$

for all $x \in D$,

$$\overline{U}(x_0, t^*) \subseteq D, \tag{11.2.9}$$

and

$$h_\delta = (\ell + \delta \ell_0) \eta \leq \delta, \tag{11.2.10}$$

where,

$$t^* = \lim_{n \to \infty} t_n, \tag{11.2.11}$$

with

$$t_0 = 0, \ t_1 = \eta, \ t_{n+2} = t_{n+1} + \frac{\ell(t_{n+1} - t_n)^2}{2(1 - \ell_0 t_{n+1})} \quad (n \geq 0). \tag{11.2.12}$$

Then

(a) sequence $\{x_n\}$ $(n \geq 0)$ generated by NK method (2.1.3) is well defined, remains in $\overline{U}(x_0, t^*)$ for all $n \geq 0$, and converges to a unique solution x^* of equation $F(x) = 0$ in $\overline{U}(x_0, t^*)$. Moreover, the following estimates hold for all $n \geq 0$:

$$\|x_{n+2} - x_{n+1}\| \leq \frac{\ell\|x_{n+1} - x_n\|^2}{2(1 - \ell_0\|x_{n+1} - x_0\|)} \leq t_{n+2} - t_{n+1}, \tag{11.2.13}$$

$$\|x_n - x^*\| \leq t^* - t_n \leq t_0^{**} - t_n \leq \alpha\eta, \tag{11.2.14}$$

where,

$$t_0^{**} = \left[1 + \frac{\delta^0}{2} + \frac{1}{1 - \frac{\delta}{2}} \frac{\delta}{2} \frac{\delta^0}{2}\right]\eta, \ t^{**} = \alpha\eta, \ \alpha = \frac{2}{2 - \delta}, \tag{11.2.15}$$

and

$$\delta^0 = \frac{\ell\eta}{1 - \ell_0\eta}. \tag{11.2.16}$$

Furthermore the solution x^* is unique in $U(x_0, R)$, $R = \frac{2}{\ell_0} - t^*$ provided that

$$U(x_0, R) \subseteq D, \ and \ \ell_0 \neq 0. \tag{11.2.17}$$

(b) If

$$H_\alpha = (\ell + 2\ell_0\alpha)\eta < 2, \tag{11.2.18}$$

then

$$0 \leq a = \frac{\ell\eta}{2(1 - \ell_0\alpha\eta)} < 1 \tag{11.2.19}$$

and

$$\|x_{n+1} - x_n\| \leq a_0\|x_n - x_{n-1}\|^2 \leq a^{2^n - 1}\eta \ for \ all \ n \geq 0, \tag{11.2.20}$$

$$\|x_n - x^*\| \leq t^* - t_n \leq a^{2^n - 1}\overline{\overline{b}}\eta \leq a^{2^n - 1}\overline{b}\eta \leq a^{2^n - 1}b\eta, \ for \ all \ n \geq 0 \tag{11.2.21}$$

where,

$$a_0 = \frac{\ell}{2(1 - \ell_0\alpha\eta)}, \ \overline{\overline{b}} = 1 + a^{2^n} + a^{2^{n+1}} + \cdots, \tag{11.2.22}$$

$$\overline{b} = 1 + \frac{a^{2^n}}{1 - a^2} \ and \ b = \frac{2 - a^2}{a^2}. \tag{11.2.23}$$

Proof. Part (a) follows immediately as a special case of Theorem 5.3.3. For part (b) note that (11.2.18) implies (11.2.19), whereas (11.2.20) is a consequence of Proposition 3 in [184, p. 193]. Finally (11.2.21) follows from (11.2.20).

Corollary 11.2.2. *If*

$$2a = \frac{\ell\eta}{1 - \ell_0\alpha\eta} \leq 1 \tag{11.2.24}$$

then under the rest of the hypotheses of Theorem 11.2.1

$$\|x_{n+1} - x_n\| \leq 2\left(\frac{1}{2}\right)^{2^n} \eta \quad (n \geq 0). \tag{11.2.25}$$

That is x_0 is an approximate zero of F and sequence $\{x_n\}$ $(n \geq 0)$ converges to x^* at least R-quadratically with R_2 factor $\frac{1}{2}$ [1], [184].

Let $\gamma \geq 0$ and $\lambda \geq 0$. It is convenient for us to define parameters

$$\beta_0 = \gamma\eta, \quad \beta_1 = \gamma\lambda, \quad \beta = \gamma\alpha\eta \tag{11.2.26}$$

and functions f, f_1, g, h by

$$f(\delta, \beta_0, \beta_1, \beta) = \frac{(1 - \beta_1)^4}{\left[2(1 - \beta_1)^2 - 1\right]^2 (1 - \beta)^2}\left[\frac{2}{1 - \beta} + \delta(2 - \beta)\right]\left(\beta_0 + \frac{\beta_1}{1 - \beta_1}\right), \tag{11.2.27}$$

$$f_1(\alpha, \beta_0, \beta_1) = \frac{\alpha(1 - \beta_1)^2}{2(1 - \beta_1)^2 - 1}\left(\beta_0 + \frac{\beta_1}{1 - \beta_1}\right) + \beta_1 - \frac{\left[2(1 - \beta_1)^2 - 1\right](1 - \beta)^2}{(1 - \beta_1)^2(2 - \beta)}, \tag{11.2.28}$$

$$g(\alpha, \beta_0, \beta_1, \beta) = \frac{2(1 - \beta_1)^4}{\left[2(1 - \beta_1)^2 - 1\right]^2 (1 - \beta)^2}\left[\frac{1}{1 - \beta} + \alpha(2 - \beta)\right]\left(\beta_0 + \frac{\beta_1}{1 - \beta_1}\right), \tag{11.2.29}$$

$$h(\alpha, \beta_0, \beta_1, \beta) = \frac{(1 - \beta_1)^4}{\left[2(1 - \beta_1)^2 - 1\right]^2 (1 - \beta)^2}\left[\frac{2}{1 - \beta} + \alpha(2 - \beta)\right]\left(\beta_0 + \frac{\beta_1}{1 - \beta_1}\right), \tag{11.2.30}$$

respectively, provided that $\beta \neq 1$, $\beta_1 \neq 1$ and $\beta_1 \neq \frac{2-\sqrt{2}}{2}$.

We can state and prove the main semilocal convergence theorem for method (6.1.3) involving analytic operators:

Theorem 11.2.3. *Let $F: D \subseteq X \to Y$ be an analytic operator. Assume there exist points $x_0, \bar{x}_0 \in D$ and parameters $\eta \geq 0, \gamma \geq 0, \lambda \geq 0, \delta \in [0, 1]$ such that*

$$\|x_0 - \bar{x}_0\| \leq \lambda \leq \alpha\eta, \tag{11.2.31}$$

$$\alpha\gamma\eta < 1, \tag{11.2.32}$$

$$2\lambda\gamma < 2 - \sqrt{2}, \tag{11.2.33}$$

$$f(\delta, \beta_0, \beta_1, \beta) \leq \delta, \tag{11.2.34}$$

$$f_1(\alpha, \beta_0, \beta_1) \leq 0, \tag{11.2.35}$$

$$\overline{U}(x_0, \alpha\eta) \subseteq D \tag{11.2.36}$$

and condition (11.2.6) hold. Then sequence $\{x_n\}$ $(n \geq 0)$ starting at $\overline{x}_0$ is well defined, remains in $\overline{U}(x_0, \alpha\eta)$ for all $n \geq 0$ and converges to a unique zero x^* of F in $\overline{U}(x_0, \alpha\eta)$, so that estimates (11.2.13) and (11.2.14) hold for η, ℓ, ℓ_0 replaced by

$$\overline{\eta} = \frac{(1 - \beta_1)^2}{2(1 - \beta_1)^2 - 1} \left[\eta + \frac{\lambda}{1 - \beta_1} \right], \tag{11.2.37}$$

$$\overline{\ell} = \frac{2\gamma(1 - \beta_1)^2}{[2(1 - \beta_1)^2 - 1](1 - \beta)^3}, \tag{11.2.38}$$

and

$$\overline{\ell}_0 = \frac{(1 - \beta_1)^2(2 - \beta)\gamma}{[2(1 - \beta_1)^2 - 1](1 - \beta)^2}, \tag{11.2.39}$$

respectively. Moreover the solution x^* is unique in $U(x_0, \overline{R})$, $\overline{R} = \frac{2}{\ell_0} - t^*$ provided that

$$U(x_0, \overline{R}) \subseteq D, \quad and \quad \gamma \neq 0. \tag{11.2.40}$$

Furthermore if

$$g(\alpha, \beta_0, \beta_1, \beta) < 2, \tag{11.2.41}$$

then estimates (11.2.20)–(11.2.21) hold (with η, ℓ, ℓ_0 replaced by $\overline{\eta}, \overline{\ell}$ and $\overline{\ell}_0$ respectively). Finally if

$$h(\alpha, \beta_0, \beta_1, \beta) \leq 1, \tag{11.2.42}$$

then $\overline{x}_0$ (and x_0) is an approximate zero of F and sequence $\{x_n\}$ $(n \geq 0)$ converges to x^* at least R-quadratically with R_2 factor $\frac{1}{2}$, so that estimates (11.2.24) and (11.2.25) hold.

Proof. We shall show that under the stated conditions, the hypotheses of Theorem 11.2.1 hold with $\overline{x}_0, \overline{\eta}, \overline{\ell}_0$ and ℓ replacing x_0, η, ℓ_0 and ℓ, respectively. We first obtain a bound on $\|x_0 - x^*\|$. For all $x \in U(x_0, \alpha\eta)$, we obtain in turn

$$\|F'(x_0)^{-1}F''(x)\| = \left\| F'(x_0)^{-1} \sum_{i=0}^{\infty} \frac{1}{i!} F^{(2+i)}(x_0)(x - x_0)^i \right\|$$

$$\leq \gamma \sum_{i=0}^{\infty} (i + 2)(i + 1) [\gamma \|x - x_0\|]^i$$

$$\leq \frac{2\gamma}{(1 - \gamma \|x - x_0\|)^3} \leq \frac{2\gamma}{(1 - \gamma\alpha\eta)^3} = \frac{2\gamma}{(1 - \beta)^3}. \tag{11.2.43}$$

According to Theorem 11.2.1, there exists a unique zero x^* of F in $U(x_0, t^*)$. Therefore NK sequence starting at $\overline{x}_0$ converges to the same zero x^*. We can have

$$\left\| F'(x_0)^{-1} \left[F'(\overline{x}_0) - F'(x_0) \right] \right\| = \left\| F'(x_0)^{-1} \sum_{i=1}^{\infty} \frac{1}{i!} F^{(i+1)}(x_0)(\overline{x}_0 - x_0)^i \right\|$$

$$\leq \sum_{i=1}^{\infty}(i+1)\left[\gamma \|\overline{x}_0 - x_0\| \right]^i \leq \frac{1}{(1 - \gamma \|\overline{x}_0 - x_0\|)^2} - 1$$

$$\leq \frac{1}{(1 - \gamma\lambda)^2} - 1 \leq \frac{1}{(1 - \beta_1)^2} - 1 < 1 \quad \text{(by (11.2.33))}. \qquad (11.2.44)$$

It follows by the identity

$$B^{-1} = \sum_{i=0}^{\infty}(-1)^i \left[A^{-1}(B - A) \right]^i A^{-1}$$

and (11.2.44) that for any $x \in X$,

$$\|F'(\overline{x}_0)^{-1}x\| \leq \frac{(1 - \beta_1)^2}{2(1 - \beta_1)^2 - 1} \|F'(x_0)^{-1}x\|. \qquad (11.2.45)$$

Hence, we can have

$$\|F'(\overline{x}_0)^{-1}F(\overline{x}_0)\| \leq \frac{(1 - \beta_1)^2}{2(1 - \beta_1)^2 - 1} \|F'(x_0)^{-1}F(\overline{x}_0)\|$$

$$\leq \frac{(1 - \beta_1)^2}{2(1 - \beta_1)^2 - 1} \left\| F'(x_0)^{-1} \sum_{i=0}^{\infty} \frac{1}{i!} F^{(i)}(x_0)(\overline{x}_0 - x_0)^i \right\|$$

$$\leq \frac{(1 - \beta_1)^2}{2(1 - \beta_1)^2 - 1} \left(\eta + \|\overline{x}_0 - x_0\| \sum_{j=0}^{\infty} \left[\gamma \|\overline{x}_0 - x_0\| \right]^j \right)$$

$$\leq \frac{(1 - \beta_1)^2}{2(1 - \beta_1)^2 - 1} \left(\eta + \frac{\lambda}{1 - \gamma\lambda} \right) \equiv \overline{\eta}.$$
$$(11.2.46)$$

Moreover, for $x \in U(\overline{x}_0, \alpha\overline{\eta})$ we have $x \in U\left(x_0, \frac{1}{\ell_0}\right)$, as

$$\|x - \overline{x}_0\| \leq \|x - x_0\| + \|x_0 - \overline{x}_0\| \leq \alpha\overline{\eta} + \lambda \leq \frac{1}{\ell_0} \quad \text{(by (11.2.35))}, \qquad (11.2.47)$$

and $\overline{\ell}_0$ is given by (11.2.39) where we have used $\overline{x}_0$ for x in (11.2.44) and (11.2.45) (to obtain $\overline{\ell}_0$). Therefore, we get

$$\|F'(\overline{x}_0)^{-1}F''(x)\| \leq \frac{(1 - \beta_1)^2}{2(1 - \beta_1)^2 - 1} \|F'(x_0)^{-1}F''(x)\|$$

$$\leq \frac{(1 - \beta_1)^2}{2(1 - \beta_1)^2 - 1} \frac{2\gamma}{(1 - \gamma\beta_1)^3} \equiv \overline{\ell}. \qquad (11.2.48)$$

Theorem 11.2.1 and Corollary 11.2.2 can now apply with $\bar{x}_0$, $\bar{\eta}$, $\bar{\ell}_0$ and $\bar{\ell}$ as hypotheses (11.2.10), (11.2.18), and (11.2.24) become (11.2.34), (11.2.41), and (11.2.42), respectively.

Remark 11.2.4. In general

$$\ell_0 \leq \ell \tag{11.2.49}$$

holds and $\frac{\ell}{\ell_0}$ can be arbitrarily large. Note that the famous Newton-Kantorovich hypothesis (2.2.17) is a special case of corresponding condition (11.2.10). Simply set $\ell_0 = \ell$ and $\delta = 1$ in (11.2.10) to obtain (2.2.17). Our conditions (11.2.10) and (11.2.24) for $\alpha = \frac{3}{2}$ become

$$\left(\ell + \frac{2}{3}\ell_0\right)\eta \leq \frac{2}{3}, \tag{11.2.50}$$

and

$$\left(\ell + \frac{3}{2}\ell_0\right)\eta \leq 1, \tag{11.2.51}$$

respectively, which are weaker than condition

$$\ell\eta \leq \frac{4}{9} \tag{11.2.52}$$

given in [184, p. 12] (for sufficiently small ℓ_0). Note that all the above benefits are obtained under the same computational cost as in practice the computation of ℓ requires that of ℓ_0. Consequently, the same benefits obtained in Section 2.2 carry over when we compare our Theorem 11.2.3 (based on Theorem 11.2.1 (or Corollary 11.2.2)) with Theorem 2 in [184, p. 12].

Let us provide an example where conditions of Theorem 11.2.3 hold but corresponding crucial condition:

$$\eta \leq \frac{1}{2}\lambda \leq \frac{1}{40\gamma} \tag{11.2.53}$$

in Theorem 2 in [184, p. 12] fails.

Example 11.2.5. Let

$$\lambda = \frac{3}{2}\eta, \quad \delta = \frac{2}{3} \tag{11.2.54}$$

and

$$\beta_0 = \gamma\eta = \frac{1}{22} = .0\overline{45}. \tag{11.2.55}$$

Then using (11.2.15) and (11.2.26) we get

$$\alpha = \frac{3}{2} \text{ and } \beta = \beta_1 = .06\overline{81}.$$

Conditions (11.2.34), (11.2.35), (11.2.41), and (11.2.42) are satisfied, as

$$f(\delta, \beta_0, \beta_1, \beta) = .651987692 < \delta = \frac{2}{3},$$

$$f_1(\alpha, \beta_0, \beta_1) = -.10652731 \le 0,$$

$$g(\alpha, \beta_0, \beta_1, \beta) = 1.50775228 < 2,$$

and

$$h(\alpha, \beta_0, \beta_1, \beta) = .943087149 < 1.$$

That is the conditions of our Theorem 11.2.3 hold but condition (11.2.53) required for the application of Theorem 2 in [184] does not hold. Hence the conclusions of our Theorem 11.2.3 hold but not the ones in Theorem 2 in [184] in this case.

These facts influence (widens) choices. Indeed by simply repeating the proofs given in Sections 2–6 in [184], with the above changes we can show:

Application 11.2.6. *"Optimal" sequences $\{t^{(i)}\}$ $(i \ge 0)$ given by*

$$t^{(i+1)} = \left(1 - \frac{1}{41\sqrt{m}}\right) t^{(i)} \quad \text{(LP Barrier Method, QP Barrier Method)},$$

$$t^{(i+1)} = \left(1 - \frac{1}{40\sqrt{m}}\right) t^{(i)} \quad \text{(Primal-dual LP Algorithm,}$$

Primal-dual QP Algorithm)

can be replaced by wider

$$\bar{t}^{(i+1)} = \left(1 - \frac{1}{23\sqrt{m}}\right) \bar{t}^{(i)}$$

and

$$\bar{t}^{(i+1)} = \left(1 - \frac{1}{22\sqrt{m}}\right) \bar{t}^{(i)}$$

respectively.

The rest of the results in [184] can also be improved if rewritten using our Theorem 11.2.3 above. However, we leave the details to the motivated reader.

The observations/improvements made here are important in computational mathematics and scientific computing.

11.3 Exercises

11.3.1. (a) [177] Use the Newton-Kantorovich theorem to show that if: $F: D \subseteq X \to Y$ is analytic on X, $x_0 \in D$ with $F'(x_0)^{-1} \in L(Y, X)$,

$$\gamma\eta \leq \frac{\alpha}{\sqrt{2}} = .11909565,$$

where α is the positive root of the cubic equation $\left(\sqrt{2}-1\right)(1-r)^3 - \sqrt{2}r = 0$, and $U(u_0, r_0) \subseteq D$, where $r_0 = \frac{\alpha}{\gamma}$, then x_0 is an approximate zero of F.

(b) If $D = X$ and $\gamma\eta \leq \alpha_1 = .15229240$, where α_1 is the positive zero of cubic equation $(1-r)^3 - 4r = 0$, then show x_0 is an approximate zero of F.

(c) If $D = X$ and $\gamma\eta = 3 - 2\sqrt{2} = .171573$ show the quadratic convergence of NK method (2.1.3) but with ratio not necessarily smaller or equal to .5.

11.3.2. [184] Let $F: D \subseteq X \rightarrow Y$ be a twice continuously Fréchet-differentiable operator. Assume there exist $x_0 \in D$, $\eta > 0$, $\ell > 0$ such that $F'(x_0)^{-1} \in L(Y, X)$, $\left\| F'(x_0)^{-1} F(x_0) \right\| \leq \eta$, $\left\| F'(x_0)^{-1} F''(x) \right\| \leq \ell$, for all $x \in D$,

$$\gamma\eta \leq \frac{4}{9} \text{ and } U\left(x_0, \frac{3}{2}\eta\right) \subseteq D.$$

Then show that the Newton-Kantorovich method is well defined, remains in $U\left(x_0, \frac{3}{2}\eta\right)$ for all $n \geq 0$, converges to a unique zero x^* of F in $D \cap U\left(x_0, \frac{1}{\ell}\right)$, and satisfies

$$\|x_n - x^*\| \leq 2\left(\frac{1}{2}\right)^{2^n} \eta, \quad (n \geq 1).$$

11.3.3. [174] Let $F: D \subseteq X \rightarrow Y$ be analytic. Assume there exist $x_0 \in D$, $\delta \geq 0$ such that

$$F'(x_0)^{-1} \in L(Y, X),$$

$$\eta \leq \frac{1}{2}\delta \leq \frac{1}{40\gamma} \text{ and}$$

$$\bar{U}(x_0, 4\delta) \subseteq D.$$

If $\|\bar{x} - x_0\| \leq \delta$, then the Newton-Kantorovich method starting at $\bar{x}$ is well defined and converges to a unique zero x^* of F in $\bar{U}\left(x_0, \frac{3}{2}\eta\right)$ so that

$$\|x_n - x^*\| \leq \frac{7}{2}\left(\frac{1}{2}\right)^{2^n} \delta.$$

11.3.4. Consider the LP barrier method [174]. Let $Int = \{x | A(x) > b\}$, A is a real matrix with a_i denoting the ith row of A, and let $h: Int \times \mathbf{R}_+ \rightarrow \mathbf{R}_+$ denote the map $h(x, t) = C^T x - t \sum \ln(a_i x - b_i)$. For fixed t, the map $x \rightarrow h(x, t)$ is strictly convex having a unique minimum. The sequence of minima as $t \rightarrow 0$ converges to the optimal solution of the LP. The algorithm simply computes a Newton-Kantorovich sequence $\{x_n\}$ $(n \geq 0)$, where

$$x_{n+1} = x_n - \nabla_x^2 h(x_n, t_{n+1})^{-1} \nabla_x h(x_n, t_{n+1})$$

and $t_n \rightarrow 0$.
Show:

(a) For suitable x_0, we can always choose $t_{n+1} = \left(1 - \frac{1}{41\sqrt{m}}\right) t_n$, and each x_n will be a "good" approximation to the minimum of the map $x \rightarrow h(x, t_n)$. Hint: Let y be the minimum of $x \rightarrow h(x, t)$, assume $\|\bar{y} - y\| \le \frac{1}{20}$ and show

$$\|\bar{y}' - y'\|' \le \frac{1}{20},$$

where $\|x\| = \|\Delta(y)^{-1} A(x)\|_2$ and for $t' > 0$, y' and $\|\cdot\|'$ are defined in the obvious way. Then use Exercise 11.3.3.

(b) An $O\left(\sqrt{m}L\right)$ iteration bound.

(c) The choice of sequence $\{t_n\}$ above can be improved if 41 is replaced by any $a \in \{20, 21, ..., 40\}$.
    Hint: Use Theorem 11.2.1. Note that this is an improvement over the choice in (a).

(d) An $O\left(\sqrt{m}L\right)$ iteration bound using choices of sequence $\{t_n\}$ given in (c).

(e) How to improve along the same lines as above the QP barrier method, the primal LP algorithm, the primal dual LP algorithm, and the primal-dual QP algorithm as introduced in [174].

References

1. Alizadeh, F., Haeberly, J.-P.A., Overton, M.L., Primal-dual interior point algorithms for semidefinite programming: stability, convergence and numerical results, SIAM J. Optim. **8** (1988), 743–768.

2. Allgower, E.L., A survey of homotopy methods for smooth mappings, Lecture Notes in Math., vol. 878, Springer-Verlag, 1980, 1–29.

3. Allgower, E.L., Böhmer, K., Potra, F.A., Rheinboldt, W.C., A mesh-independence principle for operator equations and their discretizations, SIAM J. Numer. Anal. **23** (1986), no. 1, 160–169.

4. Amat, S., Busquier, S., Gutiérrez, J.M., On the local convergence of secant-type methods, Intern. J. Comput. Math. **81** (2004), 1153–1161.

5. Anselone, P.M., Moore, R.H., An extension of the Newton-Kantorovich method for solving nonlinear equations with applications to elasticity, J. Math. Anal. Appl. **13** (1996), 476–501.

6. Appell, J., DePascale, E., Lysenko, J.V., Zabrejko, P.P., New results on Newton-Kantorovich approximations with applications to nonlinear integral equations, Numer. Funct. Anal. Optimiz. **18** (1997), no. 1 & 2, 1–17.

7. Argyros, I.K., Quadratic equations and applications to Chandrasekhar's and related equations, Bull. Austral. Math. Soc. **32** (1985), 275–297.

8. Argyros, I.K., On the cardinality of solutions of multilinear differential equations and applications, Intern. J. Math. Math. Sci. **9** (1986), no. 4, 757–766.

9. Argyros, I.K., On the approximation of some nonlinear equations, Aequationes Mathematicae **32** (1987), 87–95.

10. Argyros, I.K., On polynomial equations in Banach space, perturbation techniques and applications, Intern. J. Math. Math. Sci. **10** (1987), no. 1, 69–78.

11. Argyros, I.K., Newton-like methods under mild differentiability conditions with error analysis, Bull. Austral. Math. Soc. **37** (1987), 131–147.

12. Argyros, I.K., Improved error bounds for the modified secant method, Intern. J. Computer Math. **43** (1992), no. 1 & 2, 99–109.

13. Argyros, I.K., Some generalized projection methods for solving operator equations, J. Comp. Appl. Math. **39** (1992), no. 1, 1–6.

14. Argyros, I.K., On the convergence of generalized Newton-methods and implicit functions, J. Comp. Appl. Math. **43** (1992), 335–342.

15. Argyros, I.K., On the convergence of inexact Newton-like methods, Publ. Math. Debrecen **42** (1992), no. 1 & 2, 1–7.

16. Argyros, I.K., On the convergence of a Chebysheff-Halley-type method under Newton-Kantorovich hypothesis, Appl. Math. Lett. **5** (1993), no. 5, 71–74.

17. Argyros, I.K., Newton-like methods in partially ordered linear spaces, J. Approx. Th. Applic. **9** (1993), no. 1, 1–10.

18. Argyros, I.K., On the solution of undetermined systems of nonlinear equations in Euclidean spaces, Pure Math. Appl. **4** (1993), no. 3, 199–209.

19. Argyros, I.K., A convergence theorem for Newton-like methods under generalized Chen–Yamamato-type assumptions, Appl. Math. Comp. **61** (1994), no. 1, 25–37.

20. Argyros, I.K., On the discretization of Newton-like methods, Int. J. Computer. Math. **52** (1994), 161–170.

21. Argyros, I.K., A unified approach for constructing fast two-step Newton-like methods, Mh. Math. **119** (1995), 1–22.

22. Argyros, I.K., Results on controlling the residuals of perturbed Newton-like methods on Banach spaces with a convergence structure, Southwest J. Pure Appl. Math. **1** (1995), 32–38.

23. Argyros, I.K., On the method of tangent hyperbolas, J. Appr. Th. Appl. **12** (1996), no. 1, 78–96.

24. Argyros, I.K., On an extension of the mesh-independence principle for operator equations in Banach space, Appl. Math. Lett. **9** (1996), no. 3, 1–7.

25. Argyros, I.K., A generalization of Edelstein's theorem on fixed points and applications, Southwest J. Pure Appl. Math. **2** (1996), 60–64.

26. Argyros, I.K., Advances in the efficiency of computational methods and applications, World Scientific Publ. Co., River Edge, NJ, USA, 2000.

27. Argyros, I.K., On the convergence of a Newton-like method based on m-Fréchet-differentiable operators and applications in radiative transfer, J. Comput. Anal. Applic. **4** (2002), no. 2, 141–154.

28. Argyros, I.K., On the convergence of Newton-like methods for analytic operators and applications, J. Appl. Math. Computing **10** (2002), no. 1–2, 41–50.

29. Argyros, I.K., A unifying semilocal convergence theorem for Newton-like methods based on center Lipschitz conditions, Comput. Appl. Math. **21** (2002), no. 3, 789–796.

30. Argyros, I.K., A semilocal convergence analysis for the method of tangent hyperbolas, Journal of Concrete and Applicable Analysis **1** (2002), no. 2, 135–144.

31. Argyros, I.K., New and generalized convergence conditions for the Newton-Kantorovich method, J. Appl. Anal. **9** (2003), no. 2.

32. Argyros, I.K., On the convergence and application of Newton's method under weak Hölder continuity assumptions, Int. J. Computer Math. **80** (2003), no. 5, 767–780.

33. Argyros, I.K., On a theorem of L.V. Kantorovich concerning Newton's method, J. Comp. Appl. Math. **155** (2003), 223–230.

34. Argyros, I.K., An improved error analysis for Newton-like methods under generalized conditions, J. Comput. Appl. Math. **157** (2003), no. 1, 169–185.

35. Argyros, I.K., An improved convergence analysis and applications for Newton-like methods in Banach space, Numer. Funct. Anal. Optimiz. **24** (2003), no. 7 and 8, 653–672.

36. Argyros, I.K., On the convergence and application of generalized Newton methods, Nonlinear Studies **10** (2003), no. 4, 307–322.

37. Argyros, I.K., On the comparison of a weak variant of the Newton-Kantorovich and Miranda theorems, J. Comp. Appl. Math. **166** (2004), no. 2, 585–589.

38. Argyros, I.K., A convergence analysis and applications for the Newton-Kantorovich method in K-normed spaces, Rendiconti del Circolo Mathematico di Palermo **LIII** (2004), 251–271.

39. Argyros, I.K. On the Newton-Kantorovich hypothesis for solving nonlinear equations, J. Comput. Appl. Math. (2004).

40. Argyros, I.K., A note on a new way for enlarging the convergence radius for Newton's method, Math. Sci. Res. J. **8** (2004), no. 5, 147–153.

41. Argyros, I.K., Toward a unified convergence theory for Newton-like methods of "bounded deterioration", Adv. Nonlinear Var. Inequal. **8** (2005), no. 2, 109–120.

42. Argyros, I.K. and Szidarovszky, F., Convergence of general iteration schemes, J. Math. Anal. Applic. **168** (1992), 42–62.

43. Argyros, I.K., Szidarovszky, F., The Theory and Application of Iteration Methods, C.R.C. Press, Boca Raton, Florida, 1993.

44. Argyros, I.K., Szidarovszky, F., On the convergence of modified contractions, J. Comput. Appl. Math., **55** (1994), no. 2, 97–108.

45. Atkinson, K.E., A Survey of Numerical Methods for the Solution of Fredholm Integral Equations of the Second Kind, SIAM, Philadelphia, 1976.

46. Ben-Israel, A., A Newton-Raphson method for the solution of systems of operators, J. Math. Anal. Appl. **15** (1966), 243–252.

47. Bi, W., Ren, H., Wu, Q., Local convergence analysis of a modified secant method for finding zeros of derivatives under Argyros-type condition, to appear in the J. Comput. Appl. Math.

48. Brent, R.P., Algorithms for Minimization Without Derivatives, Prentice Hall, Englewood Cliffs, New Jersey, 1973.

49. Browder, F.E., Petryshyn, W.V., The solution by iteration of linear functional equations in Banach spaces, Bull. Amer. Math. Soc. **72** (1996), 566–570.

50. Brown, P.N., A local convergence theory for combined inexact-Newton/finite-difference projection methods, SIAM J. Numer. Anal. **24** (1987), 407–434.

51. Brown, P.N., Saad, Y., Convergence theory of nonlinear Newton-Krylov algorithms, SIAM J. Optimiz. **4** (1994), no. 2, 297–230.

52. Broyden, C.G., A class of methods for solving nonlinear simultaneous equations, Math. Comput. **19** (1965), 577–593.

53. Caponetti, D., De Pascale, E., Zabreuiko, P.P., On the Newton-Kantorovich method in K-normed spaces, Rend. Circ. Mat. Palermo, Ser. II, **49** (2000), no. 3, 545–560.

54. Cătinaş, E., On some iterative methods for solving nonlinear equations, Revue d'Analyse Numérique et de Theorie de l'Approximation **23** (1994), no. 1, 47–53.

55. Cătinaş, E., Inexact perturbed Newton methods and applications to a class of Krylov solvers, J. Optim. Theory Appl. **108** (2001), no. 3, 543–570.

56. Cătinaş, E., Affine invariant conditions for the inexact perturbed Newton method, Revue d'Analyse Numérique et de Théorie de l'Approximation **31** (2002), no. 1, 17–20.

57. Cătinaş, E., Păvăloiu, I., On a third order iterative method for solving polynomial operator equations, Revue d'Analyse Numérique et de Théorie de l'Approximation **31** (2002), no. 1, 21–28.

58. Chen, X., Yamamoto, T., Convergence domains of certain iterative methods for solving nonlinear equations, Numer. Funct. Anal. Optimiz. **10** (1989), no. 1 & 2, 37–48.

59. Chen, X., Nashed, M.Z., Convergence of Newton-like methods for singular operator equations using outer inverses, Numer. Math. **66** (1993), 235–257.

60. Chen, X., Nashed, Z., Qi, L., Convergence of Newton's method for singular smooth and nonsmooth equations using adaptive outer inverses, SIAM J. Optim. **7** (1997), no. 2, 445–462.

61. Chow, S.N., Palmer, K.J., On the numerical computation of orbits of dynamical systems: The one-dimensional case, J. Dynamics Diff. Eq. **3** (1991), 361–379.

62. Chu, M.T., On a numerical treatment for curve tracing of the homotopy method, Numer. Math. **42** (1983), 323–329.

63. Cianciaruso, F., De Pascale, E., Zabrejko, P.P., Some remarks on the Newton-Kantorovich approximations, Atti. Sem. Mat. Fis. Univ. Modena **48** (2000), 207–215.

64. Ciancaruso, F., DePascale, E., Newton-Kantorovich approximations when the derivative is Hölderian: Old and new results, Numer. Funct. Anal. Optimiz. **24** (2003), no. 7 and 8, 713–723.

65. Cohen, G., Auxiliary problem principle extended to variational inequalities, J. Optim. Theory Appl. **59** (1988), no. 2, 325–333.

66. Collatz, L., Functional Analysis and Numerisch Mathematik, Springer-Verlag, New York, 1964.

67. Danes, J., Fixed point theorems, Nemyckii and Uryson operators, and continuity of non-linear mappings, Comment. Math. Univ. Carolinae **11** (1970), 481–500.

68. Danfu, H., Xinghua, W., The error estimates of Halley's method (submitted).

69. Darbo, G., Punti uniti in trasformationa codominio non compatto, Rend. Sem. Mat. Univ. Padova **24** (1955), 84–92.

70. Daubechies, I., Ten Lectures in Wavelets, Conf. Board Math. Sci. (CBMS), vol. 61, Society for Industrial and Applied Mathematics (SIAM), Philadelphia, PA, 1992.

71. Davis, H.T., Introduction to Nonlinear Differential and Integral Equations, Dover Publications, Inc., New York, 1962.

72. Decker, D.W., Keller, H.B., Kelley, C.T., Convergence rates of Newton's method at singular points, SIAM J. Numer. Anal. **20** (1983), no. 2, 296–314.

73. Dembo, R.S., Eisenstat, S.C., Steihaug, T., Inexact Newton methods, SIAM J. Numer. Anal. **19** (1982), no. 2, 400–408.

74. Dennis, J.E., Toward a unified convergence theory for Newton-like methods, In: Nonlinear Functional Anal. and Appl. (L.B. Rall, ed.), Academic Press, New York, 1971.

75. Dennis, J.E., On the convergence of Broyden's method for nonlinear systems of equations, Math. Comput. **25** (1971), no. 115, 559–567.

76. De Pascale, E., Zabrejko, P.P., New convergence criteria for the Newton-Kantorovich method and some applications to nonlinear integral equations, Rend. Sem. Mat. Univ. Padova **100** (1998), 211–230.

77. Deuflhard, P., Pesh, H.J., Rentrop, P.A., Modified continuation method for the numerical solution of nonlinear two boundary value problems by shooting techniques, Numer. Math. **26** (1978), 327–343.

78. Deuflhard, P., Heindl, G., Affine invariant convergence theorems for Newton's method, and extensions to related methods, SIAM J. Numer. Anal. **16** (1979), no. 1, 1–10.

79. Deuflhard, P., Potra, F.A., Asymptotic mesh independence of Newton-Galerkin methods and a refined Mysovskii theorem, SIAM J. Numer. Anal., **29** (1992), no. 5, 1395–1412.

80. Diallo, O.W., On the theory of linear integro-differential equations of Barbashin type in lebesgue spaces (Russian), VINITI, 1013, 88, Minsk (1988).

81. Diaconu, A., On the approximation of solutions of equations in Banach spaces using approximant sequences, In: Analysis, Functional Analysis, Functional Equations, Approximation and Convexity, Carpatica, Cluj-Napoca, 1999, 62–72.

82. Dunford, N., Schwartz, J.T., Linear operators, Part I, Int. Publ., Leyden, 1963.

83. Edelstein, M., On fixed and periodic points under contractive mappings., J. London Math. Soc. **37** (1962), 74–79.

84. Eisenstat, S.C., Walker, H.F., Globally convergent of inexact Newton methods, SIAM J. Optim. **4** (1994), no. 2, 393–422.

85. Eisenstat, S.C., Walker, H.F., Choosing the forcing terms in an inexact Newton method, SIAM J. Sci. Comput. **17** (1996), no. 1, 16–32.

86. Ezquerro, J.A., Hernández, M.A., Avoiding the computation of the second Fréchet-derivative in the convex acceleration of Newton's method., J. Comput. Appl. Math. **96** (1998), 1–12.

87. Ezquerro, J.A., Hernández, M.A., An efficient study of convergence for a fourth order two-point iteration in Banach space (submitted).

88. Ezquerro, J.A., Hernández, M.A., On a convex acceleration of Newton's method, J. Optim. Theory Appl. **100** (1999), no. 2, 311–326.

89. Ezquerro, J.A., Hernández, M.A., On the application of a fourth order two-point method to Chandrasekhar's integral equation, Aequationes Math. **62** (2001), no. 1–2, 39–47.

90. Ezquerro, J.A., Hernández, M.A., Salanova, M.A., A discretization scheme for some conservative problems, Proceedings of the 8th International Congress on Computational and Applied Mathematics, ICCAM-98 (Leuven), J. Comput. Appl. Math. **115** (2000), no. 1–2, 181–192.

91. Ezquerro, J.A., Hernández, M.A., Salanova, M.A., Recurrence relations for the midpoint method, Tamkang J. Math. **31** (2000), no. 1, 33–41.

92. Ezquerro, J.A., Gutiérrez, J.M., Hernández, M.A., Salanova, M.A., A biparametric family of inverse free multipoint iterations, Comput. Appl. Math. **19** (2000), no. 1, 109–124.

93. Feinstauer, M., Zernicek, A., Finite element solution of nonlinear elliptic problems, Numer. Math. **50** (1987), 471–475.

94. Ferreira, O.P. and Svaiter, B.F., Kantorovich's theorem on Newton's method in Riemannian manifolds, J. Complexity **18** (2002), no. 1, 304–329.

95. Foerster, H., Frommer, A., Mayer, G., Inexact Newton methods on a vector supercomputer, J. Comp. Appl. Math. **58** (1995), 237–253.

96. Fujimoto, T., Global asymptotic stability of nonlinear difference equations I, Econ. Letters **22** (1987), 247–250.

97. Fujimoto, T., Global asymptotic stability of nonlinear difference equations II, Econ. Letters **23** (1987), 275–277.

98. Galperin, A., Kantorovich's Majorization and functional equations, Numer. Funct. Anal. Optimiz. **24** (2003), no. 7 and 8, 783–811.

99. Galperin, A., Waksman, Z., Regular smoothness and Newton's method, Numer Funct. Anal. Optimiz. **15** (1994), no. 7 & 8, 813–858.

100. Gander, W., On Halley's iteration method, Amer. Math. Monthly **92** (1985), 131–134.

101. Glowinski, R., Lions, J.L., Trémolières, R., Numerical Analysis of Variational Inequalities, North-Holland, Amsterdam, 1982.

102. Gragg, W.B., Tapia, R.A., Optimal error bounds for the Newton-Kantorovich theorem, SIAM J. Numer. Anal. **11** (1974), 10–13.

103. Graves, L.M., Riemann integration and Taylor's theorem in general analysis, Trans. Amer. Math. Soc. **29** (1927), no. 1, 163–177.

104. Gutierez, J.M., A new semilocal convergence theorem for Newton's method, J. Comput. Appl. Math. **79** (1997), 131–145.

105. Gutiérrez, J.M., Hernández, M.A., A family of Chebyshev-Halley type methods in Banach spaces, Bull. Austral. Math. Soc. **55** (1997), 113–130.

106. Gutiérrez, J.M., Hernández, M.A., Salanova, M.A., Resolution of quadratic equations in Banach spaces, Numer. Funct. Anal. Optim. **17** (1996), no. 1 & 2, 113–121.

107. Hackl, J., Wacker, Hj., Zulehner, W., An efficient step size control for continuation methods, BIT **20** (1980), no. 4, 475–485.

108. Hadeller, K.P., Shadowing orbits and Kantorovich's theorem, Numer. Math. **73** (1996), 65–73.

109. Haeberly, J.-P.A., Remarks on nondegeneracy in mixed semidefinite-quadratic programming, unpublished memorandum, available from URL: http://corcy.fordham.edu/haeberly/papers/sqldegen.ps.gz .

110. Häubler, W.M., A Kantorovich-type convergence analysis for the Gauss-Newton method, Numer. Math. **48** (1986), 119–125.

111. Hernández, M.A., A note on Halley's method, Num. Math. **59** (1991), no. 3, 273–276.

112. Hernández, M.A., Newton's Raphson's method and convexity, Zb. Rad. Prirod.-Mat. Fax. Ser. Mat. 22 **1** (1992), 159–166.

113. Hernández, M.A., Salanova, M.A., A family of Chebyshev-Halley type methods, Int. J. Comp. Math. **47** (1993), 59–63.

114. Hernández, M.A., Relaxing convergence conditions for Newton's method, J. Math. Anal. Appl. **249** (2000), no. 2, 463–475.

115. Hernández, M.A., Chebyshev's approximation algorithms and applications, Comput. Math. Appl. **41** (2001), no. 3–4, 433–445.

116. Hernández, M.A., Rubio, M.J., Ezquerro, J.A., Secant-like methods for solving nonlinear integral equations of the Hammerstein type, J. Comput. Appl. Math. **115** (2000), 245–254.

117. Hernández, M.A., Rubio, M.J., Semilocal convergence of the secant method under mild convergence conditions of differentiability, Comput. Math. Appl. **44** (2002), no. 3–4, 277–285.

118. Hernández, M.A., Salanova, M.A., Sufficient conditions for semilocal convergence of a fourth order multipoint iterative method for solving equations in Banach spaces, Southwest J. Pure Appl. Math. (1999), 29–40.

119. Hille, E. and Philips, R.S., Functional Analysis and Semigroups, Amer. Math. Soc. Coll. Publ., New York, 1957.

120. Higle, J.L., Sen, S., On the convergence of algorithms with applications to stochastic and nondifferentiable optimization, SIE Working Paper #89-027, University of Arizona, 1989.

121. Huang, Z.D., A note on the Kantorovich theorem for Newton iteration, J. Comput. Appl. Math. **47** (1993), no. 2, 211–217.

122. Jarrat, P., Some efficient fourth order multipoint methods for solving equations, BIT **9** (1969), 119–124.

123. Kanno, S., Convergence theorems for the method of tangent hyperbolas, Math. Japon. **37** (1992), no. 4, 711–722.

124. Kantorovich, L.V., The method of successive approximation for functional equations, Acta Math. **71** (1939), 63–97.

125. Kantorovich, L.V., Akilov, G.P., Functional Analysis in Normed Spaces, Pergamon Press, New York, 1964.

126. Kelley, C.T., Solving nonlinear equations with Newton's method, SIAM series: Fundamentals of Algorithmic, Philadelphia, 2003.

127. King, R.E., Tangent method for nonlinear equations, Numer. Math. **18** (1973), 298–304.

128. Krasnosel'skii, M.A., Approximate solution of operator equations, Walter Noordhoff Publ., Groningen, 1972.

129. Krasnosel'skii, M.A., Vainikko, G.M., Zabreiko, P.P., Rutiskii, Ya.B., Stetsenko, V.Ya., Approximate Solution of Operator Equations, Wolters-Noordhoff Publishing, Groningen, 1972.

130. Krein, S.G., Linear equations in Banach spaces, Birkhäuser, Boston, 1982.

131. Kung, H.T., The complexity of obtaining starting points for solving operator equations by Newton's method, Technical report, no. 044-422, Carnegie-Mellon Univ., Pittsburgh, PA, October 1975, Article in Traub, J.F., Analytic Computational Complexity.

132. Lancaster, P., Error analysis for the Newton-Raphson method, Numer. Math. **9** (1968), 55–68.

133. Laumen, M., Newton's mesh independence principle for a class of optimal design problems, SIAM J. Control. Optim. **37** (1999), no. 4, 1070–1088.

134. Laumen, M., A Kantorovich theorem for the structured PSB update in Hilbert spaces, J. Optim. Theory Appl. **105** (2000), no. 2, 391–415.

135. Lysenko, J.V., Conditions for the convergence of the Newton-Kantorovich method for nonlinear equations with Hölder linearizations, Dokl. Akad. Nauk USSR **38** (1994), 20–24 (Russian).

136. Marcotte, P., Wu, J.H., On the convergence of projection methods, J. Optim. Theory Appl. **85** (1995), no. 2, 347–362.

137. Mayer, J., A generalized theorem of Miranda and the theorem of Newton-Kantorovich, Numer. Funct. Anal. Optim. **23** (2002), no. 3–4, 333–357.

138. McCormick, S.F., A revised mesh refinement strategy for Newton's method applied to two-point boundary value problems, Lecture Notes in Mathemaics, vol. 674, Springer-Verlag, Berlin, 1978, 15–23.

139. Meyer, G.H., On solving nonlinear equations with a one parameter operator imbedding, SIAM J. Numer. Anal. **5** (1968), no. 4, 739–752.

140. Meyer, P.W., Das modifizierte Newton-Verfahren in verallgemeinerten Banach-Räumen, Numer. Math. **43** (1984), no. 1, 91–104.

141. Meyer, P.W., Newton's method in generalized Banach spaces, Numer. Funct. Anal. Optim. **9** (1987), no. 3 & 4, 244–259.

142. Miranda, C., Un osservatione su un teorema di Brower, Boll. Unione Ital. Serr. **11** (1940), no. 3, 5–7.

143. Moore, R.E., Methods and Applications of Interval Analysis, SIAM, Philadelphia, PA, 1979.

144. Nayakkankuppam, M.V., Overton, M.L., Conditioning of semidefinite programs, Math. Progr. **85** (1999), 525–540.

145. Nesterov, Y., Nemirosky, A., Interior Point Polynomial Methods in Convex Programming, SIAM, Philadelphia, PA, 1999.

146. Neumaier, A., Shen, Z., The Krawczyk operator and Kantorovich's theorem, J. Math. Anal. Appl. **149** (1990), no. 2, 437–443.

147. Nguen, D.F., Zabrejko, P.P., The majorant method in the theory of the Newton-Kantorovich approximations and the Pták error estimates, Numer. Funct. Anal. Optimiz. **9** (1987), no. 5 & 6, 671–686.

148. Noble, B., The Numerical Solution of Nonlinear Integral Equations and Related Topics, University Press, Madison, WI, 1964.

149. Noor, K.I., Noor, M.A., Iterative methods for a class of variational inequalities, In: Numerical Analysis of Singular Perturbation Problems (Hemker and Miller, eds.), Academic Press, New York, 1985, 441–448.

150. Noor, M.A., An iterative scheme for a class of quasivariational inequalities, J. Math. Anal. and Appl. **110** (1985), no. 2, 463–468.

151. Noor, M.A., Generalized variational inequalities, Appl. Math. Lett. **1** (1988), 119–122.

152. Ojnarov, R., Otel'baev, M., A criterion for a Uryson operator to be a contraction, Dokl. Akad. Nauk. SSSR, **255** (1980), 1316–1318 (Russian).

153. Okuguchi, K., Expectations and Stability in Oligopoly Models, Springer-Verlag, New York, 1976.

154. Ortega, J.M., Rheinboldt, W.C., Iterative Solution of Nonlinear Equations in Several Variables, Academic Press, New York, 1970.

155. Ostrowski, A.M., Solution of Equations in Euclidean and Banach Spaces, Academic Press, New York, 1973.
156. Paardekooper, M.H.C., An upper and a lower bound for the distance of a manifold to a nearby point, J. Math. Anal. Applic. **150** (1990), 237–245.
157. Palmer, K.J., Stuffer, D., Rigorous verification of chaotic behavior of maps using validated shadowing, Nonlinearity **12** (1999), 1683–1698.
158. Păvăloiu, I., Sur la méthode de Steffensen pour la résolution des équations opérationnelles non linéaires, Rev. Roumaine Math. Pures Appl. **13** (1968), no. 6, 857–861.
159. Păvăloiu, I., Sur une généralisation de la méthode de Steffensen, Rev. Anal. Numér. Théor. Approx. **21** (1992), no. 1, 59–65.
160. Păvăloiu, I., A converging theorem concerning the chord method, Rev. Anal. Numér. Théor. Approx. **22** (1993), no. 1, 83–85.
161. Păvăloiu, I., Bilateral approximations for the solutions of scalar equations., Rev. Anal. Numér. Théor. Approx. **23** (1994), no. 1, 95–100.
162. Potra, F.A., On the convergence of a class of Newton-like methods, In: Iterative Solution of Nonlinear Systems of Equations, Lecture Notes in Math., vol. 953, Springer-Verlag, New York, 1982.
163. Potra, F.A. On an iterative algorithm of order 1.839... for solving nonlinear operator equations, Numer. Funct. Anal. Optim. **7** (1984–1985), no. 1, 75–106.
164. Potra, F.A., Newton-like methods with monotone convergence for solving nonlinear operator equations, Nonlinear Anal., Theory Methods Appl. **11** (1987), no. 6, 697–717.
165. Potra, F.A., The Kantorovich method and interior point methods, Math. Progr. Ser. A **102** (2005), 47–50.
166. Potra, Florian-A., Pták, V., Sharp error bounds for Newton's method, Numer. Math. **34** (1980), no. 1, 63–72.
167. Potra, F.A., Pták, V., Nondiscrete induction and iterative processes, Pitman, London, 1984.
168. Pousin, J., Rappaz, J., Consistency stability a priori and a posteriori errors for Petrov-Galerkin's method applied to nonlinear problems, Numer. Math. **69** (1994), 213–231.
169. Rall, L.B., Convergence of Stirling's method in Banach spaces, Aequationes Math. **12** (1973), 12–20.
170. Rall, L.B., A quadratically convergent iteration method for computing zeros of operators satisfying autonomous differential equations, Math. Comput. **30** (1976), no. 133, 112–114.
171. Rall, L.B., A comparison of the existence theorems of Kantorovich and Moore, SIAM J. Numer. Anal. **17** (1980), no. 1, 148–161.
172. Ren, H., On the local convergence of deformed Newton's method under Argyros-type condition, to appear in J. Math. Anal. Appl.
173. Renegar, J., A polynomial-type algorithm based on Newton's method for linear programming, Math. Progr., Ser. A., **40** (1988), no. 1, 59–93.
174. Renegar, J., Shub, M., Modified complexity analysis for Newton LP methods, Math. Progr., Ser. A., **53** (1992), no. 1, 1–16.
175. Rheinboldt, W.C., An adaptive continuation process for solving systems of nonlinear equations, Publish Academy of Sciences, Banach Ctr. Publ. **3** (1977), 129–142.
176. Rheinboldt, W.C., Solution fields of nonlinear equations and continuation methods, SIAM J. Numer. Anal. **17** (1980), no. 2, 221–237.
177. Rheinboldt, W.C., On a theorem of S. Smale about Newton's method for analytic mappings, Appl. Math. Lett. **1** (1988), 69–72.

178. Robinson, S.M., Generalized equations. In: Mathematical Programming: The State of the Art (A. Bachem, M. Grötschel and B. Korte, eds.), Springer, Berlin, 1982, 346–367.
179. Robinson, S.M., An implicit function theorem for a class of nonsmooth functions, Math. Oper. Res. **16** (1991), no. 2, 292–309.
180. Robinson, S.M., Newton's method for a class of nonsmooth functions, Set-Valued Analysis **2** (1994), 291–305.
181. Schmidt, J.W., Leonhardt, H., Eingrenzung von lösungen mit hilfe der Regula-Falsi, Computing **6** (1970), 318–329.
182. Schmidt. W.F., Adaptive step size selection for use with the continuation method, Intern. J. Numer. Math. Engrg. **12** (1978), 677–694.
183. Slugin, S.N., Monotonic processes of bilateral approximation in a partially ordered convergence group, Soviet. Math. **3** (1962), 1547–1551.
184. Smale, S., Newton's method estimates from data at one point. In: The Merging of Disciplines in Pure, Applied, and Computational Mathematics, Springer-Verlag, New York, 1986, 185–196.
185. Stirling, J., Methodus differentialis: sive tractatus de summatione et interpolatione serierum infinitarum, W. Boyer, London, 1730.
186. Stoffer, D., Kirchgraber, U., Verification of chaotic behavior in the planar restricted three body problem, Appl. Numer. Math. **39** (2001), no. 3–4, 415–433.
187. Szidarovszky, F., Bahill, T., Linear Systems Theory, CRC Press, Boca Raton, FL, 1992.
188. Tapia, R.A., The weak Newton method and boundary value problems, SIAM J. Numer. Anal. **6** (1969), no. 4, 539–550.
189. Tishyadhigama, S., Polak, E., Klessig, R., A comparative study of several convergence conditions for algorithms modeled by point-to-set maps, Math. Programming Stud. **10** (1979), 172–190.
190. Törnig, W., Monoton konvergente Iterationsverfahren zür Lösung michtlinearer differenzen–randwertprobleme, Beiträge zür Numer. Math. **4** (1975), 245–257.
191. Traub, J.F., Iterative methods for the solution of equations, Prentice-Hall Series in Automatic Computation, Prentice-Hall, Inc., Englewood Cliffs, N.J., 1964.
192. Traub, J.F., Analytic Computational Complexity, Academic Press, New York–London, 1975.
193. Tricomi, F.G., Integral Equations, Interscience Publ., London, 1957.
194. Tsuchiya, T., An application of the Kantorovich theorem to nonlinear finite element analysis, Numer. Math. **84** (1999), 121–141.
195. Uko, L.U., Generalized equations and the generalized Newton method., Mathematical Programming **73** (1996), 251–268.
196. Ulm, S., Iteration methods with divided differences of the second order, Dokl. Akad. Nauk SSSR, **158** (1964), 55–58 (Russian).
197. Urabe, M., Convergence of numerical iteration in solution of equations, J. Sci. Hiroshima Univ., Ser. A, **19** (1976), 479–489.
198. Uzawa, H., The stability of dynamic processes, Econometrica **29** (1961), 617–631.
199. Vandergraft, J.S., Newton's method for convex operators in partially ordered spaces, SIAM J. Numer. Anal. **4** (1967), 406–432.
200. Varga, R.S., Matrix Iterative Analysis, Prentice-Hall, Englewood Cliffs, NJ, 1962.
201. Verma, R.U., Nonlinear variational and constrained hemivariational inequalities involving relaxed operators., Z. Angew. Math. Mech. **77** (1997), no. 5, 387–391.
202. Verma, R.U., Approximation-solvability of nonlinear variational inequalities involving partially relaxed monotone (PRM) mappings, Adv. Nonlinear Var. Inequal. **2** (1999), no. 2, 137–148.

203. Verma, R.U., A class of projection-contraction methods applied to monotone variational inequalities., Appl. Math. Lett. **13** (2000), no. 8, 55–62.
204. Verma, R.U., Generalized multivalued implicit variational inequalities involving the Verma class of mappings, Math. Sci. Res. Hot-Line **5** (2001), no. 2, 57–64.
205. Walker, H.J., A summary of the developments on imbedding methods, continuation methods, In: (H.J. Wacker, ed.) Academic Press, New York, 1978, 1–36.
206. Wang, D., Zhao, F., The theory of Smale's point estimation and its applications, J. Comput. Appl. Math. **60** (1995), 253–269.
207. Werner, W., Uber ein verfahren der ordnung $1 + \sqrt{2}$ zur Nullstellenbestimmung, Numer. Math. **32** (1970), 333–342.
208. Wu, J.W, Brown, D.P., Global asymptotic stability in discrete systems, J. Math. Anal. Appl. **140** (1989), no. 1, 224–227.
209. Yamamoto, T., A method for finding sharp error bounds for Newton's method under the Kantorovich assumptions, Numer. Math. **44**, (1986), 203–220.
210. Yamamoto, T., A convergence theorem for Newton-like methods in Banach spaces, Numer. Math. **51** (1987), 545–557.
211. Yamamoto, T., On the method of tangent hyperbolas in Banach spaces, J. Comput. Appl. Math. **21** (1988), 75–86.
212. Yamamoto, T., Chen, Z., Convergence domains of certain iterative methods for solving nonlinear equations, Numer. Funct. Anal. Optim. **10** (1989), 34–48.
213. Ypma, T.J., Numerical solution of systems of nonlinear algebraic equations, Ph.D. thesis, Oxford, 1982.
214. Ypma, T.J., Affine invariant convergence results for Newton's methods, BIT **22** (1982), 108–118.
215. Ypma, T.J., The effect of rounding error on Newton-like methods, IMA J. Numer. Anal., **3** (1983), 109–118.
216. Ypma, T.J., Convergence of Newton-like iterative methods, Numer. Math. **45** (1984), 241–251.
217. Zabrejko, P.P., K-metric and K-normed linear spaces, a survey, Collect. Math. **48** (1997), no. 4–6, 825–859.
218. Zabrejko, P.P., Nguen, D.F., The majorant method in the theory of Newton-Kantorovich approximations and the Pták error estimates, Numer. Funct. Anal. Optim. **9** (1987), no. 5–6, 671–684.
219. Zincenko, A.I., A class of approximate methods for solving operator equations with nondifferentiable operators, Dopovidi Akad. Nauk Ukrain. RSR (1963), 156–161.
220. Zuhe, S., Wolfe, M.A., A note on the comparison of the Kantorovich and Moore theorems, Nonlinear Anal. **15** (1990), no. 3, 329–332.

Glossary of Symbols

$\mathbf{R}^n$	real n-dimensional space
$\mathbf{C}^n$	complex n-dimensional space
$X \times Y, X \times X = X^2$	Cartesian product space of X and Y
$e^1, \ldots, e^n$	the coordinate vectors of $\mathbb{R}^n$
$x = (x_1, \ldots, x_n)^T$	column vector with component x_i
x^T	the transpose of x
$\{x_n\}_{n \geq 0}$	sequence of points from X
$\|\cdot\|$	norm on X
$\|\cdot\|_p$	L_p norm
$\|\cdot\|$	absolute value symbol
$/\cdot/$	norm symbol of a generalized Banach space X
$\langle x, y \rangle$	set $\{z \in X \mid z = tx + (1 - t)\,y,\ t \in [0, 1]\}$
$U(x_0, R)$	open ball $\{z \in X \mid \|x_0 - z\| < R\}$
$\bar{U}(x_0, R)$	closed ball $\{z \in X \mid \|x_0 - z\| \leq R\}$
$U(R) = U(0, R)$	ball centered at the zero element in X and of radius R
$U, \bar{U}$	open, closed balls, respectively no particular reference to X, x_0, or R
$M = \{m_{ij}\}$	matrix $1 \leq i, j \leq n$
M^{-1}	inverse of M
M^+	generalized inverse of M
$\det M$ or $|M|$	determinant of M
M^k	the kth power of M
rank M	rank of M
I	identity matrix (operator)
L	linear operator

L^{-1}	inverse
null L	null set of L
rad L	radical set of L
$F: D \subseteq X \to Y$	an operator with domain D included in X, and values in Y
$F'(x)$, $F''(x)$	first, second Fréchet derivatives of F evaluated at x
δ_{ij}	Kronecker delta
$\sum$	summation symbol
$\prod$	product of factors symbol
$\int$	integration symbol
$\in$	element inclusion
$\subset, \subseteq$	strict and nonstrict set inclusion
$\forall$	for all
$\Rightarrow$	implies
$\cup, \cap$	union, intersection
$A - B$	difference between sets A and B
$\bar{B}$	mean of a bilinear operator B
$Q(X, Z)$	set of all quadratic operators from X to Z
$Q_F^*(X)$	set of all bounded quadratic operators Q in X such that Q has finite rank
X^{2*}	set of all bounded quadratic functionals
$\oplus$	direct sum
$A^{\#}$	outer inverse of A
dim A	dimension of A
codim A	codimension of A
$]\cdot[$	K-norm

Index

Additive operator, 1
Algorithmic model, 445
Analytic operator, 127
Antitone operator, 11
Arnoldi's method, 73
Autonomous differential equation, 74, 215

Banach lemma, 4
Banach space, 3
Banach space with a convergence structure, 417
Bijective linear operator, 30
Bilinear operator, 2
Biparametric family of multipoint iterations, 255
Bounded linear operator, 2
Bounds on manifolds, 103
Broyden's method, 193

Central path, 476
Chaotic behavior, 166
Chebyshev method, 256
Chebyshev-Halley method, 256
Complementarity gap, 476
Computational complexity, 325
Continuation methods, 116
Contraction mapping, 26
Convergence on a cone, 80
Convergence radius, 73
Convergence structure, 417
Convexity, 19

Dilation measure, 114
Discretization, 187

Divided differences, 295

Embedding, 106
Euler's method, 154

Finite element analysis, 157
Fixed point, 25
Fréchet derivative, 5
Fredholm operator, 29
Functional, 2

Gâteaux derivative, 36
Gauss-Newton method, 121
Generalized conjugate residual, 73
Generalized inverse, 379
Generalized minimum residual, 73
Gershgorin's theorem, 39
Green kernel, 52

Halley method, 258
Hilbert space, 104
Hölder condition, 62
Homogeneous operator, 1
Horizontal linear complementarity problem (HLCP), 477

Inner inverse, 379
Interior point method, 475
Inverse operator, 3

j-linear operators, 2
Jacobi-Newton method, 21
Jacobi-Secant method, 21

K-normed spaces, 395

Kantorovich fixed point theorem, 12
King-Werner method, 233
Krawczyk operator, 134

Laplace operator, 29
Linear complementarity problem (LCP), 477
Linear operator, 1
Linear space, 1
LP method, 482

Mathematica, 71
Mesh independence, 170
Midpoint method, 253
Miranda theorem, 137
Moore theorem, 135

Neumann series, 4
Newton-Kantorovich theorem, 43
Newton method (inexact), 320
Newton method (weak), 102
Newton's method, 41
Newton's method modified, 42
Newton-like method, 261
Newton-like (two-point), 275
Nilpotent operator, 5

Partially ordered topological space, 9
Point-to-set-mapping, 445
Primal-dual path following algorithms, 475
PSB update, 162
Pseudo-orbit, 113

Quasivariational inequality, 362

Radius of convergence, 89
Regular smoothness, 75
Regular space, 11
Riccati operator, 32
Riemannian integration, 110

Secant method, 54
Secant-type methods, 111
Semidefinite program, 180
Semilocal convergence, 42
Shadowing lemma, 166
Shadowing orbits, 113
Steffensen's method, 207
Stirling's method, 202
Super-Halley method, 256
Symmetric operator, 24

Tangent-hyperbola, 219
Taylor's theorem, 9
Tensor product, 34
Terra incognita, 62

Undetermined system, 104
Uryson operator, 32

Variational inequality, 339

Weierstrass theorem, 28

Yamamoto, 129